Neisseria

Molecular Mechanisms of Pathogenesis

Edited by

Caroline Attardo Genco and Lee Wetzler

Boston University School of Medicine
Boston, MA
USA

Caister Academic Press

Caister Academic Press
Norfolk, UK

www.caister.com

British Library Cataloguing-in-Publication Data
A catalogue record for this book is available from the British Library

ISBN: 978-1-904455-51-6

Description or mention of instrumentation, software, or other products in this book does not imply endorsement by the author or publisher. The author and publisher do not assume responsibility for the validity of any products or procedures mentioned or described in this book or for the consequences of their use.

Cover images taken from Figure 4.2.

Printed and bound in Great Britain

Contents

Contributors

Michael A. Apicella
Department of Microbiology
The University of Iowa
51 Newton Road
Iowa City, IA
USA

michael-apicella@uiowa.edu

Kenneth Barth
Department of Microbiology and Immunology
School of Medicine and Dentistry
University of Rochester
Rochester, NY
USA

kenneth_barth@urmc.rochester.edu

Samuel Bish
National Cancer Institute/NIH
Technology Transfer Center
Rockville, MD
USA

bishse@mail.nih.gov

Sandrine Bourdoulous
Institut Cochin
Département de Biologie Cellulaire
INSERM
CNRS UMR8104
Université Paris Descartes
Paris
France

sandrine.bourdoulous@inserm.fr

Elizabeth Burrowes
Division of Infectious Diseases and Immunology
University of Massachusetts Medical School
Worcester, MA
USA

e_burrowes@yahoo.co.uk

Etienne Carbonnelle
INSERM unité 570
Université Paris Descartes
Faculté de Médecine
Paris
France

etienne.carbonnelle@inserm.fr

Virginia L. Clark
Department of Microbiology and Immunology
School of Medicine and Dentistry
University of Rochester
Rochester, NY
USA

ginny_clark@urmc.rochester.edu

John K. Davies
Department of Microbiology
Monash University
Clayton
VIC
Australia

john.davies@med.monash.edu.au

Megan Falsetta
Department of Microbiology
The University of Iowa
51 Newton Road
Iowa City, IA
USA

megan-falsetta@uiowa.edu

Sarah A. Follows
Department of Microbiology
Department of Medicine
Section of Molecular Medicine
Boston University School of Medicine
Boston, MA
USA

sarah.follows@gmail.com

Jason P. Folster
National Antimicrobial Resistance Monitoring System
Centers for Disease Control and Prevention
Atlanta, GA
USA

gux8@cdc.org

Stuart Hill
Department of Biological Sciences
Northern Illinois University
DeKalb, IL
USA

sahill@niu.edu

J. Claire Hoe
Oxford Vaccine Group
Department of Paediatrics
University of Oxford
Centre for Clinical Vaccinology and Tropical Medicine
Churchill Hospital
Oxford
UK

claire.hoe@paediatrics.ox.ac.uk

Gwenda Hughes
Sexually Transmitted Bacteria Reference Laboratory (STBRL)
Centre for Infections
Health Protection Agency
London
UK

gwenda.hughes@hpa.org.uk

Vincent M. Isabella
Department of Microbiology and Immunology
School of Medicine and Dentistry
University of Rochester
Rochester, NY
USA

vincent_isabella@urmc.rochester.edu

Catherine Ison
Sexually Transmitted Bacteria Reference Laboratory (STBRL)
Centre for Infections
Health Protection Agency
London
UK

catherine.ison@hpa.org.uk

Dominic F. Kelly
Oxford Vaccine Group
Department of Paediatrics
University of Oxford
Centre for Clinical Vaccinology and Tropical Medicine
Churchill Hospital
Oxford
UK

dominic.kelly@paediatrics.ox.ac.uk

Rosanna Leuzzi
Novartis Vaccines
Siena
Italy

rosanna.leuzzie@novartis.com

Lisa A. Lewis
Division of Infectious Diseases and Immunology
University of Massachusetts Medical School
Worcester, MA
USA

lisa.lewis@umassmed.edu

Susan Lewis
Oxford Vaccine Group
Department of Paediatrics
University of Oxford
Centre for Clinical Vaccinology and Tropical Medicine
Churchill Hospital
Oxford
UK

susan.lewis@paediatrics.ox.ac.uk

Paola Massari
Department of Medicine
Section of Infectious Diseases
Boston University School of Medicine
Boston, MA
USA

pmassari@bu.edu

Jeffrey R. Mellin
Department of Microbiology
Boston University School of Medicine
Boston, MA
USA

jrmellin@bu.edu

Xavier Nassif
INSERM unité 570
Université Paris Descartes
Faculté de Médecine
Paris
France

xavier.nassif@inserm.fr

Ryan Neil
Department of Microbiology
The University of Iowa
51 Newton Road
Iowa City, IA
USA

ryan-neil@uiowa.edu

Robert A. Nicholas
Department of Pharmacology
University of North Carolina-Chapel Hill School of Medicine
Chapel Hill, NC
USA

nicholas@med.unc.edu

Gunnstein Norheim
Oxford Vaccine Group
Department of Paediatrics
University of Oxford
Centre for Clinical Vaccinology and Tropical Medicine
Churchill Hospital
Oxford
UK

gunnstein.norheim@paediatrics.ox.ac.uk

Tim W. Overton
Biochemical Engineering
School of Chemical Engineering
The University of Birmingham
Birmingham
UK

t.w.overton@bham.ac.uk

Julia B. Patrone
The Johns Hopkins University Applied Physics Laboratory
Milton S. Eisenhower Research Center
Laurel, MD
USA

julia.patrone@jhuapl.edu

Mariagrazia Pizza
Novartis Vaccines
Siena
Italy

mariagrazia.pizza@novartis.com

Andrew J. Pollard
Oxford Vaccine Group
Department of Paediatrics
University of Oxford
Centre for Clinical Vaccinology and Tropical Medicine
Churchill Hospital
Oxford
UK

andrew.pollard@paediatrics.ox.ac.uk

Sanjay Ram
Division of Infectious Diseases and Immunology
University of Massachusetts Medical School
Worcester, MA
USA

sanjay.ram@umassmed.edu

Rino Rappuoli
Novartis Vaccines
Siena
Italy

rino.rappuoli@novartis.com

Peter A. Rice
Division of Infectious Diseases and Immunology
University of Massachusetts Medical School
Worcester, MA
USA

peter.rice@umassmed.edu

Manish Sadarangani
Oxford Vaccine Group
Department of Paediatrics
University of Oxford
Centre for Clinical Vaccinology and Tropical Medicine
Churchill Hospital
Oxford
UK

manish.sadarangani@paediatrics.ox.ac.uk

Kate L. Seib
Novartis Vaccines
Siena
Italy

kate.seib@novartis.com

Laura Serino
Novartis Vaccines
Siena
Italy

laura.serino@novartis.com

Davide Serruto
Novartis Vaccines
Siena
Italy

davide.serruto@novartis.com

William M. Shafer
Department of Microbiology and Immunology
Emory University School of Medicine
Atlanta, GA
USA

wshafer@emory.edu

Christopher Steichen
Department of Microbiology
The University of Iowa
51 Newton Road
Iowa City, IA
USA

christopher-steichen@uiowa.edu

Daniel C. Stein
University of Maryland
Department of Cell Biology and Molecular Genetics
College Park
Maryland
USA

dcstein@umd.edu

Matthew D. Snape
Oxford Vaccine Group
Department of Paediatrics
University of Oxford
Centre for Clinical Vaccinology and Tropical Medicine
Churchill Hospital
Oxford
UK

matthew.snape@paediatrics.ox.ac.uk

Caroline Trotter
Department of Social Medicine
University of Bristol
Bristol
UK

caroline.trotter@bristol.ac.uk

Lee Wetzler
Department of Medicine
Division of Infectious Diseases
Boston University School of Medicine
Boston Medical Center
Boston, MA
USA

lwetzler@bu.edu

Preface

Great progress has been made over the past two decades in our understanding of the pathogenesis of diseases caused by the pathogenic bacteria *Neisseria gonorrhoeae* and *N. meningitidis*. *N. gonorrhoeae* is the agent of gonorrhoea and *N. meningitidis* an agent of acute bacterial meningitis and/or septicaemia. *N. gonorrhoeae* infections are acquired by sexual contact and usually affect the mucous membrane of the urethra in men and the endocervix in women, although the infection may disseminate to a variety of tissues. *N. meningitidis* infection results from the blood-borne dissemination (meningococcaemia) of the meningococcus, usually following an asymptomatic or mildly symptomatic nasopharyngeal carrier state.

Interestingly, recent scientific advances have allowed a resurgence of important research regarding the pathogenic *Neisseria*. This includes the burgeoning knowledge obtained from genomics and proteomics and use of these techniques in examining the physiology and pathogenesis of these organisms, along with the development of new vaccines via 'reverse vaccinology'. Moreover, our knowledge regarding induction of innate immune responses via pattern recognition receptors (especially Toll-like receptors) have aided investigations into immunity to these pathogens.

The emphasis of this book is on the molecular and cellular mechanisms of Neisseria pathogenesis, the public health impact of the pathogenic Neisseria, and how basic and applied research are contributing to the diagnosis, treatment, and prevention of diseases caused by these bacteria. We have attempted to bring together experts in Neisseria research from around the world, whose studies have collectively contributed to our understanding of the pathogenesis of gonococcal and meningococcal diseases.

The volume is structured into four separate sections: (I) Genetic mechanisms, (II) Interaction with Host Cells, (III) Host Responses, and (IV) Clinical Correlations. We have not attempted to make this a comprehensive treatise, but wanted to have specific hotbed areas of research and knowledge covered to bring scientists interested in the pathogenic *Neisseria* up to speed with the latest information available. We feel this task has been more than accomplished by the selection of these international experts in Nesseriology as chapter authors.

The next 10 years should yield important information regarding *Neisseria* pathogenesis, disease mechanisms, and immune responses which will lead us closer to more effective therapeutic interventions. It is crucial that areas of priority in *Neisseria* research be defined, as we move ahead in our understanding of these pathogens and their interactions with the host at the cellular and molecular levels. We hope that the work described in this volume will aid scientists in setting research agendas to target and prioritize important areas for future studies.

Caroline Attardo Genco
Lee Wetzler

Part I

Genetic Mechanisms

Gene Expression Strategies of the Pathogenic Neisseriae

1

Jeffrey R. Mellin and Stuart Hill

Abstract

The pathogenic *Neisseria* consist of the species *Neisseria gonorrhoeae* and *Neisseria meningitidis*, both of which are Gram-negative diplococci and exclusive human pathogens. *Neisseria gonorrhoeae* is the causative agent of gonorrhoea and is transmitted via sexual contact and principally infects the mucosal surfaces of the cervix in women and the urethra in men. In contrast, *N. meningitidis* is transmitted via respiratory droplets leading to colonization of the nasopharynx and, in rare cases, dissemination of meningococci into the bloodstream and cerebrospinal fluid leading to meningitis and/or septicaemia. In order to effectively colonize their niches within the human host, both *N. gonorrhoeae* and *N. meningitidis* must successfully compete with significant resident flora for space and nutrients, as well as evade host immune defences. Consequently, this requires an ability to quickly sense and respond to a plethora of changing environmental stimuli. At the most fundamental level, this requires the modulation of gene expression as circumstances require. The following is a discussion of mechanisms of bacterial gene regulation with a specific focus on the strategies utilized by the pathogenic *Neisseria* to control gene expression of factors with roles in neisserial pathogenesis. Topics will include the use of alternative sigma factors, positive and negative regulation of gene transcription by specific transcription factors and adaptation via the accumulation of temporally random spontaneous mutations within chromosomal DNA. Additionally, some discussion will focus on emerging mechanisms of gene regulation such as posttranscriptional regulation via small RNA molecules.

Introduction

All bacteria have evolved mechanisms to regulate gene expression in response to environmental signals both at a transcriptional and post-transcriptional level. Commonly, gene regulatory strategies involve DNA-binding proteins which bind at the promoter to repress or activate transcription, or, alternatively, operate in conjunction with other proteins to elicit their effects. Generally, these proteins act by inhibiting or activating transcription initiation. However, in some instances, DNA-binding regulatory proteins can affect chromosomal supercoiling and local DNA structure. Other mechanisms of gene regulation, such as phase switching, can affect gene expression independent of regulatory proteins via reversible changes in chromosomal DNA. A number of gene regulatory mechanisms also act at the post-transcriptional level either through modulation of cis-acting secondary structures in mRNA molecules, trans-acting RNA regulatory molecules (termed small RNAs) or effects on RNA turnover.

In the class of DNA-binding regulatory proteins of bacteria, one of the best characterized is the sigma factor family of proteins. Sigma factors are small proteins that associate with the RNA polymerase (RNAP) holoenzyme and impart promoter specificity to the RNAP complex (Mooney *et al.*, 2005; Gruber and Gross, 2003). Thus by interchanging alternative sigma factors with the core enzyme, the RNAP complex can be directed to differing sets of genes whose promoters are recognized by only one particular sigma factor (Kazmierczak *et al.*, 2005). In the pathogenic *Neisseria*, the use of alternative sigma factors is relatively limited compared with many other bacteria, however, both *N. gonorrhoeae* and *N. meningitidis* possess an *rpoD* gene, an *ecf* sigma factor gene and a *rpoH* gene which is important for gonococcal invasion of epithelial cells (Du and Arvidson, 2006; Du *et al.*, 2005; Gunesekere *et al.*, 2006a; Gunesekere *et al.*, 2006b; Laskos *et al.*, 2004).

Sigma factors are unique amongst DNA-binding regulatory proteins in that they are a component of the RNA polymerase complex, whereas many other DNA-binding transcription factors affect RNA polymerase in trans. These proteins often work either by excluding RNA polymerase from target promoters thereby repressing transcription or by increasing the affinity of RNA polymerase for a particular promoter thus causing transcription activation. Several DNA-binding transcription factors have been well characterized and their roles in neisserial gene expression are well established. These include proteins such as OxyR, which is involved in the response to oxidative stress (Seib *et al.*, 2007), Fur, which is involved in iron adaptation (Delany *et al.*, 2006; Delany *et al.*, 2003a,b, 2004; Grifantini *et al.*, 2003; Sebastian *et al.*, 2002; Shaik *et al.*, 2007), FNR and NsrR, which are involved in anaerobic growth response and will be discussed in greater detail in Chapter 2 (Householder *et al.*, 1999; Lissenden *et al.*, 2000; Overton *et al.*, 2006; Rock *et al.*, 2007; Whitehead *et al.*, 2007), and the architectural-DNA-binding protein integration host factor (IHF), which is involved in a host of regulatory schemes (Fyfe and Davies, 1998; Goosen and van de Putte, 1995; Hill *et al.*, 1997; Lee *et al.*, 2006; Martin *et al.*, 2005). In general, the pathogenic *Neisseria* possess a relatively small number of transcription factors when compared with other bacteria such as *E. coli*, and this review will delve into some of these systems in greater detail in the main body of this chapter.

An unusual feature of the pathogenic *Neisseria* is that they utilize few two-component regulatory systems, which probably reflects the restrictive environment in which they find themselves. Two-component regulatory systems consist of a membrane-bound sensor kinase and an associated DNA-binding regulatory protein. When the sensor kinase senses an external stimulus this leads to autophosphorylation of the sensor which then transfers the phosphate moiety to the cognate transcriptional regulator. Once phosphorylated the regulatory protein becomes activated and thereby regulates transcription at target genes. In the pathogenic *Neisseria*, the number of two-component regulatory systems that have been identified is small. These include a PhoP/PhoQ system of *N. gonorrhoeae* and an orthologous MisR/MisS system of *N. meningitidis* which respond to varying magnesium levels within the environment and may play a role in serum resistance (Johnson *et al.*, 2001; Tzeng *et al.*, 2006). Another two component system is the NarQ/NarP system, which operates in conjunction with the FNR transcription factor, that responds to growth under anaerobic conditions (Overton *et al.*, 2006; Whitehead *et al.*, 2007).

The preceding examples of gene regulatory mechanisms that act at the level of transcription initiation involve DNA-binding regulatory proteins. In many bacteria, and in particular the pathogenic *Neisseria*, a distinct gene expression strategy termed phase variation works stochastically, independent of any regulatory proteins. Phase variation controls expression of many proteins particularly immunogenic surface components such as the Opacity proteins and lipooligosaccharide expression in both *N. gonorrhoeae* and *N. meningitidis* (Banerjee *et al.*, 1998; Belland *et al.*, 1997; Jordan *et al.*, 2005; Meyer *et al.*, 1986; Muralidharan *et al.*, 1987; Saunders *et al.*, 2000). At the molecular level, phase variation is mediated by multiple mechanisms which affect both transcription and translation. For example, regulation of the *porA* and *opc* genes of *N. meningitidis* is affected by subtle changes in the promoter sequence causing the −35/−10 interval to change which leads to modulation of their transcription levels (Sarkari *et al.*, 1994; van der Ende *et al.*, 1995). This is in contrast to the regulation of many genes such as the Opacity (*opa*) genes, the lipooligosaccharide glucosyl transferase (*lgt*) genes and the haemoglobin receptor genes, *hpuAB* and *hmbR*, which are also mediated by changes in the underlying chromosomal sequence, but whose effects are manifest through changes in the translational reading frame (Banerjee *et al.*, 1998; Lewis *et al.*, 1999; Muralidharan *et al.*, 1987; Richardson and Stojiljkovic, 1999). In all cases, phase variation enables the pathogenic *Neisseria* to modulate gene expression in a switch-like manner transitioning between ON and OFF states and back again.

Other translational control mechanisms have recently gained more attention in the pathogenic *Neisseria* as well as in other bacteria and include cis-acting elements within the mRNA molecules, trans-acting, small RNA regulators as well as mechanisms that affect mRNA access to the translation machinery and mRNA turnover (Condon, 2007; Gottesman, 2005; Majdalani *et al.*, 2005; Mellin *et al.*, 2007; Winkler and Breaker, 2005). Collectively, the various gene expression strategies employed by *N. gonorrhoeae* and *N. meningitidis* constitute a complex symphony of regulatory mechanisms which enable each organism to enter a human host, establish a foothold, colonize the appropriate mucosal surface, and finally, allow the organism to proliferate prior to moving on to the next host. Consequently, the orchestration of these expression strategies must ensure that the correct genes are expressed at the appropriate time and in the appropriate locale. Ultimately, understanding how these differing transcriptional and posttranscriptional regulatory mechanisms function and interact will be invaluable as these various strategies underpin the pathogenesis of both *N. gonorrhoeae* and *N. meningitidis.*

Classical regulation

Alternative sigma factors

As mentioned previously, the pathogenic *Neisseria* utilize only a small number of alternative sigma factors in order to regulate gene expression. A search of the various *Neisseria* genomes that have been sequenced has identified only *rpoD, ecf (rpoE), rpoH* and an inactive *rpoN*-like sigma factor (Fig. 1.1) (Du and Arvidson, 2006; Du *et al.*, 2005; Gunesekere *et al.*, 2006a,b; Laskos *et al.*, 2004). Of these, the RpoH sigma factor is the most well characterized and has been shown to be a mediator of the heat stress response regulon (Gunesekere *et al.*, 2006a; Laskos *et al.*, 2004), as well as of genes that are important for invasion of mucosal epithelial cells (Du and Arvidson, 2006).

RpoH is important for responding to environmental stresses and is itself regulated at three different levels: transcription, translation and protein activity (Du *et al.*, 2005; Laskos *et al.*, 2004). In gonococci, the *rpoH* gene is transcribed from an RpoD-dependent promoter, with transcription increasing in response to heat stress as well as when gonococci attach to epithelial cells (Du *et al.*, 2005). However, RpoH regulation is augmented at the post-transcriptional level via two methods: (i) *rpoH* mRNA translation is prevented by an intrinsic secondary structure motif that sequesters the ribosome binding site and evidence suggests this mechanism functions in *N. gonorrhoeae* as observed in other organisms (Laskos *et al.*, 2004; Yuzawa *et al.*, 1993), and (ii) RpoH protein activity is also modulated via binding of preformed RpoH using the DnaK chaperone system which ultimately targets RpoH for protease degradation under normal conditions (Laskos *et al.*, 2004). In both of the above scenarios, RpoH polypeptide is sequestered away from the RNAP core holoenzyme thus preventing RpoH-mediated activation. However, RpoH can become activated when bacteria are subjected to heat stress because under heat stress denatured proteins accumulate within the cell allowing the DnaK complex to preferentially bind the misformed proteins thereby releasing RpoH within the cytoplasm resulting in a rapid up-regulation of RpoH-activated genes.

The above mode of RpoH regulation was determined by overexpressing the protein from a plasmid in conjunction with microarray analysis which identified 12 genes whose expression levels were stimulated under these conditions. The proteins that were induced included the protein chaperones GroEL/GroES, DnaK/DnaJ/GrpE and ClpB as well as the proteases ClpXP, ClpAP, HslUV, Lon and ClpB (Gunesekere *et al.*, 2006a) which collectively coordinate protein homeostasis. In gonococci, RpoH also appears to be an important regulator of *ppiD* (NGO0376) which has been shown to be involved in the invasion of epithelial cells (Du *et al.*, 2006). Therefore, RpoH plays a major, yet complicated role in how gonococci respond to physical stress and epithelial cell invasion making this sigma factor an important coordinator of *N. gonorrhoeae* pathogenesis.

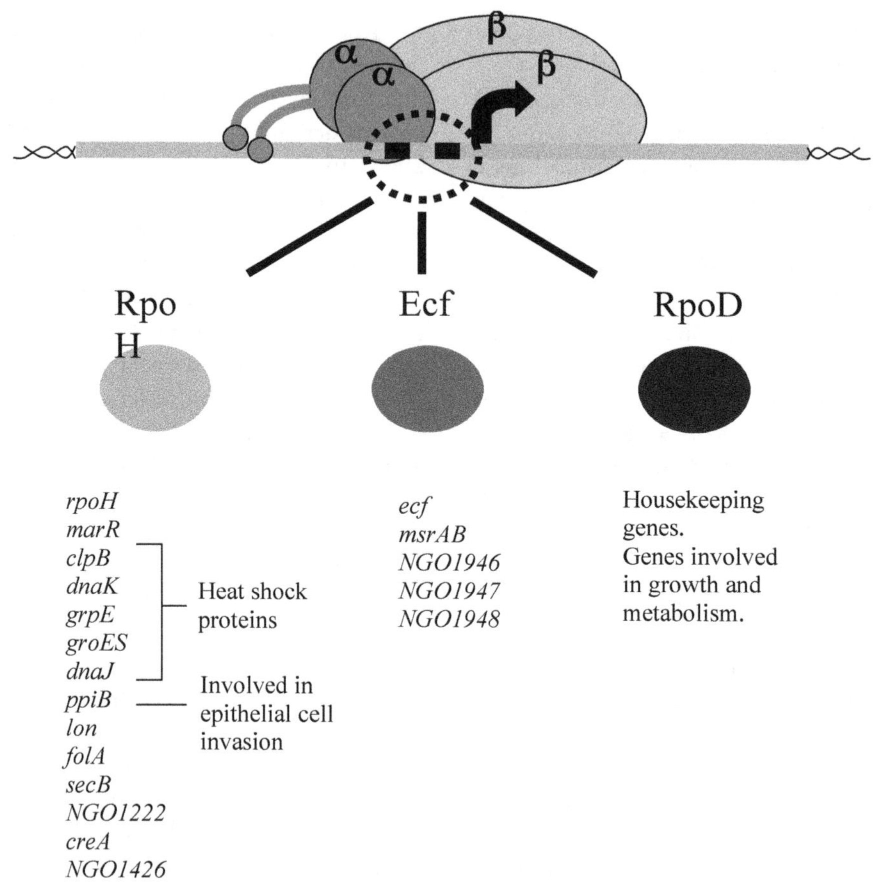

Figure 1.1 Regulation by alternative sigma factors. A schematic of the core RNA polymerase holoenzyme consisting of two β subunits, two α subunits and σ factor subunit (outlined by dashed lines) is shown. The alternative sigma factor subunits utilized by the *Neisseria gonorrhoeae* are shown below with the subset of genes known to be regulated by each respective sigma factor listed below.

Small DNA-binding proteins

Role of integration host factor as a transcription factor

Small DNA-binding proteins exert their effects on gene transcription in multiple ways. Some prevent RNA polymerase from binding to a promoter thereby repressing transcription, others directly recruit RNA polymerase and stimulate transcription, or, alternatively, other small DNA-binding protein recruit yet other regulatory proteins that either activate or repress RNAP activity. Integration host factor (IHF) is one such small DNA-binding protein that along with HU protein have long been known to be important regulatory proteins in Gram-negative bacteria(Friedman, 1988). IHF is a multi-functional, heterodimeric protein encoded by the *ihfA* (*himA*) and *ihfB* (*hip* or *himD*) genes (Flamm and Weisberg, 1985; Miller, 1984) and participates in many diverse biological systems (Drlica and Rouviere-Yaniv, 1987; Friedman, 1988; Goosen and van de Putte, 1995; Rice *et al.*, 1996). Physical analysis has shown that IHF binds to its target DNA in a sequence-specific manner, and, following binding to its target sequence, causes localized DNA bending within the vicinity of the IHF binding site (Rice *et al.*, 1996). Therefore, by bending DNA, IHF either facilitates the alignment of transcriptional regulators with their cognate promoter, recruit accessory proteins to a particular site, or, correctly juxtapose DNA segments for either genetic exchange or activation of RNA polymerase (Rice *et al.*, 1996). Gonococcal IHF is also a heterodimeric protein encoded by the *N. gonorrhoeae ihfA* and *ihfB* genes (Hill *et al.*, 1998) and likewise purified protein has also been shown to physically bend DNA when added in an *in vitro* assay (Hill *et al.*, 1997).

A slight variation on the above themes is observed with IHF regulation of the *N. gonorrhoeae pilE* gene encoding the type IV pilin subunit. Like *E. coli* IHF protein, the gonococcal homologue recognizes a similar consensus DNA binding motif which has allowed similar motifs to be identified in several other neisserial systems (Hill *et al.*, 1997, 2002; Lee *et al.*, 2006; Martin *et al.*, 2005). Involvement of IHF in the transcriptional regulation of *pilE* was initially demonstrated through gel retardation assays and DNA footprinting (coupled with chromosomal deletion analysis) of the *pilE* promoter DNA using purified gonococcal IHF which identified an IHF-binding site immediately upstream of the −35/−10 promoter (P1) of *pilE* (Hill *et al.*, 1997). When the IHF-binding site was deleted from the gonococcal *pilE* promoter, transcription from the P1 promoter was reduced approximately 10-fold when *pilE* mRNA levels were assessed by Northern blotting and primer extension analysis (Hill *et al.*, 1997). Likewise, a comparable reduction in the level of transcription was also observed when the IHF binding site was mutated in a recombinant construct where the *pilE* promoter DNA was fused to a chloramphenicol acetyl transferase (CAT) reporter gene (Fyfe and Davies, 1998). Initially, it was suggested that IHF binding to the *pilE* promoter induced a bend in the promoter DNA which would facilitate the interaction of some upstream regulatory protein with RNA polymerase (Hill *et al.*, 1997). However, owing to an appropriate regulatory protein having not been identified, such ideas are currently out of favour. The current model for *pilE* transcription now involves IHF binding in conjunction with specialized promoter sequences called UP elements (which flank the IHF binding site; see Fig. 1.2) (Fyfe and Davies, 1998). UP elements are AT-rich sequences that bind to the alpha subunits of RNAP and in doing so stimulate the formation of an active open transcription complex. Consequently, the resulting effect of IHF binding at the *pilE* promoter appears to be the establishment of the correct promoter structure which allows RNAP to engage in transcript elongation.

More conventional IHF effects have been observed at the genes encoding the FarAB efflux pump (Lee *et al.*, 2006). The FarAB efflux pump, along with the mtrCDE efflux system (Hagman *et al.*, 1995; Pan and Spratt, 1994), is a member of the major facilitator family of bacterial efflux pumps and is important for gonococcal resistance to hydrophobic antimicrobial agents such as long-chain fatty acids (Lee *et al.*, 2003; Lee and Shafer, 1999) which are present in mucosal secretions and will be discussed in more detail in Chapter 13. Transcription of *farAB* is jointly regulated by IHF and the transcriptional repressor FarR, a member of the MarR family of transcriptional regulators that is found in both *N. gonorrhoeae* and *N. meningitidis.* FarR binds as a dimer to two sites in the promoter region of the *farAB* operon that lie adjacent to the IHF binding site. FarR alone is unable to repress *farAB* transcription. However, IHF binding is proposed to induce a bend in the promoter DNA (Lee *et al.*, 2006) that orients the promoter in such a way that facilitates binding of two FarR dimers which subsequently causes *farAB* transcription to be repressed.

Iron homeostasis and the ferric uptake regulator

The ability to scavenge the host environment for essential nutrients is a key requirement for survival of bacterial pathogens (Escolar *et al.*, 1999). Many bacteria utilize transcriptional regulators which bind metal ions, and, upon binding, are triggered to regulate the transcription of target genes involved in the homeostasis of that particular ion species (Braun *et al.*, 1998). Thus, for bacteria, many ion-binding regulators act as environmental sensors of essential ions and help adjust gene expression accordingly. The ferric uptake regulator (Fur) is a master regulator of iron homeostasis in many Gram-negative and Gram-positive bacteria and the genetic response to changing iron levels has been linked to the regulation of a number of virulence factors in *Neisseria* (Delany *et al.*, 2004). The Fur protein is a well-studied transcription factor and is representative of many metal ion-responsive transcriptional regulators such as MntR, DtxR and Zur.

Neisserial Fur is a ~16-kDa protein which monitors the level of ferrous ions in the cytoplasm via two metal-binding sites within the polypeptide

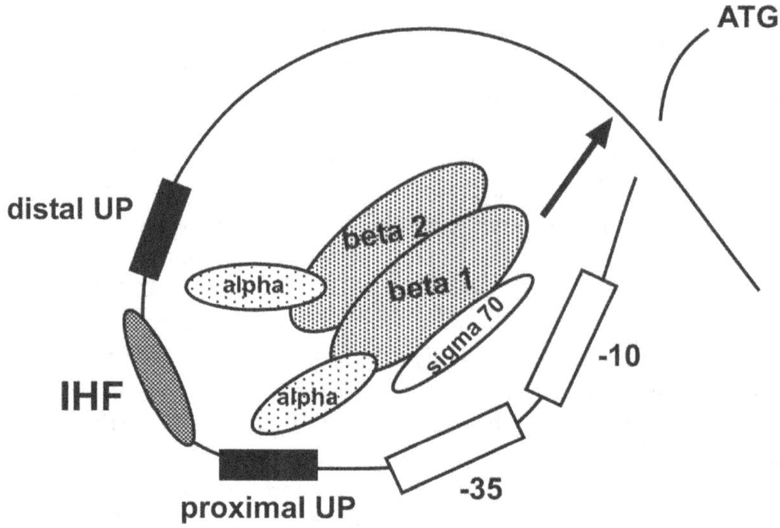

Figure 1.2 IHF effect at *pilE*. IHF has been shown to bind upstream of the –35/–10 P1 promoter at *pilE*. Flanking the IHF binding site are two UP-elements that are thought to interact with the alpha subunits of RNAP. Therefore, IHF binding to the *pilE* promoter is hypothesized to induce a DNA bend within the promoter DNA which facilitates the interaction of the UP elements with RNAP leading to the formation of an open active transcription complex.

(Escolar *et al.*, 1999; Hantke, 2001). In an iron-limited environment, Fur is present in the cytoplasm as a monomer. However, as iron concentrations in the environment increase, ferrous ions are translocated across the inner and outer membranes, which upon binding to Fur cause the protein to undergo a conformational change and homodimerize (Stojiljkovic and Hantke, 1995). In the dimeric conformation, Fur can bind DNA at specific sequences termed Furboxes (Desai *et al.*, 2006; Grifantini *et al.*, 2003; Stojiljkovic and Hantke, 1995). In most examples, when Fur binds to a Furbox within the promoter of a target gene, transcription is repressed due to the occlusion of RNA polymerase binding at the promoter. However, in one instance, Fur binding to the *norB* promoter has been shown to activate transcription in *N. meningitidis* (Delany *et al.*, 2004). Through alignment analysis of DNA fragments that presented electrophoretic mobility shifts when incubated with recombinant meningococcal Fur protein, a 24 basepair consensus binding site was identified $(NATWAT)_4$, which closely matched a neisserial consensus sequence that had been proposed after analysis using *E. coli* derived Fur (Grifantini *et al.*, 2003).

In the pathogenic *Neisseria*, Fur functions more as a global regulator rather than as a specific regulator of genes involved in iron homeostasis. Microarray analysis of the iron-regulated transcriptome of *N. gonorrhoeae* identified 203 such iron-regulated genes (Ducey *et al.*, 2005). A similar study in *N. meningitidis* identified 233 genes whose transcription was affected by changes in iron concentration within the medium (Grifantini *et al.*, 2003). Of these 233 genes, half were shown to contain a Fur-binding consensus sequences in their promoter regions (Grifantini *et al.*, 2003). A subsequent study that examined the microarray-derived transcriptome profile obtained from a meningococcal Fur mutant identified 83 genes whose transcription is apparently controlled by Fur protein (Delany *et al.*, 2006). Interestingly, transcription of 44 of these genes was shown to be repressed, with transcription of 39 genes being activated, indicating that binding of Fur protein can have both negative and positive effects on gene expression (Delany *et al.*, 2004, 2006; Shaik *et al.*, 2007).

Of the Fur-regulated genes identified in *N. meningitidis* and *N. gonorrhoeae* many as expected are involved in maintaining iron-homeostasis. Iron is both an essential cofactor for many cellular enzymes yet can be toxic at higher concentrations. Thus, iron homeostasis has to be tightly regulated in all bacteria. Fur regulation has a varied role in this process. For example, enzymes such as superoxide dismutase B (*sodB*) and succinate dehydrogenase (*sodDBAC*) both use iron as a cofactor, with Fur stimulating expression of these genes when iron is plentiful (Delany *et al.*, 2006; Mellin *et al.*, 2007). Conversely, when iron is limiting, the expression of the iron-scavenging siderophores such as the transferrin-binding proteins (*tbpA*, *tbpB*) and the lactoferrin-binding proteins (*lbpA* and *lppB*) are up-regulated, yet when iron again increases in concentration, Fur represses transcription of the genes encoding these proteins (Delany *et al.*, 2006). When one considers that some of these proteins that are regulated by Fur are considered potential vaccine candidates as well as putative virulence factors (Delany *et al.*, 2004), then one can see that Fur plays a pivotal role in not just maintaining iron homeostasis but also in regulating the expression of genes important in meningococcal pathogenesis.

The mechanism of Fur-mediated repression is well established as being via direct binding of dimerized Fur protein to a target promoter. However, the mechanism of Fur-mediated activation is less clear. In bacteria such as *Escherichia coli* and *Pseudomonas aeruginosa,* Fur has been shown to mediate indirect positive effects on gene expression, especially at the post-transcriptional level by repressing the production of small regulatory RNA molecules (Masse and Gottesman, 2002; Wilderman *et al.*, 2004). One Fur-repressed small regulatory RNA, termed NrrF, has recently been identified in *N. meningitidis* with an orthologue being present in *N. gonorrhoeae* (Mellin *et al.*, 2007) indicating that this type of Fur-mediated regulation is present in the pathogenic *Neisseria*. NrrF was shown to be required for the positive regulation of components of the Succinate dehydrogenase enzyme complex (*sdhA* and *sdhC*) which is an iron-containing enzyme of the tricarboxylic acid cycle. Potential basepairing interactions between NrrF and both the *sdhA* and *sdhC* mRNAs were demonstrated but a thorough understanding of the molecular mechanism of NrrF-mediated regulation has yet to emerge. Thus in the case of *sdhA* and *sdhC,* activation by Fur is mediated indirectly by repression of a regulatory small RNA. However, many Fur-activated genes are not regulated by NrrF, leaving the details of Fur-mediated gene activation incomplete.

Denitrification: FNR and NsrR and NarQ/NarP

Neisseria gonorrhoeae has been isolated from clinical samples along with an abundance of obligate anaerobes suggesting that *in vivo N. gonorrhoeae* can grow anaerobically (Boulanger and Murphy, 2002; Knapp and Clark, 1984). Subsequent work has established that *N. gonorrhoeae* and *N. meningitidis* are able to survive under anaerobic conditions by using a truncated denitrification pathway in which nitrite is reduced to nitrous oxide by the nitrite and nitric oxide reductases AniA and NorB respectively (Clark *et al.*, 1987; Knapp and Clark, 1984; Lissenden *et al.*, 2000). The regulation of AniA and NorB expression under anaerobic conditions is complex and will be discussed at length in Chapter 2, however it is touched on it here as an example of a how complex genetic regulation can integrate many signals via cooperating and counteracting transcription factors (Fig. 1.3).

Regulation of *aniA* and *norB* involves the oxygen-sensing transcription factor FNR, the NO-sensing repressor NsrR, the transcriptional repressor ArsR, the iron-binding ferric uptake regulator Fur, as well as the two-component signalling system NarQ/NarP (Householder *et al.*, 1999; Isabella *et al.*, 2008; Overton *et al.*, 2006; Rodionov *et al.*, 2005). Interestingly, deletion of the *fnr* gene from *N. meningitidis* strain MC58 attenuated proliferation of the organism in both an infant rat and an infant mouse model by over an order of magnitude, suggesting that adaptation to anaerobic growth is an important aspect of meningococcal pathogenesis (Bartolini *et al.*, 2006).

Both *aniA* and *norB* are divergently transcribed from promoters located within a common intergenic region which contains putative binding sites for NsrR, FNR and NarP as well as Fur and ArsR (Householder *et al.*, 1999; Isabella *et al.*, 2008; Overton *et al.*, 2006; Rodionov *et al.*, 2005). For *N. gonorrhoeae,* if nitrous oxide (NO) is absent, then NsrR binds to the *norB* and *aniA* promoters as well as to a promoter of a gene of unknown function (*dnrN*) causing transcription to be repressed (Overton *et al.*, 2006). Repression is relieved by NO which inactivates NsrR's DNA-

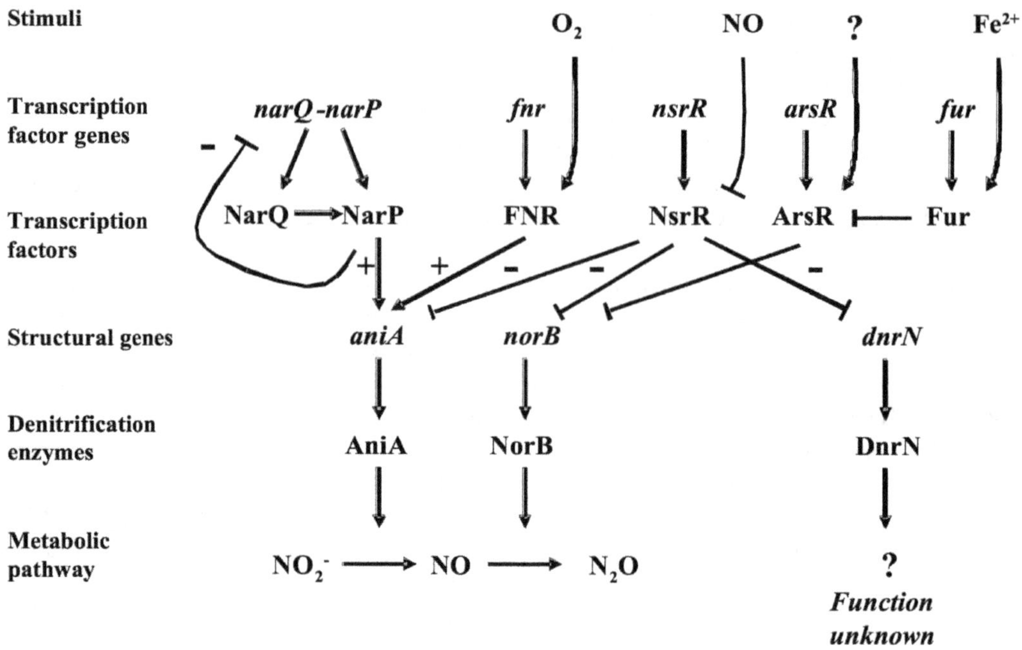

Figure 1.3 The truncated denitrification pathway of *N. gonorrhoeae* and its regulation by FNR, NarQ–NarP, and NsrR. The various stimuli, genes, and transcription factors involved in the denitrification pathway are shown and their interactions diagrammed. Arrows or bars indicate transcriptional activation or repression by the transcription factor shown, respectively. Arrows or bars from stimuli also indicate activation or repression of the indicated transcription factor, respectively.

binding activity and in doing so induces expression of *norB*. Additionally *norB* expression can be repressed by the ArsR repressor protein, although the signals ArsR responds to remain unknown. This ArsR-mediated repression in turn can be antagonized by the binding of iron-bound Fur to the same region of the promoter, thus there are many signals which can both repress and relieve repression of *norB* expression.

For *N. meningitidis*, similar events seem to occur, with *aniA* expression also being elevated in an NsrR mutant. However, in meningococci, *aniA* transcription does not appear to be induced by the presence of NO but rather is positively regulated by additional factors which include the availability of oxygen as well as the availability of nitrite (Rock *et al.*, 2007). These additional layers of *aniA* regulation appear to be mediated by the involvement of the FNR regulator in conjunction with the NarQ/NarP two component regulatory system. With this system, FNR 'senses' oxygen concentrations in the environment through direct binding to O_2 molecules. Once oxygen-activated, FNR induces the expression of nine transcriptional units in *N. meningitidis* (Bartolini *et al.*, 2006) and up to fourteen transcriptional units in *N. gonorrhoeae* including *aniA* (Whitehead *et al.*, 2007). Furthermore, microarray analysis showed that FNR also appears to repress transcription of six genes in *N. gonorrhoeae* which included *norB* and *dnrN* which is logical given that AniA reduces nitrite to NO which then derepresses NsrR repression of these genes. Finally, the NarQ/NarP system seems to be also required for full activation of *aniA* transcription as demonstrated by studies showing significantly reduced β-galactosidase activity from an *aniA–lacZ* fusion in a *narP* mutant. Unfortunately, the ligand for NarQ, which is the sensor-kinase component of the system, has not been identified, and there is some speculation that NarQ is always in a ligand-insensitive, 'locked-on' state although the details of this remain unclear (Whitehead *et al.*, 2007).

Therefore, by using these three different transcriptional regulators, *N. meningitidis* and *N. gonorrhoeae* are able to respond to both anaerobic growth conditions as well as respond to NO exposure as may be encountered following macrophage/neutrophil digestion during the course of an infection. So whereas the denitrification pathway, which is essential for anaerobic growth, requires both the nitrite reductase AniA and the nitrous oxide reductase NorB, responding to NO exposure appears to require only the expression of the *norB* gene. Thus, the pathogenic *Neisseria* have evolved a gene expression strategy that allows them to sense various environmental inputs at differing stages in the denitrification pathway and adjust the expression of various genes accordingly.

Stochastic regulation of gene expression

While transcription factors exert their regulatory effects in response to environmental cues, many bacterial genes are regulated stochastically via changes in the underlying chromosomal sequence. Generally this type of regulation, termed phase variation, results in either an ON or OFF expression state for a particular gene, although in some instances, this can result in graded levels of expression. Importantly, many immunogenic outer membrane components as well as several virulence factors are subject to phase variation in the pathogenic *Neisseria* and it appears that phase-variable expression of genes is a key element of neisserial pathogenesis. An introduction to the mechanisms of phase-variation will be discussed here.

Frameshifting mutations

The opacity proteins of *N. meningitidis* and *N. gonorrhoeae* are outer membrane adhesins that are important for intimate attachment to host mucosal surfaces as well as other lymphocytic interactions (Henderson *et al.*, 1999) (see Chapter 5). The gonococcal genome contains 11–12 *opa* genes depending upon the strain (Bhat *et al.*, 1992; Connell *et al.*, 1990) that code for related but structurally distinct proteins and their expression undergoes frequent phase transitions during the course of an infection (Swanson *et al.*, 1987). The molecular basis of this ON/OFF switching is the presence within the signal peptide encoding region of -CTTCT- pentameric repeats termed coding repeats (CRs). Polypurine and polypyrimidine stretches within the duplex (such as the -CTTCT- repeat) have the capacity to form H-DNA structures (Belland, 1991) which consist of a region of triple-stranded DNA with an extruded single-stranded region (Fig. 1.4). Such regions are thought to promote slipped-stranded mispairing events (Dybvig, 1993). During DNA replication, transient slipped-stranded mispairing within this region leads to the addition or subtraction of CRs in the resultant chromosomal copy. Depending upon the number of CRs that are either added or subtracted in the respective *opa* gene, the daughter cell will inherit an out of frame gene in two out of three instances. An out-of-frame mutation results in the expression of a truncated Opa protein in the daughter strain and

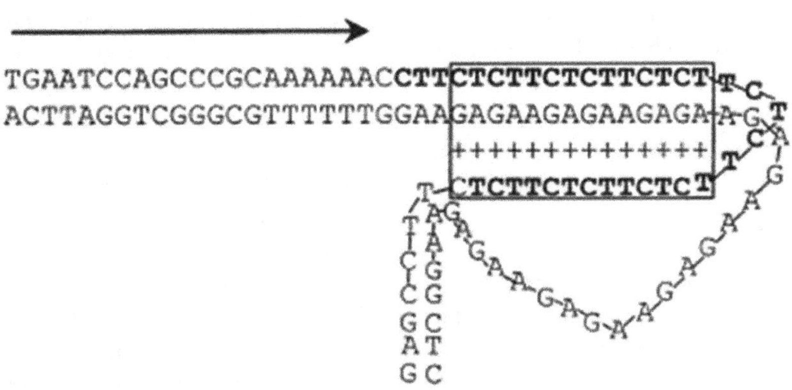

Figure 1.4 Schematic representation of H-DNA formation in a CR region. The *opaC* coding repeat region is shown in bold and is used to indicate the proposed triple-stranded and single-stranded regions. Residues within the triple-stranded region forming triple residue bonds (C:G:C or T:A:T) are shown within the box. The arrow indicates the direction of *opaC* transcription.

this Opa is not expressed on the cell surface (Fig. 1.5A). Consequently, *N. gonorrhoeae* is able to switch from an Opa^+ to Opa^- expression state by regulating expression of the Opa proteins at the level of translation via stochastic alterations in the underlying gene sequence.

The genes of the lipooligosaccharide (LOS) synthesis pathway are subject to a similar type of frameshifting-mediated phase variation but use a slightly different mechanism. LOS of *N. gonorrhoeae* consists of a lipid A moiety coupled to an oligosaccharide core with three variable oligosaccharide side branches termed the α, β and γ chains. *Neisseria* spp. encode a number of different lipooligosaccharide glucosyl transferases (*lgts*) responsible for the biosynthesis of these oligosaccharide sidechains and depending upon which glucosyl transferases is expressed, the resulting LOS molecule can have a widely different structure (Banerjee *et al.*, 1998; Yang and Gotschlich, 1996). As was the case with the *opa* genes, expression is controlled at the level of translation. However, rather than using CRs, the *lgt* genes contain either short poly(C) or poly(G) tracts in their coding regions and these short homopurine or homopyrmidine tracts allow similar slip-stranded mispairing during replication resulting in the insertion or deletion of a nucleotide(s) into the poly(C)/(G) tract and a potential frameshift in the coding region of the gene. In *N. gonorrhoeae,* this mechanism controls the expression of the LgtG glucosyl transferase (Banerjee *et al.*, 1998) which is responsible for attachment of the β chain, as well as expression of the LgtA, LgtB and LgtD glucosyl transferases which attach the α chain to LOS (Yang and Gotschlich, 1996). Thus, during the course of an infection multiple LOS variants will be present at any given time reflecting the differing *lgt* expression states from one bacterium to another.

Promoter mutations

In the case of *opa* and *lgt* regulation, nucleotides are either inserted or deleted within the coding region of the gene which ultimately influences subsequent translational outcomes. However, a different form of phase variation can moderate transcription and is mediated via small changes in repetitive stretches of nucleotides that are located within the promoter region of a gene. One such gene regulated in this fashion is the *opc* gene of *N. meningitidis* which also encodes for an outer membrane adhesin. Within the *opc* promoter resides a variable tract of repeated cytosine residues immediately upstream of the −10 region of the promoter. Addition or subtraction of nucleotides within this region during DNA replication results in the expansion or contraction of the −35/−10 interval which influences the transcriptional activity of the *opc* gene. For example, when the number of cytosine residues is 12 or 13, meningococci exhibit high levels of *opc* transcription. In contrast, when the number of cytosine residues is either 11 or 14, *opc* transcription is diminished approximately 10-fold, and is no longer observed when the number of cytosine residues in the poly(C) tract is less than 11 or greater than 15 (Henderson *et al.*, 1999; Sarkari *et al.*, 1994). Thus, by subtly altering the number of residues in the poly(C) tract within the *opc* promoter during DNA replication, *N. meningitidis* is able to transition between the Opc^{++}, Opc^+ and Opc^- expression states (Fig. 1.5B).

Similar transitions between 'low', 'medium' and 'high' transcription states are observed in the regulation of yet another adhesin (NadA) of *N. meningitidis*. NadA is expressed by three of the four hypervirulent strains of *N. meningitidis* and promotes the attachment and invasion of epithelial cells (Capecchi *et al.*, 2005). The *nadA* promoter has a region of TAAA repeats, termed a microsatellite, which can vary in number due to slip-stranded mispairing events during DNA replication. Depending upon the number of repeats that are present in the promoter, *nadA* will exhibit either a low-, medium- or high-level expression phenotype (Martin *et al.*, 2005). The repetitive sequences within the microsatellite region (as well as regions further upstream) also contain putative IHF and Fur binding sites which also appear to be involved in the repression of *nadA* expression. Phase-variable regulation of *nadA* transcription appears to depend only upon changes within the microsatellite region. However, if stochastic mutations in this microsatellite region influence IHF binding to the *nadA* promoter (Martin *et al.*, 2005) then this may provide an added layer of *nadA* regulation through IHF and Fur protein interactions at the promoter. Thus, phase-variable transcription of *nadA* potentially 'incorporates

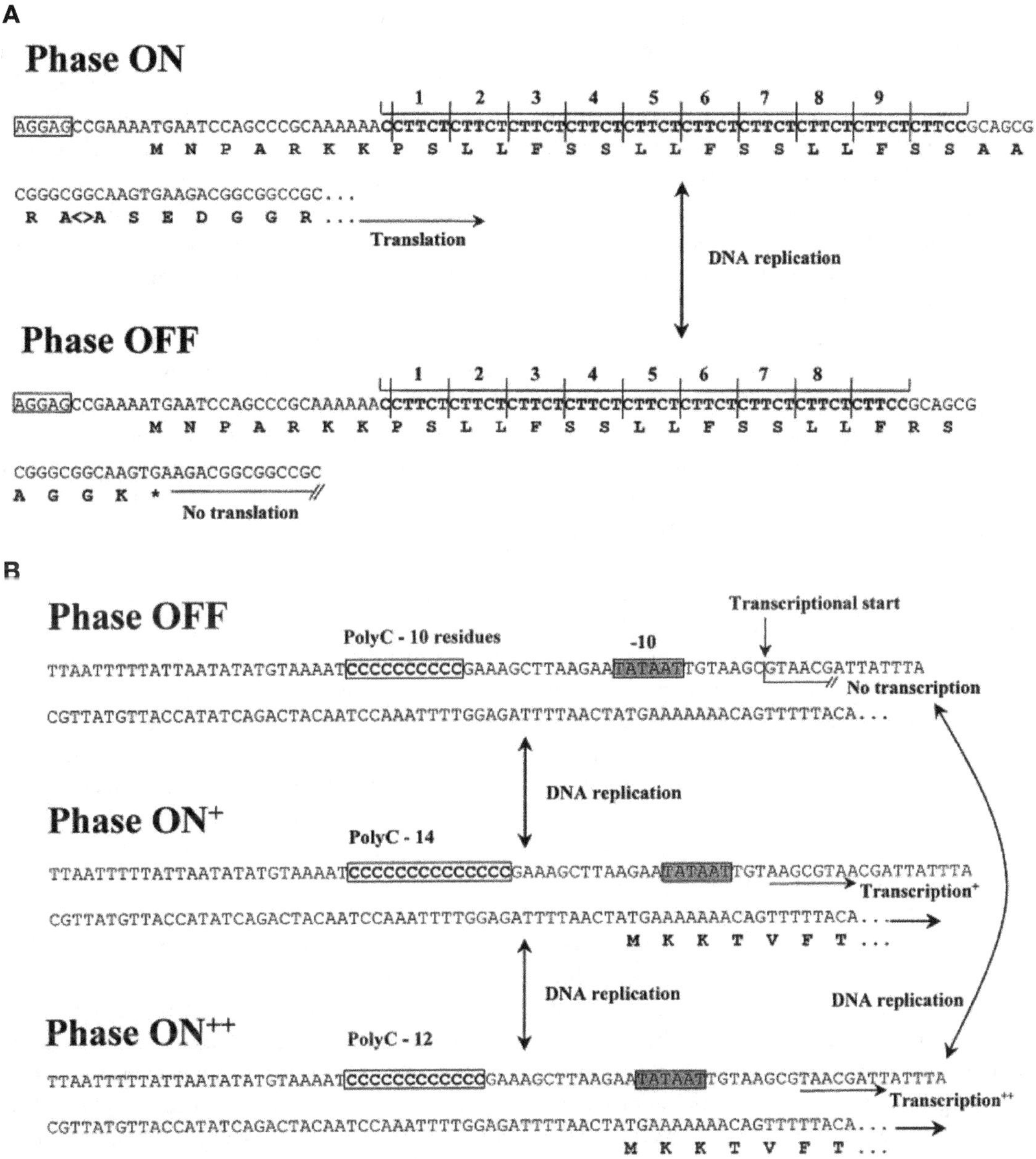

Figure 1.5 Phase variation mechanisms. **A** Schematic representation of the *opaC* gene from *N. gonorrhoeae*. The sequence contains the Shine–Dalgarno sequence (boxed) and extends to the first few residues of the coding region. The signal peptidase cleavage site is indicated (<>). Within the region encoding the signal sequence of the protein are tandem CRs (designated 1–9). The phase ON state contains nine CRs. The phase OFF state possesses eight CRs, resulting in the introduction of a stop codon (asterisk) at the end of the signal peptide encoding region and the formation of a truncated peptide. **B** Representation of the transcriptional strand-slippage mechanism of phase variation in the *opc* gene of *N. meningitidis*. The sequence of the 5′-flanking region plus the 5′-terminal portion of the *opc* gene is shown. The –10 region and a tract of contiguous C bases (polycytidine) are boxed. The mapped transcriptional start site is indicated. Transcription does not occur in the phase OFF state, which possesses 10 cytidine residues. Intermediate transcription occurs in those sequences that contain 14 C bases, and maximal transcription occurs in those containing 12 C residues.

interplay between stochastic (mutational) and prescriptive (classical) mechanisms of gene regulation' (Martin *et al.*, 2005).

As can be seen, phase variation is clearly a major means by which the pathogenic *Neisseria* respond to their environment (phase variation is discussed in greater detail in Chapter 4). This was substantiated by a genome-wide comparison of the four publicly available, sequenced *N. gonorrhoeae* genomes which showed that, excluding the *opa* and *pilC* genes, there are 64 genes which contain coding repeat sequences in either their promoter or coding region that have the potential to undergo phase variation (Jordan *et al.*, 2005). An earlier comparative genome study had identified 119 putative phase-variable genes in *N. meningitidis* (Snyder *et al.*, 2001). All told, the pathogenic *Neisseria* spp. seem to possess more potential phase-variable genes than any other bacterial species described to date (Snyder *et al.*, 2001). Consequently, any given mucosal population of pathogenic *Neisseria* spp. is going to be non-clonal with respect to their differing expression profiles arising from their array of phase-variable genes. Importantly, a study of epidemic *N. meningitidis* serogroup A isolates has revealed increased mutation rates associated with defects in the mismatch repair pathways in over half of the clones studied (Richardson *et al.*, 2002), with the elevated mutation rates correlating with significantly increased phase variation rates. Consequently, it was hypothesized that the greater phenotypic diversity of these hypermutator populations would more readily lead to the emergence of variants with the appropriate phenotype to evade immune pressure and increase epidemic spread.

Rec-mediated phase variation

Opa variation and the various other frameshifting scenarios outlined above are examples of either illegitimate recombination events (i.e. occur RecA-independently), or, alternatively, occur due to errors during DNA replication. RecA-mediated homologous recombination is designed to maintain fidelity during the recombination process. Nonetheless, an intrinsic error rate is observed in the *Neisseria*. For example, phase variation of pilus expression in the gonococcus can occur either (i) RecA-independently, when replication errors accumulate in poly-cytosine tracts located within the *pilE* open reading frame (Koomey *et al.*, 1987) or in poly-cytosine tracts that precede the *pilC* gene (Robertson and Meyer, 1992) or (ii) RecA-dependently, when *pilE/pilS* recombination events lead to the creation of missense or nonsense mutations within the *pilE* opening reading frame (Bergstrom *et al.*, 1986). In either scenario, pilus$^+$ to pilus$^-$ transitions are observed. Because RecA-mediated homologous recombination is intrinsically mutagenic, and, given the innate competency of the *Neisseria* spp. for DNA transformation, then horizontal exchange of chromosomal DNA at the mucosal surface has the capacity to phase-vary many genes due to recombination error within repeated sequences.

Future directions

Studies on gene regulation in the pathogenic *Neisseria* have predominantly focused on prescriptive, protein-mediated mechanisms as well as stochastic, mutationally driven mechanisms. However, it has become increasingly clear that this paints an incomplete picture of neisserial gene regulation strategies. Additional regulatory mechanisms, such as regulatory small RNA molecules, riboswitches and mRNA processing pathways, have garnered significant attention in other organisms, and recent work is beginning to establish a similar role for some of these mechanisms in the pathogenic *Neisseria*.

Certainly some of the most prevalent of these newly appreciated regulatory schemes involve posttranscriptional regulatory small RNAs. Small RNA molecules have been implicated in the regulation of wide-ranging processes that include iron homeostasis (Masse and Gottesman, 2002; Wilderman *et al.*, 2004), adaptation to temperature changes (Repoila and Gottesman, 2001; Repoila and Gottesman, 2003) and quorum-sensing (Lenz *et al.*, 2004) to name but a few. As mentioned previously, the first such regulatory small RNA (sRNA) of the pathogenic *Neisseria* was recently identified by a genome-wide bioinformatics screen searching for Fur-regulated sRNA molecules and was named neisserial regulatory RNA involved with iron (NrrF) (Mellin *et al.*, 2007). Interestingly, this study also identified four previously unknown transcripts that were speculated to be unknown sRNA molecules although these remain currently uncharacterized. However, their discovery does suggest that the pathogenic *Neisseria,* like other Gram negative bacteria, may eventually be shown to use a broad array of sRNA regulators in the control of gene expression.

The mechanism of NrrF-mediated repression of these genes remains unclear. Potential basepairing interactions between the NrrF and its target mRNAs have been demonstrated and since basepairing interactions are central to the function of many known sRNAs, this is likely to be crucial for NrrF as well. The authors of this study also proposed the possibility that NrrF might work in conjunction with an RNA chaperone protein termed Hfq which has been shown to be an essential cofactor for much sRNA-mediated regulation in other bacteria (Aiba, 2007; Brennan and Link, 2007; Majdalani *et al.,* 2005) and has a highly conserved orthologue in the pathogenic *Neisseria* (Zhang *et al.,* 2002). Unpublished work from the Genco laboratory has indicated that Hfq is required for the appropriate iron-responsive regulation of various genes in the pathogenic *Neisseria* such as the superoxide dismutase B gene (*sodB)* and it is presumed that Hfq will assume a central role in sRNA-mediated gene regulation in the *Neisseria* spp. as observed with other bacteria. However, the details remain to be determined.

Currently, evidence is lacking in the *Neisseria* for riboswitch regulation that manifests either through differential mRNA degradation or translational attenuation. However, given the prevalence of this type of regulation in other bacteria (Gelfand *et al.,* 1999; Johansson *et al.,* 2002; Mandal *et al.,* 2003; Winkler *et al.,* 2002; Winkler and Breaker, 2005), we anticipate that riboswitch regulation through metabolite binding within the un-translated regions of neisserial mRNAs is likely to be demonstrable in the near future.

Conclusions

The gene expression strategies of the pathogenic *Neisseria* are clearly numerous and diverse. Gene expression is regulated not only at the level of transcription but also via posttranscriptional and translational regulatory mechanisms as well. Like all bacteria, *Neisseria* spp. respond to stimuli from their environment through the actions of a number of transcriptional regulatory proteins and it is becoming increasingly clear that they also employ alternative strategies such as small RNA regulatory molecules to modulate gene expression as well. Interestingly, the pathogenic *Neisseria* appear to be unique (along with *Haemophilus* spp.) in the extent with which they utilize stochastic changes in chromosomal structure to diversify the phenotypic make-up of a given population. Consequently, this enables them to respond quickly to various environmental pressures at the population level.

Our understanding of the gene expression strategies of the pathogenic *Neisseria* is still only in its infancy, but with the publishing of the *N. meningitidis* and *N. gonorrhoeae* genome sequences this should lead to the future identification of other transcriptional regulators. Furthermore, there likely remain a bounty of undiscovered and uncharacterized sRNAs and ribozyme factors whose contributions to gene regulation may elucidate many existing conundrums regarding neisserial gene regulation. What is for sure, however, is that furthering our understanding of the gene expression strategies of the pathogenic *Neisseria* will provide invaluable insights into the basis of pathogenesis and may provide the keys to therapeutic intervention and prevention.

References

Aiba, H. (2007). Mechanism of RNA silencing by Hfq-binding small RNAs. Curr. Opin. Microbiol. *10,* 134–139.

Banerjee, A., Wang, R., Uljon, S.N., Rice, P.A., Gotschlich, E.C., and Stein, D.C. (1998). Identification of the gene (lgtG) encoding the lipooligosaccharide beta chain synthesizing glucosyl transferase from *Neisseria gonorrhoeae.* Proc. Natl. Acad. Sci. USA *95,* 10872–10877.

Bartolini, E., Frigimelica, E., Giovinazzi, S., Galli, G., Shaik, Y., Genco, C., Welsch, J.A., Granoff, D.M., Grandi, G., and Grifantini, R. (2006). Role of FNR and FNR-regulated, sugar fermentation genes in *Neisseria meningitidis* infection. Mol. Microbiol. *60,* 963–972.

Belland, R.J. (1991). H-DNA formation by the coding repeat elements of neisserial opa genes. Mol. Microbiol. *5,* 2351–2360.

Belland, R.J., Morrison, S.G., Carlson, J.H., and Hogan, D.M. (1997). Promoter strength influences phase variation of neisserial opa genes. Mol. Microbiol. *23,* 123–135.

Bergstrom, S., Robbins, K., Koomey, J.M., and Swanson, J. (1986). Piliation control mechanisms in *Neisseria gonorrhoeae.* Proc. Natl. Acad. Sci. USA *83,* 3890–3894.

Bhat, K.S., Gibbs, C.P., Barrera, O., Morrison, S.G., Jahnig, F., Stern, A., Kupsch, E.M., Meyer, T.F., and Swanson,

J. (1992). The opacity proteins of *Neisseria gonorrhoeae* strain MS11 are encoded by a family of 11 complete genes. Mol. Microbiol. *6*, 1073–1076.

Boulanger, M.J., and Murphy, M.E. (2002). Crystal structure of the soluble domain of the major anaerobically induced outer membrane protein (AniA) from pathogenic *Neisseria*: a new class of copper-containing nitrite reductases. J. Mol. Biol. *315*, 1111–1127.

Braun, V., Hantke, K., and Köster, W., (1998) Bacterial iron transport: mechanisms, genetics, and regulation. In: Sigel, A., and Sigel, H., Editors, *Metal Ions in Biological Systems, Vol. 35, Iron Transport and Storage in Microorganisms*, Marcel Dekker, New York, 67–145.

Brennan, R.G., and Link, T.M. (2007). Hfq structure, function and ligand binding. Curr. Opin. Microbiol. *10*, 125–133.

Capecchi, B., Adu-Bobie, J., Di Marcello, F., Ciucchi, L., Masignani, V., Taddei, A., Rappuoli, R., Pizza, M., and Arico, B. (2005). *Neisseria meningitidis* NadA is a new invasin which promotes bacterial adhesion to and penetration into human epithelial cells. Mol. Microbiol. *55*, 687–698.

Clark, V.L., Campbell, L.A., Palermo, D.A., Evans, T.M., and Klimpel, K.W. (1987). Induction and repression of outer membrane proteins by anaerobic growth of *Neisseria gonorrhoeae*. Infect. Immun. *55*, 1359–1364.

Condon, C. (2007). Maturation and degradation of RNA in bacteria. Curr. Opin. Microbiol. *10*, 271–278.

Connell, T.D., Shaffer, D., and Cannon, J.G. (1990). Characterization of the repertoire of hypervariable regions in the Protein II (opa) gene family of *Neisseria gonorrhoeae*. Mol. Microbiol. *4*, 439–449.

Delany, I., Grifantini, R., Bartolini, E., Rappuoli, R., and Scarlato, V. (2006). Effect of *Neisseria meningitidis* fur mutations on global control of gene transcription. J. Bacteriol. *188*, 2483–2492.

Delany, I., Ieva, R., Alaimo, C., Rappuoli, R., and Scarlato, V. (2003a). The iron-responsive regulator Fur is transcriptionally autoregulated and not essential in *Neisseria meningitidis*. J. Bacteriol. *185*, 6032–6041.

Delany, I., Rappuoli, R., and Scarlato, V. (2004). Fur functions as an activator and as a repressor of putative virulence genes in *Neisseria meningitidis*. Mol. Microbiol. *52*, 1081–1090.

Delany, I., Spohn, G., Rappuoli, R., and Scarlato, V. (2003b). An anti-repression Fur operator upstream of the promoter is required for iron-mediated transcriptional autoregulation in *Helicobacter pylori*. Mol. Microbiol. *50*, 1329–1338.

Desai, P.J., Angerer, A., and Genco, C.A. (1996). Analysis of Fur binding to operator sequences within the *Neisseria gonorrhoeae* fbpA promoter. J. Bacteriol. *178*, 5020–5023.

Drlica, K., and Rouviere-Yaniv, J. (1987). Histonelike proteins of bacteria. Microbiol. Rev. *51*, 301–319.

Du, Y., and Arvidson, C.G. (2006). RpoH mediates the expression of some, but not all, genes induced in *Neisseria gonorrhoeae* adherent to epithelial cells. Infect. Immun. *74*, 2767–2776.

Du, Y., Lenz, J., and Arvidson, C.G. (2005). Global gene expression and the role of sigma factors in *Neisseria gonorrhoeae* in interactions with epithelial cells. Infect. Immun. *73*, 4834–4845.

Ducey, T.F., Carson, M.B., Orvis, J., Stintzi, A.P., and Dyer, D.W. (2005). Identification of the iron-responsive genes of *Neisseria gonorrhoeae* by microarray analysis in defined medium. J. Bacteriol. *187*, 4865–4874.

Dybvig, K. (1993). DNA rearrangements and phenotypic switching in prokaryotes. Mol. Microbiol. *10*, 465–471.

Escolar, L., Perez-Martin, J., de Lorenzo, V., (1999). Opening the iron box: transcriptional metalloregulation by the Fur protein. J. Bacteriol. *181*, 6223–6229.

Flamm, E.L., and Weisberg, R.A. (1985). Primary structure of the hip gene of *Escherichia coli* and of its product, the beta subunit of integration host factor. J. Mol. Biol. *183*, 117–128.

Friedman, D.I. (1988). Integration host factor: a protein for all reasons. Cell *55*, 545–554.

Fyfe, J.A., and Davies, J.K. (1998). An AT-rich tract containing an integration host factor-binding domain and two UP-like elements enhances transcription from the pilEp1 promoter of *Neisseria gonorrhoeae*. J. Bacteriol. *180*, 2152–2159.

Gelfand, M.S., Mironov, A.A., Jomantas, J., Kozlov, Y.I., and Perumov, D.A. (1999). A conserved RNA structure element involved in the regulation of bacterial riboflavin synthesis genes. Trends Genet. *15*, 439–442.

Goosen, N., and van de Putte, P. (1995). The regulation of transcription initiation by integration host factor. Mol. Microbiol. *16*, 1–7.

Gottesman, S. (2005). Micros for microbes: non-coding regulatory RNAs in bacteria. Trends Genet. *21*, 399–404.

Grifantini, R., Sebastian, S., Frigimelica, E., Draghi, M., Bartolini, E., Muzzi, A., Rappuoli, R., Grandi, G., and Genco, C.A. (2003). Identification of iron-activated and -repressed Fur-dependent genes by transcriptome analysis of *Neisseria meningitidis* group B. Proc. Natl. Acad. Sci. USA *100*, 9542–9547.

Gruber, T.M., and Gross, C.A. (2003) Multiple sigma subunits and the partitioning of bacterial transcription space. Annu. Rev. Microbiol. *57*, 441–466.

Gunesekere, I.C., Kahler, C.M., Powell, D.R., Snyder, L.A., Saunders, N.J., Rood, J.I., and Davies, J.K. (2006a). Comparison of the RpoH-dependent regulon and general stress response in *Neisseria gonorrhoeae*. J. Bacteriol. *188*, 4769–4776.

Gunesekere, I.C., Kahler, C.M., Ryan, C.S., Snyder, L.A., Saunders, N.J., Rood, J.I., and Davies, J.K. (2006b). Ecf, an alternative sigma factor from *Neisseria gonorrhoeae*, controls expression of msrAB, which encodes methionine sulfoxide reductase. J. Bacteriol. *188*, 3463–3469.

Hagman, K.E., Pan, W., Spratt, B.G., Balthazar, J.T., Judd, R.C., and Shafer, W.M. (1995). Resistance of *Neisseria gonorrhoeae* to antimicrobial hydrophobic agents is modulated by the mtrRCDE efflux system. Microbiology *141 (Pt 3)*, 611–622.

Hantke, K. (2001). Iron and metal regulation in bacteria. Curr. Opin. Microbiol. *4*, 172–177.

Henderson, I.R., Owen, P., and Nataro, J.P. (1999). Molecular switches – the ON and OFF of bacterial phase variation. Mol. Microbiol. *33*, 919–932.

Hill, S.A., Samuels, D.S., Carlson, J.H., Wilson, J., Hogan, D., Lubke, L., and Belland, R.J. (1997). Integration host factor is a transcriptional cofactor of *pilE* in *Neisseria gonorrhoeae*. Mol. Microbiol. *23*, 649–656.

Hill, S.A., Belland, R.J., and Wilson, J. (1998). The *ihf* mRNA levels decline as *Neisseria gonorrhoeae* enters the stationary growth phase. Gene *215*, 303–310.

Hill, S.A., Samuels, D.S., Nielsen, C., Knight, S.W., Pagotto, F., and Dillon, J.A. (2002). Integration host factor interactions with *Neisseria* gene sequences: correlation between predicted binding sites and *in vitro* binding of *Neisseria* -derived IHF protein. Mol Cell Probes *16*, 153–158.

Householder, T.C., Belli, W.A., Lissenden, S., Cole, J.A., and Clark, V.L. (1999). *Cis-* and *trans*-acting elements involved in regulation of aniA, the gene encoding the major anaerobically induced outer membrane protein in *Neisseria gonorrhoeae*. J. Bacteriol. *181*, 541–551.

Isabella, V., Wright, L.F., Barth, K., Spence, J.M., Grogan, S., Genco, C.A., Clark, V.L., (2008). *Cis-* and *trans*-acting elements involved in regulation of norB (norZ), the gene encoding nitric oxide reductase in *Neisseria gonorrhoeae*. Mol. Microbiol. *154*, 226–239.

Johansson, J., Mandin, P., Renzoni, A., Chiaruttini, C., Springer, M., and Cossart, P. (2002). An RNA thermosensor controls expression of virulence genes in *Listeria monocytogenes*. Cell *110*, 551–561.

Johnson, C.R., Newcombe, J., Thorne, S., Borde, H.A., Eales-Reynolds, L.J., Gorringe, A.R., Funnell, S.G., and McFadden, J.J. (2001). Generation and characterization of a PhoP homologue mutant of *Neisseria meningitidis*. Mol. Microbiol. 39, 1345–1355.

Jordan, P.W., Snyder, L.A., and Saunders, N.J. (2005). Strain-specific differences in *Neisseria gonorrhoeae* associated with the phase variable gene repertoire. BMC Microbiol. *5*, 21.

Kazmierczak, M.J., Wiedmann, M., and Boor, K.J., (2005) Alternative sigma factors and their roles in bacterial virulence. Micobiol. Mol. Biol. Rev. *69*, 527–543.

Knapp, J.S., and Clark, V.L. (1984). Anaerobic growth of *Neisseria gonorrhoeae* coupled to nitrite reduction. Infect. Immun. *46*, 176–181.

Koomey, M., Gotschlich, E.C., Robbins, K., Bergstrom, S., and Swanson, J. (1987). Effects of *recA* mutations on pilus antigenic variation and phase transitions in *Neisseria gonorrhoeae*. Genetics *117*, 391–398.

Laskos, L., Ryan, C.S., Fyfe, J.A., and Davies, J.K. (2004). The RpoH-mediated stress response in *Neisseria gonorrhoeae* is regulated at the level of activity. J. Bacteriol. *186*, 8443–8452.

Lee, E.H., Hill, S.A., Napier, R., and Shafer, W.M. (2006). Integration host factor is required for FarR repression of the FarAB-encoded efflux pump of *Neisseria gonorrhoeae*. Mol. Microbiol. *60*, 1381–1400.

Lee, E.H., Rouquette-Loughlin, C., Folster, J.P., and Shafer, W.M. (2003). FarR regulates the FarAB-encoded efflux pump of *Neisseria gonorrhoeae* via an MtrR regulatory mechanism. J. Bacteriol. *185*, 7145–7152.

Lee, E.H., and Shafer, W.M. (1999). The FarAB-encoded efflux pump mediates resistance of gonococci to long-chained antibacterial fatty acids. Mol. Microbiol. *33*, 839–845.

Lenz, D.H., Mok, K.C., Lilley, B.N., Kulkarni, R.V., Wingreen, N.S., and Bassler, B.L. (2004). The small RNA chaperone Hfq and multiple small RNAs control quorum sensing in *Vibrio harveyi* and *Vibrio cholerae*. Cell *118*, 69–82.

Lewis, L.A., Gipson, M., Hartman, K., Ownbey, T., Vaughn, J., and Dyer, D.W. (1999). Phase variation of HpuAB and HmbR, two distinct haemoglobin receptors of *Neisseria meningitidis* DNM2. Mol. Microbiol. *32*, 977–989.

Lissenden, S., Mohan, S., Overton, T., Regan, T., Crooke, H., Cardinale, J.A., Householder, T.C., Adams, P., O'Conner, C.D., Clark, V.L., Smith, H., and Cole, J.A. (2000). Identification of transcription activators that regulate gonococcal adaptation from aerobic to anaerobic or oxygen-limited growth. Mol. Microbiol. *37*, 839–855.

Majdalani, N., Vanderpool, C.K., and Gottesman, S. (2005). Bacterial small RNA regulators. Crit Rev Biochem Mol Biol *40*, 93–113.

Mandal, M., Boese, B., Barrick, J.E., Winkler, W.C., and Breaker, R.R. (2003). Riboswitches control fundamental biochemical pathways in Bacillus subtilis and other bacteria. Cell *113*, 577–586.

Martin, P., Makepeace, K., Hill, S.A., Hood, D.W., and Moxon, E.R. (2005). Microsatellite instability regulates transcription factor binding and gene expression. Proc. Natl. Acad. Sci. USA *102*, 3800–3804.

Masse, E., and Gottesman, S. (2002). A small RNA regulates the expression of genes involved in iron metabolism in *Escherichia coli*. Proc. Natl. Acad. Sci. USA 99, 4620–4625.

Mellin, J.R., Goswami, S., Grogan, S., Tjaden, B., and Genco, C.A. (2007). A novel Fur- and iron-regulated small RNA, NrrF, is required for indirect Fur-mediated regulation of the *sdhA* and *sdhC* genes in *Neisseria meningitidis*. J. Bacteriol. *189*, 3686–3694.

Meyer, T.F., Haas, R., Stern, A., Fiedler, H., Frosch, M., Jahnig, F., Muralidharan, K., and Veit, S. (1986). Surface proteins of pathogenic '*Neisseria*': antigenic variation and conserved epitopes. Ann Sclavo Collana Monogr. 3, 407–414.

Miller, H.I. (1984). Primary structure of the *himA* gene of *Escherichia coli*: homology with DNA-binding protein HU and association with the phenylalanyl-tRNA synthetase operon. Cold Spring Harb. Symp. Quant. Biol. *49*, 691–698.

Mooney, R.A., Darst, S.A., and Landick, R. (2005) Sigma and RNA polymerase: an on-again off-again relationship? Mol. Cell *11*, 335–345.

Muralidharan, K., Stern, A., and Meyer, T.F. (1987). The control mechanism of opacity protein expression in the pathogenic neisseriae. Antonie Van Leeuwenhoek 53, 435–440.

Overton, T.W., Whitehead, R., Li, Y., Snyder, L.A., Saunders, N.J., Smith, H., and Cole, J.A. (2006). Coordinated regulation of the *Neisseria gonorrhoeae*-truncated denitrification pathway by the nitric oxide-sensitive repressor, NsrR, and nitrite-insensitive NarQ–NarP.J. Biol. Chem. *281*, 33115–33126.

Pan, W., and Spratt, B.G. (1994). Regulation of the permeability of the gonococcal cell envelope by the *mtr* system. Mol. Microbiol. *11*, 769–775.

Repoila, F., and Gottesman, S. (2001). Signal transduction cascade for regulation of RpoS: temperature regulation of DsrA.J. Bacteriol. *183*, 4012–4023.

Repoila, F., and Gottesman, S. (2003). Temperature sensing by the *dsrA* promoter. J. Bacteriol. *185*, 6609–6614.

Rice, P.A., Yang, S., Mizuuchi, K., and Nash, H.A. (1996). Crystal structure of an IHF–DNA complex: a protein-induced DNA U-turn. Cell *87*, 1295–1306.

Richardson, A.R., and Stojiljkovic, I. (1999). HmbR, a hemoglobin-binding outer membrane protein of *Neisseria meningitidis*, undergoes phase variation. J. Bacteriol. *181*, 2067–2074.

Richardson, A.R., Yu, Z., Popovic, T., and Stojiljkovic, I. (2002). Mutator clones of *Neisseria meningitidis* in epidemic serogroup A disease. Proc. Natl. Acad. Sci. USA *99*, 6103–6107.

Robertson, B.D., and Meyer, T.F. (1992). Genetic variation in pathogenic bacteria. Trends Genet *8*, 422–427.

Rock, J.D., Thomson, M.J., Read, R.C., and Moir, J.W. (2007). Regulation of denitrification genes in *Neisseria meningitidis* by nitric oxide and the repressor NsrR.J. Bacteriol. *189*, 1138–1144.

Rodionov, D.A., Dubchak, I.L., Arkin, A.P., Alm, E.J., and Gelfand, M.S. (2005). Dissimilatory metabolism of nitrogen oxides in bacteria: comparative reconstruction of transcriptional networks. PLoS Comput Biol *1*, e55.

Sarkari, J., Pandit, N., Moxon, E.R., and Achtman, M. (1994). Variable expression of the Opc outer membrane protein in *Neisseria meningitidis* is caused by size variation of a promoter containing poly-cytidine. Mol. Microbiol. *13*, 207–217.

Saunders, N.J., Jeffries, A.C., Peden, J.F., Hood, D.W., Tettelin, H., Rappuoli, R., and Moxon, E.R. (2000). Repeat-associated phase variable genes in the complete genome sequence of *Neisseria meningitidis* strain MC58. Mol. Microbiol. *37*, 207–215.

Sebastian, S., Agarwal, S., Murphy, J.R., and Genco, C.A. (2002). The gonococcal Fur regulon: identification of additional genes involved in major catabolic, recombination, and secretory pathways. J. Bacteriol. *184*, 3965–3974.

Seib, K.L., Wu, H.J., Srikhanta, Y.N., Edwards, J.L., Falsetta, M.L., Hamilton, A.J., Maguire, T.L., Grimmond, S.M., Apicella, M.A., McEwan, A.G., and Jennings, M.P. (2007). Characterization of the OxyR regulon of *Neisseria gonorrhoeae*. Mol. Microbiol. *63*, 54–68.

Shaik, Y.B., Grogan, S., Davey, M., Sebastian, S., Goswami, S., Szmigielski, B., and Genco, C.A. (2007). Expression of the iron-activated *nspA* and *secY* genes in *Neisseria meningitidis* group B by Fur-dependent and -independent mechanisms. J. Bacteriol. *189*, 663–669.

Snyder, L.A., Butcher, S.A., and Saunders, N.J. (2001). Comparative whole-genome analyses reveal over 100 putative phase-variable genes in the pathogenic *Neisseria* spp. Microbiology *147*, 2321–2332.

Stojiljkovic, I., and Hantke, K. (1995). Functional domains of the *Escherichia coli* ferric uptake regulator protein (Fur). Mol Gen. Genet. *247*, 199–205.

Swanson, J., Robbins, K., Barrera, O., and Koomey, J.M. (1987). Gene conversion variations generate structurally distinct pilin polypeptides in *Neisseria gonorrhoeae*. J. Exp. Med. *165*, 1016–1025.

Tzeng, Y.L., Zhou, X., Bao, S., Zhao, S., Noble, C., and Stephens, D.S. (2006). Autoregulation of the MisR/MisS two-component signal transduction system in *Neisseria meningitidis*. J. Bacteriol. *188*, 5055–5065.

van der Ende, A., Hopman, C.T., Zaat, S., Essink, B.B., Berkhout, B., and Dankert, J. (1995). Variable expression of class 1 outer membrane protein in *Neisseria meningitidis* is caused by variation in the spacing between the −10 and −35 regions of the promoter. J. Bacteriol. *177*, 2475–2480.

Whitehead, R.N., Overton, T.W., Snyder, L.A., McGowan, S.J., Smith, H., Cole, J.A., and Saunders, N.J. (2007). The small FNR regulon of *Neisseria gonorrhoeae*: comparison with the larger *Escherichia coli* FNR regulon and interaction with the NarQ–NarP regulon. BMC Genomics *8*, 35.

Wilderman, P.J., Sowa, N.A., FitzGerald, D.J., FitzGerald, P.C., Gottesman, S., Ochsner, U.A., and Vasil, M.L. (2004). Identification of tandem duplicate regulatory small RNAs in *Pseudomonas aeruginosa* involved in iron homeostasis. Proc. Natl. Acad. Sci. USA *101*, 9792–9797.

Winkler, W., Nahvi, A., and Breaker, R.R. (2002). Thiamine derivatives bind messenger RNAs directly to regulate bacterial gene expression. Nature *419*, 952–956.

Winkler, W.C., and Breaker, R.R. (2005). Regulation of bacterial gene expression by riboswitches. Annu. Rev. Microbiol. *59*, 487–517.

Yang, Q.L., and Gotschlich, E.C. (1996). Variation of gonococcal lipooligosaccharide structure is due to alterations in poly-G tracts in *lgt* genes encoding glycosyl transferases. J. Exp. Med. *183*, 323–327.

Yuzawa, H., Nagai, H., Mori, H., and Yura, T. (1993). Heat induction of sigma 32 synthesis mediated by mRNA secondary structure: a primary step of the heat shock response in *Escherichia coli*. Nucleic Acids Res. *21*, 5449–5455.

Zhang, A., Wassarman, K.M., Ortega, J., Steven, A.C., and Storz, G. (2002). The Sm-like Hfq protein increases OxyS RNA interaction with target mRNAs. Mol. Cell *9*, 11–22.

Regulation and Function of the Neisserial Denitrification Pathway: Life with Limited Oxygen

2

Virginia L. Clark, Vincent M. Isabella, Kenneth Barth and Tim W. Overton

Abstract

Neisseria gonorrhoeae and *N. meningitidis* can grow anaerobically and/or microaerobically by anaerobic respiration, using the central two enzymes of the denitrification pathway. Nitrite reductase (AniA) converts nitrite to nitric oxide (NO), which is then reduced to nitrous oxide by nitric oxide reductase (NorB). The regulation of expression of *aniA* and *norB* is complex, involving multiple transcriptional activators and/or repressors. Transcription of *aniA* only occurs when oxygen concentrations are low due to the absolute requirement for FNR, an oxygen-sensitive transcriptional activator. When nitrite is present, it is converted to nitric oxide by AniA, resulting in inactivation of the repressor, NsrR, and commensurate derepression and NarP-dependent activation. Expression of *norB* is controlled by three repressors, NsrR, Fur, and ArsR. Transcription of *norB* can occur aerobically if nitric oxide is present, by inactivation of NsrR. Fur and ArsR compete for overlapping binding sites; Fur + iron displaces ArsR, causing further derepression of *norB*. In addition to enabling growth under oxygen-limited conditions, the denitrification pathway may alter nitric oxide concentrations during infection, leading to changes in NF-κB activation and resulting in suppression of the immune response. When a long-half life NO donor is added to mimic host production of NO, gonococci reduce NO from a pro-inflammatory (> 1 μM) to an anti-inflammatory steady-state concentration (100–200 nM). In the presence of nitrite, most strains produce a high steady-state level of NO, while strains requiring arginine, hypoxanthine, and uracil (AHU) set a steady-state level < 1 μM. In meningococci the presence of the denitrification pathway protects the bacterium against the toxic effects of reactive nitrogen species and also alters cytokine production profiles by host macrophages.

Introduction

Anaerobic/microaerobic growth of *Neisseria* spp.

Neisseria gonorrhoeae was long considered to be an obligate aerobe, which presented a conundrum due to frequent co-isolation of this organism with obligate anaerobes. In 1982 Short *et al.* reported that *N. gonorrhoeae* can survive extended incubation anaerobically and that fresh clinical isolates from urethral exudates in males would occasionally grow anaerobically on initial culture, but could not be subcultured anaerobically. This suggested that the gonococcus could grow in the absence of oxygen by anaerobic respiration. In 1984 Knapp and Clark identified the terminal electron acceptor that supports anaerobic growth as nitrite and demonstrated that 204/204 different gonococcal strains could grow anaerobically under these conditions. Most meningococcal strains were found to be unable to grow under strict anaerobic conditions.

Clark *et al.* (1987) then described the induction and repression of gonococcal outer membrane proteins in response to anaerobiosis, suggesting that this is a relevant phenotype for the gonococcus. This was confirmed in 1988 (Clark *et al.*, 1987) by the demonstration of antibodies to the major anaerobically induced outer membrane protein, Pan1, in sera from women with gonococcal infections. A related protein was demonstrated to be present in commensal *Neisseria* spp. and in *N. meningitidis* when cells were cultured anaerobically (Hoehn and Clark, 1990). The gene encoding Pan1 was cloned and sequenced (Hoehn and Clark, 1992a) and renamed *aniA* (*an*aerobically *i*nduced protein A), as there was no orthologue to AniA found by Blast searches. AniA was also demonstrated to be a lipoprotein (Hoehn and Clark, 1992b), further confirming its location in the outer membrane.

The neisserial denitrification pathway

Berks *et al.* (1995) were the first to recognize that AniA has a low level of similarity to the copper-containing nitrite reductases. The function of AniA in nitrite reduction was confirmed in microaerobic gonococcal cultures by Meilles *et al.* (1997), in anaerobic gonococcal cultures and cell free extracts by Cardinale (1999), and in microaerobic meningococcal cultures by Rock and Moir (2005). The AniA protein, minus the lipid-modified N-terminus and the C-terminal proline-rich repeat region, has been crystallized and the structure determined (Boulanger and Murphy, 2002). Despite low sequence similarity (approximately 30%), AniA has a trimeric structure with copper-binding domains found between the monomers, as is found in copper-containing nitrite reductases from soil bacteria. The main structural differences are localized to two attenuated surface loops that map to deletions in the sequence alignment. It was postulated that one of the loops may be required to interact with a lipid-anchored azurin, while the other loop may facilitate a more intimate interaction with the lipid membrane.

The identification of AniA as a nitrite reductase, whose product is nitric oxide, and the identification of nitrous oxide as the end product of anaerobic respiration on nitrite (Lissenden *et al.*, 2000) led to the identification of a single component gonococcal nitric oxide reductase, NorB (NorZ) (Householder *et al*, 2000). Insertional inactivation of either *aniA* or *norB* results in a mutant unable to grow anaerobically in the presence of nitrite. An *aniA:ermC* and a *norB:kan* mutant were mixed and incubated anaerobically on nitrite-containing plates. While the *norB* mutant survived, it did not increase in number, while the *aniA* mutant increased >10-fold overnight (V. L. Clark, unpublished). Thus, it is NorB that is responsible for maintaining an oxidation/reduction balance in the absence of oxygen, allowing the organism to grow anaerobically.

AniA (nitrite reductase) and NorB (nitric oxide reductase) constitute the middle two reactions of the denitrification pathway that converts nitrate into molecular nitrogen (Table 2.1). The first reaction of this pathway is catalysed by the NarG complex. Neither *N. gonorrhoeae* nor *N. meningitidis* contains orthologues to this complex or to the periplasmic nitrate reductase, NapA, so they are incapable of converting nitrate to nitrite. Those bacteria, such as the Enterobacteriaceae, that contain only NarG or NapA are not denitrifiers, as they do not form the gaseous products, nitric oxide, nitrous oxide, or molecular nitrogen. Some bacteria, such as *Pseudomonas* spp. or the well-studied *Paracoccus denitrificans* (Baker *et al.*, 1998), contain all four reactions of the denitrification pathway. Others (see Table 2.1) lack one or two of these reactions. For more extensive discussion of the constitution and regulation of transcription of the denitrification pathway, see reviews by Zumft (1997), Baker *et al.* (1998), Richardson (2000), Philippot (2002), and Rodionov *et al.* (2005).

The last step in the denitrification pathway is the reduction of nitrous oxide, the product of NorB, to molecular nitrogen. An examination of the gonococcal genome revealed that the nitrous oxide reductase (*nosZ*), accessory/assembly proteins (*nosDFYL*), and the regulator (*norR*) are present. However, there are translational stop codons in *nosR*, *nosZ*, and *nosD*, rendering the complex non-functional. In *N. meningitidis* there is a deletion (~ 4 kb in size) that extends from the middle of *nosR* into the middle of *nosD*, completely eliminating *nosZ*. Thus the pathogenic *Neisseria* have used two different mechanisms to inactivate this last step of the denitrification pathway. This is interesting, as the only steps of the denitrification pathway that allow coupling of anaerobic respiration to ATP synthesis are the first, nitrate reductase, for which there is no orthologue in *N. gonorrhoeae* or *N. meningitidis*, and the last, nitrous oxide reductase.

Table 2.1 Distribution of nitrate, nitrite, nitric oxide and nitrous oxide reductases among selected bacteria

Species	Nitrate reduction	Nitrite reduction	Nitric oxide reduction	Nitrous oxide reduction
N. gonorrhoeae		NirK[1]	NorZ[2]	Translational stop codons in *nosZ, nosR,* and *nosD*
N. meningitidis		NirK	NorZ	Deletion from middle of *nosR* to middle of *nosD*, eliminating *nosZ*
Brucella suis	NarG[3]	NirK	NorBC	NosZ[4]
Burkholderia mallei	NarG	NirK	NorZ	NosZ
Corynebacteria diphtheriae	NarG	NirK	NorZ	
Escherichia coli	NarG, NapA	NirBD, Nrf[5]		
Mycobacterium tuberculosis	NarG			
Nitrosomonas europea		NirK	NorBC	
Paracoccus denitrificans	NarG, NapA	NirS	NorBC	NosZ
Pseudomonas aeruginosa PAO1	NarG, NapA	NirS	NorBC	NosZ
Ralstonia metallidurans	NarG, NapA	NirS	NorZ	NosZ
Salmonella typhimurium	NarG	NirBD[5]		

1 Nitrite reductase that converts nitrite to nitric oxide is one of two types: a copper-containing enzyme (NirK) or a cytochrome cd_1-containing enzyme (NirS).

2 Nitric oxide reductases are either one large subunit (NorZ or NorB) or two small subunits (NorBC) in size

3 NarG refers to the multisubunit membrane-bound dissimilatory nitrate reductase, while NapA refers to the heterodimeric periplasmic nitrate reductase.

4 NosZ refers to the homodimeric copper-containing nitrous oxide reductase.

5 NirBD and Nrf convert nitrite to ammonia, not to nitric oxide.

Gonococci can use their denitrification pathway to support aerobic as well as anaerobic growth. When the long half-life NO donor is added to wild type strains, there is no inhibition of growth, whereas growth is arrested in an isogenic *norB* mutant. However, this growth arrest is not due to NO toxicity, as similar NO levels are obtained in both strains. In both the wild type and the *norB* mutant, the addition of the NO donor causes a complete inhibition of cytochrome oxidase activity (Kenneth Barth, unpublished). Thus, the function of NorB in the wild type strain is to use nitric oxide reductase activity to maintain an oxidation/reduction balance to allow growth to continue. Meningococci have also been shown to use the denitrification pathway under aerobic conditions (Rock *et al.*, 2005).

It is not yet known what the electron donors are for the gonococcal denitrification pathway. It is probable that a variety of membrane bound dehydrogenases convert ubiquinone to ubiquinol commensurate with the production of ATP (Coupling Site I). The ubiquinol then serves as the electron donor for NorB, as is seen with orthologous nitric oxide reductases (Tavares *et al.*, 2006). It is more difficult to envisage the process of electron donation to AniA, as it is located in the outer membrane. It has been suggested that other copper-containing nitrite reductases use pseudoazurin or cytochromes as electron donors (Tavares *et al.*, 2006). A double mutant of *N. gonorrhoeae* deficient in both azurin (*laz*) and cytochrome c peroxidase (*ccp*) had ~ 75% of the nitrite reductase activity of the parent strain (K. Barth, unpublished). Whole cells of *N. gonorrhoeae* had twice the nitrite reductase activity when lactate rather than glucose was provided (Lissenden *et al.*, 2000).

Can *Neisseria gonorrhoeae* and *Neisseria meningitidis* grow under strictly anaerobic conditions?

It has recently been suggested that *N. gonorrhoeae* and *N. meningitidis* can grow microaerobically but not anaerobically, as they possess only the aerobic (class I) type of ribonucleotide reductase (RNR) (Rock *et al.*, 2005). This is in contrast to the growth curves of Knapp and Clark (1984) in which it was demonstrated that gonococcal strain NRL905 grew in prereduced media in a nitrite-limited manner, and that gonococci can be subcultured anaerobically on prereduced plates under strictly anaerobic conditions.

It is possible that gonococci possess an RNR with a very high affinity for oxygen that requires only minute concentrations of oxygen. Class I reductases use an iron centre with ferrous to ferric conversion to generate a tyrosyl free radical, and this reaction requires oxygen. An alternative explanation for how gonococci can grow under strictly anaerobic conditions is that nitric oxide, itself a free radical, can replace oxygen in this reaction. NRL905 generates a low (< 100 nM) nitric oxide steady-state level that may accomplish this reaction, while higher NO concentrations may inhibit RNR. This would explain why NRL905, but not most gonococcal strains, can grow anaerobically in broth. All strains can grow anaerobically on plates, where NO would not accumulate to RNR inhibitory levels.

Scope of this chapter

We will first describe the complex regulation of *aniA* and *norB*. FNR and NarP activate and NsrR represses *aniA* expression while *norB* is repressed by NsrR and an ArsR-like regulator and activated indirectly by Fur. The emphasis will be on regulation in *N. gonorrhoeae* genes while differences will be noted for regulation of these genes in *N. meningitidis*. We will also discuss what is known about the FNR, NarP, and NsrR regulons. We will then describe nitric oxide metabolism in *N. gonorrhoeae* and the factors that determine the NO steady-state level. This will include a discussion of how NO affects signal transduction pathways in the host, activating NF-κB when present at high concentration (> 1 μM) but inhibiting NF-κB activation at low concentration (100–200 nM). We will also discuss how NO metabolism and steady-state levels may account in part for immunosuppression by the gonococcus. We will conclude with a discussion of the effects of NO metabolism on interaction with host cells by *N. meningitidis*, resulting in alterations in cytokine production profiles.

The complex regulation of *aniA* and *norB*: transcription factors and stimuli

As *aniA* was identified as an anaerobically and nitrite-induced gene, the mechanism of regulation at the *aniA* promoter was investigated. At least five transcription factors are directly involved in the regulation of the divergent *aniA* and *norB* promoters; FNR, NarQ–NarP, NsrR, ArsR and Fur (Fig. 2.1).

The FNR protein: molecular aspects of function

FNR, the regulator of fumarate and nitrite reduction, is a transcription regulator that was first discovered in the enterobacterium *Escherichia coli* (Lambden and Guest, 1976). It is a member, along with CRP, the catabolite repressor protein, of the FNR-CRP superfamily of transcriptional regulators (Körner *et al.*, 2003). Members of the superfamily are present in both Gram-negative and -positive bacteria, respond to diverse stimuli, such as cyclic nucleotides, oxygen, nitric oxide and carbon monoxide, and regulate the expression of many different genes in response to these stimuli (Körner *et al.*, 2003).

E. coli FNR is the most well-characterized FNR protein. The structure of *E. coli* FNR has been modelled on that of the crystal structure of *E. coli* CRP (Weber and Steitz, 1987); and can be divided into three regions with specific functions: an N-terminal domain that co-ordinates an iron–sulphur centre responsible for oxygen sensing; a dimerization helix that allows FNR monomers to dimerize; and a C-terminal DNA-binding

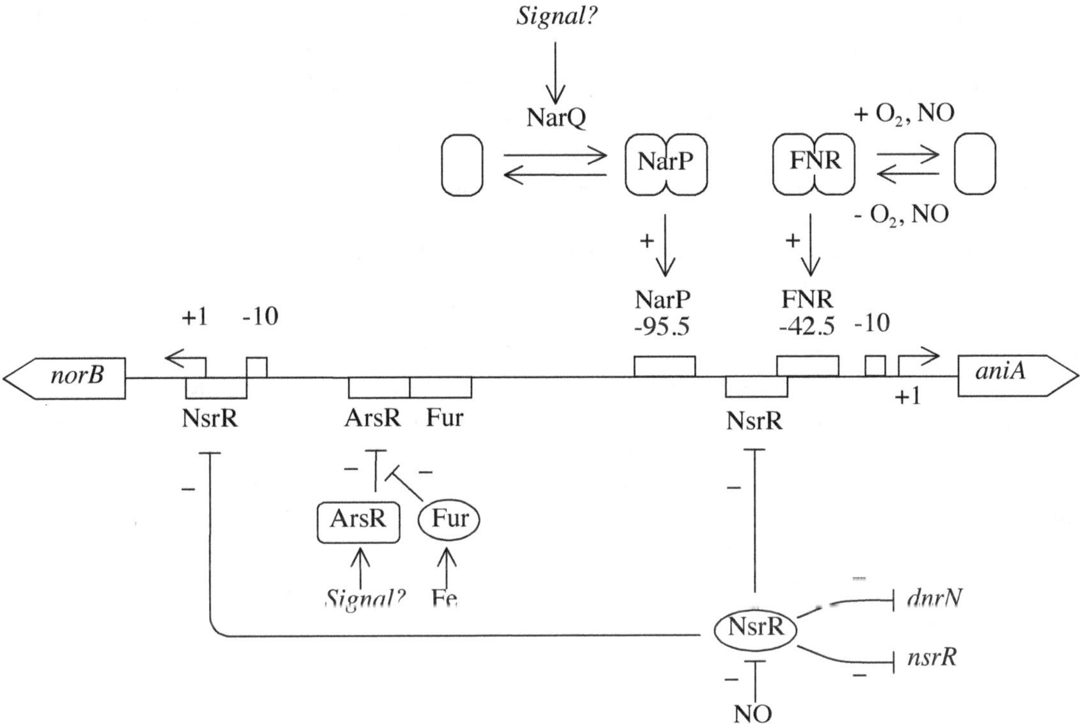

Figure 2.1 The arrangement of transcription factor binding sites at the *aniA–norB* intergenic region. Transcription activators are shown above the promoter, repressors below. The sequence of the FNR binding site in *N. gonorrhoeae* (TTGACNNNNATTTAA) differs from that of *N. meningitidis* (TTTGACNNNNATCAA) by one basepair (underlined); the sequence of the NsrR binding site in *N. gonorrhoeae* (TTTAACATTCATATTTTGTGAATTTTAAA) differs from that of *N. meningitidis* (TTTAACATTCATATTTTGTGAATTTTAAT) by one basepair (underlined); the sequence of the ArsR binding site in *N. gonorrhoeae* (CATATATGTATTTGTTCTATATG) differs from that of *N. meningitidis* (CATATATGTATTTGTTTTATATG) by one basepair (underlined); the NarP and Fur binding sites are identical in both *N. gonorrhoeae* and *N. meningitidis.*

domain. The FNR protein of *N. gonorrhoeae,* encoded by ORF NGO1579, is closely related to that of *E. coli* (Overton *et al.,* 2003). The two proteins share 38% and 50% amino acid sequence identity and similarity respectively. The *N. gonorrhoeae* FNR is believed to function in much the same way as the *E. coli* protein. The *N. meningitidis* FNR protein is 97% identical in amino acid sequence to that of *N. gonorrhoeae.* Unlike some denitrifiers, *Neisseria* express only one FNR protein.

E. coli FNR senses oxygen directly via an iron–sulphur centre. Four cysteine residues, C20, C23, C29 and C122, coordinate the FeS centre; these residues are essential for FNR function (Green *et al.,* 1993). Under anoxic conditions, the FeS centre occurs in the form [4Fe–4S]. Upon exposure to oxygen, the FeS centre dissociates to form a [2Fe–2S] centre, and after prolonged exposure, the FeS centre dissociates from FNR altogether (Kiley and Beinert 2003). Amino acid substitutions around the FeS-coordinating cysteine residues alter the stability of the centre in response to oxygen. Thus, a LH28 substitution increases the aerobic activity of *E. coli* FNR by stabilizing the [4Fe–4S] form of the FeS centre in the presence of oxygen (Bates *et al.,* 2000).

The cysteine residues that coordinate the FeS centre of *E. coli* FNR are conserved in the neisserial protein, and FNR^{Ng} function is impaired if these cysteine residues are substituted with alanine residues (Overton *et al.,* 2003). In addition, amino acid substitutions close to the cysteine residues cause alterations in oxygen sensitivity of FNR^{Ng} (Overton *et al.,* 2003).

As well as oxygen, *E. coli* FNR also responds to reactive nitrogen species such as nitric oxide (Cruz-Ramos *et al.*, 2002). The FeS centre reacts with nitric oxide to form a dinitrosyl–iron–cysteine complex, rendering the FNR molecule inactive. This mode of regulation has been shown to be important at the *E. coli hmpA* promoter, regulating expression of an NO-detoxifying flavohaemoglobin (Cruz-Ramos *et al.*, 2002).

E. coli FNR dimerizes via a dimerization helix (Moore *et al.*, 2006). Substitutions in the helix can alter the dimerization properties of FNR^{Ec}; similar substitutions in FNR^{Ng} result in similar changes in dimerization, suggesting that the mode of dimerization is similar. Active dimeric *E. coli* FNR binds to a consensus inverted repeat sequence of TTGAT-N_4-ATCAA via one helix–turn–helix motif per monomer (Spiro *et al.*, 1990). *N. gonorrhoeae* FNR is able to bind to the same DNA sequences as *E. coli* FNR. FNR^{Ng} expressed in *E. coli* was able to repress transcription from simple and complex FNR-repressible promoters (Overton *et al.*, 2003).

When bound to the DNA, FNR activates transcription initiation by making positive contacts with RNA polymerase. Surface patches of FNR (activating regions 1, 2 and 3) contact the αCTD, αNTD and σ subunits of RNAP respectively (Blake *et al.*, 2002). Studies expressing the *N. gonorrhoeae* FNR protein in *E. coli* revealed that, while FNR^{Ng} could bind to *E. coli* FNR binding sites, it could only poorly activate transcription. This deficiency was most likely due to poor alignment of the FNR^{Ng} activating regions and their cognate regions on *E. coli* RNA polymerase (Overton *et al.*, 2003); indeed, activation was restored by a second-site substitution, SF18, thought to affect coordination of the FeS centre.

FNR activates expression of *aniA*

In *N. gonorrhoeae*, the first FNR-dependent promoter to be discovered was *aniA* (Fig. 2.1; Householder *et al.*, 1999). FNR binds to the *aniA* promoter at a position 42.5 basepairs upstream of the transcription start site (that is, the centre two bases of the inverted repeat binding site are 42 and 43 bases upstream). The sequence of the binding site is <u>TTGAC</u>TTAA<u>ATTAA</u>, similar to the consensus sequence of FNR^{Ec}. FNR therefore acts as a class-II activator and is theoretically able to contact the α and σ subunits of RNAP. FNR is absolutely required for efficient transcription of *aniA* in both *N. meningitidis* and *N. gonorrhoeae*, and *fnr* strains are unable to reduce nitrite (Householder *et al.*, 1999; Rock *et al.*, 2005). *N. meningitidis* serogroup B *fnr* mutant strains were also shown to be attenuated in mouse and infant rat models (Bartolini *et al.*, 2006).

The *Neisseria gonorrhoeae* FNR regulon is small

Several studies have revealed that the FNR regulon in *E. coli* is very large; at least 100 operons are thought to be regulated by FNR (Kang *et al.*, 2005, Constantinidou *et al.*, 2006). In contrast, the *N. gonorrhoeae* FNR regulon is small. After *aniA*, the second FNR-dependent promoter discovered in *N. gonorrhoeae* was that of *ccp*, regulating expression of the outer membrane lipoprotein cytochrome *c* peroxidase, believed to be important in detoxification of reactive peroxide species generated by the host immune system (Lissenden *et al.*, 2000; Turner *et al.*, 2003). Similar to at the *aniA* promoter, FNR binds to P^{ccp} at position −42.5 with respect to the predicted transcription start. As is the case of *aniA*, expression of *ccp* is absolutely dependent upon FNR.

A transcriptomics and proteomics approach was used to define the extent of the *N. gonorrhoeae* FNR regulon (Whitehead *et al.*, 2007). This study identified only a small number of genes whose expression was different in fnr^+ and *fnr* strains. Chromatin immunoprecipitation (ChIP) was used to measure FNR–DNA interactions and so confirm direct regulation by FNR. A total of 20 transcripts were differentially expressed in fnr^+ and *fnr* strains. The transcripts most highly activated by FNR were confirmed by ChIP: these were *aniA*; a gene encoding a putative iron-uptake outer membrane protein (NGO1688); and a transcript of unknown function. A putative phosphotransferase, encoded by ORF NGO1716, was found to be repressed by FNR and confirmed by ChIP. The study demonstrated that FNR in *N. gonorrhoeae* has a very small regulon, reflecting the limited range of environmental conditions encountered by the gonococcus. It is clear that the gonococcus often encounters oxygen-limited environments during infection, as AniA, whose gene is tightly regulated by FNR, is expressed during infection in women (Clark *et al.*, 1988) and in men (Clark, unpublished).

Several genes involved in metal uptake were differentially expressed in *fnr*$^+$ and *fnr* strains; this might reflect an interaction between oxygen-limited growth and metal uptake mechanisms in the gonococcus. In addition, three genes regulated by FNR are thought to be involved in the response to reactive oxygen species. Since *N. gonorrhoeae* frequently encounters oxygen-limited conditions in the host, it is natural to expect links between regulation by FNR and mechanisms responding to other host-encountered stimuli, such as changes in metal ion concentration and exposure to reactive oxygen species. Indeed, FNR might be important in the response to reactive nitrogen species, as induction of *aniA* and reduction of NO_2^- to NO would activate expression of NorB. Differential expression of *norB* and the RNS-induced *dnrN* gene was observed in *fnr*$^+$ and *fnr* strains; this was probably due to indirect regulation by NsrR rather than FNR.

The *Neisseria meningitidis* FNR regulon

A second study used transcriptomic methods to define the FNR regulon of *N. meningitidis* (Bartolini *et al.*, 2006). Like the gonococcal study, the meningococcal FNR regulon was found to be small; 11 genes in nine operons were found to be FNR-regulated. Two of these operons were implicated in sugar metabolism and fermentation, encoding genes required for utilization of maltose and α and β hexoses. Two other FNR-dependent genes encode cytochromes c_4 and c_5; these genes are not FNR-dependent in *N. gonorrhoeae*. It is suggested that, since *N. meningitidis* does not possess a *ccp* gene, cytochromes c_4 and c_5 might fulfil an oxidative protection role. Apart from *aniA*, none of the genes activated by FNR in *N. meningitidis* were found to be regulated by FNR in *N. gonorrhoeae*. This might suggest that, although the FNR proteins of *N. meningitidis* serogroup B and *N. gonorrhoeae* share 97% amino acid sequence identity, they have subtly different roles in the two pathogenic *Neisseria* species.

The NarQ–NarP two-component regulator system

The first nitrate/nitrite-sensing two-component regulator system, NarX–NarL, was discovered in *E. coli* (Stewart and Parales, 1988). NarX is a transmembrane protein containing a sensing and a transmitter domain; NarL contains receiver and DNA-binding domains (Stewart and Bledsoe, 2003). On sensing a ligand, NarX becomes autophosphorylated on a specific histidine residue. Phosphotransfer then occurs between NarX and a specific aspartate residue on NarL. Phospho-NarL is able to bind DNA at a specific site (consensus TACYNMT) and regulate transcription initiation. *E. coli* possesses two such two-component systems; NarX-NarL and NarQ–NarP, each of which responds to different concentrations of nitrate and nitrite. Whereas NarL can bind to single heptamer binding sites, *E. coli* NarP only binds to '7–2–7' sites; two heptamers arranged in an inverted repeat with two bases between.

In contrast, *N. gonorrhoeae* possesses only one system, NarQ–NarP, encoded by two open reading frames (NGO0753 and 0752) arranged in an operon. The peptide sequence of gonococcal NarQ is 26% identical and 37% similar to that of *E. coli* NarQ, whereas gonococcal NarP is 41% identical and 54% similar in sequence to *E. coli* NarP. The gonococcal and meningococcal NarQ proteins are 88% identical, while the NarP proteins are 99% identical in sequence.

NarP also activates expression of *aniA*

N. gonorrhoeae NarQ–NarP activates expression of *aniA*; NarP binds to the *aniA* promoter at a 7–2–7 inverted repeat positioned 95.5 basepairs upstream of the transcription start site (Householder, 2000). NarQ–NarP was originally believed to respond to nitrite but not to nitrate. The evidence for this was threefold: expression of *aniA* was activated by NarQ–NarP; *aniA* expression was activated in the presence of nitrite but not nitrate; and a deletion analysis of the *aniA* promoter revealed that *aniA* transcription was activated by nitrite only when the NarP binding site was present (Householder *et al.*, 1999). However, further analysis revealed that NsrR, not NarQ–NarP, was the protein responsible for regulating *aniA* in response to nitrite (Overton *et al.*, 2006; Isabella *et al.*, 2008). The deletion analysis removed both the NarP binding site and the then-uncharacterized NsrR binding site, thus removing nitrite regulation.

The role of NarQ–NarP in *N. gonorrhoeae* is unclear. NarQ–NarP does activate expression of *aniA*, but not in response to any stimulus investigated thus far. When heterologously expressed in *E. coli*, gonococcal NarQ appears to be constitutively active and phosphorylates the *E. coli* NarL protein in the absence of ligand (Overton *et al.*, 2006). Studies with chimeric *N. gonorrhoeae–E. coli* NarQ proteins have likewise demonstrated constitutive activity (Whitehead and Cole, 2006). However, the mechanism of ligand sensing in the *E. coli* NarQ and NarX proteins is also unclear. Investigations of the region of *E. coli* NarQ thought to sense nitrate and nitrite, the 'P-box', have failed to conclusively identify the mechanism of sensing; amino acid substitutions within and outside the P-box result in NarQ proteins with altered ligand sensing properties (Chiang *et al.*, 1997).

There are two current hypotheses for the function of gonococcal NarQ. First, that gonococcal NarQ is constitutively active, its ligand sensing function having being taken over by NsrR. Gonococcal NarQ–NarP might act as an 'amplifier' at the *aniA* promoter, possibly interacting with RNA polymerase as a class-I activator to increase transcription initiation when FNR is present. NarP–NsrR interactions might be important in regulation; however, *aniA* expression is still activated in the presence of nitrite in a *narP* strain (Overton *et al.*, 2006). Alternatively, gonococcal NarQ could sense and respond to a ligand that is present in all growth conditions that have been tested to date, or responds to a ligand present at extremely low concentrations. In *N. meningitidis*, NarQP might activate *aniA* in response to nitrite; unlike the gonococcus, *aniA* expression is nitrite-dependent in an *nsrR* mutant, although the role of NarQP has not been specifically investigated (Rock *et al.*, 2007).

The *Neisseria gonorrhoeae* NarP regulon is also small

Microarray experiments revealed that many genes were differentially expressed in *narP* and *narP*$^+$ strains; however, many of these changes were due to differences in growth rate. Putative NarP binding sites were located upstream of five genes, each of which was confirmed by chromatin immunoprecipitation to be bound by NarP *in vivo* (Overton *et al.*, 2006). As well as *aniA*, *norB* was more highly expressed in a *narP*$^+$ strain; this is likely to be in indirect effect due to activation by NO, the product of AniA, via NsrR. A single putative NarP binding site is present in the divergent *aniA-norB* promoter region; it is closer to *aniA*. Three promoters were shown to be directly repressed by NarP: *narQP*, demonstrating autoregulation commonly encountered in transcription factors; and two genes of unknown function, one of which (NGO1215) is also activated by FNR (Whitehead *et al.*, 2007). Thus, as in the case of FNR, gonococcal NarQP has a small regulon focused on denitrification. By comparison, *E. coli* NarX–NarL and NarQ–NarP are thought to regulate expression of around 90 and 50 operons respectively (Constantinidou *et al.*, 2006).

NsrR: a member of a new class of transcription regulators

The third regulator acting at the *aniA-norB* promoter region is NsrR. The first NsrR protein was discovered in *Nitrosomonas europea* as an Rrf2-type nitrite-sensing repressor in the winged-helix superfamily (Beaumont *et al.*, 2004). Since then, NsrR has been demonstrated to respond to NO and to control the expression of nitric oxide detoxification systems in several pathogenic and environmental organisms. Also, genes in the NsrR regulon contribute to resistance to nitrosative stress in a number of bacteria (Rodionov *et al.*, 2005; Overton *et al.*, 2006; Filenko *et al.*, 2007; Gilberthorpe *et al.*, 2007; Rock *et al.*, 2007; Rogstam *et al.*, 2007). A closely related Rrf2 family protein, IscR, has been characterized in *E. coli* and is reported to sense the Fe-S cluster status of the cell by reversible binding of an [2Fe-2S] cluster (Schwartz *et al.*, 2001). NsrR contains three conserved cysteine residues that, in IscR, are presumed to coordinate the Fe-S cluster. It is predicted that NsrR also coordinates an Fe-S cluster and that the interaction of this cluster with NO perturbs the ability of the protein to bind DNA (Schwartz *et al.*, 2001). It has also been observed that the activity of *E. coli* NsrR is affected by iron limitation (Bodenmiller and Spiro, 2006).

Complex regulation of *aniA* by FNR, NsrR and NarP

NsrR has been demonstrated to repress *aniA* transcription in both *N. gonorrhoeae* and *N. meningitidis* (Overton *et al.*, 2006; Isabella *et al.*, 2008; Rock *et al.*, 2007). The NsrR binding site is centred around 60 bases upstream of the *aniA* transcription start, between the FNR and NarP binding sites. FNR is the 'master' regulator of *aniA* transcription; it is absolutely required, and acts as a class-II activator. The gonococcal NarQP two-component system has been shown to be nitrite insensitive, thus NarP is always poised to activate *aniA* (Overton *et al.*, 2007). NarP is thought to act as a class-I activator at p^{aniA}. The molecular mechanism by which NsrR mediates control in response to NO is unknown, but a model has been constructed using available data (Fig. 2.2).

Aerobically, p^{aniA} is inactive since FNR is unable to bind DNA in the presence of oxygen. It is not known whether NsrR and NarP bind to the promoter aerobically; either way, they do not affect transcription as FNR is absent. In oxygen-limited conditions in the absence of NO or NO_2^-, FNR binds to the promoter along with NsrR and NarP. FNR contacts the σ subunit of RNAP, but cannot contact αCTD as NsrR is bound upstream. An αCTD probably contacts NarP in these conditions, leading to the observed NarP-mediated activation. When NO or NO_2^- are present, NsrR becomes deactivated and ceases to bind the promoter; αCTD can then make productive contacts with FNR, further activating expression. NsrR is unlikely to interfere with NarP–RNAP interactions to a great extent, since in a *narP* background or at an *aniA* promoter with the NarP binding site removed, nitrite still induces expression.

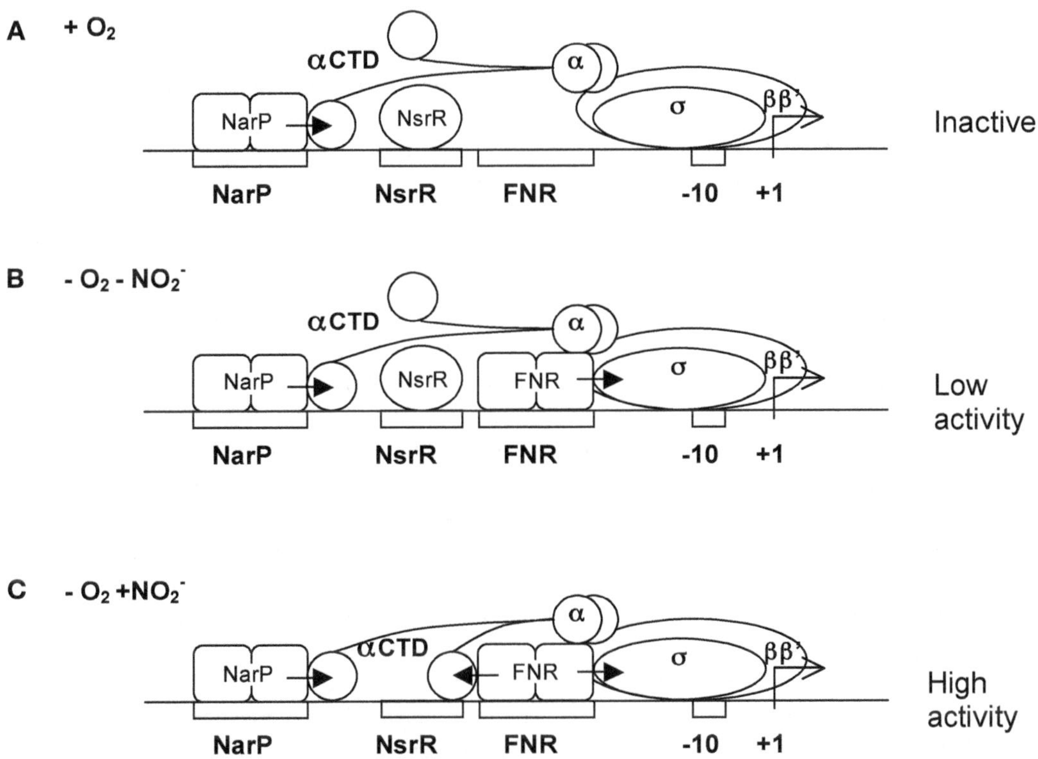

Figure 2.2 Model of transcription regulation at the gonococcal *aniA* promoter. The α, β, β′ and σ subunits of RNA polymerase are shown along with the FNR, NarP and NsrR proteins and their binding sites. **A** In the presence of oxygen, p^{aniA} is inactive as FNR cannot bind. It is not known whether NarP or NsrR are bound to the promoter. **B** In the absence of both oxygen and NO_2^-, FNR is bound and contacts the σ subunit of RNAP. NsrR is also bound, and blocks interactions between FNR and αCTD. **C** Upon exposure to nitrite, NsrR is inactive and so not bound to the promoter, allowing FNR–αCTD interactions. Thus, nitrite control is mediated by NsrR controlling the ability of FNR to interact with RNAP.

Thus, the apparent induction of *aniA* by nitrite is not a result of nitrite activation of NarQP, but rather a result of the NO-dependent derepression of NsrR, allowing optimal FNR-RNA polymerase interactions (Overton *et al.*, 2006; Isabella *et al.*, 2008). In *N. meningitidis*, the NarQP system may still respond to nitrite, as *aniA* expression was not completely dependent on NO (Rock *et al.*, 2007). A further possible level of regulation might occur via NO sensing by FNR. The *E. coli hmpA* promoter, regulating expression of an NO detoxification enzyme, is repressed by FNR (Cruz-Ramos *et al.*, 2002). Upon exposure to nitric oxide, FNR becomes deactivated as the iron–sulphur centre reacts with NO to form a dinitrosyl–iron–cysteine complex, and *hmpA* is derepressed. It is possible that a similar regulatory mechanism occurs in the gonococcus; this would explain the observation that *aniA* expression is induced by nitrite, but not by nitric oxide (Overton *et al.*, 2006).

Complex regulation of *norB* by NsrR, Fur and an ArsR-like protein

In *N. gonorrhoeae*, *aniA* transcription occurs only under anaerobic or microaerobic conditions owing to the requirement for FNR (Householder *et al.*, 1999; Overton *et al.*, 2006; Isabella *et al.*, 2008). This is not the case with *norB*, whose expression can be induced by NO even under aerobic conditions (Householder *et al.*, 2000). This is an important point when considering the host–pathogen interaction and the potential immunomodulatory effect of NO reduction on the resultant host response (Cardinale and Clark, 2005; Stevanin *et al.*, 2005; Tunbridge *et al.*, 2005; Stevanin *et al.*, 2007). The neisserial NO reductase can therefore carry out roles in both respiration and detoxification.

The *norB* upstream region contains a TG dinucleotide 3 bp upstream of the −10 motif (5′-TGNTACAAT) that is typical of extended −10 promoters that do not require a strong −35 sequence for efficient transcription (Keilty and Rosenberg, 1987). The TG dinucleotide provides critical contacts with region 3.0 of RNA polymerase, stabilizing open complex formation (Gruber and Gross, 2003; Murakami and Darst, 2003). An inverted repeat sequence spanning 29 bp and located downstream of the −10 motif was shown to be the binding site for NsrR. Thus, NsrR is thought to act as a simple repressor, preventing RNA polymerase from binding to the promoter region. Sequence substitution removing the NsrR binding site or insertional inactivation of the *nsrR* gene resulted in high level *norB* expression and a concomitant loss of NO induction capacity (Overton *et al.*, 2006; Isabella *et al.*, 2008; Rock *et al.*, 2007).

While *norB* is repressed by NsrR, it has also been reported to be activated by Fur (Delany *et al.*, 2004). Fur is an Fe^{2+}-responsive regulator of gene transcription in a wide range of bacteria whose most basic function is to regulate the transcription of genes encoding proteins that take up and metabolize iron to maintain homeostasis within the cell (Ochsner and Vasil, 1997; Escolar *et al.*, 1999; Zheleznova *et al.*, 2000). Beyond this function, Fur has been shown to act more globally in the regulation of genes important in oxidative/nitrosative stress resistance, acid resistance, and virulence (Hantke, 2001; Gilberthorpe *et al.*, 2007). The Fur protein consists of two domains. The amino-terminal domain contains the helix–turn–helix fold important for DNA binding, while the carboxy-terminal domain contains two metal binding sites and the residues involved in dimer formation (Escolar *et al.*, 1999; Hantke, 2001). By comparison with the closely related *E. coli* homologue, two cysteine residues of the neisserial Fur protein, C95 and C98, are predicted to tetrahedrically coordinate a Zn^{2+} ion and two nitrogen/oxygen atoms. The regulatory iron-binding site is less well characterized, but it is known that at least two histidines and one carboxylate bind the ferrous ion (Hantke, 2001). A 19 bp conserved motif was believed to constitute the Fur box and to be the target of binding by Fe^{2+}-occupied Fur dimers. However, recent evidence suggests that the actual Fur recognition sequence consists of 6-bp repeats (5′-NATWAT), and that at least three of these repeats are required for the formation of a functional regulatory complex between protein and DNA (Grifantini *et al.*, 2003).

Fur functions as a repressor through its Fe^{2+}-dependent DNA binding activity when its binding site overlaps the −10/−35 region, preventing RNA polymerase from gaining access to the promoter (Escolar *et al.*, 1999). Fur can also function as an activator of gene expression, though the mechanisms of positive regulation are believed to be primarily indirect and are not very well characterized. Fur has been shown to indirectly activate gene transcription through controlling expression of other regulators or by repression of antisense regulatory RNA. (Masse and Gottesman, 2002; Wennerhold and Bott, 2006; Mellin *et al.*, 2007) The gonococcal *norB* gene illustrates a novel mechanism of indirect activation by Fur, through inhibition of repressor binding (Isabella *et al.*, 2008).

Using *in vitro* transcription analysis, Delany *et al.* (2004) originally reported that Fur directly activated *norB* expression in *N. meningitidis*. However, removal of the Fur binding site in the gonococcal *norB* upstream region had no effect on *norB* expression when monitored from a translational *norB::lacZ* fusion, a more physiological *in vivo* technique. Contrary to this finding, analysis of a full-length *norB::lacZ* fusion in a gonococcal *fur* mutant showed an approximate fourfold decrease in *norB* expression, suggesting that Fur was acting as an activator (Isabella *et al.*, 2008). These contradictory results were resolved upon the detection of three copies of a 6-bp conserved motif located immediately downstream of the Fur binding site. The first 6-bp motif partially overlapped the predicted Fur binding site and had been completely removed in the *norB::lacZ* fusion with a deletion of the Fur binding site. Analysis of a *norB::lacZ* fusion in which a majority of the Fur binding site was removed but all three copies of the 6-bp motif were kept intact showed the same fourfold decrease in expression observed in a *fur* mutant, irrespective of whether in a wild-type or *fur* background. This suggested that a repressor bound to this sequence and that Fur competed with this repressor for binding to the *norB* upstream region by virtue of its overlapping binding site. Thus, Fur can indirectly activate *norB* transcription by preventing a repressor from binding. Through *in silico* analysis and DNA affinity chromatography this repressor was subsequently determined to be a small 10.6-kDa ArsR homologue (NGO1562). The first two repeats of the putative ArsR binding site are centred at –71.5. Presumably too far away to prevent RNAP binding, this repressor probably functions by tethering the RNAP and preventing promoter clearance though direct interactions with the RNAP αCTD (Isabella *et al.*, 2008).

It is unlikely that Fur would directly activate *norB* expression in *N. meningitidis* as proposed by Delany *et al.* (2004) while indirectly activating *norB* in *N. gonorrhoeae*. The very high homology between the meningococcal and gonococcal Fur and ArsR proteins (99% and 92% amino acid identity respectively), as well as the perfect conservation of their respective binding sites, suggest that meningococcal *norB* is regulated in a manner analogous to that seen in the gonococcus, though this remains to be investigated. The apparent results of Delany *et al.* (2004) may be due to non-physiological concentrations of protein used in the *in vitro* transcription assay, use of *E. coli* RNAP, or due to the fact that the plasmid template used in the assay was isolated from *E. coli* and may have been contaminated with a tightly bound *E. coli* ArsR protein, whose binding site is remarkably similar to that predicted for the gonococcus (Xu *et al.*, 1996; Isabella *et al.*, 2008).

The ArsR family of transcriptional regulators generally controls the expression of genes required for metal ion detoxification and efflux (Busenlehner *et al.*, 2003; Penella and Giedroc, 2005). Unlike Fur, ArsR family proteins bind to their operator regions in the absence of their inducing metal ligands. Derepression occurs by the coordination of metal ions through key Cys, His, Glu, or Asp residues (Busenlehner *et al.*, 2003). The metal ligand potentially coordinated by this gonococcal ArsR homologue is unknown, and the protein's amino acid sequence shows little similarity to any proposed metal binding sites in functionally characterized ArsR proteins from other organisms (Busenlehner *et al.*, 2003). This presents the possibility that the gonococcal ArsR homolog is degenerate in its metal sensory domain or that the ligand it recognizes is novel. However, it would be expected that this regulator plays an important role in determining the level of *norB* expression during infection, when iron is limiting and Fur may not be bound to its operator.

norB transcription is subject to 'fine-tuning' during infection

In its natural environment as an obligate human pathogen, *N. gonorrhoeae* will be presented with a variety of environmental factors that may act as stimuli for gene expression. Regulation of *norB* is a prime example of a gene whose expression is regulated by the integration of multiple signals. Importantly, this demonstrates that bacterial response to varying stimuli *in vivo* does not simply result in an on or off phenomena, but actually involves a more complex network of gene regulation and the ability to 'fine-tune' gene expression. Fur's competition for operator binding with ArsR modulates *norB* transcription. Even on chemically defined medium lacking metal ions and in the absence of Fur, ArsR still only represses *norB* expression about fourfold (Vincent Isabella, unpublished data). The ability of ArsR to regulate transcription of *norB* depends on both the protein's affinity for its binding site and the strength of its interaction with RNAP. It is interesting to note that in the presence of NO, when NsrR is inactive, the *norB* promoter retains considerable activity even when ArsR is presumed to be bound (Isabella *et al.*, 2008). Under physiological conditions, NsrR and ArsR are probably the major regulators of *norB* expression, with Fur acting to alter *norB* expression in response to changing concentrations of iron (Fig. 2.3).

The *Neisseria gonorrhoeae* NsrR regulon

No comprehensive transcriptomic or proteomic analysis has been completed for the gonococcal NsrR protein. Locating genes in the NsrR regulon has primarily relied on *in silico* analysis of the putative binding site. Aside for *aniA* and *norB*, NsrR has been shown to regulate three other genes, *dnrN*, *mobA*, and *nsrR*.

The gonococcal *dnrN* gene is homologous to the *ytfE* gene in *E. coli* and the *norA* gene of *Ralstonia eutropha*. The *ytfE* gene was initially identified in a transcriptomic screen of

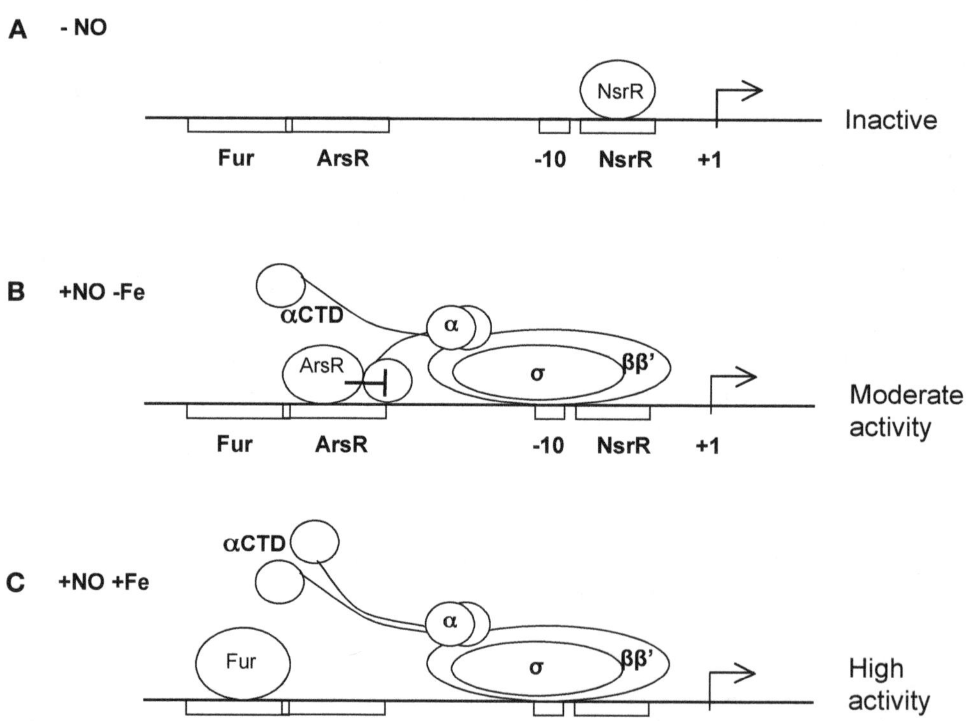

Figure 2.3 Model of transcription regulation at the gonococcal *norB* promoter. The α, β, β′ and σ subunits of RNA polymerase are shown along with the NsrR, ArsR, and Fur proteins and their binding sites. **A** In the absence of nitric oxide, p^{norB} is inactive as NsrR prevents RNAP binding. **B** In iron-limiting conditions and in the presence of nitric oxide, NsrR is inactivated and allows RNAP binding. The ArsR repressor presumably tethers RNAP through contacts with αCTD, preventing RNAP from clearing the promoter. **C** When iron is abundant, Fur can bind to its operator and prevent ArsR binding. Maximal induction of p^{norB} occurs in the presence of NO and Fe.

NO-activated genes (Justino *et al.*, 2005); this regulation was later shown to be NsrR-dependent (Bodenmiller and Spiro, 2006; Filenko *et al.*, 2007). Although the mechanism of YtfE action is unknown, is thought to have a role in the repair of iron–sulphur centres damaged by reactive nitrogen species (Justino *et al.*, 2007); as such, it is thought to be important in the physiological response to reactive nitrogen species (RNS). In *R. eutropha*, *norA* is believed to be a NO-binding cytoplasmic protein that regulates gene transcription by lowering the free cytoplasmic concentration of NO (Strube *et al.*, 2007). NsrR regulation of *dnrN* was confirmed by quantitative real-time PCR and translational *lacZ* fusion, though the precise role of the *dnrN* gene product in NO or RNS metabolism has not yet been elucidated (Overton *et al.*, 2006; Isabella *et al.*, 2008). As at the *norB* promoter, the putative NsrR binding site at p^{dnrN} is located downstream of the predicted −10 site and NsrR is believed to repress transcription by hindering promoter binding by RNA polymerase.

A putative NsrR binding site was originally reported to be located upstream of *narQP*, but not to be involved in the regulation of those genes (Rodionov *et al.*, 2005; Overton *et al.*, 2006; Isabella *et al.*, 2008). This binding site lies in the intergenic region between the *narQP* and *mobA* genes, and further characterization established that this site is actually responsible for regulation of *mobA* (Vincent Isabella, unpublished data). The *mobA* gene (NGO0754) encodes a protein required for molybdopterin guanine dinucleotide (MGD) synthesis, an essential cofactor for many bacterial molybdoenzymes (Schrag *et al.*, 2001; Guse *et al.*, 2003). MobA catalyses the synthesis of MGD from molybdopterin and GTP. In *E. coli*, molybdopterin is synthesized in a complicated enzymatic process involving the gene products of the *moa*, *moe*, and *mog* operons (Schwarz, 2005). Neisserial species do not contain orthologues to any of these operons, so it is a mystery why *mobA*, utilized in the very last step of MGD synthesis, is maintained in the genome and is regulated. Perhaps these bacteria possess some novel mechanism to hijack molybdopterin from the host or from other commensal organisms. Another possibility is that the neisserial *mobA* orthologue has evolved a novel function. Regulation of this gene by NO, and the lack of neisserial proteins with homology to molybdoenzymes from other organisms, alludes to this possibility. In any case, further work is needed to characterize the role of this protein on anaerobiosis or NO metabolism.

A translational *lacZ* fusion to *nsrR* confirmed that this gene was autoregulated. β-Galactosidase activities of the fusion in the Δ*nsrR* mutant were 95-fold higher aerobically than were observed in the wild type, however, the putative NsrR binding site was shown to have low sequence similarity to the *norB* NsrR binding site (Isabella *et al.*, 2008). The *iscR* gene in *E. coli* was also found to be autoregulated, and the IscR binding site in the *iscR* gene had little similarity to the IscR binding sequence found in other genes (Giel *et al.*, 2006). This presents the possibility that NsrR may also recognize a different binding site at its own promoter.

An *in silico* position weight matrix search of the gonococcal genome identified potential NsrR binding sites located upstream of a number of genes (T. W. Overton, unpublished data). Quantitative PCR was used to determine whether these genes were regulated by NsrR; expression of the genes with the three best potential NsrR sites was measured in *nsrR* and *nsrR*$^+$ strains. However, none of these genes were found to be differentially expressed in the two strains.

Comparison of the currently defined gonococcal NsrR regulon with that of *E. coli*, which has been determined using a transcriptomic approach, reveals a great deal of overlap; in both organisms, NsrR regulates genes involved in the nitrate or nitrite reduction process (the *E. coli* *nap* and *nrf* operons) and *dnrN*/*ytfE* (Filenko *et al.*, 2007). The *E. coli* NsrR regulon is larger than that currently defined in the gonococcus, but the additional NsrR-regulated genes in *E. coli* are not present in the gonococcus.

Summary of *aniA* and *norB* regulation

We have proposed a model for anaerobic regulation of *aniA* and *norB* (Isabella *et al.*, 2008). When gonococci are aerobically grown, there are only basal levels of expression of *aniA* and *norB* due to the inactivity of FNR in the presence of oxygen and the activity of NsrR in the absence of NO respectively. When cells are shifted to an anaerobic environment in the absence of nitrite,

FNR becomes active and *aniA* is partially induced while *norB* is still repressed due to the lack of NO. When nitrite is added to anaerobically growing cells, AniA reduces nitrite to NO, which in turn inactivates NsrR. This results in high level expression of *norB* and an increase in expression of *aniA* as NsrR-mediated repression is relieved. The presence of NO also derepresses *nsrR*, resulting in an increase in NsrR. This explains the lower level of anaerobic *norB* expression in the wild type as compared to the *nsrR* mutant, and raises the possibility that NsrR may act as a detoxifying NO sink. This model also proposes that while *aniA* cannot be expressed aerobically due to the absolute requirement for FNR (Householder *et al.*, 1999), *norB* can be expressed aerobically in the presence of NO, as might be encountered *in vivo* during infection. The *norB* gene is subject to a second level of regulation by Fur and a gonococcal ArsR homolog. In the presence of iron, Fur binds to its operator and inhibits the binding of the ArsR. However, when iron is limiting, Fur is not bound to its operator allowing the ArsR-like protein to bind and repress *norB* expression.

Nitric oxide metabolism by *Neisseria gonorrhoeae*

The ability of the gonococcus to produce and degrade NO suggests that the immunosuppressive effects of gonococcal infection may be due in part to NO effects on cytokine induction. NO has been shown to be an important modulator of cellular events, including the immune response (Deora and Lander, 2000; Liaudat *et al.*, 2000; Stefano *et al.*, 2000; Mateo and de Artinano, 2000; Davis *et al.*, 2001). Low basal levels of NO (nM) are anti-inflammatory while the high levels of NO (μM) generated by inducible nitric oxide synthase (iNOS) or absence of NO are proinflammatory (Stefano *et al.*, 2000; Mateo and de Artinano, 2000; Davis *et al.*, 2001).

When NO levels are low (< 1 μM) the direct effects of NO predominate and most often involve the interaction of NO with metal complexes in proteins (Stefano *et al.*, 2000; Mateo and de Artinano, 2000; Davis *et al.*, 2001). NO activation of soluble guanylyl cyclase (sGC), a haem-containing protein, leads to translocation of sGC to the plasma membrane (Ignarro *et al.*, 01984). This activation requires low levels of NO (EC_{50} = 100 nM), and it occurs as a result of NO produced by one of the two constitutive NOS (cNOS) enzymes, eNOS or nNOS. Activation of sGC leads to an increase in cyclic GMP, a second messenger, which in turn affects an array of modulators, including ion channels, phosphodiesterases, and protein kinases.

The transcriptional regulator, NF-κB, is required for transcription of genes for the immunoglobulin κ chain, cytokines, cytokine receptors, cell adhesion molecules, and iNOS. NO has complex effects on NF-κB, with low levels of NO, such as that produced by cNOS or NO donors, inhibiting NF-κB activation, while high levels of NO, such as that produced by iNOS, activate NF-κB (Deora and Lander, 2000; Liaudat *et al.*, 2000; Stefano *et al.*, 2000). Thus low levels of NO are anti-inflammatory while high levels of NO are proinflammatory (Connelly *et al.*, 2001). NF-κB is activated by bacterial LPS and other bacterial products through the action of Toll-like receptors, resulting in cytokine production and inflammation. Pretreatment of human cells with NO donors that produce low levels of NO prevents LPS-, IL-1β-, and TNF-α-inducible cytokine and iNOS synthesis, by inhibiting NF-κB activation (Peng *et al.*, 1995; Stefano *et al.*, 2000). The low levels of NO increase IκB-α promoter activity (Peng *et al.*, 1995) and prevent IκB-α degradation (Peng *et al.*, 1995; Katsuyama *et al.*, 1998; Reynaert *et al.*, 2004), thus preventing NF-κB activation due to the inhibitory effect of IκB-α. Degradation of IκB-α is blocked by NO inhibition of IκB kinase (Reynaert *et al.*, 2004), which prevents phosphorylation of IκB-α and subsequent ubiquitination and degradation by the proteosome (Peng *et al.*, 1995; Katsuyama *et al.*, 1998; Reynaert *et al.*, 2004).

Thus, if *N. gonorrhoeae* can modify the NO steady-state level during infection from a pro- to an anti-inflammatory level, it can prevent NF-κB activation and subsequent cytokine production. Steady-state NO concentrations during denitrification have been investigated in three different species, *Paracoccus denitrificans, Pseudomonas stutzeri,* and *Achromobacter cycloclastes* (Goretski *et al.*, 1990). In all three cases NO was detected only during denitrification and, once established, NO concentrations did not vary with time, ranging between 1 and 65 nM, depending on the species. The steady-state NO level was relatively independent of cell concentration, nitrate or nitrite concentration, and growth medium composition.

Kinetic parameters for gonococcal nitrite reductase (Nir) and nitric oxide reductase (Nor)

Anaerobically grown whole cells (strain F62) mutated in *norB* or with *norB* under the control of the *tac* promoter were used to determine the affinity of Nir and Nor, respectively, for their substrates. These strains were found to have a Nir K_m of 20 μM nitrite and a Nor K_m of 1.2 μM nitric oxide. The V_{max} values for Nir and Nor were 135 nmol nitrite reduced/min-OD_{600} (pH 6.7) and 270 nmol nitric oxide reduced/min-OD_{600} (pH 7.5) respectively (Cardinale and Clark, 2005). The K_m values represent concentrations likely to occur during infection, while the V_{max} values represent activities that will rapidly reduce the nitrite and NO that the organism will encounter. In seven gonococcal strains the nitrite reductase and nitric oxide reductase activities were similar to the specific activities seen in F62 (Barth and Clark, 2008).

NO steady-state levels during denitrification by *Neisseria gonorrhoeae* F62

N. gonorrhoeae F62 was found to establish a NO steady-state level when either nitrite or NO was provided to anaerobically grown cells. A steady-state concentration of NO after nitrite addition was dependent on the nitrite concentration until saturation was reached at 1 mM nitrite. The NO steady-state level decreased as the pH increased and the ratio of the activities of Nir and Nor correlated with the NO steady-state level in the presence of nitrite (Cardinale and Clark, 2005).

When the long half-life NO donor, DETA/NO, was used as the electron acceptor, gonococci also established a NO steady-state level. The concentration of NO at steady state was found to be a function of the concentration of NO generated by DETA/NO, with gonococci rapidly reducing the NO from a pro-inflammatory (> 1 μM) to an anti-inflammatory (~100 nM) concentration (Cardinale and Clark, 2005).

NO steady-state levels during denitrification by other gonococcal strains

When the long half-life NO donor, DETA/NO, was added to mimic host production of NO, all gonococcal strains reduced the NO steady-state level from a pro-inflammatory concentration (1.5 μM) to a new steady-state level in the anti-inflammatory range (100–200 nM) (Barth and Clark, 2008).

When nitrite was added, gonococcal strains fell into two categories in terms of the NO steady-state level that was established. Most gonococcal strains produced a NO steady-state level in the pro-inflammatory range (> 1 μM) in the presence of 2 mM nitrite. AHU strains (require arginine, hypoxanthine, and uracil), however, produced a NO steady-state level in the anti-inflammatory range (< 1 μM) (Barth and Clark, 2008). AHU strains are clonal and have been associated with asymptomatic infection in males.

Implications of gonococcal NO metabolism on host responses

Local gonococcal infections in women are generally asymptomatic. Cervical mucus from women who were uninfected or had gonococcal infections was examined for the presence of local anti-gonococcal antibody (Hedges *et al.*, 1999; Russell *et al.*, 1999) and for the presence of cytokines (Hedges *et al.*, 1998; Russell *et al.*, 1999). The antibody response to infection was weak and the anti-gonococcal antibody levels did not rise at 2 weeks after treatment. In subjects with concomitant rectal infection, the local rectal antibody response was also poor, despite the presence of organized lymphoid tissue at this site (Hedges *et al.*, 1999). This suggests that the lack of natural immunity to GC is not due solely to antigenic variation of the organism.

Local cytokine responses were not elevated in women infected with *N. gonorrhoeae* as compared to uninfected women; they were also not elevated if a woman was also infected with *Chlamydia trachomatis* or *Trichomonas vaginalis* (Hedges *et al.*, 1998). These data suggest that the presence of gonococci at the cervix not only suppresses an immune response to itself, but also to other organisms. Modulation of nitric oxide levels by the gonococcus to maintain anti-inflammatory levels may account, at least in part, for this immunosuppression.

Gonococcal infection in male patients is almost always symptomatic, with the most common clinical manifestation being urethritis. While females would produce NO from one or more of the three NOS enzymes, males would also come in contact with an intermittent supply of nitrite through the flushing of urine. The majority of gonococcal isolates tested showed generation of a high NO steady-state level, representing over 1 μM NO, in the presence of 2 mM nitrite. This could lead to μM levels of NO being produced from nitrite by gonococci rather than from iNOS, enhancing inflammation. Gonococcal strains requiring arginine, hypoxanthine and uracil are correlated with asymptomatic infections in men (Crawford *et al.*, 1977). Even in the presence of high nitrite, the NO steady-state level during denitrification by AHU strains remained far in the anti-inflammatory range. Thus, these strains may suppress an inflammatory response in males by setting a low NO level in the presence of nitrite.

The mean number of gonococci recovered from cervical aspirates in infected females was found to be 1.45×10^5 (Lowe and Krause, 1976) and 1×10^6 CFU (Young, *et al.*, 1983), while semen samples were found to contain an average of 6×10^6 CFU (Isbey, *et al.*, 1997). The higher number of gonococci in infected males may overwhelm the ability of host cells to produce NO, resulting in an elimination of NO rather than a steady-state anti-inflammatory level.

Neisseria meningitidis denitrification and subsequent effects on the host immune response

Neisseria meningitidis is likely exposed to varying degrees of stresses during the stages of colonization and subsequent disease. Upon growth within the nasopharynx, this organism must endure a multitude of killing mechanisms employed by resident macrophages, including not only oxidative stresses from the generation of reactive oxygen species, but also the direct and indirect effects of reactive nitrogen species; specifically, nitric oxide.

Genetic analysis of the *N. meningitidis* MC58 genome has revealed that similarly to *N. gonorrhoeae,* this other pathogenic neisserial species contains a partial denitrification pathway, which has been shown to be functional, allowing for microaerobic growth (Anjum *et al.,* 2002). Nitrite utilization occurs through the activity of a nitrite reductase (AniA), resulting in production of nitric oxide as an intermediate, which is then reduced to N_2O through the activity of a nitric oxide reductase (NorB) (Anjum *et al.,* 2002). NorB in addition to its role in denitrification aids in defence to NO. When challenged with nitrosative-stress generating compounds, SNP, GSNO, and SNAP, a *norB* mutant strain showed greatly attenuated growth compared to that of the wild-type. Cytochrome *c′*, encoded by *cycP,* also functions in defence to RNS, as it binds and removes NO within *Rhodobacter capulatus* (Cross *et al.,* 2000), thus eliminating the potential to undergo further reactions that generate more damaging species such as peroxynitrite ($ONOO^-$). Within the meningococcus, cytochrome *c′* has been shown to be involved in nitrosative stress protection to a lesser extent (Anjum *et al.,* 2002). It is believed that cytochrome *c′* is able to bind NO until NorB levels reach sufficient expression to reduce the NO. These types of detoxification mechanisms are independent of oxygen (Cross *et al.,* 2001), which differs from *E. coli* and *S. enterica,* which require oxygen (Poole and Hughes, 2000).

The presence of both *norB* (nitric oxide reductase) and *cycP* (cytochrome *c′*) within *N. meningitidis* offers potential of this bacterium to resist the normal microbicidal activity of nitric oxide, which can be generated at micromolar concentrations via the inducible NO synthase, expressed within macrophages, neutrophils, and epithelial cells (Aktan, 2004). Upon infection of primary human macrophages or within a nasopharyngeal mucosa model, *norB* and to a lesser degree *cycP,* enhanced survival of *N. meningitidis* (Stevanin *et al.,* 2005). This enhanced intracellular survival occurred in parallel with a depletion of NO, which required viable bacteria capable of protein synthesis. These studies indicate the requirement of active NO detoxification systems for optimal survival during nasopharyngeal colonization and interactions with human macrophages.

Recent studies highlight the fact that bacterial–host interactions must consider not only the proper mechanisms that occur during an immune response, but activities of the pathogen, which may lead to direct effects on the host cells. Subsequently, *Neisseria* metabolism of nitric oxide functions to undermine the immune system to a more favourable environment in which the pathogen can continue to colonize. Variations in NO levels can alter the activation state of NF-κB, a global transcriptional regulator that controls the expression of several cytokines, receptors, chemokines, cell adhesion molecules, and iNOS, as discussed above.

This is apparent when examining the effects of NO levels and NO metabolism on the cytokine and chemokine profile of isolated human monocyte-derived macrophages. In the absence of bacteria, macrophages had a differential cytokine expression in response to NO levels when a NO donor was introduced, causing up-regulation of both IL-12 and TNF-α. Infection with wild-type *N. meningitidis* MC58 strain causes a typical cytokine and chemokine response, showing elevated expression of IFN-γ, IL-1β, MIP-1α, IL-6, and TNF-α (Stevanin *et al.*, 2007), many of whose expression has previously been shown to be linked to variations in local NO concentrations in animal models (Gardner *et al.*, 2002; Niedbala *et al.*, 2002). A *norB* mutant instead resulted in a 4-fold reduction in a pro-inflammatory cytokine expression, IL-12, while an anti-inflammatory cytokine IL-10 had a 2-fold increase in expression compared to infection with wild-type (Stevanin *et al.*, 2007). However, expression of TNF-α was increased two-fold in the *norB* mutant. Thus, studies suggest that infection with *N. meningitidis* favours a pro-inflammatory response, and consumption of NO affects in part the extent of the immune response. This is directly linked to the ability of the wild-type strain to metabolize NO through the activity of *norB*, as a *cycP* mutant strain gives a similar cytokine response as infection with wild-type (Stevanin *et al.*, 2007).

It would appear that these pathogens have evolved the denitrification system to function as an alternative respiratory pathway, which is beneficial for the bacteria under appropriate growth conditions. However, the metabolism of nitrogen oxides, through the activity of a nitric oxide reductase has additional implications on the effects it may have on the host immune response, which are only now beginning to be elucidated.

Concluding remarks

It is likely that both *N. gonorrhoeae* and *N. meningitidis* exist in an environment that is limited in oxygen, using anaerobic respiration to support growth. The two genes comprising the neisserial denitrification pathway are highly regulated and respond to multiple signals, including low oxygen levels (FNR), NO (NsrR), and iron concentrations (Fur/ArsR). Thus, the expression and activity of nitrite reductase and nitric oxide reductase may vary considerably during infection. The activities of these two enzymes, in turn, determines the NO steady-state level, which is also affected by environmental factors such as pH and the concentration of nitrite and NO. The pathogenic *Neisseria* use NO as a terminal electron acceptor to support growth, but expression of the denitrification pathway may also greatly influence the host immune response by affecting the signalling properties of NO.

Note added in proof

NsrR and its interaction with its DNA binding site has been further studied *in vitro*. Purified NsrR was shown to contain a [2Fe–2S] cluster that was required for DNA binding and for sensing and responding to NO (Isabella *et al.*, 2008). Three conserved cysteine residues in NsrR were demonstrated to be involved in coordination of this [2Fe–2S] cluster. The 29-bp inverted repeat sequence presumed to make up the NsrR operator sequence was confirmed by DNase I footprinting. EMSA analysis was used to calculate the affinity of NsrR for various binding sites in order to determine nucleotides important in the protein / DNA interaction.

The role of the gonococcal DnrN protein in iron–sulphur centre repair has been further elucidated (Overton *et al.*, 2008). A *dnrN* mutant was shown to be more sensitive to nitrite than a *dnrN*+ strain, and a *dnrN* mutation increased the sensitivity of a catalase-deficient strain to hydrogen peroxide. Following RNS-mediated damage, the FeS centers of both FNR and NsrR were found to be repaired slower in a *dnrN* strain than in the parental strain. The extent of the NsrR regulon in *N. meningitidis* has also been defined; Heurlier *et al.* (2008) used a transcriptomic approach and identified that the meningococcal NsrR regulon is similar to that currently defined in the gonococcus. The gene downstream of *aniA*, designated *nirV*, was shown to be in the same operon as *aniA*. This study also confirmed that meningococcal FNR senses NO as well as oxygen.

References

Aktan, F. (2004) iNOS-mediated nitric oxide production and its regulation. Life Sci. 75, 639–653.

Anjum, M.F., Stevanin, T.M, Read, R.C, and Moir, J.W. (2002). Nitric oxide metabolism in *Neisseria meningitidis*. J. Bacteriol. *184*, 2987–2993.

Baker, S.C., Ferguson, S.J., Ludwig, B., Page, M.D., Richter, O.-M.H., and van Spanning, R.J.M. (1998) Molecular genetics of the genus *Paracoccus*: metabolically versatile bacteria with bioenergetic flexibility. Microbiol. Mol. Biol. Rev. *62*, 1046–1078.

Barth, K., and Clark, V.L. (2008) Differences in nitric oxide steady-states between arginine, hypoxanthine, uracil auxotrophs (AHU) and non-AHU strains of *Neisseria gonorrhoeae* during anaerobic respiration in the presence of nitrite. Can. J. Microbiol. *54*, 639–646.

Bartolini, E., Frigimelica, E., Giovinazzi, S., Galli, G., Shaik, Y., Genco, C., Welsch, J.A., Granoff, D.M., Grandi, G., and Grifantini, R. (2006). Role of FNR and FNR-regulated, sugar fermentation genes in *Neisseria meningitidis* infection. Mol. Microbiol. *60*, 963–972.

Bates, D.M., Popescu, C.V., Khoroshilova, N., Vogt, K., Beinert, H., Münck, E., and Kiley, P.J. (2000). Substitution of leucine 28 with histidine in the *Escherichia coli* transcription factor FNR results in increased stability of the [4Fe–4S](2+) cluster to oxygen. J. Biol. Chem. *275*, 6234–6240.

Beaumont, H.J., Lens, S.I., Reijnders, W.N., Westerhoff, H.V., and van Spanning, R.J. (2004). Expression of nitrite reductase in *Nitrosomonas europaea* involves NsrR, a novel nitrite-sensitive transcription repressor. Mol. Microbiol. *54*, 148–158.

Berks, B.C., Fergusen, S.J., Moir, J.W.B., and Richardson, D.J. (1995). Enzymes and associated electron transport systems that catalyze the respiratory reduction of nitrogen oxides and oxyanions. Biochim. Biophys. Acta *1232*, 97–173.

Blake, T., Barnard, A., Busby, S.J., and Green, J. (2002). Transcription activation by FNR: evidence for a functional activating region 2. J. Bacteriol. *184*, 5855–5861.

Bodenmiller, D.M., and Spiro, S. (2006). The *yjeB* (*nsrR*) gene of *Escherichia coli* encodes a nitric oxide-sensitive transcriptional regulator. J. Bacteriol. *188*, 874–881.

Bogdan, C. (2001). Nitric oxide and the immune response – review. Nat. Immun. 2, 907–916.

Boulanger, M.J., and Murphy, M.E. (2002). Crystal structure of the soluble domain of the major anaerobically induced outer membrane protein (AniA) from pathogenic *Neisseria*: a new class of copper-containing nitrite reductases. J. Mol. Biol. *315*, 1111–1127.

Busenlehner, L.S., Pennella, M.A., and Giedroc, D.P. (2003). The SmtB/ArsR family of metalloregulatory transcriptional repressors: structural insights into prokaryotic metal resistance. FEMS Microbiol. Rev. *27*, 131–143.

Cardinale, J.A. (1999). Structural and functional analysis of aniA, the major anaerobically induced outer membrane protein of *Neisseria gonorrhoeae*. PhD thesis, University of Rochester, Rochester, NY.

Cardinale, J.A., and Clark, V.L. (2005). Determinants of nitric oxide steady state levels during anaerobic respiration by *Neisseria gonorrhoeae*. Mol. Microbiol. *58*, 177–188.

Chiang, R.C., Cavicchioli, R., and Gunsalus, R.P. (1997). 'Locked-on' and 'locked-off' signal transduction mutations in the periplasmic domain of the *Escherichia coli* NarQ and NarX sensors affect nitrate- and nitrite-dependent regulation by NarL and NarP. Mol. Microbiol. *24*, 1049–1060.

Clark, V.L., Campbell, L.A., Palermo, D.A., Evans, T.M., and Klimpel, K.W. (1987). Induction and repression of outer membrane proteins by anaerobic growth of *Neisseria gonorrhoeae*. Infect. Immun. *55*, 1359–1364.

Clark, V.L., Knapp, J.S., Thompson, S. and Klimpel, K.W. (1988). Presence of antibodies to the major anaerobically induced gonococcal outer membrane protein in sera from patients with gonococcal infections. Microbial Pathogen. *5*, 381–390.

Connelly L., Palacios-Callender M., Ameixa C., Moncada S., and Hobbs A.J. (2001) Biphasic regulation of NF-kappa B activity underlies the pro- and anti-inflammatory actions of nitric oxide. J. Immunol. *166*, 3873–3881.

Constantinidou, C., Hobman, J.L., Griffiths, L., Patel, M.D., Penn, C.W., Cole, J.A., and Overton, T.W. (2006). A reassessment of the FNR regulon and transcriptomic analysis of the effects of nitrate, nitrite, NarXL, and NarQP as *Escherichia coli* K12 adapts from aerobic to anaerobic growth. J. Biol. Chem. *281*, 4802–4815.

Cross, R., Aish, J., Paston, S.J., Poole, R.K., and Moir, J.W. (2000). Cytochrome *c'* from *Rhodobacter capsulatus* confers increased resistance to nitric oxide. J Bacteriol. *182*, 1442–1447.

Cross, R., Lloyd, D., Poole, R.K., and Moir, J.W. (2001). Enzymatic removal of nitric oxide catalyzed by cytochrome c; in *Rhodobacter capsulatus*. J. Bacteriol. *183*, 3050–3054.

Cruz-Ramos, H., Crack, J., Wu, G., Hughes, M.N., Scott, C., Thomson, A.J., Green, J., and Poole, R.K. (2002). NO sensing by FNR: regulation of the *Escherichia coli* NO-detoxifying flavohaemoglobin, Hmp. EMBO J. *21*, 3235–3244.

Davis,, K.L., Martin, E., Turko, I.V., and Murad, F. (2001). Novel effects of nitric oxide. Annu. Rev. Pharmacol. Toxicol. *41*, 203–236.

Delany, I., Rappuoli, R., and Scarlato, V. (2004). Fur functions as an activator and as a repressor of putative virulence genes in *Neisseria meningitidis*. Mol. Microbiol. *52*, 1081–1090.

Deora, A.A., and Lander, H.M. (2000). Regulation of signal transduction and gene expression by reactive nitrogen species. In Sen, C.K., Sies, H., and Baeurle, P.A., eds. Antioxidant and Redox Regulation of Genes. Academic Press, San Diego, CA.

Deora, A.A., Win, T., Vanhaesebroeck, B., and Lander., H.M. (1998). A redox-triggered ras-effector interaction. Recruitment of phosphotidylinositol-3-kinase to Ras by redox stress. J. Biol. Chem. *273*, 29923–29928.

Escolar, L., Perez-Martin, J., and de Lorenzo, V. (1999). Opening the iron box: transcriptional metalloregulation by the Fur protein. J. Bacteriol. *181*, 6223–6229.

Filenko, N., Spiro, S., Browning, D.F., Squire, D., Overton, T.W., Cole, J., and Constantinidou, C. (2007). The NsrR regulon of *Escherichia coli* K-12 includes genes encoding the hybrid cluster protein and the periplasmic, respiratory nitrite reductase. J. Bacteriol. *189*, 4410–4417.

Gardner, R.C., Laskin, J.D., Dambach, M.D., Sacco, M., Durham, S.K., Bruno, M.K., Cohen, S.D., Gordon, M.K., Gerecke, D.R., Zhou, P., and Laskin, D.L. (2002). Reduced hepatotoxicity of acetaminophen in mice lacking inducible nitric oxide synthase: potential role of tumor necrosis factor-alpha and interleukin-10. Toxicol. Appl. Pharmacol. *184*, 27–36.

Giel, J.L., Rodionov, D., Liu, M., Blattner, F.R., and Kiley, P.J. (2006). IscR-dependent gene expression links iron–sulphur cluster assembly to the control of O2-regulated genes in *Escherichia coli*. Mol. Microbiol. *60*, 1058–1075.

Gilberthorpe, N.J., Lee, M.E., Stevanin, T.M., Read, R.C., and Poole, R.K. (2007). NsrR: a key regulator circumventing *Salmonella enterica* serovar Typhimurium oxidative and nitrosative stress *in vitro* and IFN-γ-stimulated J774.2 macrophages. Microbiology. *153*, 1756–1771.

Goretski, J., Zafiriou, O.C., and Hollocher, T.C. (1990). Steady-state nitric oxide concentrations during denitrification. J. Biol. Chem. *265*, 11535–11538.

Grifantini, R., Sebastian, S., Frigimelica, E., Draghi, M., Bartolini, E., Muzzi, A., Rappuoli, R., Grandi, G., and Genco, C.A. (2003). Identification of iron-activated and -repressed Fur-dependent genes by transcriptome analysis of *Neisseria meningitidis* group B. Proc. Natl. Acad. Sci. USA *100*, 9542–9547.

Green. J., Sharrocks. A.D., Green, B., Geisow, M., and Guest, J.R. (1993). Properties of FNR proteins substituted at each of the five cysteine residues. Mol. Microbiol. *8*, 61–8.

Gruber, T.M., and Gross, C.A. (2003). Multiple sigma factors and the partitioning of bacterial transcription space. Annu. Rev. Microbiol. *57*, 441–466.

Guse, A., Stevenson, C.E., Kuper, J., Buchanan, G., Schwarz, G., Giordano, G., Magalon, A., Mendel, R.R., Lawson, D.M., and Palmer, T. (2003). Biochemical and structural analysis of the molybdenum cofactor biosynthesis protein MobA.J. Biol. Chem. *278*, 25302–25307.

Hantke, K. (2001). Iron and metal regulation in bacteria. Curr. Opin. Microbiol. *4*, 172–177.

Hedges S.R., Sibley D.A., Mayo M.S., Hook, III E.W., Russell M.W. (1998) Cytokine and antibody responses in women infected with *Neisseria gonorrhoeae*: effects of concomitant infections. J. Infect. Dis. *178*, 742–751.

Hedges, S.R., Mayo M.S., Mestecky J., Hook III E.W., and Russell M.W. (1999) Limited local and systemic antibody responses to *Neisseria gonorrhoeae* during uncomplicated genital infections. Infect. Immun. *67*, 3937–3946.

Hoehn, G.T., and Clark, V.L. (1990). A protein antigenically related to the major gonococcal anaerobically induced outer membrane protein is present in commensal *Neisseria* but poorly expressed in *N. meningitidis* strains. Infect. Immun. *58*, 3929–3933.

Hoehn, G.T., and Clark, V.L. (1992a). Isolation and nucleotide sequence of the gene (*aniA*) encoding the major anaerobically induced outer membrane protein of *Neisseria gonorrhoeae*. Infect. Immun. *60*, 4695–4703.

Hoehn, G.T., and Clark, V.L. (1992b). The major anaerobically induced outer membrane protein of *Neisseria gonorrhoeae*, Pan1, is a lipoprotein. Infect. Immun. *60*, 4704–4708.

Householder, T.C., Belli, W.A., Lissenden, S., Cole, J.A., and Clark, V.L. (1999). *cis*- and *trans*-acting elements involved in regulation of *aniA*, the gene encoding the major anaerobically induced outer membrane protein in *Neisseria gonorrhoeae*. J. Bacteriol. *181*, 541–551.

Householder, T.C. (2000). Regulation of gene expression in *Neisseria gonorrhoeae aniA* and *norB*, examples from the denitrification pathway. PhD thesis, University of Rochester, Rochester, NY.

Householder, T.C., Fozo, E.M., Cardinale, J.A., and Clark, V.L. (2000). Gonococcal nitric oxide reductase is encoded by a single gene, *norB*, which is required for anaerobic growth and is induced by nitric oxide. Infect. Immun. *68*, 5241–5246.

Ignarro, L.J., Wood K.S., and Wolin M.S. (1984) Regulation of purified soluble guanylate cyclase by porphyrins and metalloporphyrins: a unifying concept. Adv. Cyclic Nucleotide Protein Phosphorylation Res. *17*, 267–274.

Isabella, V.M., Wright, L.F., Barth, K., Spence, J.M., Grogan, S., Genco, C.A., and Clark, V.L. (2008). *cis*- and *trans*-acting elements involved in regulation of *norB (norZ)*, the gene encoding nitric oxide reductase in *Neisseria gonorrhoeae*. Microbiol. *154*, 226–239.

Isbey S.F., Alcorn T.M., Davis R.H., Haizlip J., Leone P.A., and Cohen M.S. (1997) Characterisation of *Neisseria gonorrhoeae* in semen during urethral infection in men. Genitourin. Med. *73*, 378–382.

Justino, M.C., Vicente, J.B., Teixeira, M., and Saraiva, L.M. (2005). New genes implicated in the protection of anaerobically grown *Escherichia coli* against nitric oxide. J. Biol. Chem. *280*, 2636–2843.

Justino, M.C., Almeida, C.C., Teixeira, M., and Saraiva, L.M. (2007). *Escherichia coli* di-iron YtfE protein is necessary for the repair of stress-damaged iron-sulfur clusters. J. Biol. Chem. *282*, 10352–10359.

Kang, Y., Weber, K.D., Qiu, Y., Kiley, P.J., and Blattner, F.R. (2005). Genome-wide expression analysis indicates that FNR of *Escherichia coli* K-12 regulates a large number of genes of unknown function. J. Bacteriol. *187*, 1135–1160.

Katsuyama K., Shichiri M., Marumo F., and Hirata Y. (1998) NO inhibits cytokine-induced iNOS expression and NF-kappaB activation by interfering with phosphorylation and degradation of IkappaB-alpha. Arterioscler. Thromb. Vasc. Biol. *18*, 1796–1802.

Keilty, S., and Rosenberg, M. (1987). Constitutive function of a positively regulated promoter reveals new sequences essential for activity. J. Biol. Chem. *262*, 6389–6395.

Kiley, P.J., and Beinert, H. (2003). The role of Fe-S proteins in sensing and regulation in bacteria. Curr. Opin. Microbiol. *6*, 181–185.

Knapp, J.S., and Clark, V.L. (1984). Anaerobic growth of *Neisseria gonorrhoeae* coupled tonitrite reduction. Infect. Immun. *46*, 176–181.

Körner, H., Sofia, H.J., and Zumft, W.G. (2003). Phylogeny of the bacterial superfamily of Crp-Fnr transcription regulators: exploiting the metabolic spectrum by controlling alternative gene programs. FEMS Microbiol. Rev. *27*, 559–592.

Lambden, P.R., and Guest, J.R. (1976). Mutants of *Escherichia coli* K12 unable to use fumarate as an anaerobic electron acceptor. J. Gen. Microbiol. *97*, 145–160.

Liaudet, L., S.F. Garcia, and S. Csaba. (2000). Biology of nitric oxide signaling. Crit. Care Med. *28*, N37–N52.

Lissenden, S., Mohan, S., Overton, T., Regan, T., Crooke, H., Cardinale, J.A., Householder, T.C., Adams, P., O'Conner, C.D., Clark, V.L., Smith, H., and Cole, J.A. (2000). Identification of transcription activators that regulate gonococcal adaptation from aerobic to anaerobic or oxygen-limited growth. Mol. Microbiol. *37*, 839–855.

Lowe T.L. and Kraus S.J. (1976) Quantitation of *Neisseria gonorrhoeae* from women with gonorrhea. J. Infect. Dis. *133*, 621–626.

Masse, E., and Gottesman, S. (2002). Small RNA regulates the expression of genes involved in iron metabolism in *Escherichia coli.* Proc. Natl. Acad. Sci. USA *99*, 4620–4625.

Mateo, A.O., and de Artinano, M.A.A. (2000). Nitric oxide reactivity and mechanisms involved in it biological effects. Pharm. Res. *42*, 421–427.

Mellies, J., Jose, J., and Meyer, T.F. (1997). The *Neisseria gonorrhoeae* gene aniA encodes an inducible nitrite reductase. Mol. Gen. Genet. *256*, 525–532.

Mellin, J.R., Goswami, S., Grogan, S., Tjaden, B., and Genco, C.A. (2007). A novel fur- and iron- regulated small RNA, NrrF, is required for indirect fur-mediated regulation of the sdhA and sdhC genes in *Neisseria meningitidis*. J. Bacteriol. *189*, 3686–3694.

Murakami, K.S., and Darst, S.A. (2003). Bacterial RNA polymerases: the wholo story. Curr. Opin. Struct. Biol. *13*, 31–39.

Moore, L.J., Mettert, E.L., and Kiley, P.J. (2006). Regulation of FNR dimerization by subunit charge repulsion. J. Biol. Chem. *281*, 33268–33275.

Nakano, M.M., Geng, H., Nakano, S., and Kobayashi, K. (2006). The nitric oxide-responsive regulator NsrR controls ResDE-dependent gene expression. J. Bacteriol. *188*, 5878–5887.

Niedbala, W., Wei, X.Q., Campbell, C., Thomas, D., Koma, M.K., and Liew, F.Y. (2002). Nitric oxide preferentially induces type 1 T-cell differentiation by selectively up-regulating IL-12 receptor β2 expression via cGMP. Proc. Natl. Acad. Sci. USA *99*, 16186–16191.

Ochsner, U.A., and Vasil, M.L. (1997). Gene repression by the ferric uptake regulator in *Pseudomonas aeruginosa*: cycle selection of iron-regulated genes. Proc. Natl. Acad. Sci. USA *93*, 4409–4414.

Overton, T., Reid, E.G., Foxall, R., Smith, H., Busby, S.J., and Cole, J.A. (2003). Transcription activation at *Escherichia coli* FNR-dependent promoters by the gonococcal FNR protein: effects of a novel S18F substitution and comparisons with the corresponding substitution in *E. coli* FNR. J. Bacteriol. *185*, 4734–4747.

Overton, T.W., Whitehead, R., Li, Y., Snyder, L.A., Saunders, N.J., Smith, H., and Cole, J.A. (2006). Coordinated regulation of the *Neisseria gonorrhoeae*-truncated denitrification pathway by the nitric oxide-sensitive repressor, NsrR, and nitrite-insensitive NarQ–NarP. J. Biol. Chem. *281*, 33115–33126.

Peng H.B., Libby P., and Liao J.K. (1995) Induction and stabilization of I kappa B alpha by nitric oxide mediates inhibition of NF-kappa B. J. Biol. Chem. *270*, 14214–14219.

Pennella, M.A., and Giedroc, D.P. (2005). Structural determinants of metal selectivity in prokaryotic metal-responsive transcriptional regulators. Biometals *18*, 413–428.

Philippot, L. (2002) Denitrifying genes in bacterial and Archaeal genomes. Biochim. Biophys. Acta *1577*, 355–376.

Poole, R.K., and M.N. Hughes. (2000). New functions for the ancient globin family: bacterial responses to nitric oxide and nitrosative stress. Mol. Microbiol. *36*, 775–783.

Reynaert N.L., Ckless K., Korn S.H., Vos N., Guala A.S., Wouters E.F., van der Vliet A., and Janssen-Heininger Y.M.. (2004) Nitric oxide represses inhibitory kappaB kinase through S-nitrosylation. Proc. Natl. Acad. Sci. USA *101*, 8945–8950.

Richardson, D.J. (2000) Bacterial respiration: a flexible process for a changing environment. Microbiology *146*, 551–571.

Rock, J.D., and Moir, J.W. (2005). Microaerobic denitrification in *Neisseria meningitidis*. Biochem. Soc. Trans. *33*, 134–136.

Rock, J.D., Mahnane, M.R., Anjum, M.F., Shaw, J.G., Read, R.C., and Moir, J.W. (2005). The pathogen *Neisseria meningitidis* requires oxygen, but supplements growth by denitrification. Nitrite, nitric oxide and oxygen control respiratory flux at genetic and metabolic levels. Mol. Microbiol. *58*, 800–809.

Rock, J.D., Thomson, M.J., Read, R.C., and Moir, J.W. (2007). Regulation of denitrification genes in *Neisseria meningitidis* by nitric oxide and the repressor NsrR.J. Bacteriol. *189*, 1138–1144.

Rodionov, D.A., Dubchak, I.L., Arkin, A.P., Alm, E.J., and Gelfand, M.S. (2005). Dissimilatory metabolism of nitrogen oxides in bacteria: comparative reconstruction of transcriptional networks. PLoS Comput. Biol. *1*, 415–431.

Rogstam, A., Larsson, J.T., Kjelgaard, P., and von Wachenfeldt, C. (2007). Mechanisms of adaptation to nitrosative stress in *Bacillus subtilis*. J. Bacteriol. *189*, 3063–3071.

Russell M.W., Hedges S.R., Wu H.-Y., Hook, III E.W., and Mestecky J. (1999) Mucosal Immunity in the genital tract: prospects for vaccines against sexually transmitted diseases – a review. Am. J. Reprod. Immunol. *42*, 58–63.

Schrag, J.D., Huang, W., Sivaraman, J., Smith, C., Plamondon, J., Larocque, R., Matte, A., and Cygler, M. (2001). The crystal structure of *Escherichia coli* MoeA, a protein from the molybdopterin synthesis pathway. J. Mol. Biol. *310*, 419–431

Schwartz, C.J., Giel, J.L., Patschkowski, T., Luther, C., Ruzicka, F.J., Beinert, H., and Kiley, P.J. (2001). IscR, an Fe-S cluster-containing transcription factor, represses expression of *Escherichia coli* genes encoding Fe-S cluster assembly proteins. Proc. Natl. Acad. Sci. USA *98*, 14895–14900.

Schwarz, G. (2005). Molybdenum cofactor biosynthesis and deficiency. Cell Mol. Life Sci. *62*, 2792–2810.

Short, H.B., Clark, V.L.. Kellogg, Jr., D.S., and Young, F.E. (1982). Anaerobic survival of clinical isolates and laboratory strains of *Neisseria gonorrhoeae*: use in transfer and storage. J. Clin. Microbiol. *15*, 915–919.

Spiro, S., Gaston, K.L., Bell, A.I., Roberts, R.E., Busby, S.J., and Guest, J.R. (1990). Interconversion of the DNA-binding specificities of two related transcription regulators, CRP and FNR. Mol. Microbiol. *4*, 1831–1838.

Stefano, G.B., Gouman, Y., Bilfinger, T.V., Welters, I.D., and Cadet, P. (2000). Basal nitric oxide limits immune, nervous, and cardiovascular excitation: human endothelia express a mu opiate receptor. Prog. Neurobiol. *60*, 513–530.

Stevanin, T.M., Moir, J.W., and Read, R.C. (2005). Nitric oxide detoxification systems enhance survival of *Neisseria meningitidis* in human macrophages and in nasopharyngeal mucosa. Infect. Immun. *73*, 3322–3329.

Stevanin, T.M., Laver, J.R., Poole, R.K., Moir, J.W., and Read, R.C. (2007). Metabolism of nitric oxide by *Neisseria meningitidis* modifies release of NO-regulated cytokines and chemokines by human macrophages. Microbes Infect. *9*, 981–987.

Stewart, V., and Bledsoe, P.J. (2003). Synthetic *lac* operator substitutions for studying the nitrate- and nitrite-responsive NarX–NarL and NarQ–NarP two-component regulatory systems of *Escherichia coli* K-12. J. Bacteriol. *185*, 2104–2111.

Stewart, V., and Parales, J. Jr. (1988). Identification and expression of genes *narL* and *narX* of the *nar* (nitrate reductase) locus in *Escherichia coli* K-12. J. Bacteriol. *170*, 1589–1597.

Strube, K., de Vries, S., and Cramm, R. (2007). Formation of a dinitrosyl iron complex by NorA, a nitric oxide-binding di-iron protein from *Ralstonia eutropha* H16. J. Biol. Chem. *282*, 20292–20300.

Tavares, P., Pereira, A.S., Moura, J.J., and Moura, I. (2006). Metalloenzymes of the denitrification pathway. J. Inorg. Biochem. *100*, 2087–100.

Tunbridge, A.J., Stevanin, T.M., Lee, M., Marriott, H.M., Moir, J.W., Read, R.C., and Dockrell, D.H. (2005). Inhibition of macrophage apoptosis by *Neisseria meningitidis* requires nitric oxide detoxification mechanisms. Infect. Immun. *74*, 729–733.

Turner, S., Reid, E., Smith, H., and Cole, J. (2003). A novel cytochrome *c* peroxidase from *Neisseria gonorrhoeae*: a lipoprotein from a Gram-negative bacterium. Biochem. J. *373*, 865–873.

Umansky, V.S., Hehner, P., Dumont, A., Hofmann, T.G., Schirrmacher, V., Droge, W., and Schmitz, M.L. (1998). Co-stimulatory effect of nitric oxide on endothelial NFκB implies a physiological self-amplifying mechanism. Eur. J. Immunol. *28*, 2276–2282.

Weber, I.T., and Steitz, T.A. (1987). Structure of a complex of catabolite gene activator protein and cyclic AMP refined at 2.5 Å resolution. J. Mol. Biol. *198*, 311–326.

Wennerhold, J. and Bott, M. (2006). The DtxR regulon of *Corynebacterium glutamicum*. J. Bacteriol. *188*, 2907–2918.

Whitehead, R.N., and Cole, J.A. (2006). Different responses to nitrate and nitrite by the model organism *Escherichia coli* and the human pathogen *Neisseria gonorrhoeae*. Biochem. Soc. Trans. *34*, 111–114.

Whitehead, R.N., Overton, T.W., Snyder, L.A., McGowan, S.J., Smith, H., Cole, J.A., and Saunders, N.J. (2007). The small FNR regulon of *Neisseria gonorrhoeae*: comparison with the larger *Escherichia coli* FNR regulon and interaction with the NarQ–NarP regulon. BMC Genomics. *8*, 35.

Xu, C., Shi, W., and Rosen, B. (1996). The chromosomal *arsR* gene of *Escherichia coli* encodes a *trans*-acting metalloregulatory protein. J. Biol. Chem. *271*, 2427–2432.

Young H., Sarafian S.K., Harris A.B., and McMillan A. (1983) Non-cultural detection of *Neisseria gonorrhoeae* in cervical and vaginal washings. J. Med. Microbiol. *16*, 183–191.

Zheleznova, E.E., Crosa, J.H., and Brennan, R.G. (2000). Characterization of the DNA- and metal-binding properties of *Vibrio anguillarum fur* reveals conservation of a structural Zn^{2+} ion. J. Bacteriol. *182*, 6264–6267.

Zumft, W.G. (1997) Cell biology and molecular basis of denitrification. Microbiol. Mol. Biol. Rev. *61*, 533–616.

Genomics and Recombination

3

John K. Davies

Abstract

Analyses of the available genome sequences of *Neisseria* species and strains have considerably increased our knowledge of the recombination processes occurring within this genus. The accumulation of multiple copies of mobile genetic elements such as insertion sequences and bacteriophages has obviously contributed to genomic diversity. However these elements have also served as substrates for homologous recombination reactions that, at a low frequency, have resulted in various genomic rearrangements. Numerous examples of the horizontal acquisition of DNA segments are also obvious. The largest of these acquired DNA segments have been termed Islands of Horizontal Transfer, and appear to have been acquired at a low frequency via an as yet unknown process. A second class of horizontally acquired DNA segments, involving smaller fragments termed minimal mobile elements, seem to be moving through the bacterial population at a higher frequency. These elements consist of a series of mutually exclusive cassettes, containing just a few genes, which target specific positions on the genomes. It is unclear how these elements evolved, but their movement appears to be facilitated by the natural transformation system, and integration occurs through homologous recombination involving conserved flanking genes. Each sequenced genome also contains a remarkable amount of non-coding repetitive DNA sequences. The most common of these repeats is involved in recognition and uptake of genus-specific DNA during transformation. Other repeats appear to be the target for site-specific recombinases or, especially when occurring in repeat arrays, the homologous recombination system. Finally, the genome sequences have also revealed numerous, previously unknown, examples of genes that appear to be subject to phase or antigenic variation. While the systems operating on a few of these genes have been known for some time, the magnitude of the variable gene repertoire was much greater than expected.

Introduction

The number of publicly available neisserial genome sequences has lagged behind those available for many other bacterial species. Analyses of the complete genome sequences of *Neisseria meningitidis* serogroup A strain Z2491, serogroup B strain MC58 and serogroup C strains FAM18 and 053442 have been published (Bentley *et al.*, 2007; Parkhill *et al.*, 2000; Peng *et al.*, 2008; Tettelin *et al.*, 2000). Recently the compete genome sequence of strain α14, a capsule null carrier isolate of *N. meningitidis*, was published along with the near complete genomes for two additional carrier isolates, serogroup 29E strain α153 and serogroup W135 strain α275 (Schoen *et al.*, 2008). The genome sequence of *Neisseria gonorrhoeae* strain FA1090 has also been completed (GenBank accession number AE004969), and although an analysis of the sequence has not been published several annotations are available (see Web Resources below). In addition, the complete genome sequence of *Neisseria lactamica* has been completed (see Web Resources below), although an annotation is not yet publicly available.

The genome sequences provide a series of 'snapshots' of individual genomes at a particular point in time. The completion of each meningococcal genome sequence has resulted in the identification of novel genes, and the number of new genes discovered has not decreased with each successive project. This is indicative of what has been termed an 'open pan-genome', where large DNA segments are moving horizontally through the species in a relatively efficient manner (Muzzi *et al.*, 2008; Schoen *et al.*, 2008). This is almost certainly also true of *N. gonorrhoeae* and the commensal *Neisseria* species, but more genome sequences are needed for these species before this can be confirmed.

All of these genomes are approximately 2.2 Mb in size, have a GC content of about 52%, and contain approximately 2000 coding sequences (Bentley *et al.*, 2007; Parkhill *et al.*, 2000; Peng *et al.*, 2008; Schoen *et al.*, 2008; Tettelin *et al.*, 2000). A surprisingly large portion (18–20%) of each of the annotated genomes appears to be non-coding, and contains numerous repetitive sequences (Bentley *et al.*, 2007; Parkhill *et al.*, 2000; Peng *et al.*, 2008; Schoen *et al.*, 2008; Tettelin *et al.*, 2000). The papers analysing the genome sequences, and a recent review, have described and compared the genetic content of these strains (Bentley *et al.*, 2007; Parkhill *et al.*, 2000; Peng *et al.*, 2008; Schoen *et al.*, 2008; Snyder *et al.*, 2005a; Tettelin *et al.*, 2000). That analysis will not be repeated here, and this chapter will concentrate on what the genome sequences reveal about the remarkable variety of recombination processes within *Neisseria* spp. It is clear that there is considerable potential within these species for genomic rearrangements, horizontal acquisition of DNA from related and unrelated species, and generation of nucleotide sequence diversity at specific loci.

Homologous recombination

Homologous recombination is a process that can involve genetic exchange between similar DNA sequences, but is also central to the processes of DNA replication and repair (Kline *et al.*, 2003). It has been known for some time that *N. gonorrhoeae* and *N. meningitidis* lack the SOS regulon that controls expression of genes involved in recombination and DNA repair in many species (Black *et al.*, 1998). In agreement with this, each of the genomes lack a gene encoding the equivalent of a LexA repressor, and recombination and DNA repair genes appear to be constitutively expressed (Black *et al.*, 1998). In addition to *recA*, genes equivalent to those encoding components of the RecBCD and RecF recombination pathways of *E. coli* have been identified in *Neisseria* spp. (Kline *et al.*, 2003). The RecBCD pathway, encoded by the *recB*, *recC* and *recD* genes, has been shown to be involved, to at least some extent, in pilin antigenic variation, transformation and DNA repair (Hamilton and Dillard, 2006; Hill *et al.*, 2007; Mehr and Seifert, 1998). Although *recO*, *recR*, *recQ* and *recJ* genes are present, neisserial genomes lack a *recF* homologue, leading to the designation of this recombination pathway as 'RecF-like' (Kline *et al.*, 2003). Inactivation of this pathway leads to severe defects in DNA repair and pilin antigenic variation (Mehr and Seifert, 1998). Both recombination pathways are reliant on a functional RecA protein.

Natural transformation system

The natural transformation system of *N. gonorrhoeae* has been the subject of a recent review (Hamilton and Dillard, 2006), and will therefore only be described briefly here. Unlike other naturally transformable species, *Neisseria* spp. are competent for DNA uptake in all growth phases and no soluble competence factor appears necessary. *Neisseria* spp. preferentially take up DNA that contains a genus-specific DNA uptake sequence (DUS), although the method by which this sequence is recognized by the transformation system remains unclear. The extracellular DNA needed for transformation presumably results from autolysis of sibling cells, or co-infecting strains of the same genus. A type IV secretion system, which actively exports chromosomal DNA from some strains within the genus, can also act to provide donor DNA (Hamilton *et al.*, 2005). The incoming DNA appears to be processed as part of the uptake process, with at least some of the DNA being converted to a single-stranded form. In addition, plasmid DNA from foreign backgrounds appears to be converted to a linear form, presumable through interaction with restriction enzymes. Chromosomal DNA does not appear to be subject to this restriction barrier.

It has been suggested that this is because it is converted to a single-stranded form before entering the cytoplasm, although the widespread distribution of methylases within the genus might also provide some level of protection from restriction. It has been known for a long time that piliation is linked to competence, although there is no direct evidence that the pilin subunit PilE is directly involved in DNA uptake. Rather it seems that an intact and functional pilus assembly apparatus is needed for DNA import. A model has been proposed that partially explains how this interaction between pilus assembly and transformation might work (Hamilton and Dillard, 2006). Integration of the incoming DNA into the recipient cell chromosome requires the RecA protein for homologous recombination. Integration appears to be independent of the 'RecF-like' pathway of recombination, whilst there is still some debate around the involvement of the RecBCD pathway. Mutants in this pathway display reduced, but still easily quantitated, rates of transformation. The transformation system is incredibly efficient, with transformation frequencies of 20% reported at high DNA concentrations (Gunn and Stein, 1996).

Large repetitive sequences and homologous recombination

Each of the sequenced neisserial genomes contains intact or truncated mobile genetic elements. Movement of these elements is assumed to involve the action of site-specific recombinases, but they are sufficiently large to serve another role, as substrates for homologous recombination reactions. This has resulted in rearrangement, deletion or amplification of segments of the chromosome.

Bacteriophages

There is a tendency with bacteriophage to think primarily of lytic bacteriophage, and *Neisseria* spp. have long been regarded as unusual in that they seemed to be immune to bacteriophage attack. However there are situations, particularly with filamentous bacteriophage, where viral particles can be extruded from the host cell without lysis. In other situations the viral genome can be integrated into the host chromosome in a quiescent state known as a prophage. In fact each sequenced neisserial genome contains multiple apparently intact, or truncated, prophages. These can be classified into two major groups based on whether the genetic organization suggests they resemble filamentous bacteriophage that package single-stranded DNA, or tailed bacteriophage that contain double-stranded DNA.

One study identified 11 intact and 12 partial copies of filamentous bacteriophages in searches of the *N. gonorrhoeae* and *N. meningitidis* Z2491, MC58 and FAM18 genomes (Kawai *et al.*, 2005). Each intact and some partial sequences were flanked on their right hand side by a gene (termed either *piv* or *irg*) encoding an unusual site-specific recombinase (Skaar *et al.*, 2005). It has been suggested that the *piv* gene is part of the bacteriophage, and the encoded recombinase allows excision of a circular form of the bacteriophage genome. The filamentous bacteriophage could be classified into 4 subtypes based on sequence similarity, and one of these sequence types appears to target dRS3 repeats (see below) as integration sites (Kawai *et al.*, 2005). Subsequently, extrachromosomal circular copies of this bacteriophage, and an extracellular nuclease-resistant form that contains single-stranded DNA, have been detected (Bille *et al.*, 2005). This bacteriophage initially appeared to be preferentially associated with disease-associated isolates of *N. meningitidis*. It was found in all 29 disease-associated isolates examined but in only two of 20 non-disease associated isolates (Bille *et al.*, 2005). Subsequently it was suggested that the presence of this bacteriophage somehow alters transmission dynamics (Moxon and Jansen, 2005). However, recent more detailed studies have detected this bacteriophage in 60% of virulent and 42% of carriage isolates, suggesting that it is not a characteristic of all invasive meningococcal isolates (Hotopp *et al.*, 2006).

The double-stranded DNA bacteriophages detected in *N. meningitidis* have been described as Mu-like (Bentley *et al.*, 2007), while those described in *N. gonorrhoeae* more closely resemble P2 (Piekarowicz *et al.*, 2007). *N. gonorrhoeae* contains two intact and three partial copies of these bacteriophages, constituting about 5% of the gonococcal genome, and they were variably present in the genomes of other strains screened (Piekarowicz *et al.*, 2007). The two intact copies and one of the partial copies show a high degree of sequence similarity (Piekarowicz *et al.*, 2007). As with the filamentous bacteriophages, extrachromosomal copies of the double-stranded DNA bacteriophage genomes have been detected (Piekarowicz *et al.*, 2007).

Insertion sequences

The genome of *N. meningitidis* MC58 contains 22 intact and 29 remnant insertion sequences (IS), including 14 copies of IS4351 (Tettelin *et al.*, 2000). Likewise, strain Z2491 contains 43 copies of complete or partial IS, including 14 sequences related to IS1016, 22 sequences related to IS1106, and 7 sequences related to IS1655 (Parkhill *et al.*, 2000). IS1655 appears to be specific to *N. meningitidis* (Schoen *et al.*, 2008). The available annotations of the genomes of *N. meningitidis* FAM18 and *N. gonorrhoeae* FA1090 also suggest the presence of more than 40 IS in each genome, with multiple copies of individual IS, particularly IS1016.

Gross genomic rearrangements

The meningococcal genomes appear to be largely co-linear, but in each strain large inversion events have occurred relative to the other genomes. The inversion event in Z2491 has occurred by recombination between non-coding repeat arrays (see below), whilst the separate inversion in FAM18 appears to involve copies of IS1106. In strain MC58 an inversion has occurred between duplicated genes (NMB0497 and NMB1779) encoding Fha, a large repetitive surface protein. Both strains Z2491 and FAM18 contain a single copy of this gene (Bentley *et al.*, 2007). The genome sequences of strains 053442 and α14 also provide evidence of inversions and translocations involving insertion sequences and repetitive sequences (Peng *et al.*, 2008; Schoen *et al.*, 2008). A comparison of the genomes of *N. gonorrhoeae* and *N. meningitidis* revealed much more complex rearrangements (Snyder *et al.*, 2005a).

Amplification of genome segments

The genome of meningococcal strain MC58 contains a perfect tandem duplication of 32kb. This involves genes NMB1124 to NMB1159, which are duplicated as NMB1162 to NMB1197 (Snyder *et al.*, 2005a).

Several examples of amplification events on the gonococcal genome have been reported. Pulse-field gel electrophoresis studies of DNA from a defined lineage of *N. gonorrhoeae* MS11 piliation variants has identified an *Nhe*I fragment which appears to expand and contract in 26-kb increments (Gibbs and Meyer, 1996). It was also noted that a 1.4-kb repeat was present on this *Nhe*I fragment, and it was suggested that the repeat might facilitate the size alterations seen. More recently, others have described the amplification of several different segments of the *N. gonorrhoeae* genome (Barten and Meyer, 2001). In each case the amplified region was flanked by directly repeated sequences. However, in contrast to what has been observed in other species, it was suggested that amplification could occur in the absence of a functional *recA* gene (Barten and Meyer, 2001).

The repeat surrounding the 26-kb amplicon (Gibbs and Meyer, 1996) was termed ISNg1 and shows most sequence similarity to IS*150* from *E. coli*. ISNg1 is, in fact, defective as the *orfA* and *orfB* needed for transposition have been disrupted by frameshift mutations (Hamilton and Davies, unpublished data). In contrast to the findings described above, this laboratory has found that this particular amplification is RecA-dependent, and that it occurs at a frequency similar to that predicted for an amplicon of this nature (Hamilton and Davies, unpublished data).

Horizontal transfer of DNA segments

A recent three-way coding sequence comparison of three *N. meningitidis* genomes revealed less than 100 unique genes per genome. The majority of these encode hypothetical proteins of unknown function, and many are associated with prophages, IS elements, or minimal mobile elements (see below) (Bentley *et al.*, 2007).

Islands of horizontal transfer

Many of these unique genes are clustered together on the chromosome, but neither the genes they contain, the structure of the DNA segment nor the site of integration suggest that they are classical pathogenicity islands, which contain identifiable virulence genes, are flanked by direct repeats, and target tRNA genes. These segments of the MC58 genome were termed islands of horizontal transfer (IHT), the name reflecting the fact that they display an atypical nucleotide composition, and do not contain non-coding repeats, that are common elsewhere on the genome (Tettelin *et al.*, 2000). IHT-A consists of two subregions. The first of 8.5 kb contains genes involved in capsule production, and another gene encoding an adenine rRNA methylase (Tettelin *et al.*, 2000). The second subregion of 5.4 kb is flanked by copies of IS1016, and encodes mainly hypothetical proteins of unknown function (Tettelin *et al.*, 2000). IHT-B is 17.1 kb and exclusively encodes hypothetical proteins, although some of these proteins display sequence similarity to plasmid replication proteins (Hotopp *et al.*, 2006; Tettelin *et al.*, 2000). IHT-C is larger at 32.6 kb, and contains genes that encode toxin-related proteins as well as some bacteriophage proteins. Based on genomic sequence comparisons, none of these IHTs were identified in strain Z2491, which however appeared to contain its own different IHT (Tettelin *et al.*, 2000). Subsequently, a large study using comparative genomic hybridization, suggested that many genes from IHT-A are found in all meningococcal strains (Hotopp *et al.*, 2006). As has been noted previously, it may not be coincidental that the duplicate copies of the *fhaA* and *fhaB* genes found in meningococcal strain MC58 are located adjacent to both IHT-B and IHT-C (Snyder *et al.*, 2007b).

Additional IHT were identified during the comparative genomic hybridization study involving a diverse collection of 48 meningococcal isolates, and a few strains of *N. gonorrhoeae* and *N. lactamica* (Hotopp *et al.*, 2006). IHT-D and IHT-E both contain prophage elements, and in addition IHT-E contains a transposon containing genes encoding part of a type I secretion system (Hotopp *et al.*, 2006). Recently additional IHT (termed IHT-F through IHT-K) were identified in strain 053442 (Peng *et al.*, 2008). Partial copies of IHT-G, which involves a *maf* locus, can also be found in strains Z2491 and MC58. Hierarchical clustering using the data from comparative genomic hybridization placed meningococcal strains into five groups, and it was noted that strains within each group had similar multilocus sequence type (MLST) profiles (Hotopp *et al.*, 2006). The groups could also be distinguished by the presence or absence of the IHTs and two bacteriophages (Table 3.1). The fact that these genomic changes parallel the MLST profiles of the strains examined suggests that the acquisition of IHT is an infrequent event (Hotopp *et al.*, 2006).

Table 3.1 Grouping of *Neisseria* species and strains based on the presence or absence of bacteriophage and IHT[a]

Species	CGH group[b]	Genome sequence	MuMenB/ PNM2[c]	PNM1	IHT-A	IHT-B	IHT-C	IHT-D	IHT-E
N. meningitidis	1		Variable	–	+	–	–	–	–
N. meningitidis	2	Z2491	+	+	+	–	–	–	–
N. meningitidis	3	FAM18	–	–	+	–	–	–	+
N. meningitidis	4		+	–	+	+	+	–	Partial
N. meningitidis	5	MC58	+	–	+	Variable	Variable	+	Partial
N. gonorrhoeae		FA1090	–	–		–	–	–	–
N. lactamica			–	–		–	–	+	–

a + refers to the presence, and – to the absence, of bacteriophage and IHT.

b Grouped using comparative genomic hybridization (Hotopp *et al.*, 2006).

c These two bacteriophage with high nucleotide sequence identity could not be distinguished.

Gonococcal genetic island

A large island, termed the gonococcal genetic island (GGI), has been found in most, but not all, strains of *N. gonorrhoeae* (Dillard and Seifert, 2001). The GGI can vary in gene content from strain to strain, but encodes a non-contact induced type IV secretion system that is capable of exporting DNA into the environment, where it can subsequently be used in the transformation of other cells (Dillard and Seifert, 2001). The GGI has always been found integrated at a specific site near the *ung* gene, close to the replication terminus (Hamilton *et al.*, 2005; Snyder *et al.*, 2005b). The specific site of integration appears to be a *dif* site, which is targeted by the site-specific recombinase XerCD (Hamilton *et al.*, 2005; Snyder *et al.*, 2005b). This raises the possibility that GGI integration (and presumably excision) could occur through site-specific, rather than homologous recombination. In support of this suggestion a model plasmid containing a *dif* site, but little additional homologous DNA, could be shown to integrate into, and excise from the chromosome (Hamilton *et al.*, 2005). Subsequently, both integrated and extrachromosomal plasmid forms of the GGI have been detected (Snyder *et al.*, 2005b). The GGI was originally described as being specific to the gonococcus (Dillard and Seifert, 2001), but subsequent studies have suggested that at least some strains of *N. meningitidis* contain specific variants of the GGI (Snyder *et al.*, 2005b).

Acquisition of islands

The question arises as to the means by which *Neisseria* spp. acquired these islands in the first place. The almost complete absence of DUS in the islands would suggest that they would be inefficiently processed by the transformation system. In addition, reanalysis of early genetic experiments suggests that processing of the incoming DNA ensures that fragments larger than 25 kb are rarely or never imported by the transformation system (Hamilton and Dillard, 2006). Some of the islands contain genes that are bacteriophage-related. This, and the presence of apparently intact prophages integrated into neisserial genomes suggests that transduction is a theoretically possible means by which DNA might be acquired from a foreign source. However, transduction of *Neisseria* spp. has never been demonstrated in the laboratory. Finally, some plasmid-related genes are found within islands, and some strains of *N. gonorrhoeae* are known to carry conjugative plasmids (Roberts and Falkow, 1977). However, the conjugative plasmids do not mobilize chromosomal sequences (Norlander *et al.*, 1979). This does not exclude the possibility that neisserial cells might act as rare recipients in cross-species conjugation events. Again, it would theoretically possible that some of the islands were acquired through conjugation.

Minimal mobile elements

The discovery of cassettes containing different genes, inserted at identical locations on the genomes of different strains of *Neisseria* spp., led to the concept of the Minimal Mobile Element (MME) (Saunders and Snyder, 2002). For instance, a gene encoding a hypothetical protein (NMB1020) is found between the *trpE* and *purK* genes in *N. meningitidis*. At the same position in the *N. gonorrhoeae* are genes encoding a restriction-modification system (Zhu *et al.*, 1999). In *N. polysaccharea* there is only a short intergenic region between the *trpE* and *purK* genes, a so-called 'empty' site (Snyder *et al.*, 2007a). Around 40 known or candidate MME sites have now been identified in *Neisseria* spp., along with more than 100 cassettes (Saunders and Snyder, 2002; Snyder *et al.*, 2004; Snyder *et al.*, 2007a). It has been suggested that 'empty' sites are the native configuration and the starting point for the evolution of MMEs, as no example of an empty site containing the remnants of a deletion event has been observed (Snyder *et al.*, 2007a). All MMEs share the common characteristics that they are flanked by conserved house-keeping genes, often similarly orientated and with related functions, and that unlike other islands the vast majority of the cassettes or flanking genes contain copies of the DUS (Snyder *et al.*, 2007a). It is envisaged that initial event, where a foreign cassette is inserted into an 'empty' site, is a rare event (Snyder *et al.*, 2007a). However, what is not clear is why a cassette containing an alternative gene or genes should target an identical 'empty' site in another strain. The combination of two or more such rare events involving identical sites would seem highly unlikely. No matter how they evolved, the end results are a series of cassettes that can be inserted at specific positions in the neisserial chromosomes. Given the presence of homologous flanking DNA and the DUS, and the efficiency of the natural transformation system in these species, it is likely that MMEs are highly mobile and readily move from strain to strain through the bacterial population. The species specific DNA uptake displayed by the natural transformation system would dictate that the majority of this horizontal exchange would occur within the genus *Neisseria,* but there is some evidence that MMEs have crossed species barriers (Snyder *et al.,* 2007a).

Non-coding repetitive sequences

All of the characterized neisserial genomes contain numerous non-coding repetitive sequences. These can occur singly, or in arrays of different repeats, and occupy more than 6% of each neisserial genome.

DNA uptake sequence (DUS)

The most numerous of the repeats is the DUS that is essential for the specific recognition of neisserial DNA by the natural transformation system (Goodman and Scocca, 1988). The DUS was originally described as a 10-bp repeat, but recent compilations suggest that in 76% of cases the sequence is a 12-mer 5′-ATGCCGTCTGAA-3′, that appears to be more efficiently recognized than the 10-bp sequence by the transformation system (Ambur *et al.,* 2007). Each neisserial genome contains approximately 1900 copies of the DUS, and by itself it therefore comprises approximately 1% of the genome (Ambur *et al.,* 2007). The DUS is not evenly distributed around the genomes with only 35% being found in coding sequences (Ambur *et al.,* 2007; Davidsen *et al.,* 2004). Those intragenic sequences are also not evenly distributed between coding sequences, and a bias towards genome maintenance genes has been noted (Davidsen *et al.,* 2004). In 65% of cases the repeat is intergenic, and most often (50% of all DUS) occurs as an inverted repeat downstream of genes where it acts as a transcriptional terminator (Ambur *et al.,* 2007). It has been estimated that 25% of all neisserial genes use an inverted repeat, involving the DUS, as a rho-independent terminator (Ambur *et al.,* 2007).

Neisserial intergenic mosaic elements (NIMEs)

Each meningococcal genome also contains 600–700 copies of a family of repetitive sequences termed the neisserial Intergenic Mosaic Elements (NIMEs). NIMEs consist of two components, dRS3 repeats and repetitive sequence (RS) elements (Bentley *et al.,* 2007; Parkhill *et al.,* 2000). The dRS3 repeat is a 20-bp sequence that, as inverted repeat, flanks the RS elements. The RS elements are a series of 50–150 nucleotide repeats that can be grouped into approximately 100 sequence families. In the FAM18 genome, the dRS3 repeat has also served as an integration site for a filamentous bacteriophage (Bentley *et al.,* 2007).

NIMEs, in particular, occur in arrays that seem to be preferentially associated with genes encoding cell surface proteins. It has been suggested that these NIME arrays promote sequence variation in neighbouring genes by increasing the frequency of recombination (Bentley *et al.,* 2007). The suggestion is that the NIME arrays promote the exchange of flanking sequences by facilitating homologous recombination, or by acting as targets for a site-specific recombinase. A candidate for the latter role is an integrase/recombinase encoded by the filamentous bacteriophage, which has been shown to bind to dRS3 within the arrays (Bille *et al.,* 2005; Kawai *et al.,* 2005).

Correia repeat enclosed elements (CREEs)

Another common repetitive sequence has been termed the Correia repeat enclosed element (CREEs). This same repetitive sequence has also been called nemis (for *Neisseria* miniature insertion sequence) and the neisserial repeat (Black *et al.,* 1995; Mazzone *et al.,* 2001). It has recently been suggested that CREE are part of a class of repeat sequences termed MITEs (miniature inverted repeat transposable elements) that are widespread in bacterial, archaeal and eukaryotic species (Delihas, 2008). CREEs contain a 26–27 bp sequence, the Correia repeat (CR), which as an inverted repeat flanks a central core region (Buisine *et al.,* 2002; Liu *et al.,* 2002). Two versions of the core sequence exist, a full-length sequence of approximately 104 bp, and a smaller 51-bp version of that same sequence.

Each meningococcal genome contains approximately 170 full-length CREEs, and about half that number of deleted versions, occupying 1.6–1.7% of the genome (Liu *et al.*, 2002). In addition, each genome contains approximately 30 identifiable CREE fragments. CREEs appear to be less common in the genome of *N. gonorrhoeae*, in which there are fewer than 100 CREE (Liu *et al.*, 2002). CREEs located upstream of specific genes have been shown to affect transcription (Black *et al.*, 1995; De Gregorio *et al.*, 2002; De Gregorio *et al.*, 2003). The full-length versions also contain a characterized IHF binding site, and it is feasible that IHF binding could affect transcription of nearby genes (Buisine *et al.*, 2002). The CREE structure is reminiscent of an IS element in that it is flanked by an inverted repeat sequence, and although they do not contain a transposase gene, there is evidence to suggest they are mobile. For instance, many of the meningococcal CREE are inserted at sites that do not contain a CREE at the equivalent position in the gonococcal genome. When the nucleotide sequences are compared it is apparent that CREE insertion is accompanied by a dinucleotide (TA) target site duplication (Buisine *et al.*, 2002). Again this is characteristic of mobile elements.

Other repeats

A series of other non-coding repeats have been termed REP2 through REP5. Each meningococcal genome contains approximately 20 copies of REP2 and REP4, and 10 copies of REP3 and REP5 (Bentley *et al.*, 2007). Finally, nineteen copies of a 183 bp AT-rich repeat (ATR), with imperfect 40-bp inverted repeats, has been noted in the Z2491 genome (Parkhill *et al.*, 2000).

Variable genes

The phase and antigenic variation systems of *Neisseria* species have been well studied. Phase variation involves a process of slipped-strand mispairing at tracts of short direct nucleotide repeats. Variations in the number of repeats can switch gene expression on or off by creating frameshift mutations if the repeats are located within the coding sequence, or by altering transcription levels when the repeats are located in the promoter region. These systems are discussed elsewhere and will not be considered in detail here, other than to note that genome sequencing has revealed that such systems are much more widespread in *Neisseria* than was previously suspected (Bentley *et al.*, 2007; Peng *et al.*, 2008).

Antigenic variation in pilin production has also been well studied, especially in *N. gonorrhoeae*. Pilin is encoded by the pilin expression locus, *pilE*. Upstream of the *pilE* gene and elsewhere are a series of promoter-less silent loci, *pilS*, that contain partial variant pilin coding sequences. A unidirectional recombination process can move variant pilin sequences into *pilE*, resulting in the production of antigenically distinct pilin molecules. This process of antigenic variation is also discussed elsewhere, and will not be examined in detail here. Meningococcal strains MC58 and Z2491 have a pilin expression locus similar to that observed in *N. gonorrhoeae*, except that the silent loci in *N. meningitidis* are flanked by NIME arrays (Parkhill *et al.*, 2000; Tettelin *et al.*, 2000). The pilin expression locus of *N. meningitidis* strain FAM18 has largely been deleted, and has been functionally replaced by insertion of a class 2 pilin gene elsewhere on the chromosome (Bentley *et al.*, 2007).

However, the genome sequences have revealed additional loci that may be subject to similar antigenic variation systems. Loci encoding a putative haemagglutinin (*hrpA*, also termed *fhaA*) and adhesin (*mafB*) contain, downstream of the intact gene, what appear to be silent loci encoding alternative C-terminal ends for the relevant protein. Each of the meningococcal genomes contain three *maf* loci, while a single *fha/hrp* locus is found in strains Z2491 and FAM18 (Bentley *et al.*, 2007; Parkhill *et al.*, 2000). Meningococcal strain MC58 contains two *fha* loci, and, as mentioned above, recombination between these two copies appears to have resulted in the strain-specific inversion of a segment of the genome (Tettelin *et al.*, 2000).

Another example of variable genes was described well before the genome sequences became available. *N. meningitidis* contains three or four copies of *opa* genes, while *N. gonorrhoeae* contains up to 11 *opa* gene copies. These genes encode outer membrane proteins that act as adhesins, and their expression is subject to both phase and antigenic variation. The antigenic variation in Opa expression differs from that seen with PilE in that it occurs at a much slower rate, and involves a straightforward homologous recombination reaction between variant regions of the participating *opa* genes (Aho *et al.*, 1991). Although there would seem to be the potential for recombination between the *opa* genes found within a single cell, the diversity of Opa proteins that can be expressed cannot be explained solely by intracellular recombination reactions. For instance, in an investigation of Opa protein diversity in a collection of related *N. meningitidis* strains, 26 different Opa proteins were identified, encoded by just four genes per genome (Hobbs *et al.*, 1998). Such diversity would more likely be generated by recombination following transformation with DNA from other strains, generating what have been termed mosaic genes (Snyder *et al.*, 2005a).

Conclusions

In terms of the recombination processes occurring within this genus, the genome sequence data have yielded findings that might have been expected, and those that were unexpected. Given that the same thing has been observed with many other species, it is not surprising that *Neisseria* species have accumulated multiple IS and bacteriophages within their genomes. Furthermore, given that the genomes contain multiple copies of individual IS it is not surprising that they have served as substrates for homologous recombination, resulting in genome rearrangements and amplification of specific genome segments. Nor was it particularly surprising to find large DNA segments that appear to have been horizontally acquired. Again, such islands have been noted in many other species.

The surprises include the existence of MMEs, which were previously unknown and were first described in this genus, although similar systems have since been discovered in other genera (Snyder *et al.*, 2007a). The existence of a highly efficient genus-wide natural transformation system had always suggested the possibility of widespread horizontal transfer of genetic material, but both the number of gene cassettes and the integration sites so far discovered suggest a genus-wide pool of genetic material much wider than previously suspected. Also a surprise was the sheer volume of non-coding repetitive sequences present in the genomes. Although the role of many of these repeats remains unknown, those such as the NIMEs that occur in arrays have also served as substrates for the homologous recombination system, resulting in genomic rearrangements. Finally, the genome sequences have also revealed a previously unappreciated repertoire of variable genes, subject to high frequency antigenic and phase variation.

Future trends

The introduction of new massively parallel sequencing technologies, and the resultant lowering of the cost of sequencing a bacterial genome, will undoubtedly lead to multiple additional neisserial genome sequences becoming available in the near future. For a species with highly efficient systems for horizontal transfer of DNA, it is probable that we will need many additional genome sequences from diverse strains within a species before we can come close to an adequate knowledge of what has been termed the 'open pan-genome' – the total genetic information circulating within a naturally transformable species (Muzzi *et al.*, 2008; Schoen *et al.*, 2008). The most pressing need would seem to be for additional genome sequences from *N. gonorrhoeae*, where only a single sequence is available for this pathogen. However, an equally convincing case can be made for sequencing multiple genomes from the commensal *Neisseria* species. Given the highly efficient natural transformation system operating in this genus, the commensal species have the potential to act as reservoirs of genetic information that could easily cross species barriers.

Evidence is also beginning to accumulate as to how the pathogenic *Neisseria* species evolved, and additional genome sequence data will assist in understanding this process. The *N. meningitidis* strain α14 is the closest meningococcal strain to *N. gonorrhoeae* and *N. lactamica* in a number of different phylogenetic analyses (Schoen *et al.*, 2008). It has been suggested that α14 resembles an unencapsulated common ancestor to both *N. meningitidis* and *N. gonorrhoeae*, with *N. meningitidis* evolving relatively recently by acquiring a capsule locus from *Pasteurella* via horizontal transfer (Schoen *et al.*, 2008).

On a more sobering note, the genome comparisons so far possible have not been helpful in identifying genes directly involved in virulence. Apart from the previously known difference in capsule production, comparisons of the genetic content of *N. gonorrhoeae* and *N. meningitidis* have not explained how these species cause such different diseases. Comparisons between the pathogenic and commensal *Neisseria* species, and between carriage and disease isolates of *N. meningitidis* have also been uninformative (Schoen *et al.*, 2008; Snyder and Saunders, 2006). It is clear that there is still much to be learnt about the genetic, recombination and pathogenic potential of these species.

Web resources

Several annotations of the *N. gonorrhoeae* FA1090 genome sequence are available at http://stdgen.northwestern.edu/; www.molbiol.ox.ac/uk/data.shtml; http://wishart.biology.ualberta.ca/BacMap/index.html; http://cmr.tigr.org/tigr-scripts/CMR/CmrHomePage.cgi. The complete genome sequence of *Neisseria lactamica* is available at www.sanger.ac.uk/Projects/N_lactamica.

References

Aho, E.L., Dempsey, J.A., Hobbs, M.M., Klapper, D.G., and Cannon, J.G. (1991). Characterization of the *opa* (class 5) gene family of *Neisseria meningitidis*. Mol. Microbiol. *5*, 1429–1437.

Ambur, O.H., Frye, S.A., and Tonjum, T. (2007). New functional identity for the DNA uptake sequence in transformation and its presence in transcriptional terminators. J. Bacteriol. *189*, 2077–2085.

Barten, R., and Meyer, T.F. (2001). DNA circle formation in *Neisseria gonorrhoeae*: a possible intermediate in diverse genomic recombination processes. Mol Gen Genet *264*, 691–701.

Bentley, S.D., Vernikos, G.S., Snyder, L.A., Churcher, C., Arrowsmith, C., Chillingworth, T., Cronin, A., Davis, P.H., Holroyd, N.E., Jagels, K., Maddison, M., Moule, S., Rabbinowitsch, E., Sharp, S., Unwin, L., Whitehead, S., Quail, M.A., Achtman, M., Barrell, B., Saunders, N.J., and Parkhill, J. (2007). Meningococcal genetic variation mechanisms viewed through comparative analysis of serogroup C strain FAM18. PLoS Genet *3*, e23.

Bille, E., Zahar, J.R., Perrin, A., Morelle, S., Kriz, P., Jolley, K.A., Maiden, M.C., Dervin, C., Nassif, X., and Tinsley, C.R. (2005). A chromosomally integrated bacteriophage in invasive meningococci. J. Exp. Med. *201*, 1905–1913.

Black, C.G., Fyfe, J.A., and Davies, J.K. (1995). A promoter associated with the neisserial repeat can be used to transcribe the uvrB gene from *Neisseria gonorrhoeae*. J. Bacteriol. *177*, 1952–1958.

Black, C.G., Fyfe, J.A., and Davies, J.K. (1998). Absence of an SOS-like system in *Neisseria gonorrhoeae*. Gene *208*, 61–66.

Buisine, N., Tang, C.M., and Chalmers, R. (2002). Transposon-like Correia elements: structure, distribution and genetic exchange between pathogenic *Neisseria* sp. FEBS Lett. *522*, 52–58.

Davidsen, T., Rodland, E.A., Lagesen, K., Seeberg, E., Rognes, T., and Tonjum, T. (2004). Biased distribution of DNA uptake sequences towards genome maintenance genes. Nucleic Acids Res. *32*, 1050–1058.

De Gregorio, E., Abrescia, C., Carlomagno, M.S., and Di Nocera, P.P. (2002). The abundant class of nemis repeats provides RNA substrates for ribonuclease III in neisseriae. Biochim. Biophys. Acta *1576*, 39–44.

De Gregorio, E., Abrescia, C., Carlomagno, M.S., and Di Nocera, P.P. (2003). Ribonuclease III-mediated processing of specific *Neisseria meningitidis* mRNAs. Biochem. J. *374*, 799–805.

Delihas, N. (2008). Small mobile sequences in bacteria display diverse structure/function motifs. Mol. Microbiol. *67*, 475–481.

Dillard, J.P., and Seifert, H.S. (2001). A variable genetic island specific for *Neisseria gonorrhoeae* is involved in providing DNA for natural transformation and is found more often in disseminated infection isolates. Mol. Microbiol. *41*, 263–277.

Gibbs, C.P., and Meyer, T.F. (1996). Genome plasticity in *Neisseria gonorrhoeae*. FEMS Microbiol. Lett. *145*, 173–179.

Goodman, S.D., and Scocca, J.J. (1988). Identification and arrangement of the DNA sequence recognized in specific transformation of *Neisseria gonorrhoeae*. Proc. Natl. Acad. Sci. USA *85*, 6982–6986.

Gunn, J.S., and Stein, D.C. (1996). Use of a non-selective transformation technique to construct a multiply restriction/modification-deficient mutant of *Neisseria gonorrhoeae*. Mol. Gen. Genet. *251*, 509–517.

Hamilton, H.L., and Dillard, J.P. (2006). Natural transformation of *Neisseria gonorrhoeae*: from DNA donation to homologous recombination. Mol. Microbiol. *59*, 376–385.

Hamilton, H.L., Dominguez, N.M., Schwartz, K.J., Hackett, K.T., and Dillard, J.P. (2005). *Neisseria gonorrhoeae* secretes chromosomal DNA via a novel type IV secretion system. Mol. Microbiol. *55*, 1704–1721.

Hill, S.A., Woodward, T., Reger, A., Baker, R., and Dinse, T. (2007). Role for the RecBCD recombination pathway for *pilE* gene variation in repair-proficient *Neisseria gonorrhoeae*. J. Bacteriol. *189*, 7983–7990.

Hobbs, M.M., Malorny, B., Prasad, P., Morelli, G., Kusecek, B., Heckels, J.E., Cannon, J.G., and Achtman, M. (1998). Recombinational reassortment among *opa* genes from ET-37 complex *Neisseria meningitidis* isolates of diverse geographical origins. Microbiology *144*, 157–166.

Hotopp, J.C., Grifantini, R., Kumar, N., Tzeng, Y.L., Fouts, D., Frigimelica, E., Draghi, M., Giuliani, M.M., Rappuoli, R., Stephens, D.S., Grandi, G., and Tettelin, H. (2006). Comparative genomics of *Neisseria meningitidis*: core genome, islands of horizontal transfer and pathogen-specific genes. Microbiology *152*, 3733–3749.

Kawai, M., Uchiyama, I., and Kobayashi, I. (2005). Genome comparison *in silico* in *Neisseria* suggests integration of filamentous bacteriophages by their own transposase. DNA Res. *12*, 389–401.

Kline, K.A., Sechman, E.V., Skaar, E.P., and Seifert, H.S. (2003). Recombination, repair and replication in the pathogenic neisseriae: the 3 R's of molecular genetics of two human-specific bacterial pathogens. Mol. Microbiol. *50*, 3–13.

Liu, S.V., Saunders, N.J., Jeffries, A., and Rest, R.F. (2002). Genome analysis and strain comparison of correia repeats and correia repeat-enclosed elements in pathogenic *Neisseria*. J. Bacteriol. *184*, 6163–6173.

Mazzone, M., De Gregorio, E., Lavitola, A., Pagliarulo, C., Alifano, P., and Di Nocera, P.P. (2001). Whole-genome organization and functional properties of miniature DNA insertion sequences conserved in pathogenic neisseriae. Gene *278*, 211–222.

Mehr, I.J., and Seifert, H.S. (1998). Differential roles of homologous recombination pathways in *Neisseria gonorrhoeae* pilin antigenic variation, DNA transformation and DNA repair. Mol. Microbiol. *30*, 697–710.

Moxon, E.R., and Jansen, V.A. (2005). Phage variation: understanding the behaviour of an accidental pathogen. Trends Microbiol *13*, 563–565.

Muzzi, A., Masignani, V., and Rappuoli, R. (2008). The pan-genome: towards a knowledge-based discovery of novel targets for vaccines and antibacterials. Drug Discov. Today *12*, 429–439.

Norlander, L., Davies, J., and Normark, S. (1979). Genetic exchange mechanisms in *Neisseria gonorrhoeae*. J. Bacteriol. *138*, 756–761.

Parkhill, J., Achtman, M., James, K.D., Bentley, S.D., Churcher, C., Klee, S.R., Morelli, G., Basham, D., Brown, D., Chillingworth, T., Davies, R.M., Davis, P., Devlin, K., Feltwell, T., Hamlin, N., Holroyd, S., Jagels, K., Leather, S., Moule, S., Mungall, K., Quail, M.A., Rajandream, M.A., Rutherford, K.M., Simmonds, M., Skelton, J., Whitehead, S., Spratt, B.G., and Barrell, B.G. (2000). Complete DNA sequence of a serogroup A strain of *Neisseria meningitidis* Z2491. Nature *404*, 502–506.

Peng, J., Yang, L., Yang, F., Yang, J., Yan, Y., Nie, H., Zhang, X., Xiong, Z., Jiang, Y., Cheng, F., Xu, X., Chen, S., Sun, L., Li, W., Shen, Y., Shao, Z., Liang, X., Xu, J.,and Jin, Q. (2008). Characterization of ST-4821 complex, a unique *Neisseria meningitidis* clone. Genomics *91*, 78–87.

Piekarowicz, A., Klyz, A., Majchrzak, M., Adamczyk-Poplawska, M., Maugel, T.K., and Stein, D.C. (2007). Characterization of the dsDNA prophage sequences in the genome of *Neisseria gonorrhoeae* and visualization of productive bacteriophage. BMC Microbiol *7*, 66.

Roberts, M., and Falkow, S. (1977). Conjugal transfer of R plasmids in *Neisseria gonorrhoeae*. Nature *266*, 630–631.

Saunders, N.J., and Snyder, L.A. (2002). The minimal mobile element. Microbiology *148*, 3756–3760.

Schoen, C., Blom, J., Claus, H., Schramm-Gluck, A., Brandt, P., Muller, T., Goesmann, A., Joseph, B., Konietzny, S., Kurzai, O., Schmitt, C., Friedrich, T., Linke, B., Vogel, U., and Frosch, M. (2008). Whole-genome comparison of disease and carriage strains provides insights into virulence evolution in *Neisseria meningitidis*. Proc. Natl. Acad. Sci. USA *105*, 3473–3478.

Skaar, E.P., Lecuyer, B., Lenich, A.G., Lazio, M.P., Perkins-Balding, D., Seifert, H.S., and Karls, A.C. (2005). Analysis of the Piv recombinase-related gene family of *Neisseria gonorrhoeae*. J. Bacteriol. *187*, 1276–1286.

Snyder, L.A., Davies, J.K., and Saunders, N.J. (2004). Microarray genomotyping of key experimental strains of *Neisseria gonorrhoeae* reveals gene complement diversity and five new neisserial genes associated with Minimal Mobile Elements. BMC Genomics *5*, 23.

Snyder, L.A., Davies, J.K., Ryan, C.S., and Saunders, N.J. (2005a). Comparative overview of the genomic and genetic differences between the pathogenic *Neisseria* strains and species. Plasmid *54*, 191–218.

Snyder, L.A., Jarvis, S.A., and Saunders, N.J. (2005b). Complete and variant forms of the 'gonococcal genetic island' in *Neisseria meningitidis*. Microbiology *151*, 4005–4013.

Snyder, L.A., McGowan, S., Rogers, M., Duro, E., O'Farrell, E., and Saunders, N.J. (2007a). The repertoire of minimal mobile elements in the *Neisseria* species and evidence that these are involved in horizontal gene transfer in other bacteria. Mol. Biol. Evol. *24*, 2802–2815.

Snyder, L.A., and Saunders, N.J. (2006). The majority of genes in the pathogenic *Neisseria* species are present in non-pathogenic *Neisseria lactamica*, including those designated as 'virulence genes'. BMC Genomics *7*, 128.

Snyder, L.A.S., Jordan, P.W., and Saunders, N.J. (2007b). *Neisseria*: a postgenomic view. In Bacterial pathogenomics, M.J. Pallen, K.E. Nelson, and G.M. Preston, eds. (Washington, DC, ASM Press), pp. 90–119.

Tettelin, H., Saunders, N.J., Heidelberg, J., Jeffries, A.C., Nelson, K.E., Eisen, J.A., Ketchum, K.A., Hood, D.W., Peden, J.F., Dodson, R.J., Nelson, W.C., Gwinn, M.L., DeBoy, R., Peterson, J.D., Hickey, E.K., Haft, D.H., Salzberg, S.L., White, O., Fleischmann, R.D., Dougherty, B.A., Mason, T., Ciecko, A., Parksey, D.S., Blair, E., Cittone, H., Clark, E.B., Cotton, M.D., Utterback, T.R., Khouri, H., Qin, H., Vamathevan, J., Gill, J., Scarlato, V., Masignani, V., Pizza, M., Grandi, G., Sun, L., Smith, H.O., Fraser, C.M., Moxon, E.R., Rappuoli, R., and Venter, J.C. (2000). Complete genome sequence of *Neisseria meningitidis* serogroup B strain MC58. Science *287*, 1809–1815.

Zhu, P., Morelli, G., and Achtman, M. (1999). The *opcA* and (psi)*opcB* regions in *Neisseria*: genes, pseudogenes, deletions, insertion elements and DNA islands. Mol. Microbiol. 33, 635–650.

Part II

Interactions with Host Cells

Gonococcal Biofilms

4

Michael A. Apicella, Megan Falsetta, Ryan Neil and Christopher Steichen

Abstract

Neisseria gonorrhoeae is an exclusive human pathogen. Our studies have demonstrated that it utilizes two distinct mechanisms for entry into human urethral and cervical epithelial cells involving different bacterial surface ligands and host receptors. Our recent studies have demonstrated that the gonococcus can form biofilms on glass surfaces and over human cells. In this chapter, we provided details of evidence for formation of gonococcal biofilms on human cervical epithelial cells during natural disease and provide further evidence that outer membrane blebbing by the gonococcus is crucial in biofilm formation over human cervical epithelial cells.

Introduction

Neisseria gonorrhoeae is a human-adapted, Gram-negative diplococcus that infects the human male and female reproductive tract. *N. gonorrhoeae* infections in women frequently go unnoticed and can persist for prolonged periods of time. These infections can either go completely unnoticed until they are identified through transmission to a symptomatic male partner or they can become symptomatic and eventually lead to serious upper genital tract infections. This can ultimately lead to infertility (Paris *et al.*, 1999). We have recently begun to study the role that biofilms could play in infection of the cervical surface. Our interest in the capability of *N. gonorrhoeae* to form a biofilm came about by observations made in our laboratory during 4- and 8-day infections of primary, human, urethral and cervical epithelial cells (Edwards *et al.*, 2000; Harvey *et al.*, 1997). These studies show that the gonococcus was forming microcolonies on these surfaces and eventually these transitioned into structures that resembled bacterial biofilms.

Numerous bacterial species have been have been shown to be capable of producing biofilms. Bacterial biofilms have been shown to be important commercially and medically. A bacterial biofilm is defined as a community of bacteria intimately associated with each other and included within an exopolymer matrix substance. This biological unit exhibits its own properties, quite different in comparison with those showed by the single species in planktonic form (Donlan, 2000). Studies have indicated that bacteria that reside within biofilms are more resistant to biocides and antibacterial agents than planktonic bacteria. Genetic analysis of biofilm formation has led to the proposal that extracellular signals and quorum-sensing regulatory systems are important for differentiated biofilms (Whiteley *et al.*, 2001). This model conforms to the concept of density-driven cell–cell communication. Bacterial attachment to a surface is divided into several distinct phases, including primary and reversible adhesion, secondary and irreversible adhesion, and biofilm formation (Dunne, 2002). Each of these phases is ultimately controlled by the expression of one or more gene products. Ultrastructurally, the mature bacterial biofilm resembles an underwater coral reef containing pyramidal or mushroom-shaped microcolonies of organisms embedded within an extracellular glycocalyx, with channels and cavities to allow the exchange of nutrients and waste (Dunne,

2002). The biofilm protects its inhabitants from predators, dehydration, biocides, and other environmental extremes while regulating population growth and diversity through primitive cell signals. From a physiological standpoint, surface-bound bacteria behave quite differently from their planktonic counterparts. In general, Gram-negative bacteria use acylated homoserine lactones as autoinducers, and Gram-positive bacteria use processed oligopeptides to communicate. Furthermore, there is mounting data suggesting that bacterial autoinducers elicit specific responses from host organisms. Although the nature of the chemical signals, the signal relay mechanisms, and the target genes controlled by bacterial quorum sensing systems differ, in every case the ability to communicate with one another allows bacteria to coordinate the gene expression, and therefore the behaviour, of the entire community (Miller and Bassler, 2001). Others believe that biofilm formation is multifactorial and complex. Hydrodynamics, nutrient load and intracellular carbon flux have major impacts, presumably by altering the expression of cellular traits essential for bacterial adaptation during the different stages of biofilm formation. Hence, differentiated biofilms may also be the net result of many independent interactions, rather than being determined by a particular global quorum-sensing system (Kjelleberg and Molin, 2002).

Biofilms are dynamic structures in which cells can switch from a weak interaction with the surface to an almost permanent binding through extracellular polymers. In addition, the bacteria form channels and pores to provide nutrients to bacteria within the biofilm structure. The maintenance of a biofilm can be attributed to the development and preservation of a matrix (Davies *et al.*, 1993). As more bacterial species which form biofilms are identified, the nature of the matrix structure can vary from exopolysaccharides to proteins to DNA and to complex mixtures of these components. In some bacterial species, over three hundred proteins can be detected in bacteria from mature biofilms that are not detected in planktonic bacteria (Sauer *et al.*, 2002). These proteins fall into the functional classes of metabolism, phospholipid and lipopolysaccharide (LPS) biosynthesis, membrane transport and secretion, and adaptation and protective mechanisms. In addition, biofilm bacteria are considered to be in the arrested-phase (of growth) partly because of the accumulation of acylhomoserine lactone within cell clusters (Hall-Stoodley and Stoodley, 2002). Some bacteria eventually detach from the mature biofilm to enter the surrounding fluid phase. Detachment is a physiologically regulated event in which a bacterium will release from the biofilm as a planktonic organism to move on to attach to other surfaces where they initiate biofilm development (O'Toole *et al.*, 2000). Many different mechanisms may contribute to the detachment process. O'Toole and Kolter (O'Toole *et al.*, 2000; O'Toole and Kolter, 1998) demonstrate that starvation may lead to detachment by an unknown mechanism. *Streptococcus mutans* produce a surface protein-releasing enzyme that mediates their release from biofilms (Lee *et al.*, 1996). A possible trigger for the release of this matrix-degrading enzyme could be cell density. In addition, the presence of homoserine lactones may cause biofilm reduction, as has been demonstrated with *Rhodobacter sphaeroides* (Puskas et al., 1997).

Biofilm formation during natural infection

We verified that *N. gonorrhoeae* can produce a biofilm during natural infection and can form biofilms in flow chambers over viable human cervical epithelial cells in culture. The second objective was to gain information about the structure of the gonococcal biofilm and transcriptional characteristics of the gonococcal biofilm. Using archival biopsies from women with culturally confirmed gonococcal cervicitis, we examine 10 biopsies for evidence of gonococcal biofilm formation during natural infection. In three individuals, we identified gonococci arrayed on the surface of cervical cells in formations consistent with biofilms and microcolonies. Fig. 4.1 shows an example of transmission electron micrographs of studies from a patient demonstrating evidence of microcolony/biofilm formation over cervical epithelia. Immunostaining with antibodies specific for epitopes found within the outer membrane or LOS of *N. gonorrhoeae* confirmed that these organisms were *N. gonorrhoeae*. Gonococcal infection in men has a much more intense inflammatory component than occurs in cervical infection. Extensive shedding of urethral epithelial cells is

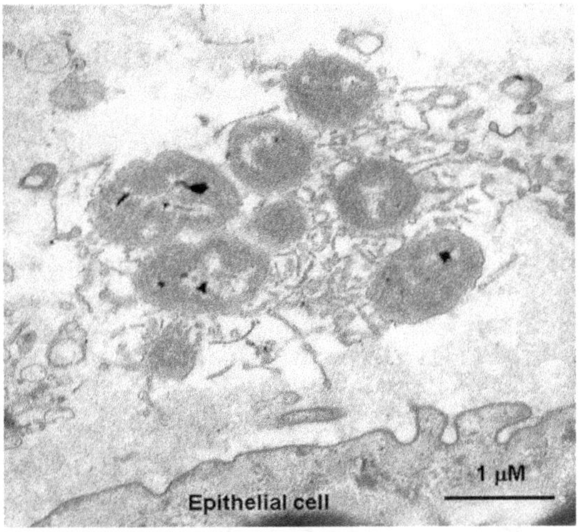

Figure 4.1 Electron micrograph of a 100-nm-thick section taken from a cervical biopsy with culture proven *N. gonorrhoeae* infection. The image demonstrates the clustering of bacteria above the cervical epithelial cell surface and the profusion of membranes between the bacteria typically seen in gonococcal biofilms. The bar represents 1 micron.

a early hallmark of urethral infection (Apicella *et al.*, 1996) and evidence of biofilm formation on these cells has not been recognized.

In vitro models of gonococcal biofilm formation over human epithelial cells

In order to develop an *in vitro* model which would simulate growth on cervical cells, we developed a special flow chamber which permitted growth of the biofilm on a confluent layer of cervical epithelia cells. E6/E7 transformed cervical epithelial were cultured in T-75 flasks (Corning, Corning, NY, USA) in MEM medium + 10% heat-inactivated fetal calf serum (FCS) (v/v) + 2 mM L-glutamine (all from Gibco, Grand Island, NY). Cells were trypsinized and placed onto 22 mm × 50 mm cover slips coated with 0.5 mg/ml bovine collagen for use in biofilms. Cell monolayers were used for biofilm experiments once they had reached at least 90% confluency as determined by visual examination.

The *N. gonorrhoeae* used to inoculate the biofilm chambers were collected after 16 hours of growth on Isovitalex, supplemented Gc Medium base (Difco). Organisms were suspended to an absorbance of 0.3 at 600 nm in 0.1 × HBE tissue culture media diluted in 1 × phosphate-buffered saline (PBS) pH 7.4 and supplemented with Isovitalex (1:100 v/v) and 100 μM sodium nitrite. Bacterial suspensions were added to biofilm chambers containing the cell monolayers and allowed to incubate at 37°C for 1 hour before the start of media flow. The flow rate through the chamber was 9 ml per hour.

To assess biofilm growth on tissue culture cells, we constructed a flow-through chamber which permitted the insertion of a cover slip coated with live transformed cervical cells. Briefly, the chamber was composed of three pieces. The bottom was a 35 mm × 70 mm × 4 mm piece of acrylic with a 17-mm-wide, 45-mm-long, and 2-mm-deep chamber machined out of the block. Media intake and outtake ports were centred on the ends of the chamber at 2 mm depth. A 1-mm-thick rubber gasket the size of the acrylic block with a 17 mm × 45 mm section removed from the centre was placed on top of the chamber bottom. The top piece was acrylic with the same 17 mm × 45 mm middle section removed. The underside of the top piece has a 22 mm × 50 mm section machined out to the same depth as the cover slip to accommodate the tissue culture cells. The whole apparatus was held together by 10 screws evenly spaced around the cover slip.

This device has allowed us to perform an array of microscopy studies in the context of human cells including light microscopy, transmission electron microscopy (TEM) and scanning electron microscopy SEM. It has also allowed visualization of the biofilm during fluid flow through the chamber. *Haemophilus influenzae* has been shown to form biofilms on glass surfaces, in the chinchilla middle ear and humans with otitis media (Greiner *et al.*, 2004; Jurcisek *et al.*, 2005; Post, 2001). Unlike the *H. influenzae* biofilms we have observed (Greiner *et al.*, 2004; Jurcisek *et al.*, 2005), gonococcal biofilms are much more easily detached from the cell surface and can be seen moving back and forth in fluid stream when the pump is turned on and off. These studies demonstrated that *N. gonorrhoeae* could form a biofilm in continuous flow over primary human genital tract epithelial cells in culture. Cryosections viewed by light microscopy indicated that gonococci were surrounded by a matrix-like material. Cryo-field emission scanning electron microscopy and TEM have shown that this biofilm matrix material consists of membranous structures that appeared to be surrounding gonococci. Binding studies with lectins and monoclonal antibody 6B4 showed that LOS-like structures predominate on the biofilm and that these structures could be identified separated from organisms within the biofilm. These membranous extensions can reach 10–15 mm in length. We presume that they are derived from the bacterial outer membrane, which is shed from the gonococcus in the form of blebs. Our studies have shown that human, cervical tract epithelial cells can serve as surfaces for gonococcal biofilm formation with minimal, if any effect, on cell viability. This may reflect the ability of the gonococcus to induce anti-apoptotic factors in these cells (Binnicker *et al.*, 2003, 2004).

The role of outer membrane blebbing in gonococcal biofilm formation

Immunoelectron microscopy experiments on primary urethral epithelial cells show membrane structures resembling those seen in Fig. 4.2A. These structures stained heavily with monoclonal antibody 6B4 which recognizes the Gala1–4Glc-NAc lipooligosaccharide epitope. It is interesting to speculate that these membranous structures are derived in large part from the fusions of blebs shed by the gonococcus and the meningococcus during growth (Devoe and Gilchrist, 1973). SEM analysis of gonococcal biofilms would confirm this observation (Fig. 4.3). These appear to encase the gonococci within sac-like structures into which a matrix may released. A recent study

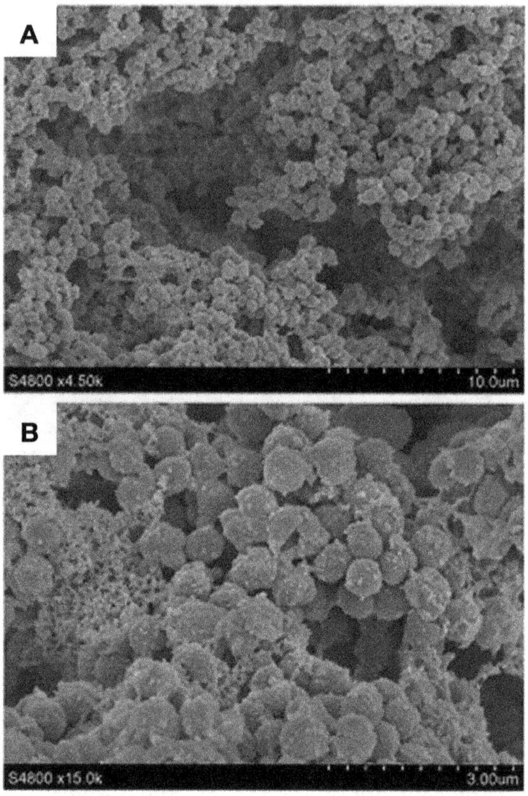

Figure 4.3 Scanning electron micrographs of a biofilm formed by *N. gonorrhoeae* strain 1291 at magnifications of **A** 4500 and **B** 15,000x. The water channels within the biofilm can be seen in panel **A** and the extensive coalesced blebbing can be seen in **B**.

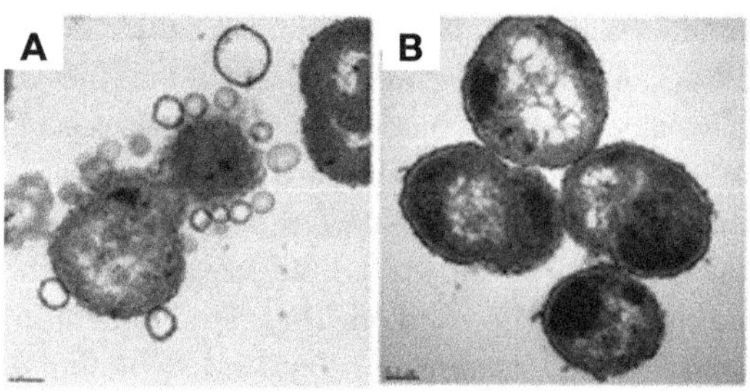

Figure 4.2 Images of surface blebbing in **A** a strain *N. gonorrhoeae* 1291*mod* mutant and **B** a 1291mod/on phase variant. As can be seen, the modoff mutant shows a hyperblebbing phenotype. This mod mutant makes a significantly great biofilm.

of biofilms produced by 39 *Neisseria meningitidis* strains show that 30% of the carriage isolates and 12.5% of the invasive disease isolates formed biofilms in microtitre wells (Yi *et al.*, 2004). These studies would support the concept that hydrophobic interactions of surfaces with pathogenic *Neisseria* enhance their ability to form biofilms. This study lends support to the theory that the membranous bacterial-derived structures may facilitate biofilm formation. *N. gonorrhoeae* strain 1291*msbB*, which has a severely compromised ability to form outer membrane blebs, produced biofilms on cultured transformed cervical cells that were significantly reduced in thickness and overall biomass compared with 1291 wt biofilms. However, the 1291*msbB* bacteria were able to attach to cultured transformed cervical cells just as well as parent bacteria were, and the growth rate in GC broth was the same for mutant and parent strains. In addition, *N. gonorrhoeae* strain FA1090 *modA13* which has an increased blebbing phenotype compared to the FA1090 parent strain (Fig. 4.2A and B). This mutant forms a biofilm of significantly greater thickness and biomass than the wild-type strain. These results further suggest that outer membranes released by *N. gonorrhoeae* within a biofilm are an integral component of biofilm formation, and hypo- or hyper-blebbing of the outer membrane directly affects biofilm thickness and density. Although it is well accepted that biofilms in general have an exopolysaccharide framework for their matrix, in addition to a host of other products produced by the cell or acquired from the surrounding environment, we have not yet explored what this framework is composed of in *N. gonorrhoeae* biofilms (Kolter, 2005; Sutherland, 2001). However, our findings suggest that incorporation of outer membranes resulting from blebbing of the gonococcus is crucial to proper biofilm formation.

The resistance of *N. gonorrhoeae* to antibiotic treatment in recent years has increased to a point where many previously used antibiotics are no longer efficacious in treating the disease (Bauer *et al.*, 2005, 2008). The ability *of N. gonorrhoeae* to form biofilms in the female genital tract further exacerbates this dilemma. Biofilm formation in bacteria has been shown to prevent complete access of antibiotics to all regions of the biofilm. The diverse microenvironments found in biofilms as a result of pH differences, oxygen variance, water channels, and varied growth rates among the embedded bacteria also contribute to the difficulty of treating biofilms with antimicrobial agents (Musk and Hergenrother, 2006). Furthermore, horizontal gene transfer among bacteria in a biofilm can increase the spread of antibiotic resistance genes (Fux *et al.*, 2005). In addition to antibiotic resistance, biofilm formation by *N. gonorrhoeae* in the female urogenital tract may also contribute to the high incidence of asymptomatic infections in women (Paris *et al.*, 1999). Whereas in men gonococcal infections tend to be of a more inflammatory nature leading to a robust polymorphonuclear leucocyte response, in most instances this does not occur in women and the majority unaware they are infected. This can lead to a greater likelihood of the bacteria being spread to additional sexual partners as well as to more severe complications of *N. gonorrhoeae* infection, including pelvic inflammatory disease when symptoms occur in women.

Owing to the ability of *N. gonorrhoeae* to form biofilms and the refractive nature of biofilms to antibiotic treatment, alternative strategies to resolve *N. gonorrhoeae* infection should be pursued. Recent interesting research has been done attempting to interfere with quorum sensing in bacteria to prevent biofilm formation from occurring. Much of this research has focused on inhibition of N-acyl homoserine lactones and LuxR-type proteins, which would prevent a quorum sensing cascade necessary for biofilm formation(Hentzer *et al.*, 2003).

Meningococcal biofilms

Recent studies have shown that *N. meningitidis* can form biofilms in plastic tubing and continuous flow chambers (Lappann *et al.*, 2006; Yi *et al.*, 2004). The studies have demonstrated that meningococci within biofilms are more resistant to penicillin but not to rifampin or ciprofloxacin (Lappann *et al.*, 2006). These studies suggested that *pilE* was not necessary for biofilm formation and that PilX, a pilus associated protein, which is involved in interbacterial aggregation, indirectly supported microcolony formation. The role that biofilms may play in meningococcal nasopharyngeal colonization and disease is not known at this time.

Conclusions

Neisseria gonorrhoeae and *N. meningitidis* can form biofilms. Our studies indicate that the basic nature of the biofilm structure in each is similar with membrane blebbing playing a major role in maintaining the integrity of the biofilm. What additional materials are essential for biofilm formation in pathogenic *Neisseria* is currently an area of intense study. It is interesting to speculate on the role that biofilms may play in asymptomatic infection (colonization by meningococci) and in disease. It is clear from studies with other organisms that microbes within biofilms are significantly more antibiotic resistant and potentially involved in increased DNA exchange. The rapid evolution of antibiotic resistance in the gonococcus suggests that a similar mechanism may be occurring during cervical infection.

References

Apicella, M.A., Ketterer, M., Lee, F.K., Zhou, D., Rice, P.A., and Blake, M.S. (1996). The pathogenesis of gonococcal urethritis in men: confocal and immunoelectron microscopic analysis of urethral exudates from men infected with *Neisseria gonorrhoeae*. J. Infect. Dis. *173*, 636–646.

Bauer, H.M., Mark, K.E., Samuel, M., Wang, S.A., Weismuller, P., Moore, D., Gunn, R.A., Peter, C., Vannier, A., DeAugustine, N., *et al.* (2005). Prevalence of and associated risk factors for fluoroquinolone-resistant *Neisseria gonorrhoeae* in California, 2000–2003. Clin. Infect. Dis. *41*, 795–803.

Bauer, H.M., Wohlfeiler, D., Klausner, J.D., Guerry, S., Gunn, R.A., and Bolan, G. (2008). California guidelines for expedited partner therapy for Chlamydia trachomatis and *Neisseria gonorrhoeae*. Sex. Transm. Dis. *35*, 314–319.

Binnicker, M.J., Williams, R.D., and Apicella, M.A. (2003). Infection of human urethral epithelium with *Neisseria gonorrhoeae* elicits an upregulation of host anti-apoptotic factors and protects cells from staurosporine-induced apoptosis. Cell. Microbiol. *5*, 549–560.

Binnicker, M.J., Williams, R.D., and Apicella, M.A. (2004). Gonococcal porin IB activates NF-kappaB in human urethral epithelium and increases the expression of host antiapoptotic factors. Infect. Immun. *72*, 6408–6417.

Davies, D.G., Chakrabarty, A.M., and Geesey, G.G. (1993). Exopolysaccharide production in biofilms: substratum activation of alginate gene expression by Pseudomonas aeruginosa. Appl. Environ. Microbiol. *59*, 1181–1186.

Devoe, I.W., and Gilchrist, J.E. (1973). Release of Endotoxin in the Form of Cell Wall Blebs During In Vitro Growth of *Neisseria Meningitidis*. J. Exp. Med. *138*, 1156–1166.

Donlan, R.M. (2000). Role of biofilms in antimicrobial resistance. ASAIO J. *46*, S47–52.

Dunne, W.M., Jr. (2002). Bacterial adhesion: seen any good biofilms lately? Clin. Microbiol. Rev. *15*, 155–166.

Edwards, J.L., Shao, J.Q., Ault, K.A., and Apicella, M.A. (2000). *Neisseria gonorrhoeae* elicits membrane ruffling and cytoskeletal rearrangements upon infection of primary human endocervical and ectocervical cells. Infect. Immun. *68*, 5354–5363.

Fux, C.A., Costerton, J.W., Stewart, P.S., and Stoodley, P. (2005). Survival strategies of infectious biofilms. Trends Microbiol. *13*, 34–40.

Greiner, L.L., Watanabe, H., Phillips, N.J., Shao, J., Morgan, A., Zaleski, A., Gibson, B.W., and Apicella, M.A. (2004). Nontypeable *Haemophilus influenzae* strain 2019 produces a biofilm containing N-acetylneuraminic acid that may mimic sialylated O-linked glycans. Infect. Immun. 72, 4249–4260

Hall-Stoodley, L., and Stoodley, P. (2002). Developmental regulation of microbial biofilms. Curr. Opin. Biotechnol. *13*, 228–233.

Harvey, H.A., Ketterer, M.R., Preston, A., Lubaroff, D., Williams, R., and Apicella, M.A. (1997). Ultrastructural analysis of primary human urethral epithelial cell cultures infected with *Neisseria gonorrheae*. Infect. Immun. *65*, 2420–2427.

Hentzer, M., Eberl, L., Nielsen, J., and Givskov, M. (2003). Quorum sensing : a novel target for the treatment of biofilm infections. BioDrugs *17*, 241–250.

Jurcisek, J., Greiner, L., Watanabe, H., Zaleski, A., Apicella, M.A., and Bakaletz, L.O. (2005). Role of sialic acid and complex carbohydrate biosynthesis in biofilm formation by nontypeable *Haemophilus influenzae* in the chinchilla middle ear. Infect. Immun. *73*, 3210–3218.

Kjelleberg, S., and Molin, S. (2002). Is there a role for quorum sensing signals in bacterial biofilms? Curr. Opin. Microbiol. *5*, 254–258.

Kolter, R. (2005). Surfacing views of biofilm biology. Trends Microbiol. *13*, 1–2.

Lappann, M., Haagensen, J.A., Claus, H., Vogel, U., and Molin, S. (2006). Meningococcal biofilm formation: structure, development and phenotypes in a standardized continuous flow system. Mol. Microbiol. *62*, 1292–1309.

Lee, S.F., Li, Y.H., and Bowden, G.H. (1996). Detachment of *Streptococcus mutans* biofilm cells by an endogenous enzymatic activity. Infect. Immun. *64*, 1035–1038.

Miller, M.B., and Bassler, B.L. (2001). Quorum sensing in bacteria. Annu. Rev. Microbiol. *55*, 165–199.

Musk, D.J., Jr., and Hergenrother, P.J. (2006). Chemical countermeasures for the control of bacterial biofilms: effective compounds and promising targets. Curr. Med. Chem. *13*, 2163–2177.

O'Toole, G., Kaplan, H.B., and Kolter, R. (2000). Biofilm formation as microbial development. Annu. Rev. Microbiol. *54*, 49–79.

O'Toole, G.A., and Kolter, R. (1998). Initiation of biofilm formation in Pseudomonas fluorescens WCS365 proceeds via multiple, convergent signalling pathways: a genetic analysis. Mol. Microbiol. *28*, 449–461.

Paris, M., Gotuzzo, E., Goyzueta, G., Aramburu, J., Caceres, C.F., Castellano, T., Jordan, N.N., Vermund, S.H., and Hook, E.W., 3rd (1999). Prevalence of gonococcal and chlamydial infections in commercial sex workers in a Peruvian Amazon city. Sex. Transm. Dis. *26*, 103–107.

Post, J.C. (2001). Direct evidence of bacterial biofilms in otitis media. Laryngoscope *111*, 2083–2094.

Puskas, A., Greenberg, E.P., Kaplan, S., and Schaefer, A.L. (1997). A quorum-sensing system in the free-living photosynthetic bacterium *Rhodobacter sphaeroides*. J. Bacteriol. *179*, 7530–7537.

Sauer, K., Camper, A.K., Ehrlich, G.D., Costerton, J.W., and Davies, D.G. (2002). *Pseudomonas aeruginosa* displays multiple phenotypes during development as a biofilm. J. Bacteriol. *184*, 1140–1154.

Sutherland, I.W. (2001). The biofilm matrix--an immobilized but dynamic microbial environment. Trends Microbiol. *9*, 222–227.

Whiteley, M., Bangera, M.G., Bumgarner, R.E., Parsek, M.R., Teitzel, G.M., Lory, S., and Greenberg, E.P. (2001). Gene expression in *Pseudomonas aeruginosa* biofilms. Nature *413*, 860–864.

Yi, K., Rasmussen, A.W., Gudlavalleti, S.K., Stephens, D.S., and Stojiljkovic, I. (2004). Biofilm formation by *Neisseria meningitidis*. Infect. Immun. 72, 6132–6138.

Newly Described Surface Structures and Adhesion Strategies of the Pathogenic Neisseriae

5

Rosanna Leuzzi, Laura Serino, Davide Serruto and Mariagrazia Pizza

Abstract

N. meningitidis and *N. gonorrhoeae* have evolved numerous surface-exposed proteins that mediate interaction with the host. Host–pathogen interaction is a multifactorial process and several surface structures have been shown to play a role in this process, suggesting that interactions mediated by different molecules are important in determining tissue tropism and invasive versus carriage potentials.

Porins, type IV pili and opacity proteins are the major objects of investigation of neisserial pathogenesis and recent studies have improved our understanding of their functionality at a molecular level of detail.

In the last few years, availability of neisserial genome sequences has enabled the identification of new surface proteins with a potential role as adhesins and virulence factors. Characterization of these new proteins have extended our knowledge on the multiple strategies evolved by pathogenic neisseriae to infect the host.

Introduction

N. meningitidis and *N. gonorrhoeae* are Gram-negative obligate human pathogens with a typical cell envelope organization composed by outer and inner cell membranes that are separated by the periplasm containing the peptidoglycan layer. The meningococcus outer membrane is surrounded by a polysaccharide capsule that is essential for pathogenicity as it confers resistance to phagocytosis and complement-mediated lysis and offers protection against environmental insults. The outer membrane consists of phospholipids, lipopolysaccharides and membrane proteins, which play a crucial role in the contact of pathogenic *Neisseria* with the outside world.

In the last few decades, the major outer membrane proteins of *Neisseria* have been identified and extensively investigated from a functional and structural point of view. They include porin, Rmp, iron-regulated proteins such as transferrin, lactoferrin and haemoglobin receptors and the major adhesins such as pili and opacity proteins (Derrick *et al.*, 2006).

More recently, thanks to the availability of the complete genome sequence of *Neisseria* strains, it has been possible to get access to the entire repertoire of the surface-exposed proteins potentially encoded by *Neisseria* and never identified by classical approaches. Further analysis from the '*in silico*' to the bench has revealed not only that many of these proteins are potential vaccine antigens, but that they could also act as potential adhesins/virulence factors.

In this chapter, we will describe the more recent findings on the best known adhesins of *Neisseria* and the properties of the newly identified surface-exposed proteins, with special emphasis on those with adhesive properties (Fig. 5.1). Finally, we will describe some of the antigens which, despite having no role in the adhesion process, have additional functions supporting a potential role in virulence and pathogenesis.

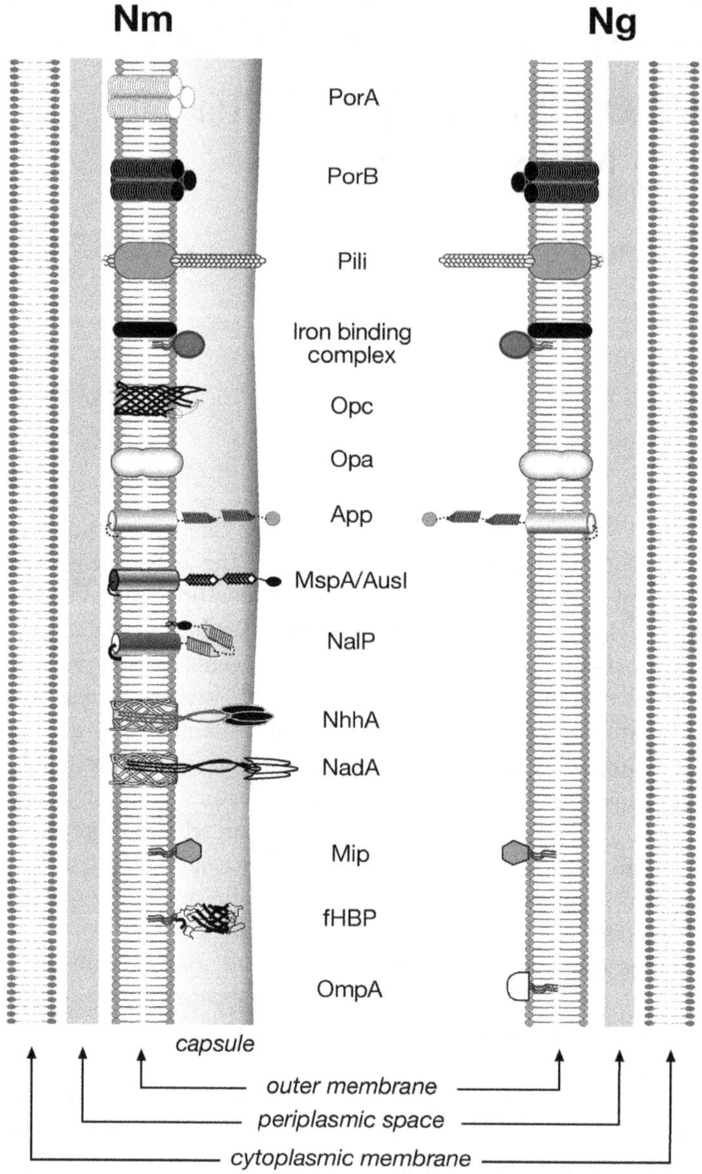

Figure 5.1 Schematic representation of the different bacterial compartments in *N. meningitidis* (Nm) and *N. gonorrhoeae* (Ng) and comparison of the main outer membrane components characterized in the two species.

Proteins involved in interactions of pathogenic *Neisseria* with host cells: new findings in known components

Adhesion of *N. meningitidis* and *N. gonorrhoeae* to mucosal epithelia of nasopharynx and genitourinary tract represents the primary event of colonization and consequently the first step of pathogenesis.

The major adhesins of *Neisseria* are pili and the opacity proteins Opa and Opc. Pili are regarded as the main adhesins in the initial attachment to the mucosal epithelia; the elongation of these polymeric filaments on the bacterial surface allows the first contact with the host cells. This is of particular importance for capsulated *N. meningitidis,* in which the presence of the capsule can influence the exposure of the surface adhesins.

Once the primary bacteria–cells contact is established, the synthesis of pili and capsule is down-regulated (Deghmane *et al.*, 2002) and a different repertoire of adhesins operates a deeper interaction with the host cell surface. Among them, the integral adhesins Opa and Opc are the major and most abundant determinants of intimate adhesion and bind to the cellular receptors in absence of capsule.

The mechanisms of pathogenesis orchestrated by pili and Opa proteins, as well as the multiple functions operated by porins, have been extensively described in the last decades. However, new insights have been provided by recent studies, as reviewed in the next sections.

Type IV pili

Type IV pili, composed of pilin subunits, are filamentous structures essential in mediating the adhesion to both epithelial and endothelial cells, and are also associated with other phenotypes such as bacterial motility, microcolony formation and natural transformation (Virji *et al.*, 1991; Rudel *et al.*, 1995; Pujol *et al.*, 1997; Merz and So, 2000; Aas *et al.*, 2002; Morand *et al.*, 2004). The pilus has been suggested to bind to the membrane cofactor protein CD46, which is expressed on all human cells (Kallstrom *et al.*, 1997).

In addition to the pilin subunit, PilE, other components are essential for the functionality of pili. Among them, PilC1 and PilC2 proteins have a crucial role in pilus assembly and in adhesion (Nassif *et al.*, 1994) and PilT is an inner membrane ATPase necessary for the retraction of pili, a function required for twitching motility and localized adherence to host cells (Merz and So, 2000).

A recent study reported the detailed structure of the gonococcus pili (Craig *et al.*, 2006). A high-resolution crystal structure of the pilin subunit and a 3D cryo-EM reconstruction of the pilus filament has helped to address the mechanism of pili assembly and the multifunctionality of this apparatus.

In the structure, the filament core is formed by spiralling three-helix bundle, which anchors the globular heads and provide strength and flexibility to the pilus. Positive charged grooves have been identified, which might act as domains involved in the binding with the negatively charged backbone of DNA, providing a mechanism for the role of pili in natural transformation. The alternation of positively and negatively charged surface patches can be important in pilus–pilus interactions, microcolony formation and adhesion to cell surfaces. Therefore, the general aspect of pili is a charged and corrugated structure in which surface protrusions are responsible for antigenic variation and for pilus–pilus interaction, whereas internal grooves contain binding sites for DNA, proteins and carbohydrates. Although pili bind to host cells by the tip-associated adhesin PilC, the pilin subunits can act directly on the host cells to drive adhesion and microcolony formation. This reconstruction allows the attribution of single functions to the architecture of pili and suggests a way by which these structures contribute to diverse binding events, including adherence to host cell surfaces, to other bacterial cells and to double-stranded DNA. Overall, this structure gives an explanation to the paradox of a conserved molecular machinery with an extremely high level of multifunctionality (Craig *et al.*, 2006).

Pelicic and co-workers recently brought new insights into pili-mediated attachment of pathogenic *Neisseria* species to human cells. They performed a signature-tagged mutagenesis screening of a collection of meningococcus mutants for those presenting reduced attachment to human endothelial cells. This led to the identification of a protein essential for adherence, named PilX, which encodes a pilin-like protein that co-purifies with pilin fibres and has intrinsic aggregative but no adhesive properties (Helaine *et al.*, 2005). Moreover, the same group solved the X-ray crystal structure of PilX that reveals α/β roll fold shared by all pilins. This suggests that PilX is a minor, or low-abundant, pilin that assembles within the filaments in a similar way to pilin. Surface-exposed motifs in PilX subunits might stabilize bacterial aggregates against the disruptive force of pilus retraction (Helaine *et al.*, 2007).

Opa proteins

The outer membrane opacity proteins Opa and Opc are so called since they are responsible for an opaque phenotype of agar-grown colonies. The Opa proteins are expressed both by pathogenic

Neisseria spp. as well as by some commensal *Neisseria* strains. A single gonococcal strain can harbour up to twelve *opa* genes, whereas meningococci usually encode three or four Opa proteins.

Despite variation in amino acid sequence, almost all currently characterized meningococcal and gonococcal Opa proteins can be grouped into two major classes, according to their binding specificity for human surface receptors: the large group Opa_{CEA}-type (specific for CEACAMs, carcinoembryonic antigen-related adhesion molecules) and the Opa_{HS}-type (specific for the glycosamicoglycan side-chains of heparan sulphate proteoglycans, HSPGs) (Hauck and Meyer, 2003) (see Chapter 7).

CEACAMs are host surface receptors represented by multiple isoforms, each with a different specificity for various epithelial and immune cells. This differential expression and the subsequent Opa recognition may explain the tissue tropism exhibited by neisserial Opa_{CEA} variants. Opa_{CEA}-positive bacteria can bind different human cell types, triggering subsequent cellular events. Interactions with epithelial cells are followed by rearrangements of the actin cytoskeleton and bacterial internalization; contact with neutrophils results in opsonin-independent phagocytosis accompanied by activation of intracellular signal cascade; binding to activated T-cells inhibits T-cell proliferation and activation. Moreover, bacteria-induced cytokines and bacterial LOS lead to increased expression of CEACAM on endothelial cells, allowing enhanced adhesion of Opa_{CEA}-expressing neisseriae. Since CEACAM recognition is not limited to pathogenic *Neisseria*, the biological events primed by Opa_{CEA} binding could also be common to other non-pathogenic and pathogenic bacteria sharing the same ecological niche.

A second target for gonococcal Opas are HSPGs. Variants of Opa with highly cationic surface-exposed loops bind strongly to HSPGs, resulting in invasion of certain cellular types. Moreover, interactions are potentiated by serum proteins such as vitronectin and fibronectin. Also in this case, Opa_{HS}-mediated binding results in stimulation of intracellular signalling pathways and cytoskeletal rearrangement, events that facilitate gonococcal invasion.

Opc is functional in meningococcus and, contrarily to Opa, is antigenically invariant and encoded by a single gene. Opc expression promotes the adherence and invasion of meningococcus independently of pili and Opa functions. This functionality is explicated in absence of capsule, suggesting that the protein is properly accessible to cell contact during the intimate phase of adhesion, when the capsule synthesis is down-regulated (Virji *et al.*, 1993) (see Chapter 6).

Interestingly, Opc is a functional homologue of gonococcal Opa_{HS} since it is able to bind the HSPG receptors and to mediate the attachment to human endothelial cells by binding to human fibronectin and vitronectin (Virji *et al.*, 1994).

Porin

The porin proteins are the predominant proteins present on the bacterial surface. Neisserial porins function as pores and control the diffusions of small metabolites. Moreover, they have been implicated in pathogenic mechanisms such as serum resistance, immune stimulation, host cell survival, and invasion (see Chapter 7) (Massari *et al.*, 2003). Recently, the role of PorB1A (whose expression correlates with systemic gonococcal infection) in the interaction with host cells has been characterized: the human heat shock glycoprotein Gp96 and the scavenger receptor SREC are PorBIA-specific receptors. Bacterial adherence is increased, whereas invasion into epithelial cells is inhibited when Gp96 is present. The scavenger receptor can function in the uptake of gonococcus by epithelial cells. Rechner *et al.* suggest a two-step process initiated by the binding of Gp96 to PorBIA, which then competes for efficient uptake of bacteria via SREC receptors; Gp96 might have a role as an anti-invasion factor while SRECs work as receptors mediating host cell entry of highly invasive gonococci (Rechner *et al.*, 2007).

Novel identified adhesin autotransporters

Classical autotransporters: App and MspA/AusI

Secreted proteins enable bacteria to interface with the external milieu. Besides their fundamental role in nutrient acquisition, secreted proteins play a crucial role in establishing interactions with hosts and consequently in determining disease.

Among the multiple secretion machineries evolved by Gram-negative bacteria, the predominant strategy in *Neisseria* is the autotransporter pathway. Autotransporter pathway, also known as type V secretion system, represents the simplest form of protein export in bacteria.

IgA protease from *N. gonorrhoeae* was the first autotransporter to be described and defined as the prototype of the autotransporters superfamily (Pohlner *et al.*, 1987). Subsequent studies on IgA protease and other autotransporters have spotlighted the wide complexity of this class of proteins, differing from each other in their structure, biogenesis and mechanism of secretion (reviewed in van Ulsen and Tommassen, 2006).

The minimal and necessary attributes defining the family of autotransporters are three structural features: (1) a signal peptide which targets the protein to the Sec machinery for the translocation across the inner membrane; (2) a C-terminal translocator or β-domain integrated in the outer membrane; and (3) an N-terminal passenger domain, often exceeding 100 kDa, whose secretion in the extracellular space is mediated by the β-domain. Once in the extracellular space, the passenger domain explicates its biological activity, which ranges from cytotoxicity, adhesion to the host surface, enzymatic activity degradation and other virulence functions.

In the majority of the Gram-negative autotransporters, the functional activity resides in the secreted passenger domain, whereas the un-cleaved form inserted in the outer membrane is only an intermediate deputed to the secretion of the functional passenger domain. Examples are IgA1 proteases of *Neisseria* and *Haemophilus*, whose secreted forms are capable of cleaving the hinge region of human secretory IgA1 (Pohlner *et al.*, 1987; Poulsen *et al.*, 1989). In addition, the passenger domain of neisserial IgA1 proteases cleaves LAMP1 (a major integral membrane glycoprotein of lysosomes), a property which contributes to the intracellular life of *Neisseria* (Hauck and Meyer, 1997) (see Chapter 6).

App of *Neisseria meningitidis* is a member of the autotransporter family. The protein has been named App for adhesion and penetration protein (Hadi *et al.*, 2001) because of its homology to Hap (*Haemophilus* adhesion and penetration protein (St. Geme *et al.*, 1994).

The structural organization of App reflects the classical architecture of IgA protease-like autotransporters, including the serine protease motif in the β-domain, which pilots the autoproteolytic cleavage of the passenger domain and its subsequent release in the extracellular space. The autoproteolysis is operated by the aminoacidic triad $H^{115}D^{158}S^{267}$, (Serruto *et al.*, 2003) reproducing the catalytic mechanism demonstrated for the cleavage of human IgA by IgA protease and for the auto-processing of Hap.

App is efficiently secreted in the extracellular space although the un-cleaved form of the protein is still present on the outer surface to some extent. The full and secreted forms of App are likely to be deputed to distinct biological activities. The full-length protein is able to mediate the adhesion of *N. meningitidis* to human conjunctiva-derived cell lines (Chang cells) but not to HUVEC endothelial cells, suggesting a cell type-specific binding (Serruto *et al.*, 2003). The autoproteolytic cleavage of App can be abolished by mutation of the serine located in the catalytic triad. The uncleaved App shows improved adhesive and aggregative properties, suggesting that the full-length protein is responsible for the functional activity, which is in agreement to what showed for the homologue Hap of *Haemophilus influenzae* (Fink *et al.*, 2003). The binding domain is likely to be located in the region of the passenger domain proximal to OM-embedded β-domain. To date, no function has been ascribed to the secreted form of App, although the evolution of this secretion pathway in *Neisseria* would suggest a relevant biological significance.

App is present in all *Neisseria* species with high sequence conservation. App of *N. meningitidis* shares, indeed, 73% and 95% identity to *N. lactamica* and *N. gonorrhoeae*, respectively.

Antibodies recognizing App are detected in sera deriving from convalescent patients and from healthy carrier population; on the contrary, antibodies are absent in sera from non-carrier population (van Ulsen *et al.*, 2001; Hadi *et al.*, 2001). These data suggest a possible implication of App both in colonization and in infection.

App, as well as IgA protease, are secreted in two forms of different molecular weights, reflecting a dual processing of the proteins. Studies on neisserial autotransporter NalP have elucidated these mechanisms of secretion (van Ulsen *et al.*, 2003). Indeed, deletion of *nalP* gene changed the secretion profile of *N. meningitidis*, resulting in the presence of a unique secreted form of both App and IgA protease. This evidence led to a picture of cooperative functionality of neisserial autotransporters, in which NalP is implicated in modulating the processing of other autotransporters. NalP-mediated processing produces the release of a larger form of the passenger domain of App and IgA protease, which includes the so called 'α-peptide', containing in its sequence a nuclear localization motif possibly responsible for an interaction with proteins located in the nucleus of eukaryotic cells.

The two secreted forms of App or, alternatively, the two anchored domains, could perform distinct biological activities. A competition between the autoproteolysis and NalP-mediated process could be a mechanism of modulation of App functions. The biological significance of such a mechanism needs to be elucidated.

MspA/AusI or NMB1998 is a novel putative autotransporter protein, recently identified in the available genomic sequence of meningococcal strain MC58 (Turner *et al.*, 2006; van Ulsen *et al.*, 2006). The gene is absent from the genome of meningococcal serogroup A strain Z2491 and from *N. gonorrhoeae* FA1090. It was identified as a phase-variable gene as it contains a poly-C stretch in the coding region (Martin *et al.*, 2003). The full-length gene codes for a predicted protein of 157 kDa, designated MspA (meningococcal serine protease A) or AusI (autotransporter serine protease I) for its homology to App, IgA1 protease and other known autotransporter proteins. Indeed, the MspA/AusI protein possesses the characteristic 3D structure of a classical autotransporter protein and is secreted and released in the external milieu. MspA has a role in adhesion to host cells. *E. coli* expressing recombinant MspA is able to adhere to epithelial cell lines, such as BEAS-2B (human bronchial epithelial cells) and HBMEs (brain microvascular endothelial cells), but not to HUVECs (endothelial cells) (Turner *et al.*, 2006). In another study, van Ulsen and colleagues showed that AusI was not able to mediate adhesion to Chang, H292 and HeLa epithelial cell lines (van Ulsen *et al.*, 2006). These data suggest a differential expression of eukaryotic surface ligands on different cell lines, as already shown for other adhesins such as NadA (which will be described later in the text) and App.

MspA is expressed *in vivo*, being recognized by sera of convalescent patients from invasive serogroup B meningococcal disease (Turner *et al.*, 2006). Furthermore, anti-MspA antibodies are able to kill the homologous MC58 strain, and this bactericidal response indicates that, analogously to App, MspA is accessible to antibodies on the bacterial surface. However, in an other study, none of the sera from convalescent patients reacted with the AusI protein on Western blot (van Ulsen *et al.*, 2006). It was then hypothesized that either AusI expression *in vivo* is too low to elicit an immune response or the infecting strains did not contain the *ausI* gene. Considering the high frequency of its presence in the strains tested, this latter hypothesis is unlikely; rather, it is possible that antibodies against AusI are directed against conformational epitopes that are not detected in the Western blotting procedure.

Trimeric autotransporters: NadA and NhhA

Recent studies on autotransporters have identified a subfamily of this class of proteins presenting the prerogative of a short translocator domain able to form an intermolecular trimeric β-barrel that pilots the secretion of the passenger domain (reviewed in Cotter *et al.*, 2005). This mechanism of secretion defines the subfamily of trimeric autotransporters. The architecture of trimeric β-domain shows significant similarities to the translocator domains of classical autotransporters in that both orchestrate the formation of a β-barrel with several transmembrane β-strands forming an internal hydrophilic channel. In classical autotransporters, all the β-strands belong to a single large translocator domain, whereas in trimeric autotransporters the membrane channel is organized by three proximal distinct subunits.

An additional distinctive element of trimeric autotransporters is the absence of autoproteolysis; whereas in all classical autotransporters the passenger domain is cleaved from the translocator domain and released extracellularly, in all identified trimeric autotransporters, the passenger domain remains un-cleaved and surface-exposed.

With respect to functionality, trimeric autotransporters are generally specialized in adhesive activities promoting bacterial interaction with either host cells or extracellular matrix components. In this context, they are functionally more homogenous compared to conventional autotransporters that evolved a wider diversity in their effector functions. Two proteins are considered prototypes of this subfamily of autotransporters: YadA of *Yersinia* species (Roggenkamp *et al.*, 2003) and Hia of *H. influenzae* (Surana *et al.*, 2004). To date, two trimeric autotransporters, NadA and NhhA, have been characterized in *Neisseria*, both showing adhesive properties.

NadA of *N. meningitidis* has been identified as member of the trimeric autotransporter family of proteins, and is homologous to YadA of *Yersinia* (van Ulsen *et al.*, 2001) and UspA2 of *Moraxella catarrhalis* (Hoiczyk *et al.*, 2000). In particular, NadA belongs to the Oca (Oligomeric coiled-coil adhesins) (Desvaux *et al.*, 2004) family, a group of autotransporters characterized by the presence of a coiled-coil motif that possibly mediates the oligomerization process. Despite the weak sequence conservation to its corresponding homologues, NadA shares substantial structural similarity converging to similar functions. NadA, similarly to the other member of the Oca family, shows a typical structural organization with three main domains: (i) a short C-terminal region with a predicted propensity to form transmembrane β-strands similarly to the β-domain of other Oca members; (ii) an intermediate coiled-coil domain stalk containing a leucine zipper; and (iii) an N-terminal globular head region. Consistently to the predicted structural features, NadA forms high-molecular-weight oligomers, stable in denaturing conditions (Comanducci *et al.*, 2002).

The role of NadA in mediating the binding to human epithelial cells was initially reported by using the recombinant purified protein (Comanducci *et al.*, 2002). Further characterization of NadA revealed that *E. coli* strains expressing the oligomeric form of NadA are able to interact with Chang conjunctiva epithelial cells, but failed to bind to HUVECs (human umbilical vein endothelial cells). It was speculated that the protein receptor molecule, which mediates the binding to Chang cells might be absent in HUVECs and differentially expressed in various epithelial cells. The use of a *nadA* isogenic knockout mutant confirmed the role of this protein in the intimate association of the bacteria with host cells, which is associated with a cytoskeletal rearrangement and the formation of membrane protrusions at the contact sites. This close association and membrane rearrangement then result in internalization of the bacteria (Capecchi *et al.*, 2005). Unlike UspA, whose binding region has been defined in the coiled-coil region (Hill and Virji, 2003; Hill *et al.*, 2005), or YadA, which has been recently shown to bind collagen, laminin and fibronectin, when in trimeric form (Nummelin *et al.*, 2004), the NadA binding region appears to be located in the N-terminal globular 'head' domain.

Analysis of sequence conservation reveals that the gene is present in about 50% of *N. meningitidis* strains including hypervirulent lineages, such as ET37, ET5 and cluster A4. However, the gene is absent in lineage III, as well as in *N. gonorrhoeae* and in the commensal *N. lactamica* (Comanducci *et al.*, 2002) (see Chapter 13). The wide conservation of NadA supports the hypothesis that it is a disease-related gene essential for full meningococcal virulence. Furthermore, from a study performed on 154 carrier strains of *N. meningitidis*, NadA resulted to segregate differently in strains isolated from healthy individuals (carriers) and in strains isolated from patients (clinical isolates). Only 5.1% of the genuine carrier population harbours the *nadA* gene. This form, named NadA4, has the same overall features of other NadA forms. It is also able to form high-molecular-weight aggregates, is exposed on the surface, is able to bind to epithelial cells and is bactericidal (Comanducci *et al.*, 2004). This evidence supports the hypothesis that NadA is an important co-factor in the pathogenesis of *Neisseria*.

NhhA is an additional member of trimeric autotrasporters identified in *Neisseria*. NhhA is the acronym of *Neisseria hia/hsf* homologue, because of its close homology to Hia and Hsf adhesins of *H. influenzae* (Peak *et al.*, 2000). In accordance with the definition of trimeric autotransporter, NhhA has a short and conserved translocator domain able to trimerize and form a transmembrane β-barrel. The functionality of the translocator domain has been confirmed with the definition of the minimal translocation unit able to allow trimerization, insertion in the outer membrane and export of the passenger domain (Scarselli *et al.*, 2006).

NhhA shares an adhesive role with other members of trimeric autotransporters. The adhesive property is intrinsic to the protein itself, since purified NhhA is able to bind human cell lines; moreover, the recombinant protein expressed in *E. coli* confers to the strain the ability to bind human cell lines (Scarselli *et al.*, 2006). The contribution of NhhA in the interaction of *N. meningitidis* with human cells has been evaluated by inactivation of *nhhA* gene in a virulent strain of *N. meningitidis*; these studies indicated that the NhhA-deficient strain showed a largely reduced ability to adhere to human cells (Scarselli *et al.*, 2006).

The adhesive properties of NhhA are extended to components of extracellular matrix, since recombinant NhhA is able to bind, besides human cells, also laminin and heparan sulphate (Scarselli *et al.*, 2006). This evidence is in accordance with the presence in the passenger domain of a repeated amino acid motif, which is peculiar to eukaryotic adhesion molecules (Scarselli *et al.*, 2001).

NhhA is expressed in *N. meningitidis* and in *N. lactamica*, but is absent in *N. gonorrhoeae*. The expression of NhhA during colonization and infection is attested by the presence of anti-NhhA antibodies in sera of patients and healthy carriers (van Ulsen *et al.*, 2001; Litt *et al.*, 2004).

Expression of autotransporters in *Neisseria* species and implications in interaction with host

The sequenced genomes of *N. meningitidis* contain up to eight genes coding for putative autotransporters, which are presumably implicated in the virulence of the bacterium. Besides the five autotransporters described above implicated in adhesion of *N. meningitidis* to the host (Fig. 5.1), the list is completed by IgA protease showing multiple virulence functions and AutA and B, which are homologues to AIDA-I of *E. coli* (van Ulsen *et al.*, 2001) and whose function in *Neisseria* is still unknown. The distribution of autotransporters in pathogenic *Neisseria* species, including *N. meningitidis* serogroups A, B and C and *N. gonorrhoeae* has been recently reviewed and discussed (van Ulsen and Tommassen, 2006).

A remarkable finding emerging from genome analysis and functional studies is the relatively low number of secreted proteins evolved by pathogenic *Neisseria* and their preferential specialization in adhesive activity. With the exception of App, this new class of adhesins is a prerogative of *N. meningitidis*. Genes encoding for NadA, NhhA and MspA/AusI are absent in *N. gonorrhoeae*; App is expressed in *N. gonorrhoeae* but the processing differs from *N. meningitidis* because of the absence of NalP. Moreover, genome analysis of *N. gonorrhoeae* reveals the presence of additional pseudogenes showing similarity to autotransporters but the ORFs are disrupted by stop codons and deletions (van Ulsen and Tommassen, 2006). This evidence suggests a propensity of evolution pressure to eliminate autotransporter proteins from the genome of *N. gonorrhoeae*. The findings of a new class of adhesin autotransporters in *N. meningitidis* suggest the evolution of a supplementary strategy to interact with the host.

It is worth noting that App, NadA and NhhA are demonstrated to be functional in capsulated strains (Serruto *et al.*, 2003; Capecchi *et al.*, 2005; Scarselli *et al.*, 2006). Attachment to nasopharyngeal epithelium is the first step of colonization operated by capsulated meningococci. While the determinant of first attachment is the expression of type IV pili, this new class of adhesins could operate as auxiliary factors potentiating the ability of meningococci to colonize the respiratory tract.

It is interesting to note that many bacteria that colonize the upper respiratory tract in humans, such as *N. meningitidis*, *H. influenzae*, *M. catarrhalis* and *B. pertussis*, possess multiple autotransporter proteins with a significant homology. For example, *H. influenzae* IgA1 protease, Hap and Hia/Hsf, which share homology to their respective meningococcal IgA1 protease, App and NhhA, all have an important role in adhesion and invasion of host cells.

Experimental evidence demonstrates that App, MspA, NadA and NhhA are expressed and surface-exposed even in presence of capsule during colonization and invasive disease. Indeed, antibodies recognizing MspA and NadA were detected in sera of patients convalescent from meningococcal disease and antibodies against App and NhhA were detected in sera from both convalescent patients and from carrier population. Moreover, App, NadA and NhhA were also identified in a whole screening of *Neisseria* genome as possible candidates for a vaccine development (Pizza *et al.*, 2000); antibodies directed against these proteins induce a bactericidal response.

Two-partner secretion system: TpsA and TpsB

Sequence analysis of available genomes of *N. meningitidis* serogroup A, B and C revealed the presence of genes encoding for a putative two-partner secretion system (or TPS). The TPS is a two-step mechanism exporting large proteins (named TpsA) across the outer membranes using a partner channel-forming protein (named TpsB). Genes coding for TpsA and TpsB are often located in the same operon. TpsB proteins are typically ~500–800 amino acids and are highly conserved; by contrast, TpsA are large proteins with more than 3000 amino acids and are commonly more variable. TpsA and TpsB proteins are both synthesized with N-terminal signal sequences for translocation across the inner membrane by the Sec secretion system. The TpsB protein is then inserted into the outer membrane where it forms a β-barrel channel that mediates the secretion of the TpsA protein. TpsA protein can remain stably anchored to the cell surface, whereas others are released into the extracellular milieu.

The two putative TPS proteins of *N. meningitidis* have been denoted HrpA and HrpB for haemagglutinin/haemolysin-related proteins, which are the prototypes of TPS in *Bordetella pertussis*. The first identification of a TPS in *Neisseria* has been disclosed by subtractive hybridization of genes specific for *N. meningitidis* and absent in *N. gonorrhoeae*. In particular, strain MC58 of *N. meningitidis* contains two copies of TPS and three additional genes of *hrpA* with different degrees of homology.

The functionality of TPS in *N. meningitidis* has been recently described (Schmitt *et al.*, 2007). HrpB operates the proteolytic processing of HrpA, leading to secretion of HrpA in the extracellular milieu. In an effort to investigate a possible role of TPS in adhesion of *Neisseria* to epithelial cells, the authors reported a contribution in adhesion of both HrpA and HrpB only in unencapsulated and LPS-truncated strain of *N. meningitidis* serogroup C. The authors failed to report evidence of a similar function in serogroup B and ascribe this difference to the presence of multiple genes coding for HrpA proteins that could share redundant functions and mask the contribution of single proteins. Interestingly, TPS is widely expressed in a large panel of carrier isolates and there is no preferential distribution in virulent strains, suggesting a role of this secretion system in colonization processes.

Newly identified virulence factors of *Neisseria gonorrhoeae*: OmpA and MIP

Ng-OmpA is a recently identified surface protein, specific for *N. gonorrhoeae* and absent in *N. meningitidis*, belonging to the OmpA-like family of proteins (Serino *et al.*, 2007). This 23-kDa protein (NG1559) was identified by screening the available *N. gonorrhoeae* FA1090 genome for potential surface exposed factors involved in adhesion and invasion. Indeed, this specific gonococcal protein has extensive similarity with a family of OmpA proteins, conserved in many Enterobacteriaceae, and shares many functional characteristics common to those proteins. A mutation in the *ompA* gene decreased the adhesive and invasive phenotype of *N. gonorrhoeae* in epithelial cells, such as human cervical carcinoma cells, ME-180 and Hec-1B. This suggests that a specific receptor might be expressed on the surface of these cells and can be utilized by *N. gonorrhoeae* to facilitate attachment and internalization. However, further studies will be required to evaluate whether the same receptor used by *E. coli* OmpA, the 95-kDa glycoprotein gp96, also mediates attachment of *N. gonorrhoeae* through the Ng-OmpA. In addition, the gonococcal OmpA protein could also serve as a colonization factor for murine epithelial cells, since the *ompA* mutant was significantly attenuated in mouse colonization experiments using a mouse model of lower genital tract infection. Considering the high conservation of this protein in many clinical isolates, it is likely that Ng-OmpA plays a crucial role in gonococcal pathogenesis. All these findings support the hypothesis that the *N. gonorrhoeae* OmpA represents a novel virulence factor, critical for the establishment of lower genital tract colonization.

NG1225 or Ng-MIP is another virulence factor identified from genome analysis of *N. gonorrhoeae*. NG1225 is homologous to several members of the macrophage infectivity potentiator (MIP) family, which consists of secreted or surface-exposed virulence factors expressed by a variety of intracellular pathogens such as *Legionella pneumophila*, *Chlamydia trachomatis* and *Trypanosoma cruzi*. Because of this homology, the protein was named Ng-MIP (Leuzzi *et al.*, 2005).

Ng-MIP is a dimeric lipoprotein showing functional similarities to the counterpart in *Legionella*, where it is implicated in intracellular survival of *L. pneumophila* in human macrophages and in protozoa. Infection assays carried out with a mutant *mip*-deficient strain and its parental wild-type strain demonstrated a substantial role of Ng-MIP in promoting intracellular survival of *N. gonorrhoeae* in human macrophages. Indeed, loss of Ng-MIP rendered gonococci more susceptible to intracellular macrophage-mediated killing and affected the bacterial persistence into phagosomes.

In accordance with other members of this family, Ng-MIP is a PPIase (peptidyl-prolyl-*cis/trans* isomerase), an enzyme catalysing the slow *cis/trans* isomerization of prolyl peptide bonds in oligopeptides involved in the proline-mediated folding of proteins. PPIases are widely expressed in bacteria but the potentially innumerable substrates made it difficult to identify the corresponding targets and no substrates have been attributed so far. However, the conservation of PPIase activity in all identified MIP proteins leads to suppose a correlation between enzymatic activity and bacterial survival in host cells. Interestingly, MIP-like proteins differ from other PPIases in their localization, in that they are exposed on bacterial membrane or secreted, implying an extracellular target for their enzymatic activity. Recently, the ability of MIP of *L. pneumophila* to bind collagen and to promote bacterial transmigration has been shown; this functionality is concerted by PPIase activity and a serine protease activity (Wagner *et al.*, 2007). It would be of interest to investigate whether this functionality could be extended to *N. gonorrhoeae.*

It is interesting to note that functionality of Ng-MIP is specific for phagocytic cells and is not extended to epithelial cells. This aspect opens new questionings on the relevance of interaction with macrophages in gonococcal infection. Infection with gonococci in women is frequently asymptomatic and evolves with low frequency to pelvic inflammatory disease (PID), characterized by a persistent inflammation. Hypothetically, the latency could be explained by persistence of *N. gonorrhoeae* in macrophages resident in genital mucosa. In this perspective, macrophages could represent a reservoir of infection leading to spreading of infection following damage of mucosa or changes in host surveillance.

The genomes of *N. meningitides* serogroup A and B possess *mip* genes encoding for proteins with 97% identity to Ng-MIP, but the functional role of these proteins needs to be explored.

Identification of new potential virulence factors by genome analysis

During the search for potential vaccine candidates in the *Neisseria* genomes, most of the putative surface-exposed or exported proteins were identified *in silico* and tested for their ability to raise an immune response in mice. In the case of meningococcus B, antigens were evaluated for their ability to induce complement-mediated bactericidal activity (summarized in Table 5.1). Some of them are under investigation as components of a new MenB vaccine, which will be described in Chapter 11. Here, we will describe the additional antigens identified during genome screening and so far characterized, which, although not involved in adhesion, could have a potential role in the *in vivo* virulence and pathogenesis of *Neisseria.*

The genome-derived Neisseria antigen GNA33 is a lipoprotein homologous to a membrane-bound lytic transglycolsylase (MltA) of *E. coli*, an enzyme involved in murein enlargement during bacterial growth and division

(Jennings *et al.*, 2002). Indeed, it was shown to be able to degrade both insoluble murein sacculi and unsubstituted glycan strands. Studies on the strain of *N. meningitidis* deficient in GNA33 gene indicate a relevant role of the corresponding protein in meningococcal growth and shaping: the GNA33 mutant exhibits a marked decreased viability, an abnormal cellular morphology and a perturbed membrane profile (Adu-Bobie *et al.*, 2004), suggesting an important contribution of this protein in meningococcal membrane architecture and in replication. These alterations cause a virulence-attenuated phenotype since GNA33 mutant is unable to cause bacteraemia in an infant rat animal model (Adu-Bobie *et al.*, 2004).

Therefore, GNA33 plays an important role in peptidoglycan metabolism, cell separation, membrane architecture and virulence. It was found to elicit bactericidal antibodies by mimicking an epitope localized in the loop 4 of PorA and to confer protection in an infant rat model by passive immunization (Granoff *et al.*, 2001).

The genome-derived Neisseria antigen GNA1870 was selected for its ability to induce bactericidal antibodies in mice (see Chapter 13) and, more recently, it has been shown to be an important factor for the survival of *Neisseria* in human blood. This is due to the ability of GNA1870 to bind human factor H (fH), the key fluid-phase regulator of the alternative complement pathway (see Chapter 8) (Ngampasutadol *et al.*, 2005). Binding of fH to GNA1870 correlates well with the level of complement C3 deposition and the degree of serum resistance. These data support the notion that inhibition by specific antibodies of the binding of a complement down-regulator to the surface of bacteria (via fHBP) contributes to the intrinsic bactericidal activity of those antibodies. To reflect the critical function of this molecule, GNA1870 was renamed fH-binding protein (fHBP) (Madico *et al.*, 2006).

Finally, the first neisserial ADP-ribosyltransferase NarE (Masignani *et al.*, 2003) was identified by homology with a group of ADP ribosyltransferases, a functionally conserved class of enzymes playing a key role in pathogenicity of various bacteria (e.g. pertussis toxin, cholera toxin, diphtheria toxins etc.) known to be the etiological agent of the relative diseases. The *narE* gene is not ubiquitous in *Neisseria* species and is specific for serogroup B (always present in ET-5 and Lineage 3 but absent in ET-37 and Cluster A4 hypervirulent lineages), since it is absent in the sequenced genomes of serogroup A and C. In the gonococcus, the *narE* gene has a frameshift resulting in a premature interruption of the gene after eight codons. Despite the absence of a predicted leader peptide, the protein is efficiently exported in the periplasm of meningococcus. *In vitro*, the protein has ADP-ribosylation and NAD-glycohydrolase activities, indicating that the structural homology with this important class of bacterial toxins is conserved at the functional level. However, the target of this enzyme *in vivo* is not known. The possible role of NarE in the virulence and pathogenesis of meningococcus needs further investigation.

Conclusions and future directions

Adhesion of pathogenic *Neisseria* to the mucosa of the human nasopharynx and genitourinary tract is the first step of host colonization and, consequently, the primary event leading to disease manifestation. Adhesion to the host is a dynamic process resulting from a balance of interactions between bacterial surface structures and host receptors; the success of such interactions is determined by the cooperation of multiple adhesins during the different stages of infection (Fig. 5.2).

Although the presence of pili and opacity protein is a necessary condition for colonization by pathogenic neisseriae, a new class of adhesins have been disclosed by genome analysis. Searching for sequence homologies to known virulence factors has highlighted a new class of adhesin autotransporters in *N. meningitidis*. App, MspA, NadA and NhhA are autotransporters involved in adhesion to epithelial cells.

In the absence of an animal model able to mirror the disease in its whole complexity, the effectors of host colonization have been studied by *in vitro* models that rely on the use of cultured cell lines. The contribution of potential adhesins to interaction with human cells has been extrapolated by using pathogenic strains mutated in single genes. Alternatively, proteins of interest have been overexpressed in *E. coli* to verify whether the

Table 5.1 Novel autotransporters and lipoproteins characterized in Nm and Ng.

	Name	Homologue	Genomes		Description	Reference
			N. meningitidis (MC58)	*N. gonorrhoeae* (FA1090)		
Autotransporters						
App	Adhesion and penetration protein	Hap adhesin of *H. influenzae*	NMB1985	NG2105	Autotransporter adhesin with autocatalytic serine protease activity Able to induce bactericidal antibody and stimulate human T-cells	Hadi *et al.* (2001), Serruto *et al.* (2003)
NalP	*Neisseria* autotransporter lipoprotein	Secreted serine protease (PrtS) of *Serratia marcescens*	NMB1969	NP	Autotransporter lipoprotein with serine protease activity involved in the processing of Neisseria surface proteins	van Ulsen *et al.* (2003)
MspA/AusI	Meningococcal serine protease A/autotransporter serine protease I		NMB1998	NP	Serine protease autotransporter with adhesive activity Able to induce bactericidal antibody	van Ulsen *et al.* (2006), Turner *et al.* (2006)
NhhA	*Neisseria hia/hsf* homologue	Hsf and Hia adhesins of *H. influenzae*	NMB0992	NP	Trimeric autotransporter adhesin involved in interaction with epithelial cells and extracellular matrix proteins Able to induce bactericidal antibody	Peak *et al.* (2000), Pizza *et al.* (2000), Scarselli *et al.* (2001, 2006)
NadA	*Neisseria* adhesin A	UspA2 of *M. catarrhalis* and YadA of enteropathogenic *Yersinia* species	NMB1994	NP	Trimeric autotransporter adhesin involved in interaction with epithelial cells Able to induce bactericidal antibody	Comanducci *et al.* (2002), Capecchi *et al.* (2005)

Lipoproteins						
Ng-MIP	Macrophage infectivity potentiator	MIP, *Legionella pneumophila*	NMB1567	NG1225	Dimeric lipoprotein with PPIase activity. Involved in the Ng persistence into phagosomes	Leuzzi *et al.* (2005)
fHBP	Factor H binding protein	Low similarity with transferrin binding proteins	NMB1870	NG0033 (the leader peptide is absent – uncertain localization)	Surface-exposed lipoprotein able to induce high levels of bactericidal antibodies Binds human factor H	Masignani *et al.* (2003), Giuliani *et al.* (2005), Madico *et al.* (2006)
OmpA	Outer membrane protein A		NP	NG1559	Colonization factor able to facilitate attachment and internalization of Ng	Serino *et al.* (2007)
GNA33	Genome-derived *Neisseria* antigen 33	MltA, *E. coli*	NMB0033	NG2048	Lipoprotein involved in peptidoglycan metabolism, cell separation and membrane architecture	Jennings *et al.* (2002), Adu-Bobie *et al.* (2004)

NP, not present.

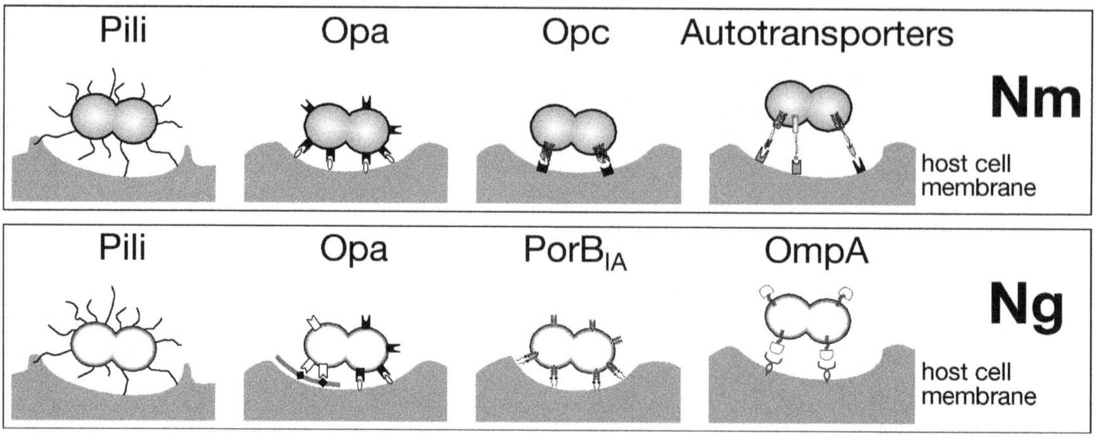

Figure 5.2 Comparison of the adhesion strategies of pathogenic neisseriae. The outer membrane proteins involved in the interaction with the host cell membranes are reported for Nm and Ng. The host receptors of Opa, Opc and PoB1A are described in the text.

presence of potential adhesins on the membrane confers an adhesive phenotype. Although these models cannot represent the multifactorial mechanisms occurring during infections, they allow to simplify a single event and to bring to light the contribution of individual adhesins.

In this context, studies on autotransporter adhesins have provided new knowledge on the strategies of colonization. Notably, the evidence that App, NadA and NhhA explicate their functional role in a capsulated strain leads to the hypothesis that they are involved in the primary stage of *Neisseria* colonization. Autotransporter adhesins and pili could be present simultaneously and cooperate to potentiate the binding to the host cells, favouring the subsequent intimate adhesion and internalization into host cells.

Since pili undergo phase and antigenic variation, colonization is carried out by a mixed population of meningococci either piliated or non-piliated or expressing structurally different pili; this heterogeneity could be compensated by the presence of multiple adhesins in the site of colonization. Importantly, expression of adhesin autotransporters has been verified *in vivo* by testing the presence of specific antibodies in sera from convalescent patients and from healthy population carrying commensal *Neisseria*. This evidence establishes a link between the *in vitro* model and the manifestation of disease and strongly support the biological relevance of new autotransporter adhesins in potentiating disease.

A major issue in understanding neisserial pathogenesis is to unravel the differential strategies evolved by the meningococcus and gonococcus. It is evident that the two species share common surface structures to target different human tissues and to lead to distinct diseases. The presence of the capsule is undoubtedly a major contradistinction, not only for a direct involvement in several virulent functions but also for the influence on the whole membrane organization and the functionality of all the surface proteins.

Expression of a different repertoire of proteins is another evolutionary strategy to adapt to specific biological niches. Several studies have been addressed in this direction in order to identify specific disease-related effectors (Tinsley and Nassif, 1996; Hotopp *et al.*, 2006; van Ulsen and Tommassen, 2006).

The new additional adhesins recently characterized provide new insights in the complex picture of *Neisseria*–host interactions. Autotransporters, with the exception of App, and two-partner secretion systems are specific for *N. meningitidis*. On the other hand, attachment of *N. gonorrhoeae* to epithelial cells is supported by an OmpA-like adhesin, which is peculiar to this species (Fig. 5.2). These specific adhesion strategies suggest that the two pathogens evolved different mechanism to colonize mucosal surfaces in different body parts. Analysis of neisserial genome sequences was instrumental to identify these new surface-exposed adhesive structures and to define new adhesion strategies of pathogenic neisseriae.

References

Aas, F.E., Lovold, C., and Koomey, M. (2002). An inhibitor of DNA binding and uptake events dictates the proficiency of genetic transformation in *Neisseria gonorrhoeae*: mechanism of action and links to Type IV pilus expression. Mol. Microbiol. *46*, 1441–1450.

Adu-Bobie, J., Lupetti, P., Brunelli, B., Granoff, D., Norais, N., Ferrari, G., Grandi, G., Rappuoli, R., and Pizza, M. (2004). GNA33 of *Neisseria meningitidis* is a lipoprotein required for cell separation, membrane architecture, and virulence. Infect. Immun. *72*, 1914–1919.

Capecchi, B., Adu-Bobie, J., Di Marcello, F., Ciucchi, L., Masignani, V., Taddei, A., Rappuoli, R., Pizza, M., and Arico, B. (2005). *Neisseria meningitidis* NadA is a new invasin which promotes bacterial adhesion to and penetration into human epithelial cells. Mol. Microbiol. *55*, 687–698.

Comanducci, M., Bambini, S., Brunelli, B., Adu-Bobie, J., Arico, B., Capecchi, B., Giuliani, M.M., Masignani, V., Santini, L., Savino, S., Granoff, D.M., Caugant, D.A., Pizza, M., Rappuoli, R., and Mora, M(2002). NadA, a novel vaccine candidate of *Neisseria meningitidis*. J. Exp. Med. *195*, 1445–1454.

Comanducci, M., Bambini, S., Caugant, D.A., Mora, M., Brunelli, B., Capecchi, B., Ciucchi, L., Rappuoli, R., and Pizza, M. (2004). NadA diversity and carriage in *Neisseria meningitidis*. Infect. Immun. *72*, 4217–4223.

Cotter, S.E., Surana, N.K., and St Geme, J.W., 3rd (2005). Trimeric autotransporters: a distinct subfamily of autotransporter proteins. Trends Microbiol *13*, 199–205.

Craig, L., Volkmann, N., Arvai, A.S., Pique, M.E., Yeager, M., Egelman, E.H., and Tainer, J.A. (2006). Type IV pilus structure by cryo-electron microscopy and crystallography: implications for pilus assembly and functions. Mol. Cell *23*, 651–662.

Deghmane, A.E., Giorgini, D., Larribe, M., Alonso, J.M., and Taha, M.K. (2002). Down-regulation of pili and capsule of *Neisseria meningitidis* upon contact with epithelial cells is mediated by CrgA regulatory protein. Mol. Microbiol. *43*, 1555–1564.

Derrick, J., Heckels, J.E., and Virji, M. (2006). Major Outer Membrane Proteins of Meningococci. In Handbook of Meningococcal Disease, M. Frosch, and M.C. Maiden, eds. (Chichester, John Wiley & Sons).

Desvaux, M., Parham, N.J., and Henderson, I.R. (2004). The autotransporter secretion system. Res. Microbiol. *155*, 53–60.

Fink, D.L., Buscher, A.Z., Green, B., Fernsten, P., and St Geme, J.W., 3rd (2003). The *Haemophilus influenzae* Hap autotransporter mediates microcolony formation and adherence to epithelial cells and extracellular matrix via binding regions in the C-terminal end of the passenger domain. Cell. Microbiol. *5*, 175–186.

Granoff, D.M., Moe, G.R., Giuliani, M.M., Adu-Bobie, J., Santini, L., Brunelli, B., Piccinetti, F., Zuno-Mitchell, P., Lee, S.S., Neri, P., Bracci, L., Lozzi, L., and Rappuoli, R.(2001). A novel mimetic antigen eliciting protective antibody to *Neisseria meningitidis*. J. Immunol. *167*, 6487–6496.

Giuliani, M.M., Santini, L., Brunelli, B., Biolchi, A., Arico, B., Di Marcello, F., Cartocci, E., Comanducci, M., Masignani, V., Lozzi, L., Savino, S., Scarselli, M., Rappuoli, R., and Pizza, M. (2005) The region comprising amino acids 100 to 255 of *Neisseria meningitidis* lipoprotein GNA 1870 elicits bactericidal antibodies. Infect. Immun. *73*, 1151–1160.

Hadi, H.A., Wooldridge, K.G., Robinson, K., and Ala'Aldeen, D.A. (2001). Identification and characterization of App: an immunogenic autotransporter protein of *Neisseria meningitidis*. Mol. Microbiol. *41*, 611–623.

Hauck, C.R., and Meyer, T.F. (1997). The lysosomal/phagosomal membrane protein h-lamp-1 is a target of the IgA1 protease of *Neisseria gonorrhoeae*. FEBS Lett. *405*, 86–90.

Hauck, C.R., and Meyer, T.F. (2003). 'Small' talk: Opa proteins as mediators of *Neisseria*-host-cell communication. Curr. Opin. Microbiol. *6*, 43–49.

Helaine, S., Carbonnelle, E., Prouvensier, L., Beretti, J.L., Nassif, X., and Pelicic, V. (2005). PilX, a pilus-associated protein essential for bacterial aggregation, is a key to pilus-facilitated attachment of *Neisseria meningitidis* to human cells. Mol. Microbiol. *55*, 65–77.

Helaine, S., Dyer, D.H., Nassif, X., Pelicic, V., and Forest, K.T. (2007). 3D structure/function analysis of PilX reveals how minor pilins can modulate the virulence properties of type IV pili. Proc. Natl. Acad. Sci. USA *104*, 15888–15893.

Hill, D.J., Edwards, A.M., Rowe, H.A., and Virji, M. (2005). Carcinoembryonic antigen-related cell adhesion molecule (CEACAM)-binding recombinant polypeptide confers protection against infection by respiratory and urogenital pathogens. Mol. Microbiol. *55*, 1515–1527.

Hill, D.J., and Virji, M. (2003). A novel cell-binding mechanism of *Moraxella catarrhalis* ubiquitous surface protein UspA: specific targeting of the N-domain of carcinoembryonic antigen-related cell adhesion molecules by UspA1. Mol. Microbiol. *48*, 117–129.

Hoiczyk, E., Roggenkamp, A., Reichenbecher, M., Lupas, A., and Heesemann, J. (2000). Structure and sequence analysis of *Yersinia* YadA and *Moraxella* UspAs reveal a novel class of adhesins. Embo J. *19*, 5989–5999.

Hotopp, J.C., Grifantini, R., Kumar, N., Tzeng, Y.L., Fouts, D., Frigimelica, E., Draghi, M., Giuliani, M.M., Rappuoli, R., Stephens, D.S., Grandi, G., and Tettelin, H.(2006). Comparative genomics of *Neisseria meningitidis*: core genome, islands of horizontal transfer and pathogen-specific genes. Microbiology *152*, 3733–3749.

Jennings, G.T., Savino, S., Marchetti, E., Arico, B., Kast, T., Baldi, L., Ursinus, A., Holtje, J.V., Nicholas, R.A., Rappuoli, R., and Grandi, G. (2002). GNA33 from *Neisseria meningitidis* serogroup B encodes a membrane-bound lytic transglycosylase (MltA). Eur. J. Biochem. *269*, 3722–3731.

Kallstrom, H., Liszewski, M.K., Atkinson, J.P., and Jonsson, A.B. (1997). Membrane cofactor protein (MCP or CD46) is a cellular pilus receptor for pathogenic *Neisseria*. Mol. Microbiol. *25*, 639–647.

Leuzzi, R., Serino, L., Scarselli, M., Savino, S., Fontana, M.R., Monaci, E., Taddei, A., Fischer, G., Rappuoli, R., and Pizza, M. (2005). Ng-MIP, a surface-exposed lipoprotein of *Neisseria gonorrhoeae*, has a peptidyl-prolyl cis/trans isomerase (PPIase) activity and is involved in persistence in macrophages. Mol. Microbiol. *58*, 669–681.

Litt, D.J., Savino, S., Beddek, A., Comanducci, M., Sandiford, C., Stevens, J., Levin, M., Ison, C., Pizza, M., Rappuoli, R., and Kroll, J.S. (2004). Putative vaccine antigens from *Neisseria meningitidis* recognized by serum antibodies of young children convalescing after meningococcal disease. J. Infect. Dis. *190*, 1488–1497.

Madico, G., Welsch, J.A., Lewis, L.A., McNaughton, A., Perlman, D.H., Costello, C.E., Ngampasutadol, J., Vogel, U., Granoff, D.M., and Ram, S. (2006). The meningococcal vaccine candidate GNA1870 binds the complement regulatory protein factor H and enhances serum resistance. J. Immunol. *177*, 501–510.

Martin, P., van de Ven, T., Mouchel, N., Jeffries, A.C., Hood, D.W., and Moxon, E.R. (2003). Experimentally revised repertoire of putative contingency loci in *Neisseria meningitidis* strain MC58: evidence for a novel mechanism of phase variation. Mol. Microbiol. *50*, 245–257.

Masignani, V., Balducci, E., Di Marcello, F., Savino, S., Serruto, D., Veggi, D., Bambini, S., Scarselli, M., Arico, B., Comanducci, M., Adu-Bobie, J., Giuliani, M.M., Rappuoli, R., and Pizza, M. (2003). NarE: a novel ADP-ribosyltransferase from *Neisseria meningitidis*. Mol. Microbiol. *50*, 1055–1067.

Massari, P., Ram, S., Macleod, H., and Wetzler, L.M. (2003). The role of porins in neisserial pathogenesis and immunity. Trends Microbiol. *11*, 87–93.

Merz, A.J., and So, M. (2000). Interactions of pathogenic neisseriae with epithelial cell membranes. Annu. Rev. Cell. Dev. Biol. *16*, 423–457.

Morand, P.C., Bille, E., Morelle, S., Eugene, E., Beretti, J.L., Wolfgang, M., Meyer, T.F., Koomey, M., and Nassif, X. (2004). Type IV pilus retraction in pathogenic *Neisseria* is regulated by the PilC proteins. Embo J. *23*, 2009–2017.

Nassif, X., Beretti, J.L., Lowy, J., Stenberg, P., O'Gaora, P., Pfeifer, J., Normark, S., and So, M. (1994). Roles of pilin and PilC in adhesion of *Neisseria meningitidis* to human epithelial and endothelial cells. Proc. Natl. Acad. Sci.USA *91*, 3769–3773.

Ngampasutadol, J., Ram, S., Blom, A.M., Jarva, H., Jerse, A.E., Lien, E., Goguen, J., Gulati, S., and Rice, P.A. (2005). Human C4b-binding protein selectively interacts with *Neisseria gonorrhoeae* and results in species-specific infection. Proc. Natl. Acad. Sci. USA *102*, 17142–17147.

Nummelin, H., Merckel, M.C., Leo, J.C., Lankinen, H., Skurnik, M., and Goldman, A. (2004). The *Yersinia* adhesin YadA collagen-binding domain structure is a novel left-handed parallel beta-roll. Embo J. *23*, 701–711.

Peak, I.R., Srikhanta, Y., Dieckelmann, M., Moxon, E.R., and Jennings, M.P. (2000). Identification and characterisation of a novel conserved outer membrane protein from *Neisseria meningitidis*. FEMS Immunol. Med. Microbiol. *28*, 329–334.

Pizza, M., Scarlato, V., Masignani, V., Giuliani, M.M., Arico, B., Comanducci, M., Jennings, G.T., Baldi, L., Bartolini, E., Capecchi, B., Galeotti, C.L., Luzzi, E., Manetti, R., Marchetti, E., Mora, M., Nuti, S., Ratti, G., Santini, L., Savino, S., Scarselli, M., Storni, E., Zuo, P., Broeker, M., Hundt, E., Knapp, B., Blair, E., Mason, T., Tettelin, H., Hood, D.W., Jeffries, A.C., Saunders, N.J., Granoff, D.M., Venter, J.C., Moxon, E.R., Grandi, G., and Rappuoli, R. (2000). Identification of vaccine candidates against serogroup B meningococcus by whole-genome sequencing. Science *287*, 1816–1820.

Pohlner, J., Halter, R., and Meyer, T.F. (1987). *Neisseria gonorrhoeae* IgA protease. Secretion and implications for pathogenesis. Antonie Van Leeuwenhoek *53*, 479–484.

Poulsen, K., Brandt, J., Hjorth, J.P., Thogersen, H.C., and Kilian, M. (1989). Cloning and sequencing of the immunoglobulin A1 protease gene (*iga*) of *Haemophilus influenzae* serotype b. Infect. Immun. *57*, 3097–3105.

Pujol, C., Eugene, E., de Saint, M.L., and Nassif, X. (1997). Interaction of *Neisseria meningitidis* with a polarized monolayer of epithelial cells. Infect. Immun. *65*, 4836–4842.

Rechner, C., Kuhlewein, C., Muller, A., Schild, H., and Rudel, T. (2007). Host glycoprotein Gp96 and scavenger receptor SREC interact with PorB of disseminating *Neisseria gonorrhoeae* in an epithelial invasion pathway. Cell Host Microbe 2, 393–403.

Roggenkamp, A., Ackermann, N., Jacobi, C.A., Truelzsch, K., Hoffmann, H., and Heesemann, J. (2003). Molecular analysis of transport and oligomerization of the *Yersinia enterocolitica* adhesin YadA.J. Bacteriol. *185*, 3735–3744.

Rudel, T., Facius, D., Barten, R., Scheuerpflug, I., Nonnenmacher, E., and Meyer, T.F. (1995). Role of pili and the phase-variable PilC protein in natural competence for transformation of *Neisseria gonorrhoeae*. Proc. Natl. Acad. Sci.USA *92*, 7986–7990.

Scarselli, M., Rappuoli, R., and Scarlato, V. (2001). A common conserved amino acid motif module shared by bacterial and intercellular adhesins: bacterial adherence mimicking cell cell recognition? Microbiology *147*, 250–252.

Scarselli, M., Serruto, D., Montanari, P., Capecchi, B., Adu-Bobie, J., Veggi, D., Rappuoli, R., Pizza, M., and Arico, B. (2006). *Neisseria meningitidis* NhhA is a multifunctional trimeric autotransporter adhesin. Mol. Microbiol. *61*, 631–644.

Schmitt, C., Turner, D., Boesl, M., Abele, M., Frosch, M., and Kurzai, O. (2007). A functional two-partner secretion system contributes to adhesion of *Neisseria meningitidis* to epithelial cells. J. Bacteriol. *189*, 7968–7976.

Serino, L., Nesta, B., Leuzzi, R., Fontana, M.R., Monaci, E., Mocca, B.T., Cartocci, E., Masignani, V., Jerse, A.E., Rappuoli, R., and Pizza, M. (2007). Identification of a new OmpA-like protein in *Neisseria gonorrhoeae* involved in the binding to human epithelial cells and *in vivo* colonization. Mol. Microbiol. *64*, 1391–1403.

Serruto, D., Adu-Bobie, J., Scarselli, M., Veggi, D., Pizza, M., Rappuoli, R., and Arico, B. (2003). *Neisseria meningitidis* App, a new adhesin with autocatalytic serine protease activity. Mol. Microbiol. *48*, 323–334.

St. Geme, J.W., III, de la Morena, M.L., and Falkow, S. (1994). A *Haemophilus influenzae* IgA protease-like protein promotes intimate interaction with human epithelial cells. Mol. Microbiol. *14*, 217–233.

Surana, N.K., Cutter, D., Barenkamp, S.J., and St. Geme, J.W., 3rd (2004). The *Haemophilus influenzae* Hia autotransporter contains an unusually short trimeric translocator domain. J. Biol. Chem. *279*, 14679–14685.

Tinsley, C.R., and Nassif, X. (1996). Analysis of the genetic differences between *Neisseria meningitidis* and *Neisseria gonorrhoeae*: two closely related bacteria expressing two different pathogenicities. Proc. Natl. Acad. Sci. USA 93, 11109–11114.

Turner, D.P., Marietou, A.G., Johnston, L., Ho, K.K., Rogers, A.J., Wooldridge, K.G., and Ala'Aldeen, D.A. (2006). Characterization of MspA, an immunogenic autotransporter protein that mediates adhesion to epithelial and endothelial cells in *Neisseria meningitidis*. Infect. Immun. *74*, 2957–2964.

van Ulsen, P., Adler, B., Fassler, P., Gilbert, M., van Schilfgaarde, M., van der Ley, P., van Alphen, L., and Tommassen, J. (2006). A novel phase-variable autotransporter serine protease, AusI, of *Neisseria meningitidis*. Microbes Infect. *8*, 2088–2097.

van Ulsen, P., and Tommassen, J. (2006). Protein secretion and secreted proteins in pathogenic Neisseriaceae. FEMS Microbiol. Rev. *30*, 292–319.

van Ulsen, P., van Alphen, L., Hopman, C.T., van der Ende, A., and Tommassen, J. (2001). In vivo expression of *Neisseria meningitidis* proteins homologous to the *Haemophilus influenzae* Hap and Hia autotransporters. FEMS Immunol. Med. Microbiol. *32*, 53–64.

van Ulsen, P., van Alphen, L., ten Hove, J., Fransen, F., van der Ley, P., and Tommassen, J. (2003). A neisserial autotransporter NalP modulating the processing of other autotransporters. Mol. Microbiol. *50*, 1017–1030.

Virji, M., Kayhty, H., Ferguson, D.J., Alexandrescu, C., Heckels, J.E., and Moxon, E.R. (1991). The role of pili in the interactions of pathogenic *Neisseria* with cultured human endothelial cells. Mol. Microbiol. *5*, 1831–1841.

Virji, M., Makepeace, K., Ferguson, D.J., Achtman, M., and Moxon, E.R. (1993). Meningococcal Opa and Opc proteins: their role in colonization and invasion of human epithelial and endothelial cells. Mol. Microbiol. *10*, 499–510.

Virji, M., Makepeace, K., and Moxon, E.R. (1994). Distinct mechanisms of interactions of Opc-expressing meningococci at apical and basolateral surfaces of human endothelial cells; the role of integrins in apical interactions. Mol. Microbiol. *14*, 173–184.

Wagner, C., Khan, A.S., Kamphausen, T., Schmausser, B., Unal, C., Lorenz, U., Fischer, G., Hacker, J., and Steinert, M. (2007). Collagen binding protein Mip enables *Legionella pneumophila* to transmigrate through a barrier of NCI-H292 lung epithelial cells and extracellular matrix. Cell. Microbiol. *9*, 450–462.

Mechanisms of Cellular Invasion of *Neisseria meningitidis*

6

Etienne Carbonnelle, Xavier Nassif and Sandrine Bourdoulous

Abstract

Neisseria meningitidis is an extracellular human specific pathogen responsible for septicaemia and meningitis. Like most bacterial pathogens, *N. meningitidis* exploits host cell signalling pathways in order to promote its uptake by host cells. *N. meningitidis* does not have a type III or a type IV secretion system. The signalling leading to bacterial internalization is induced by the type IV pili which are the main attribute mediating bacterial adhesion onto cells. The signalling induced following Type IV pilus mediated adhesion is responsible for the formation of microvilli like structures at the site of the bacteria–cell interaction. These microvilli trigger the internalization of the bacteria inside the cells. A major consequence of these signalling events is a reorganization of the actin cytoskeleton leading to the formation of membrane protrusion engulfing bacterial pathogens into intracellular vacuoles. Efficient internalization of *N. meningitidis* also requires the activation of an alternative signalling pathway coupled to the activation of the tyrosine kinase receptor ErbB2. Besides Type IV pili, other outer membrane proteins may be involved in other mechanisms of bacteria internalization inside cells.

Introduction

Neisseria meningitidis is a Gram-negative bacterium that is an obligate commensal of the human nasopharyngeal mucosa, the only known reservoir of this bacterium. The ability to colonize the nasopharynx is necessary for the propagation and the survival of the bacterium. Furthermore when the bacteria become pathogenic, the nasopharynx is the port-of-entry of the infection.

N. meningitidis then interacts with many different cell types to colonize the nasopharynx, enter the systemic circulation, and cause disease. However, the strict tropism of *N. meningitidis* for humans has prevented the development of any suitable animal model for studying the mechanisms by which this bacterium colonizes and traverses host barriers. Most of the knowledge of the pathway of infection by *N. meningitidis* has been derived from studies of its interactions with organ cultures, primary cell cultures, and immortalized cell lines. Specific models have been used to reflect individual stages in the progression of infection: the interactions between the bacterium and isolated, immortalized epithelial cells have been studied to understand the process of colonization, while rodent models have been adapted to investigate the dissemination of the bacterium in the systemic circulation. Experiments with endothelial and meningeal cells have been used to represent the subsequent stages of attachment to vascular cells and entry into the cerebrospinal fluid (CSF), leading to meningitis. While none of these models is ideal, they have revealed that *N. meningitidis* is able to adhere, enter, and traffic through human epithelial and endothelial cells, and that *N. meningitidis* can also replicate in cultured cell lines. These *in vitro* models have provided valuable insights into the bacterial factors required for pathogenesis and their cellular targets.

To mediate association with a range of host cells, this bacterium expresses a number of adhesins. However, virulent strains of *N. meningitidis* express a polysaccharide capsule, which can affect the accessibility of some bacterial ligands to host receptors. This is especially relevant for serogroups of *N. meningitidis* (B and C) that express a capsule consisting of polymers of negatively charged sialic acid which is also present on human cells. The main mean used by pathogenic *Neisseria* to interact with host cells are their Type IV pili (TFP). This adhesin is required for the initial attachment of bacteria to host cells, which may be further facilitated by the lipo-oligosaccharide (LOS), opacity (Opa) proteins and glycolipid adhesins.

The purpose of this chapter is to review our current knowledge on the interaction between the bacterium and host cells, focusing on the invasive mechanism particularly epithelial and endothelial cells, which are critical to colonization and represent the probable route of spread to the CSF. Considering the importance of Type IV pili in the pathogenesis of *N. meningitidis*, invasion mediated by this fibres will first be addressed. On the other hand, the focus of Chapter 8 is the interaction of *Neisseria gonorrhoeae* with host cells.

Type IV pili-mediated internalization

The Type IV pili

Type IV pili are polymeric filaments found on many Gram-negative bacteria. These structures, which correspond to a multimer of a pilin subunit protein encoded by the *pilE* gene, play a crucial role in the pathogenesis of *N. meningitidis* infections in humans. Non piliated mutants (PilE defective mutants) of encapsulated strains of *N. meningitidis* do not efficiently interact with host cells (Nassif *et al.*, 1994; Virji *et al.*, 1991), indicating that virulent encapsulated meningococci mainly interact with host cells by means of Type IV pili (Nassif *et al.*, 1997).

The pilin subunit (PilE) is synthesized as a preprotein, and is produced by cleavage of a N-terminal sequence from prepilin by a prepilin peptidase PilD (Lory and Strom, 1997). In addition, the pilin protein PilE undergoes several posttranslational modifications including glycosylation at serine 63 (Forest *et al.*, 1999; Marceau *et al.*, 1998). The nature of the sugar was initially reported to be a trisaccharide composed of Gal(β1–4) Gal(α1–3) 2,4-diacetamido-2,4,6-trideoxyhexose (DATDH). An alternative pilin glycosylation with a glyceramido acetamido trideoxyhexose residue was recently reported (Chamot-Rooke *et al.*, 2007). The importance of glycosylation of bacterial proteins is becoming increasingly evident, however in the case of *N. meningitidis*, pilin glycosylation has not been shown to be responsible for a major effect on the function of the Type IV pili; non-glycosylated mutants have slightly enhanced binding to epithelial cells compared with glycosylated bacteria (Marceau *et al.*, 1998).

Fully functional TFP are dynamic structures. Pilin subunits are constantly being assembled into a fibre from a platform in the inner-membrane. The fibre is then extruded through the outer membrane. A specificity of TFP is their ability to retract. Retraction is a consequence of the disassembly of pilin subunits which are then stored in the cytoplasmic membrane. Two inner membrane-associated ATPases, PilF and PilT, are thought to antagonistically promote this extension and retraction, with PilF being involved in pilus elongation (Freitag *et al.*, 1995) and PilT being necessary for pilus retraction (Merz *et al.*, 2000; Wolfgang *et al.*, 1998). The components of the pilus machinery have been identified (Fig. 6.1). Depending on their function they can be divided into three groups (Fig. 6.2) (Carbonnelle *et al.*, 2006):

1. The first group includes PilE, PilD, PilM, PilN, PilO, PilP, PilF and PilQ. This group is absolutely required for the assembly of the pilin subunit into fibre (PilE, PilD, PilM, PilN, PilO, PilP, PilF) and its extrusion through the outer membrane (PilQ).
2. The second group of components is required to counteract pilus retraction, i.e. when these components are missing the bacteria is non piliated except if pilus retraction is abolished by a PilT mutation. These components are PilC1/PilC2, PilG, PilH, PilI, PilJ, PilK, PilW. Some are located in the outer membrane (PilC1/C2, PilW) or in the cytoplasmic membrane (PilG). The others have a

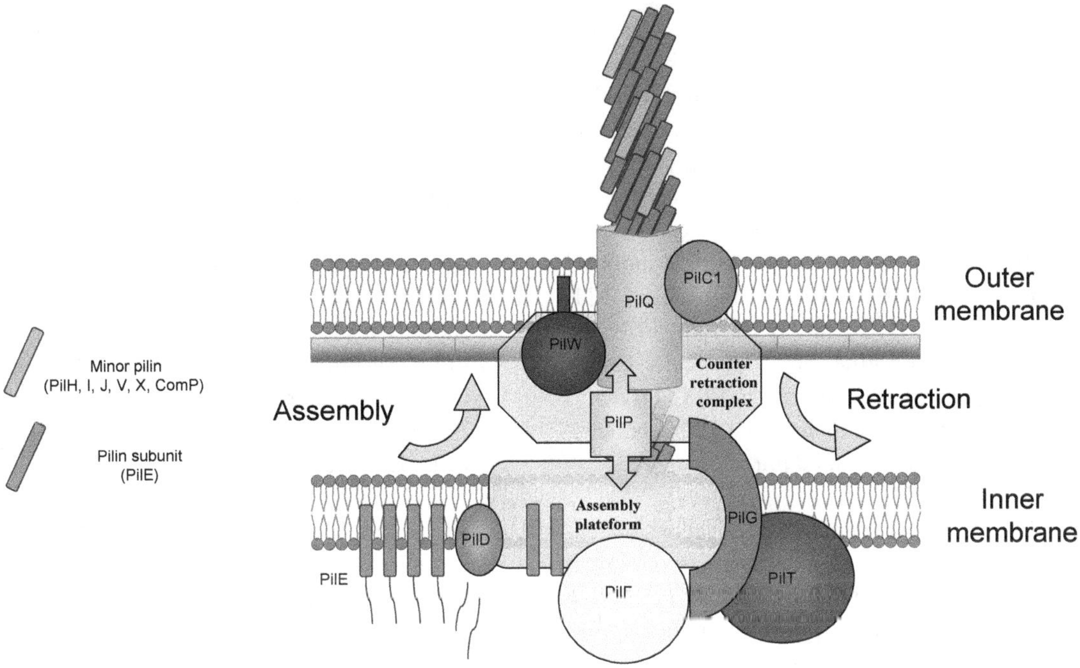

Figure 6.1 Components and dynamics of Tfp biogenesis. At the inner membrane–periplasm interface, the pilin subunits (PilE) are assembled from a plateform complex (PilD, F, M, N, O, P) and the growing fibre is translocated to the cell surface via the secretin (PilQ). It has been recently demonstrated that PilP copurifies with the inner membrane and interacts with PilQ. Two inner membrane ATPases promote pilus elongation with PilF and pilus retraction with PilT. At the periplasm-outer membrane interface, different components are required to counteract pilus retraction (PilC1/C2, G, H, I, J, K W). PilC1/C2 and PilW are located at the outer membrane whereas PilG is in the inner membrane. Some components are implicated for one of the TFP function (PilC1, W, I, J, K, X, V and ComP). Among the minor pilin(PilH, I, J, V, X and ComP), some of them are incorporated in the fibre (PilV, X and ComP).

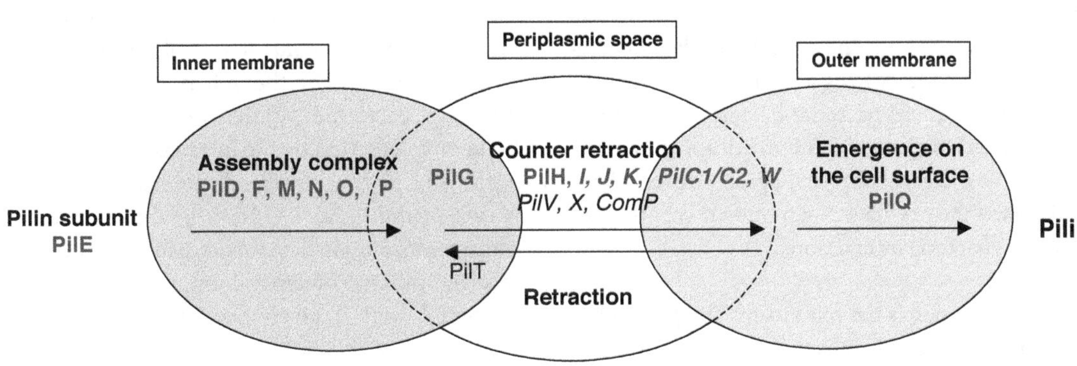

Figure 6.2 Implication of the *N. meningitidis* Pil components in the different steps of Tfp biogenesis, fibre assembly, functional maturation and counter retraction of fibre, retraction powered by the PilT protein and emergence of the fibres on the cell surface.

sequence corresponding to a signal peptide that is cleaved by PilD and are likely to be associated with the fibre (PilH, PilI, PilJ, PilK) and by analogy with the type II secretion system they may be localized only with the periplasmic part of the fibre (Durand *et al.*, 2005; Sauvonnet *et al.*, 2000).

3 The third group of components is required for one of the function of the TFP. PilC1, PilW, PilI, PilJ, PilK, PilX and PilV are important for adhesion to host cells. ComP is required for competency. PilX, PilV and ComP are likely to be minor components present in the extruded pili.

TFP-mediated meningococcal adhesion occurs in two steps. The first one involves an initial attachment of the bacteria onto the apical surface of the host cells. This step is very inefficient and only very few *Neisseria* adhere onto the host cell either as single bacterium or as diplococcus. Following this initial attachment the bacteria divide and spread onto the apical surface of the cells to form colonies and at later stage a monolayer of bacteria covering the apical surface of the host cells.

The components of the TFP that interact with the eukaryotic cells remain unknown. Initial studies suggested that the PilC1 protein was the adhesin. This hypothesis relied mostly on the fact that PilC1 was required for pili to become adhesive. Since then a similar phenotype was observed for numerous other pili components (see above and Fig. 6.2), and the nature of the bacterial component that bind to host cells remain unknown. On the other hand, adhesiveness has been correlated with the ability to establish bacteria–bacteria interactions. These bacteria–bacteria interactions are required to allow the initial adhering bacteria to grow as a colony onto the apical surface of the cells. Two factors have been shown to promote bacteria–bacteria interactions.

1 The first one is the pilin itself. The pilin subunit undergoes extensive antigenic variation, some variants are associated with hyperadhesiveness whereas others are responsible for a low adhesive phenotype (Nassif *et al.*, 1993, 1994). High adhesiveness has been correlated with the ability of pilin to form bundles that connect bacteria together. On the other hand, low adhesive pili have a poor tendency to form bundles of pili (Marceau *et al.*, 1995).

2 The second component promoting bacteria–bacteria interaction is PilX, a minor component of the fibre that assembles within the fibre in a similar manner to pilin. Recent data support a model in which surface exposed motifs of PilX stabilize bacteria–bacteria interaction via homotypic interaction between PilX molecules from pili of different bacterial cells (Helaine *et al.*, 2005).

The proposed host cell receptor for the Type IV pili of *N. meningitidis*, and of the closely related pathogen *N. gonorrhoeae*, is CD46 or membrane cofactor protein (MCP) (Kallstrom *et al.*, 1997). This receptor is a member of a superfamily of complement resistance proteins, of which at least six isoforms have been described, that are found expressed in nearly all human cell types, except erythrocytes. CD46 is a well characterized receptor protein for a number of viruses such as measles (Dorig *et al.*, 1993) and human herpes virus-6 (Santoro *et al.*, 1999). Using Chinese hamster ovary (CHO) cells transfected with CD46 isoforms, Kallström and colleagues showed that piliated neisseriae bind to BC1 and BC2 isoforms but not to C1 or C2 isoforms, and that this binding was blocked by recombinant CD46 or by anti-CD46 antibodies (Kallstrom *et al.*, 1997). Mice expressing human CD46 were more susceptible to systemic challenge by *N. meningitidis* than were control animals, although this was independent of the piliation status of the infecting bacterium (Johansson *et al.*, 2003). Following infection by the intranasal route, which is the site of the proposed interaction between *N. meningitidis* and CD46, a low proportion of transgenic animals developed disease if they had received piliated meningococci, whereas all mice remained well if given non-piliated organisms (Johansson *et al.*, 2003).

However, most of the recent published data support the notion that pilus-mediated infection of host cells occur in a CD46-independent manner:

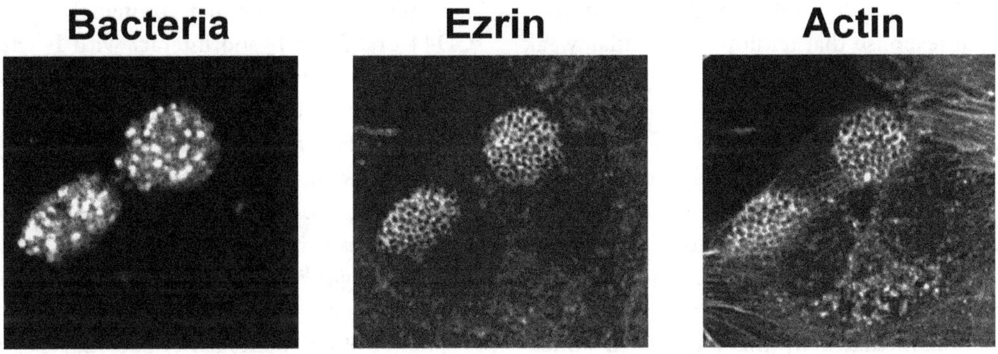

Figure 6.3 Infection of human endothelial cells by *N. meningitidis* induces the formation of cortical plaques beneath bacterial colonies. Human endothelial cells infected for 3 hours with the 2C43 strain of *N. meningitidis* were stained for bacteria, ezrin and actin and were analysed by confocal microscopy.

1 Adhesion of *N. meningitidis* to CD46 expressed in otherwise non-permissive cells was weak compared with the adhesion supported by human epithelial cells (Kallstrom *et al.*, 1997), and was not reproduced by other groups (Kirchner *et al.*, 2005). Inhibition of pilus-mediated adhesion to epithelial cells using anti-CD46 antibodies was not reproduced by several other groups (Edwards and Apicella, 2005; Edwards *et al.*, 2002; Gill *et al.*, 2003; Kirchner *et al.*, 2005).

2 In primary epithelial cells, CD46 is expressed on the basolateral surface, and is not transcytosed to the apical surface, therefore rendering most unlikely its availability to serve as initial, predominant pilus receptor on the mucosal surface of epithelial cells (Edwards and Apicella, 2005).

3 No correlation has been observed between the level of CD46 expression by cell lines and the degree of adhesion with piliated neisseriae (Kirchner *et al.*, 2005; Tobiason and Seifert, 2001). Moreover, a specific down-regulation of CD46 expression in human epithelial cell lines by RNA interference did not alter the binding efficiency of piliated gonococci (Kirchner *et al.*, 2005). These data therefore call into question the function of CD46 as an essential pilus receptor for pathogenic neisseriae.

New studies to identify pilus receptors have highlighted a potential interaction with the I-domain of some integrin alpha chains (α_M, a_1 and a_2), since adhesion of gonococci on primary urethral cells may be blocked by recombinant, soluble I domain or anti-α-chain antibodies (Edwards and Apicella, 2005; Edwards *et al.*, 2002). However, only 20–40% adherence inhibition was observed on some epithelial cell lines (Chang, ME180), while no inhibition was observed on others (T84), suggesting that another, yet undefined, pilus receptor exists.

Signalling induced by the Type IV pili

Initial attachment to host cells

The initial attachment of bacteria to the host cell by the TFP is characterized by a transient release of Ca^{2+} from intracellular stores. Semi-purified pili also trigger a cytosolic Ca^{2+} flux in human epithelial cells (Kallstrom *et al.*, 1998). This Ca^{2+} signal has a long latency, occurring ~10 min after the addition of bacteria or pili, can be blocked by anti-CD46 antibodies, and depends on the presence of PilC1, which, as mentioned above, is required for strong pilus-mediated adherence. These data suggest that pili are necessary for the Ca^{2+} signal but do not exclude the possibility that a copurified component is also required. Depletion of intracellular Ca^{2+} stores or treatment with protein kinase inhibitors results in diminished bacterial adherence (Kallstrom *et al.*,

1998). *N. meningitidis* pili may therefore elicit a host cell response that reinforces an initial, weak attachment.

Cortical plaque formation and invasion

As mentioned above, after their initial attachment to host cells, encapsulated meningococci proliferate, locally forming a colony at their site of attachment on the cell surface. The formation of these colonies is driven by bacteria–bacteria interaction mediated by the pili. When adhering to the apical membrane of epithelial cells, *N. meningitidis* induces the local elongation of microvilli towards the bacteria, leading to their engulfment and internalization (Merz *et al.*, 1996; Pujol *et al.*, 1997). Adhesion of *N. meningitidis* to endothelial cells promotes the local formation of membrane protrusions reminiscent of epithelial microvilli structures that surround bacteria and initiate their internalization within intracellular vacuoles (Eugene *et al.*, 2002). Interestingly, the formation of such protrusions was also observed *ex vivo*, by transmission electron microscopy analysis of brain sections from a child who died from fulminant meningitis (Nassif *et al.*, 2002). These observations strongly suggest that such morphological modifications of the host cell membrane may be essential for *N. meningitidis* to cross human vascular endothelium via a transcytosis pathway.

The formation of membrane protrusions by encapsulated *N. meningitidis* stems from the organization of specific molecular complexes, referred to as cortical plaques, beneath bacterial colonies (Fig. 6.3). Cortical plaques result from the recruitment of the molecular linkers ezrin and moesin that cluster several membrane integral proteins such as CD44 or ICAM-1 and from the localized polymerization of cortical actin (Eugene *et al.*, 2002; Merz *et al.*, 1999; Merz and So, 1997). Ezrin and moesin are members of the ezrin/radixin/moesin (ERM) family, known to play a major structural and regulatory role in many of the morphogenic changes of the plasma membrane, including the formation of microvilli (Bretscher *et al.*, 2002). These proteins control the organization of the cortical cytoskeleton by acting as linkers between the plasma membrane and the actin cytoskeleton. ERM proteins interact by their amino-terminal domain with the cytoplasmic domain of transmembrane proteins (so-called ERM binding proteins), such as CD44 or ICAM-1, and interact with F-actin by their carboxy-terminal domain. The expression of the truncated N-terminal domain of ezrin, which displays a dominant negative activity on actin-dependent ezrin function, largely prevented cortical actin polymerization induced by *N. meningitidis*, thus confirming the role of this molecule in *N. meningitidis*-induced cytoskeletal modifications (Eugene *et al.*, 2002).

Like other pathogenic bacteria, *N. meningitidis* requires the participation of the actin cytoskeleton to enter into non-phagocytic host cells. Cortical actin polymerization induced by *N. meningitidis* relies on the activation of both Rho and Cdc42 Rho GTPases, along with the activation of a PI3-K/Rac1GTPase signalling pathway involved in cortactin recruitment at the bacterial entry site (Eugene *et al.*, 2002; Lambotin *et al.*, 2005). Cortactin is an actin binding protein involved in the reorganization of the cortical actin cytoskeleton, which plays a major role in multiple essential cellular functions including endocytosis, neuronal morphogenesis, immune synapse formation (Gomez *et al.*, 2006; Martinez *et al.*, 2003; Schafer, 2002), and in the cellular invasion process of several bacterial pathogens (Dehio *et al.*, 1995; Selbach *et al.*, 2003). The recruitment and phosphorylation of cortactin at the *N. meningitidis* entry site is required for the formation of the actin-rich cell projections that promote efficient bacterial uptake, since the inhibition of either cortactin translocation or phosphorylation leads to structurally altered cortical actin polymerization and decreased invasion (Lambotin *et al.*, 2005). Interestingly, tyrosine phosphorylation of cortactin induced by *N. meningitidis* results from the clustering and activation of the host cell tyrosine kinase receptor ErbB2 and the downstream activation of the kinase src (Hoffmann *et al.*, 2001). ErbB2, which belongs to the family of epidermal growth factor (EGF) receptors, has no known soluble ligands, and is normally activated by heterodimerization with the other members of the family. However, the interaction of *N. meningitidis* with human endothelial cells leads to the activation of ErbB2, most likely via formation of ErbB2 homodimers, since no recruitment nor tyrosine phosphorylation of any other known ErbB family members was detected. ErbB2 activation by *N. meningitidis* infection apparently results from bacteria-induced clustering of this receptor; however, how *N. meningitidis* interaction leads to ErbB2 clustering is still unknown.

The above signalling is observed only with piliated bacteria, thus demonstrating that pili are required for these rearrangements to occur. Furthermore this signalling is greatly enhanced when mechanical force is exerted on the host cell membrane. The force exerted by pilus retraction is thought to be the motive force that induces elongation of microvilli (Merz and So, 2000). Moreover, recent data indicate that LOS integrity is required for efficient bacterial uptake. Bacteria expressing a truncated LOS structure elicit the formation of altered cortical plaques (Bonnah *et al.*, 2005; Lambotin *et al.*, 2005). Indeed, LOS provides a co-stimulatory signal leading to the activation of the PI3-K/Rac1 GTPase signalling pathway, which is required for cortactin recruitment to the bacterial adhesion sites and its subsequent phosphorylation downstream of the ErbB2/src pathway. By controlling cortactin localization, LOS contributes to the formation of actin structures and membrane protrusions associated with the internalization of *N. meningitidis* (Lambotin *et al.*, 2005).

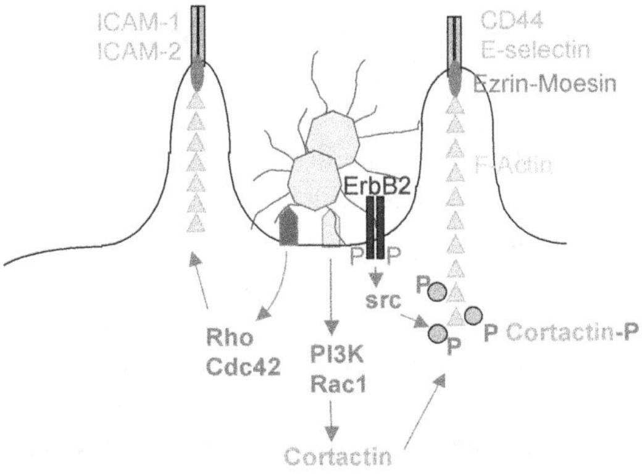

Figure 6.4 Schematic representation of the signalling pathways activated by *N. meningitidis* leading to their internalization into human endothelial cells. Type IV pili initiate the interaction of virulent, encapsulated *N. meningitidis* with human endothelial cells by interacting with a cellular receptor, possibly CD46 (Kallstrom *et al.*, 1997). This pili-dependent adhesion induces the recruitment of ezrin and the clustering of several transmembrane proteins: the ErbB2 tyrosine kinase receptor and the ezrin binding proteins CD44 and ICAM-1. The activation of both Rho and Cdc42 GTPases induces a local polymerization of cortical actin. ErbB2 clustering leads to the activation of src tyrosine kinase. In parallel, the LOS structure of *N. meningitidis*, by a mechanism which remains to be identified, provides a co-stimulatory signal leading to PI3-K and Rac1 activation, and the subsequent translocation of cortactin to site of cortical actin rearrangements. When localized at the cell plasma membrane, cortactin is tyrosine phosphorylated by src kinase and contributes to the formation of dynamic actin structures leading to the formation of membrane projections which surround bacteria and induce their internalization within endothelial intracellular vacuoles.

It is interesting to note that the signalling pathways promoted by *N. meningitidis*, summarized in Fig. 6.4, are similar to those induced by leucocyte adhesion on endothelial cells. Leucocyte adhesion promotes the remodelling of the apical endothelial plasma membrane into projections that surround adherent leucocytes (Barreiro *et al.*, 2005; Carman and Springer, 2004; Shaw *et al.*, 2004). These structures, referred to as 'endothelial docking structures' or 'transmigratory cups', are essential to promote firm adhesion and extravasation of leucocytes through paracellular as well as transcellular routes. These docking structures result from the dynamic redistribution of ICAM-1, VCAM-1 and CD44 at the endothelial-leucocyte contact area, accompanied by the recruitment of activated ERM proteins and by cortical actin polymerization. Therefore, the same set of endothelial proteins is present in the membrane protrusions induced by *N. meningitidis* and in the docking structures promoted by leucocyte adhesion. In addition, *N. meningitidis* colonization on the endothelial surface, by inducing the massive recruitment of ezrin and moesin together with ICAM-1, ICAM-2, VCAM-1, CD44, prevents their accumulation at leucocyte/endothelial cell contact area and the formation of docking structures (Doulet *et al.*, 2006). These data strongly suggest that, when adhering to endothelial cells, *N. meningitidis* hijack the cellular machinery normally used for leucocyte trans-endothelial migration to promote its own bacterial internalization and transcytosis.

Intimate adhesion

After several hours of *N. meningitidis* interaction with host cells, the bacterial colonies formed at the apical surface of the cells disperse and diffuse adherence occurs, during which process bacteria retract their pili, and adhere tightly to the host cell plasma membrane (Pujol *et al.*, 1997; Pujol *et al.*,

1999). At this stage, the host cell membrane protrusions observed during localized adhesion disappeared. Two main processes are responsible for diffuse adherence. The first is bacterial dispersal over the cell surface, which is mediated by PilT, and a reduction in pilus expression (Pujol *et al.*, 1999). The second is the intimate attachment of the bacteria, characterized by PilC1 up-regulation and capsule down-regulation (Deghmane *et al.*, 2002; Deghmane *et al.*, 2000; Taha *et al.*, 1998). This coordinated series of events is under genetic control in the bacterium which responds to host cell binding by regulation of gene expression. The region of the *pilC1* promoter involved in PilC1 up-regulation in initial adhesion has been identified and named the CREN region (contact regulatory element of *Neisseria*). Homology searches revealed that the *crgA* gene, encoding a transcriptional regulator, also possesses a CREN region and is induced upon host cell contact (Deghmane *et al.*, 2000). Insertional inactivation of the crgA gene decreases meningococcal adhesion to epithelial cells and abolishes intimate adhesion (Deghmane *et al.*, 2002; Deghmane *et al.*, 2000). CrgA protein binds to the *pilE* and *sia* promoters and induces the down-regulation of the expression of *pilE* (the main structural unit of the pilus) and of sia (the principal capsule biosynthesis protein). CrgA may then play a central regulatory role in meningococcal adhesion, particularly in switching from initial to intimate adhesion by down-regulating the bacterial surface structures that hinder this adhesion.

Twelve other genes that are also up-regulated following host cell contact have been identified (Morelle *et al.*, 2003), and possess a REP2 region corresponding to the CREN region. The functions of most of these genes are unknown but their presence confirms that the genetic processes involved in initial adhesion to epithelial and endothelial cells involve many complicated biochemical and regulatory pathways.

Other means of meningococcal interaction with host cells

For unencapsulated meningococci, several mechanisms of bacterial internalization have been described involving the interaction of cellular receptors with bacterial surface components, such as Opa and Opc outer membrane proteins (Dehio *et al.*, 1998; Merz and So, 2000; Nassif, 1999).

Opa proteins consist of eight transmembrane β-strands and four surface-exposed loops. The transmembrane and periplasmic sections are highly conserved, while the first three surface exposed loops display antigenic variation. There are 4–5 loci encoding Opa proteins in *N. meningitidis*, and expression varies between strains. Thus a single bacterium may express any number of Opa proteins at any one time, and variation is enhanced by horizontal transfer of *opa* alleles promoting the formation of hybrid recombinant *opa* loci (Dehio *et al.*, 1998). This has important functional consequences, as expression of different Opa proteins confers tropism to different cell lines (Makino *et al.*, 1991; Virji *et al.*, 1993).

Several groups have shown that many different Opas bind to members of the large CD66/CEACAM family of proteins (Dehio *et al.*, 1998). CD66-related proteins can mediate cell–cell adhesion, are encoded by different genes, and are often produced as multiple splice variants (Hammarstrom, 1999). Different CD66 family members are variably expressed on different cell types. Most family members encode transmembrane proteins, whereas a few encode peripheral membrane proteins with GPI (glycosylphosphatidylinositol) membrane anchors. Opa proteins have been observed to bind to a number of variants of CD66, and Opa structure can be altered to determine CD66 specificity (Virji *et al.*, 1999). Depending on the particular CD66 variant(s) and Opa protein(s) expressed, different host cell responses occur, including binding, uptake, and the activation of different signal transduction pathways.

In the gonococcus, one class of Opas established binding with heparan sulphate proteoglycans (HSPG) present on certain epithelial cells (van Putten and Paul, 1995). Two pathways then mediate bacterial entry into the cell, either through localized recruitment of lipid hydrolysis (Grassme *et al.*, 1997), or through a vitronectin-dependent mechanism (Gomez-Duarte *et al.*, 1997). Interaction between Opa proteins (via vitronectin) and hence integrins can be sufficient to facilitate bacterial uptake in certain cell lines. However, no interaction between HSPG and meningococcal Opa proteins has yet been described.

Opc proteins share physiochemical properties and a weak homology with the Opa group, but differ in their structure and genetic control. The *opc* gene is present in many but not all *N. meningitidis* strains and is associated with virulence (Olyhoek *et al.*, 1991; Seiler *et al.*, 1996). The expression of Opc proteins is governed by alterations in the length of a homopolymeric tract in the promoter region (Sarkari *et al.*, 1994). Expression of *opc* in non-encapsulated *N. meningitidis* confers on the bacteria the ability to adhere to and invade endothelial cells independently of Opa and pili (Virji *et al.*, 1993; Virji *et al.*, 1992; Virji *et al.*, 1994; Virji *et al.*, 1995). The action of Opc follows interaction with the serum glycoprotein, vitronectin, which acts as a molecular bridge to αvβ3 integrins on the surface of umbilical vein endothelial cells, facilitating entry of the bacterium into these cells (Virji *et al.*, 1994). Opc also affects internalization of bacteria into human brain microvascular endothelial cells; this however is mediated through the interaction of Opc with fibronectin in serum and then with αvβ integrins (Unkmeir *et al.*, 2002b).

Other molecules expressed at the bacterial cell surface have also been implicated in adhesion. Recently, a novel antigen of *N. meningitidis*, NadA, contributing to meningococcal adhesion to and invasion into epithelial cells has, been described (Capecchi *et al.*, 2005). NadA is present in three of the four known hypervirulent lineages (Comanducci *et al.*, 2002) and is a bacterial invasin which, when expressed on the surface of *Escherichia coli*, promotes adhesion to and invasion of Chang epithelial cells (Capecchi *et al.*, 2005). Deletion of the N-terminal globular domain of recombinant NadA or pronase treatment of human cells abrogated the adhesive phenotype. Moreover, a hypervirulent strain of *N. meningitidis* where the NadA gene was inactivated had a reduced ability to adhere to and invade epithelial cells *in vitro*.

The structure of lipooligosaccharide (LOS) of the meningococcus is similar to the lipopolysaccharide (LPS) of the Enterobacteriaceae, but lacks a repeating O-antigen. Differences in side chain composition underlie the individual immunotypes of different strains (Kahler and Stephens, 1998). Some immunotypes express the lacto-N-neotetraose epitope that mimics human asialocarbohydrates; the contribution of LOS to *N. gonorrhoeae* binding to the host cell may be through interaction with asialocarbohydrate receptors (Porat *et al.*, 1995). In *N. meningitidis* LOS can be further modified by the addition of terminal sialic acid moieties. LOS sialylation is a phenotypic variable, increasing the bacterium's negative surface charge density, which may produce an anti-adhesive effect (Merz and So, 2000).

The effects of the capsule composition on meningococcal adhesion have been mostly studied in strains expressing the polysialic acid, serogroup B polysaccharide. The capsule is antiphagocytic and antibactericidal due to surface-exposed polysaccharides (Estabrook *et al.*, 1998; Unkmeir *et al.*, 2002a). Thus the capsule is commonly seen in strains isolated from the blood and CSF. The capsule and sialylated LOS inhibit Opa- and Opc-mediated adhesion to host cells possibly due to negatively charged molecules that repel host cells, but do not strongly inhibit type IV pilus-mediated adhesion (van Putten, 1993; Virji *et al.*, 1995), presumably because pili extend far enough from the bacterial cell surface that electrostatic repulsion between bacterial sialic acids and the negatively charged host cell surface is negligible (Tzeng and Stephens, 2000).

Role of meningococcal host cell invasion in the pathogenesis and the dissemination of meningococci

The invasion of host cells by *N. meningitidis* appears to be essential for meningococcal pathogenesis since *in vitro* and *in vivo* evidence indicates that *N. meningitidis* uses a transcytosis pathway to cross cell monolayers (Nassif *et al.*, 2002). *N. meningitidis* appears to replicate within host cells and is capable of traversing polarized epithelial or endothelial monolayers without disrupting barrier function (Eugene *et al.*, 2002; Merz *et al.*, 1996; Pujol *et al.*, 1997). It is unknown to what extent the intracellular bacteria egress through basolateral versus apical cellular membranes. Moreover, the mechanisms by which intracellular bacteria obtain nutrients are just beginning to be addressed. However considering that pathogenesis is an accident and is not the usual outcome of the colonization of a host by *N. meningitidis*. One question is how cellular invasion helps this bacterium to disseminate. It is likely that invading the host cell at the site of colonization may help *N. meningitidis* to persist inside its host and to evade the mucosal innate immune system.

Conclusions and future trends

Recent progress has opened the way to a molecular understanding of neisserial attachment and invasion in the pathogenesis of neisserial infection. A number of adhesive factors have been characterized, and their cognate receptors defined. Recent exciting findings have considerably expanded our understanding of the cellular events involved in *N. meningitidis* invasion processes, underlining the complex sequence of signalling events induced by *N. meningitidis* to elicit its uptake in non phagocytic cells. However, in spite of recent advances in our understanding of these molecular mechanisms, much remains to be discovered about the complex molecular networks involved. Among major issues is the identification of the receptor for meningococcal pili, which would constitute a significant breakthrough in the field and pave the way to future developments of novel vaccinal strategies.

Future research should also focus on the identification of key proteins that may functionally link cortical actin cytoskeletal reorganization to the endocytic machinery responsible for the formation of intracellular vacuoles containing bacteria and their subsequent transcytosis through cellular barriers. In addition to the importance of the elucidation of this pathway for our current knowledge of *N. meningitidis* pathogenesis, these results would also certainly bring new insights in the understanding of fundamental cellular mechanisms of infection.

Targeting bacterial virulence is an alternative approach to the development of new antimicrobials that can be used to disarm pathogens in the host. The overall strategy is to inhibit specific mechanisms that promote infection and are essential to persistence in a pathogenic cascade. New approaches and weapons are needed to combat bacteria that are, owing to evolution and selection, moving targets.

References

Barreiro, O., Yanez-Mo, M., Sala-Valdes, M., Gutierrez-Lopez, M.D., Ovalle, S., Higginbottom, A., Monk, P.N., Cabanas, C., and Sanchez-Madrid, F. (2005). Endothelial tetraspanin microdomains regulate leukocyte firm adhesion during extravasation. Blood *105*, 2852–2861.

Bonnah, R.A., Hoelter, J., Steeghs, L., Enns, C.A., So, M., and Muckenthaler, M.U. (2005). Lipooligosaccharide-independent alteration of cellular homeostasis in *Neisseria meningitidis*-infected epithelial cells. Cell. Microbiol. *7*, 869–885.

Bretscher, A., Edwards, K., and Fehon, R.G. (2002). ERM proteins and merlin: integrators at the cell cortex. Nature Rev. *3*, 586–599.

Capecchi, B., Adu-Bobie, J., Di Marcello, F., Ciucchi, L., Masignani, V., Taddei, A., Rappuoli, R., Pizza, M., and Arico, B. (2005). *Neisseria meningitidis* NadA is a new invasin which promotes bacterial adhesion to and penetration into human epithelial cells. Mol. Microbiol. *55*, 687–698.

Carbonnelle, E., Helaine, S., Nassif, X., and Pelicic, V. (2006). A systematic genetic analysis in *Neisseria meningitidis* defines the Pil proteins required for assembly, functionality, stabilization and export of type IV pili. Mol. Microbiol. *61*, 1510–1522.

Carman, C.V., and Springer, T.A. (2004). A transmigratory cup in leukocyte diapedesis both through individual vascular endothelial cells and between them. J. Cell. Biol. *167*, 377–388.

Chamot-Rooke, J., Rousseau, B., Lanternier, F., Mikaty, G., Mairey, E., Malosse, C., Bouchoux, G., Pelicic, V., Camoin, L., Nassif, X., *et al.* (2007). Alternative *Neisseria* spp. type IV pilin glycosylation with a glyceramido acetamido trideoxyhexose residue. Proc. Natl. Acad. Sci. USA *104*, 14783–14788.

Comanducci, M., Bambini, S., Brunelli, B., Adu-Bobie, J., Arico, B., Capecchi, B., Giuliani, M.M., Masignani, V., Santini, L., Savino, S., *et al.* (2002). NadA, a novel vaccine candidate of *Neisseria meningitidis*. J. Exp. Med. *195*, 1445–1454.

Deghmane, A.E., Giorgini, D., Larribe, M., Alonso, J.M., and Taha, M.K. (2002). Down-regulation of pili and capsule of *Neisseria meningitidis* upon contact with epithelial cells is mediated by CrgA regulatory protein. Mol. Microbiol. *43*, 1555–1564.

Deghmane, A.E., Petit, S., Topilko, A., Pereira, Y., Giorgini, D., Larribe, M., and Taha, M.K. (2000). Intimate adhesion of *Neisseria meningitidis* to human epithelial cells is under the control of the crgA gene, a novel LysR-type transcriptional regulator. EMBO J. *19*, 1068–1078.

Dehio, C., Gray-Owen, S.D., and Meyer, T.F. (1998). The role of neisserial Opa proteins in interactions with host cells. Trends Microbiol. *6*, 489–495.

Dehio, C., Prevost, M.C., and Sansonetti, P.J. (1995). Invasion of epithelial cells by Shigella flexneri induces tyrosine phosphorylation of cortactin by a pp60c-src-mediated signalling pathway. EMBO J. *14*, 2471–2482.

Dorig, R.E., Marcil, A., Chopra, A., and Richardson, C.D. (1993). The human CD46 molecule is a receptor for measles virus (Edmonston strain). Cell *75*, 295–305.

Doulet, N., Donnadieu, E., Laran-Chich, M.P., Niedergang, F., Nassif, X., Couraud, P.O., and Bourdoulous, S. (2006). *Neisseria meningitidis* infection of human endothelial cells interferes with leukocyte transmigration by preventing the formation of endothelial docking structures. J. Cell. Biol. *173*, 627–637.

Durand, E., Michel, G., Voulhoux, R., Kurner, J., Bernadac, A., and Filloux, A. (2005). XcpX controls biogenesis of the *Pseudomonas aeruginosa* XcpT-containing pseudopilus. J. Biol. Chem. *280*, 31378–31389.

Edwards, J.L., and Apicella, M.A. (2005). I-domain-containing integrins serve as pilus receptors for *Neisseria gonorrhoeae* adherence to human epithelial cells. Cell. Microbiol. *7*, 1197–1211.

Edwards, J.L., Brown, E.J., Uk-Nham, S., Cannon, J.G., Blake, M.S., and Apicella, M.A. (2002). A co-operative interaction between *Neisseria gonorrhoeae* and complement receptor 3 mediates infection of primary cervical epithelial cells. Cell. Microbiol. *4*, 571–584.

Estabrook, M.M., Zhou, D., and Apicella, M.A. (1998). Nonopsonic phagocytosis of group C *Neisseria meningitidis* by human neutrophils. Infect. Immun. *66*, 1028–1036.

Eugene, E., Hoffmann, I., Pujol, C., Couraud, P.O., Bourdoulous, S., and Nassif, X. (2002). Microvilli-like structures are associated with the internalization of virulent capsulated *Neisseria meningitidis* into vascular endothelial cells. J. Cell Sci. *115*, 1231–1241.

Forest, K.T., Dunham, S.A., Koomey, M., and Tainer, J.A. (1999). Crystallographic structure reveals phosphorylated pilin from *Neisseria*: phosphoserine sites modify type IV pilus surface chemistry and fibre morphology. Mol. Microbiol. *31*, 743–752.

Freitag, N.E., Seifert, H.S., and Koomey, M. (1995). Characterization of the pilF-pilD pilus-assembly locus of *Neisseria gonorrhoeae*. Mol. Microbiol. *16*, 575–586.

Gill, D.B., Koomey, M., Cannon, J.G., and Atkinson, J.P. (2003). Down-regulation of CD46 by piliated *Neisseria gonorrhoeae*. J. Exp. Med. *198*, 1313–1322.

Gomez, T.S., McCarney, S.D., Carrizosa, E., Labno, C.M., Comiskey, E.O., Nolz, J.C., Zhu, P., Freedman, B.D., Clark, M.R., Rawlings, D.J., *et al.* (2006). HS1 functions as an essential actin-regulatory adaptor protein at the immune synapse. Immunity *24*, 741–752.

Gomez-Duarte, O.G., Dehio, M., Guzman, C.A., Chhatwal, G.S., Dehio, C., and Meyer, T.F. (1997). Binding of vitronectin to opa-expressing *Neisseria gonorrhoeae* mediates invasion of HeLa cells. Infect. Immun. *65*, 3857–3866.

Grassme, H., Gulbins, E., Brenner, B., Ferlinz, K., Sandhoff, K., Harzer, K., Lang, F., and Meyer, T.F. (1997). Acidic sphingomyelinase mediates entry of N. *gonorrhoeae* into nonphagocytic cells. Cell *91*, 605–615.

Hammarstrom, S. (1999). The carcinoembryonic antigen (CEA) family: structures, suggested functions and expression in normal and malignant tissues. Seminars in cancer biology *9*, 67–81.

Helaine, S., Carbonnelle, E., Prouvensier, L., Beretti, J.L., Nassif, X., and Pelicic, V. (2005). PilX, a pilus-associated protein essential for bacterial aggregation, is a key to pilus-facilitated attachment of *Neisseria meningitidis* to human cells. Molecular microbiology *55*, 65–77.

Hoffmann, I., Eugene, E., Nassif, X., Couraud, P.O., and Bourdoulous, S. (2001). Activation of ErbB2 receptor tyrosine kinase supports invasion of endothelial cells by *Neisseria meningitidis*. The Journal of cell biology *155*, 133–143.

Johansson, L., Rytkonen, A., Bergman, P., Albiger, B., Kallstrom, H., Hokfelt, T., Agerberth, B., Cattaneo, R., and Jonsson, A.B. (2003). CD46 in meningococcal disease. Science *301*, 373–375.

Kahler, C.M., and Stephens, D.S. (1998). Genetic basis for biosynthesis, structure, and function of meningococcal lipooligosaccharide (endotoxin). Critical reviews in microbiology *24*, 281–334.

Kallstrom, H., Islam, M.S., Berggren, P.O., and Jonsson, A.B. (1998). Cell signaling by the type IV pili of pathogenic *Neisseria*. J. Biol. Chem. *273*, 21777–21782.

Kallstrom, H., Liszewski, M.K., Atkinson, J.P., and Jonsson, A.B. (1997). Membrane cofactor protein (MCP or CD46) is a cellular pilus receptor for pathogenic *Neisseria*. Mol. Microbiol. *25*, 639–647.

Kirchner, M., Heuer, D., and Meyer, T.F. (2005). CD46 independent binding of neisserial type IV pili and the major pilus adhesin, PilC, to human epithelial cells. Infection and immunity 73, 3072–3082.

Lambotin, M., Hoffmann, I., Laran-Chich, M.P., Nassif, X., Couraud, P.O., and Bourdoulous, S. (2005). Invasion of endothelial cells by *Neisseria meningitidis* requires cortactin recruitment by a phosphoinositide-3-kinase/Rac1 signalling pathway triggered by the lipo-oligosaccharide. J. Cell Sci. *118*, 3805–3816.

Lory, S., and Strom, M.S. (1997). Structure–function relationship of type-IV prepilin peptidase of *Pseudomonas aeruginosa* – a review. Gene *192*, 117–121.

Makino, S., van Putten, J.P., and Meyer, T.F. (1991). Phase variation of the opacity outer membrane protein controls invasion by *Neisseria gonorrhoeae* into human epithelial cells. The EMBO journal *10*, 1307–1315.

Marceau, M., Beretti, J.L., and Nassif, X. (1995). High adhesiveness of encapsulated *Neisseria meningitidis* to epithelial cells is associated with the formation of bundles of pili. Mol. Microbiol. *17*, 855–863.

Marceau, M., Forest, K., Beretti, J.L., Tainer, J., and Nassif, X. (1998). Consequences of the loss of O-linked glycosylation of meningococcal type IV pilin on piliation and pilus-mediated adhesion. Mol. Microbiol. *27*, 705–715.

Martinez, M.C., Ochiishi, T., Majewski, M., and Kosik, K.S. (2003). Dual regulation of neuronal morphogenesis by a delta–catenin–cortactin complex and Rho. J. Cell. Biol. *162*, 99–111.

Merz, A.J., Enns, C.A., and So, M. (1999). Type IV pili of pathogenic neisseriae elicit cortical plaque formation in epithelial cells. Mol. Microbiol. *32*, 1316–1332.

Merz, A.J., Rifenbery, D.B., Arvidson, C.G., and So, M. (1996). Traversal of a polarized epithelium by pathogenic neisseriae: facilitation by type IV pili and maintenance of epithelial barrier function. Mol. Med. *2*, 745–754.

Merz, A.J., and So, M. (1997). Attachment of piliated, Opa- and Opc- gonococci and meningococci to epithelial cells elicits cortical actin rearrangements and clustering of tyrosine-phosphorylated proteins. Infect. Immun. *65*, 4341–4349.

Merz, A.J., and So, M. (2000). Interactions of pathogenic neisseriae with epithelial cell membranes. Annu. Rev. Cell. Dev. Biol. *16*, 423–457.

Merz, A.J., So, M., and Sheetz, M.P. (2000). Pilus retraction powers bacterial twitching motility. Nature *407*, 98–102.

Morelle, S., Carbonnelle, E., and Nassif, X. (2003). The REP2 repeats of the genome of *Neisseria meningitidis* are associated with genes coordinately regulated during bacterial cell interaction. J. Bacteriol. *185*, 2618–2627.

Nassif, X. (1999). Interactions between encapsulated *Neisseria meningitidis* and host cells. Int. Microbiol. *2*, 133–136.

Nassif, X., Beretti, J.L., Lowy, J., Stenberg, P., O'Gaora, P., Pfeifer, J., Normark, S., and So, M. (1994). Roles of pilin and PilC in adhesion of *Neisseria meningitidis* to human epithelial and endothelial cells. Proc. Natl. Acad. Sci. USA *91*, 3769–3773.

Nassif, X., Bourdoulous, S., Eugene, E., and Couraud, P.O. (2002). How do extracellular pathogens cross the blood–brain barrier? Trends Microbiol. *10*, 227–232.

Nassif, X., Lowy, J., Stenberg, P., O'Gaora, P., Ganji, A., and So, M. (1993). Antigenic variation of pilin regulates adhesion of *Neisseria meningitidis* to human epithelial cells. Mol. Microbiol. *8*, 719–725.

Nassif, X., Marceau, M., Pujol, C., Pron, B., Beretti, J.L., and Taha, M.K. (1997). Type-4 pili and meningococcal adhesiveness. Gene *192*, 149–153.

Olyhoek, A.J., Sarkari, J., Bopp, M., Morelli, G., and Achtman, M. (1991). Cloning and expression in *Escherichia coli* of opc, the gene for an unusual class 5 outer membrane protein from *Neisseria meningitidis* (meningococci/surface antigen). Microbial Pathogen. *11*, 249–257.

Porat, N., Apicella, M.A., and Blake, M.S. (1995). A lipooligosaccharide-binding site on HepG2 cells similar to the gonococcal opacity-associated surface protein Opa. Infect. Immun. *63*, 2164–2172.

Pujol, C., Eugene, E., de Saint Martin, L., and Nassif, X. (1997). Interaction of *Neisseria meningitidis* with a polarized monolayer of epithelial cells. Infect. Immun. *65*, 4836–4842.

Pujol, C., Eugene, E., Marceau, M., and Nassif, X. (1999). The meningococcal PilT protein is required for induction of intimate attachment to epithelial cells following pilus-mediated adhesion. Proc. Natl. Acad. Sci. USA *96*, 4017–4022.

Santoro, F., Kennedy, P.E., Locatelli, G., Malnati, M.S., Berger, E.A., and Lusso, P. (1999). CD46 is a cellular receptor for human herpesvirus 6. Cell *99*, 817–827.

Sarkari, J., Pandit, N., Moxon, E.R., and Achtman, M. (1994). Variable expression of the Opc outer membrane protein in *Neisseria meningitidis* is caused by size variation of a promoter containing poly-cytidine. Mol. Microbiol. *13*, 207–217.

Sauvonnet, N., Vignon, G., Pugsley, A.P., and Gounon, P. (2000). Pilus formation and protein secretion by the same machinery in *Escherichia coli*. EMBO J. *19*, 2221–2228.

Schafer, D.A. (2002). Coupling actin dynamics and membrane dynamics during endocytosis. Current opinion in cell biology *14*, 76–81.

Seiler, A., Reinhardt, R., Sarkari, J., Caugant, D.A., and Achtman, M. (1996). Allelic polymorphism and site-specific recombination in the opc locus of *Neisseria meningitidis*. Mol. Microbiol. *19*, 841–856.

Selbach, M., Moese, S., Hurwitz, R., Hauck, C.R., Meyer, T.F., and Backert, S. (2003). The *Helicobacter pylori* CagA protein induces cortactin dephosphorylation and actin rearrangement by c-Src inactivation. EMBO J. *22*, 515–528.

Shaw, S.K., Ma, S., Kim, M.B., Rao, R.M., Hartman, C.U., Froio, R.M., Yang, L., Jones, T., Liu, Y., Nusrat, A., *et al.* (2004). Coordinated redistribution of leukocyte LFA-1 and endothelial cell ICAM-1 accompany neutrophil transmigration. J. Exp. Med. *200*, 1571–1580.

Taha, M.K., Morand, P.C., Pereira, Y., Eugene, E., Giorgini, D., Larribe, M., and Nassif, X. (1998). Pilus-mediated adhesion of *Neisseria meningitidis*: the essential role of cell contact-dependent transcriptional up-regulation of the PilC1 protein. Mol. Microbiol. *28*, 1153–1163.

Tobiason, D.M., and Seifert, H.S. (2001). Inverse relationship between pilus-mediated gonococcal adherence and surface expression of the pilus receptor, CD46. Microbiology *147*, 2333–2340.

Tzeng, Y.L., and Stephens, D.S. (2000). Epidemiology and pathogenesis of *Neisseria meningitidis*. Microbes and Infection/Institut Pasteur *2*, 687–700.

Unkmeir, A., Kammerer, U., Stade, A., Hubner, C., Haller, S., Kolb-Maurer, A., Frosch, M., and Dietrich, G. (2002a). Lipooligosaccharide and polysaccharide capsule: virulence factors of *Neisseria meningitidis* that determine meningococcal interaction with human dendritic cells. Infect. Immun. *70*, 2454–2462.

Unkmeir, A., Latsch, K., Dietrich, G., Wintermeyer, E., Schinke, B., Schwender, S., Kim, K.S., Eigenthaler, M., and Frosch, M. (2002b). Fibronectin mediates Opc-dependent internalization of *Neisseria meningitidis* in human brain microvascular endothelial cells. Mol. Microbiol. *46*, 933–946.

van Putten, J.P. (1993). Phase variation of lipopolysaccharide directs interconversion of invasive and immuno-resistant phenotypes of *Neisseria gonorrhoeae*. EMBO J. *12*, 4043–4051.

van Putten, J.P., and Paul, S.M. (1995). Binding of syndecan-like cell surface proteoglycan receptors is required for *Neisseria gonorrhoeae* entry into human mucosal cells. EMBO J. *14*, 2144–2154.

Virji, M., Evans, D., Hadfield, A., Grunert, F., Teixeira, A.M., and Watt, S.M. (1999). Critical determinants of host receptor targeting by *Neisseria meningitidis* and *Neisseria gonorrhoeae*: identification of Opa adhesiotopes on the N-domain of CD66 molecules. Mol. Microbiol. *34*, 538–551.

Virji, M., Kayhty, H., Ferguson, D.J., Alexandrescu, C., Heckels, J.E., and Moxon, E.R. (1991). The role of pili in the interactions of pathogenic *Neisseria* with cultured human endothelial cells. Mol. Microbiol. *5*, 1831–1841.

Virji, M., Makepeace, K., Ferguson, D.J., Achtman, M., and Moxon, E.R. (1993). Meningococcal Opa and Opc proteins: their role in colonization and invasion of human epithelial and endothelial cells. Mol. Microbiol. *10*, 499–510.

Virji, M., Makepeace, K., Ferguson, D.J., Achtman, M., Sarkari, J., and Moxon, E.R. (1992). Expression of the Opc protein correlates with invasion of epithelial and endothelial cells by *Neisseria meningitidis*. Mol. Microbiol. *6*, 2785–2795.

Virji, M., Makepeace, K., and Moxon, E.R. (1994). Distinct mechanisms of interactions of Opc-expressing meningococci at apical and basolateral surfaces of human endothelial cells; the role of integrins in apical interactions. Mol. Microbiol. *14*, 173–184.

Virji, M., Makepeace, K., Peak, I.R., Ferguson, D.J., Jennings, M.P., and Moxon, E.R. (1995). Opc- and pilus-dependent interactions of meningococci with human endothelial cells: molecular mechanisms and modulation by surface polysaccharides. Mol. Microbiol. *18*, 741–754.

Wolfgang, M., Lauer, P., Park, H.S., Brossay, L., Hebert, J., and Koomey, M. (1998). PilT mutations lead to simultaneous defects in competence for natural transformation and twitching motility in piliated *Neisseria gonorrhoeae*. Mol. Microbiol. *29*, 321–330.

Part III

Host Response

Innate Immune Recognition of *Neisseria meningitidis* and *Neisseria gonorrhoeae*

7

Daniel C. Stein, Julia B. Patrone and Samuel Bish

Abstract

The existing paradigm in neisserial infections is that these pathogens present a diverse, varying cell surface, and that one of the functions of this variability is that it allows them to avoid the killing actions of the hosts' innate immune system. In this chapter, we will define the important features of the innate immune system as it relates to neisserial disease, review the role of the variable cell surface antigens of *N. gonorrhoeae* and *N. meningitidis* in eliciting responses from various host cells, and provide some insights into the molecular basis of disease. We will describe what features are essential on the bacterium for initiating colonization in various anatomical locations and in the different sexes, and review how the host responds to these colonization events (disease vs. asymptomatic colonization). We will also review the molecular mechanisms by which cells of the innate immune system become activated, and how different activation mechanisms can lead to different disease outcomes.

Introduction

The innate immune system includes anatomical barriers (skin and mucous membranes), protective cells (leucocytes) and secreted proteins (complement, antimicrobial peptides, etc.). In order to cause disease, the pathogenic *Neisseria* must overcome this first line of defence. The microbial strategies employed for meningococcal disease probably differ somewhat from those of gonococcal disease, due to the different anatomical sites in which the organisms colonize. However, given the high degree of genomic DNA sequence similarities between the two pathogens (Guibourdenche *et al.*, 1986), and the significant structural similarities in their surface antigens (Aho *et al.*, 2000; Black *et al.*, 1985; Dehio *et al.*, 1998; Jose *et al.*, 2003; Lavitola *et al.*, 1987; Mandrell *et al.*, 1988; Plante *et al.*, 1999; Snapper *et al.*, 1997; Turner *et al.*, 2001; Virji *et al.*, 1999), one can predict that many of the disease symptoms at the molecular level are probably mediated by similar mechanisms.

Fig. 7.1 represents a schematic of the classes of various types of surface structures that are found on the pathogenic *Neisseria* (Note: one significant difference between gonococci and meningococci is that meningococci can express a capsule.) Each of these components can interact with different host-cell ligands to promote colonization (see Chapter 5), and many of these structures are variably expressed on the bacterial cell surface. The *Neisseria* utilize a variety of genetic mechanisms to regulate the surface expression of most of the components shown in Fig. 7.1. Variation in expression of these phase-variable antigens occurs at such a high frequency that during infection, an individual is exposed to a variety of bacterial cell surfaces (Alcala *et al.*, 2004; Apicella *et al.*, 1987; Banerjee *et al.*, 2002; Carson *et al.*, 2000; Chen *et al.*, 1998; Danaher *et al.*, 1995; Hammerschmidt *et al.*, 1996b; Jonsson *et al.*, 1991; Lambden *et al.*, 1979; Murphy *et al.*, 1989; Sarkari *et al.*, 1994; Segal *et al.*, 1985). This suggests that even when two individuals are exposed to the same strain, the initial signalling to trigger the innate immune response can vary, due to differences in the surface antigens presented to

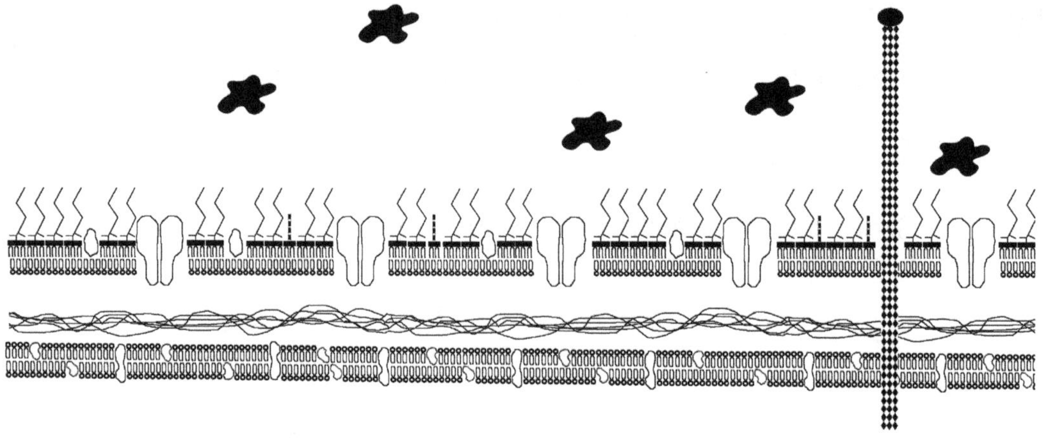

Figure 7.1 Neisserial surface structures involved in virulence. The names in parenthesis are representative protein class. The symbols in the figure represent gonococcal virulence determinants. The protein descriptions in the parenthesis are an example of each class: Pili (PilE), ⁞; Porin (PorA), ⟩⟨; LOS, ⸸; Outer Membrane proteins (PilC), ●; Adhesins (Opa), ◌; Lipoproteins (H8), ⸸; secreted proteins (IgA protease), ✱.

each host. Because of the surface variability of the pathogenic *Neisseria* strains, it is to be expected that infections caused by these organisms could elicit a broad array of symptoms. The ability of the *Neisseria* to antigenically vary their surface at high frequency suggests that their interaction with the innate immune system may have significant differences from other bacterial pathogens that present invariant surfaces to the host.

Innate immunity

The innate immune response to bacteria occurs through the nonspecific activation of many different types of cells, including polymorphonuclear leucocytes (PMNs), monocytes and macrophages, dendritic cells (DCs), and natural killer (NK) cells. These cells can be activated directly by interacting with a pathogen, or indirectly by receiving secondary signals from infected cells. In order to understand how gonococci and meningococci stimulate the innate immune system, a brief review of the innate immune system as it relates to neisserial pathogenesis is warranted.

Anatomical barriers to infections

All metazoans have mucosal epithelia, and together with the skin, these cells are the main interface between the host and the microbial world. The mucosal surface is the initial site of colonization by pathogenic *Neisseria*. Newly acquired bacteria must first compete with commensal organisms for sites of attachment on mucosal cells. Commensal microbes, such as lactobacilli, compete with exogenous microbes by: using nutrients; establishing and maintaining a low pH; generating antimicrobial bacteriocins, peroxides, and organic acids; and blocking attachments sites (Boris and Barbés, 2000). Mucosal cells are also bathed by fluids that have multiple roles: they protect the mucosa from adhesion by microorganisms; they contain antimicrobial substances; they provide a nutrient source; and they produce a matrix on which bacteria can proliferate. For example, the mucosal surface of the female reproductive tract is bathed with vaginal fluid, which is selectively antimicrobial due to the presence of lactic acid, antimicrobial proteins such as lysozyme, lactoferrin, secretory leukoprotease inhibitor (SLPI), calprotectin, human α-defensins human neutrophil peptides 1 through 3 (HNP1–3) and the β-defensins 1 and 2 (Valore *et al.*, 2002). Newly acquired bacteria must overcome all of these barriers in order to successfully establish colonization on the mucosal surface (see Chapter 6). Since the hallmark of gonococcal disease symptoms (pain, swelling, purulent exudates) can result solely from the interactions on the mucosal surface, colonization appears to be sufficient to initiate inflammation without dissemination of the pathogen. However, not all colonization events elicit symptoms (i.e. asymptomatic carriage of *N. meningitidis*). Gonococcal infections in women tend to be asymptomatic, with up to 70% of infected women experiencing no symptoms (Roy *et al.*, 2005). Hence, one challenge in the field of neisserial pathogenesis is to understand how pathogenic *Neisseria* produce this differential signalling across mucosal surfaces to yield some infections that elicit symptoms and others that do not. Other challenges include correlating these signalling events to disease and developing molecular explanations to explain why symptomatic infections do not produce significant levels of antibody (see Chapter 10).

Cellular barriers to infection

As part of the inflammatory response, PMNs and monocytes/macrophages are recruited to the site of infection. These cells are the main line of defence in the innate immune response. The role of PMNs in neisserial infections is to phagocytose and kill invading organisms. *Neisseria* strains appear to have evolved mechanisms to avoid phagocytosis and/or phagocytic killing (Bjerknes *et al.*, 1995; Estabrook *et al.*, 1992; Knepper *et al.*, 1997; Porter *et al.*, 2005; Rest and Shafer, 1989; Simons *et al.*, 2005; Virji and Heckels, 1986). However, as part of their activation/avoidance of the innate immune response, collateral tissue damage can occur. This host tissue damage seems to result from the activation of various cytokine cascades.

PMNs recognize pathogens through a variety of pattern recognition receptors (PRRs) which are capable of recognizing bacterial pathogen-associated molecular patterns (PAMPs). PAMPs are often components of the cell wall, such as lipopolysaccharide, peptidoglycan, lipoteichoic acids and cell-wall lipoproteins. In most bacterial species, the PAMPs of a given strain are invariant. In *Neisseria* infections, the structure of most of these PAMPS can change at a high frequency. Hence, another challenge to neisserial researchers is to determine how changes in PAMPs effect changes in innate immune recognition of infecting *Neisseria*.

Adaptive immunity, mediated by T and B lymphocytes, provides a highly specific immune response to pathogens and contributes to cellular memory of a microbial exposure. This allows the host to resist repeated infections by a particular pathogen. A functional adaptive immune response requires that the cells of the innate immune response communicate efficiently and effectively with T and B lymphocytes. The effects of the adaptive immune response on neisserial infection are not well-understood, and differ significantly between *N. gonorrhoeae* and *N. meningitidis*. Consistent with the absence of protective immunity resulting from previous infection with *N. gonorrhoeae*, the genital mucosal immune response in human gonorrhoea is weak: only low levels of immunoglobulin G (IgG) and immunoglobulin A (IgA) antibodies are detectable against gonococci (Russell *et al.*, 1999). This indicates that communication between the innate and adaptive immune systems is disassociated in gonococcal infections. This is contrasted with meningococcal infections where individuals who recover from infection establish long-lived immunity to reinfection. Since gonococci and meningococci both possess similar surface molecules, an ongoing challenge in the field is to explain why protective antibodies raised against meningococcal structures fail to protect against gonococcal infection (see Chapter 10).

Non-cellular barriers to infection

While anatomical barriers are somewhat effective in preventing invasion into tissues by bacteria, the pathogenic *Neisseria* can damage these barriers as part of the colonization process, or direct their entry into the cells that make up this barrier (See Chapter 6). Once bacteria have penetrated into epithelial tissues, they are exposed to humoral factors. Some of the non-cellular barriers to infection found in serum or at the site of infection that are important in neisserial disease are complement (reviewed in Chapter 8), lactoferrin and transferrin, and a variety of antimicrobial peptides (Tosi, 2005). In addition, sera may also contain antibodies specific for invading microorganisms. The *Neisseria* have evolved unique mechanisms to evade killing or to overcome the growth inhibitory properties of each of these factors (Cooper *et al.*, 1984; Lin *et al.*, 1997; McKenna *et al.*, 1988;

Schneider *et al.*, 2007; Schryvers and Morris, 1988; Vitovski *et al.*, 1999). Bacterial viability is required for gonococcal entry into host cells (Bish *et al.*, 2008), yet dead cells are able to induce cytokine expression by human monocytes (Patrone *et al.*, 2006). It is not clear if humoral components directly kill invading *Neisseria*, or if they simply limit the ability of the organism to grow.

Mucosal tissues and other cells of the innate immune response can recognize when bacterial pathogens are present. Various signalling cascades become activated and can result in the induction of the expression of a variety of cytokines. For a recent review, see Medzhitov (2007). Cytokines, often called interleukins (IL), are proteins which are produced by cells involved in host innate and adaptive immune responses. Production of these proteins occurs quickly in response to pathogens, serving to elicit an effective host response. Therefore, cytokine production is critical for efficient microbial clearance. These proteins can act in concert or alternately, one cytokine may have many functions. They can also promote the synthesis of other cytokines, thus initiating various signalling cascades. Finally, cytokines can act both locally and systemically. The lack of animal model systems in gonococcal disease has hampered our development of a complete understanding of the interplay between cells and components of the innate immune response, because studies must be performed *ex vivo*.

Cytokine activation is an important event in the pathogenesis of meningococcal disease, and has considerable bearing on the clinical outcome. A fine balance must be achieved, because limited activation will result in an inadequate immune response, whereas excess activation can be extremely destructive (Casey, 2000). During the acute phase of meningococcal infection, plasma concentrations of TNF-α, IL-1, IL-6, and IL-10 are significantly elevated, and high concentrations of these cytokines are associated with an increased risk of mortality (Riordan *et al.*, 1996). The relative roles of proinflammatory and anti-inflammatory cytokines are not clearly defined. Current research is directed at elucidating genetic determinants in both the host and the pathogen that contribute to variation in the cytokine response.

The pathogenic *Neisseria* are specifically adapted to life in their human host. A fundamental requirement for bacterial survival is the avoidance of complement-mediated bacteriolysis (reviewed in Chapter 8). Persons deficient in terminal complement components are predisposed to recurrent neisserial infections (Ross and Densen, 1984). The complement cascade comprises the classical, lectin, and alternative pathways. All three pathways converge at the level of C3b deposition. C3b is an integral component of both the classical and alternative pathway C5 convertases (C4bC3b,C2a and C3bC3b,Bb, respectively), which cleave C5 to initiate membrane attack complex (C5b-9) formation. While activation of complement on a microbial surface is beneficial in combating infection, unimpeded activation causes excessive inflammation and results in detrimental effects on host tissues. Therefore, activation of the complement system is tightly controlled (see Walport 2001 for recent review). The *Neisseria* employ several, often redundant, mechanisms to escape killing by complement. Both *N. gonorrhoeae* and *N. meningitidis* bind factor H and C4bp, which results in the inhibition of complement killing or enhanced serum resistance (Jarva *et al.*, 2005; Ram *et al.*, 2003; Ram *et al.*, 1999; Ram *et al.*, 1998). It is probable that further regulators are co-opted by the bacterium, and identifying the sites where they bind might identify antigens that could be included in much-needed vaccines to prevent disease caused by these important human pathogens.

Other considerations

The *Neisseria* can engage a variety of receptors found on macrophages and other cells to facilitate phagocytosis and/or signalling, providing the trigger for adaptive immune responses. Innate immune receptors play a key role in the recognition of neisserial pathogens and initiate the crucial innate immune response. These receptors include scavenger receptors, C-type lectins, integrins, the NOD (nucleotide-binding oligomerization domain) proteins NOD1 and NOD2, and Toll-like receptors (TLRs). In humans, there are at least 10 TLR proteins that recognize a broad range of microbe-derived ligands (Takeda *et al.*, 2003) and various neisserial products have been shown to activate some TLRs. TLR2 recognizes lipoproteins and lipid-modified sugars (i.e. lipoarabinomannan) from bacterial and fungal cell walls. TLR4 is primarily responsible for inflammatory responses to bacterial lipopolysaccharide (LPS) (Fitzgerald *et al.*, 2004). TLR5 recognizes the bacterial protein flagellin (Hayashi *et al.*, 2001). Still other TLRs recognize nucleic acids. For example, TLR9 recognizes CpG motifs in bacterial DNA (Hemmi *et al.*, 2000). Hence, an individual bacterium is likely to be recognized by multiple TLRs simultaneously. A Gram-negative bacterium such as *N. gonorrhoeae* or *N. meningitidis* will be recognized by TLR4 through its lipooligosaccharide (LOS). Lipoproteins expressed on the cell surface will signal through TLR2.

Epithelial cells derived from normal human vagina, ectocervix, and endocervix expressed mRNA for TLR1, -2, -3, -5, and -6, but do not express TLR4 or MD2. Consistent with this, endocervical epithelial cells are unresponsive to protein-free preparations of LOS isolated from the gonococcus, demonstrating that the response to *N. gonorrhoeae* at the mucosal surface of the female genital tract occurs in the absence of endotoxin recognition and TLR4-mediated signalling (Fichorova *et al.*, 2002). Massari and coworkers have demonstrated that immune stimulation by neisserial porins is TLR 2 and MyD88 dependent (Massari *et al.*, 2002), indicating that Por has significant immunostimulatory properties. PorB has also been shown to induce the costimulatory capacity of murine dendritic cells (DCs) and stimulate B-cell proliferation *in vitro* as well as enhancing the humoral response to meningococcal capsule *in vivo* (Mackinnon *et al.*, 1999; Massari *et al.*, 2002; Singleton *et al.*, 2005). PorB has also been shown to have adjuvant activity which is dependent on stimulation through TLR2 *in vitro* (Latz *et al.*, 2004).

TLRs initiate signalling through a cytoplasmic domain called the Toll/IL-1 receptor (TIR) domain (for a recent review of TLR signalling, see Strober and coworkers (Strober *et al.*, 2006)). TIR domains interact with other TIR domains. TLRs can also interact with each other as homo- or heterodimers, changing the nature of the activation response. Furthermore, TLR signalling adapter molecules (myeloid differentiation factor 88 (MyD88)), TIR domain-containing adapter protein (TIRAP), TIR domain-containing adapter-inducing interferon-β (TRIF), and TRIF-related adapter molecule (TRAM)) also contain TIR domains and can engage TLRs. The use of these adapter molecules varies between TLRs, and as a result, the inflammatory consequences of TLR signalling can depend on the specific TLR that is activated. *Neisseria* that contain a lipoprotein or Por will activate TLR2 on macrophages, via TIRAP and MyD88, triggering a signalling pathway culminating in activation of NF-κB and production of cytokines such as tumour necrosis factor-α (TNF-α). The Lip lipoprotein stimulates cytokine release and NF-κB activation in epithelial cells in a TLR2-dependent manner (Fisette *et al.*, 2003). LOS stimulation of TLR4 also leads to the activation of NF-κB via TIRAP and MyD88 but additionally activates the transcription factor interferon (IFN) regulatory factor-3 (IRF3), leading to the production of type I interferons. If the organism gains entry into the cytoplasm, peptidoglycan fragments could signal through NOD, resulting in the synergistic activation with TLR signalling. This activation is further complicated by the recent discoveries that NOD receptors recognize organisms that enter the cytoplasm of the cell or detect products that may be released or transported into the cytoplasm by processes such as phagocytosis and degradation of microbes. Thus, the inflammatory response activated by TLR stimulation may be defined not only by which TLR is activated but also by which combination of TLRs is activated, and in what temporal order (Strober *et al.*, 2006).

Because most experimental methodologies are not sensitive enough to detect events in single cells, most researchers have used very high multiplicities of infection (MOI) in order to generate reproducible experimental data. However, high MOIs really do not reflect natural neisserial infections. Hence the reliance on high MOIs to generate experimental data has the potential to generate experimenter-derived phenotypes that do not happen *in vivo*. This was clearly demonstrated by Patrone and coworkers (Patrone *et al.*, 2006) who showed that alterations in bacterial challenge dose caused a differential human monocytic cytokine response. Hence it is important to recognize that induction of a signalling cascade *in vitro* does not always correlate with *in vivo* events.

Gender and disease

Gonococcal disease severity and symptoms differ based upon host gender. As published data are limited to male clinical trials and tissue culture models of infection, factors affecting colonization of women (uterine pH, hormonal cycles, vaginal secretions, normal flora, etc.) and their impact on infectious dose and subsequent disease sequelae have not been fully explored. Specific physiological differences between genders have been shown to contribute to the complicated nature of gonococcal disease. Cervical epithelia in women express complement receptor type 3 (CR3) (Edwards *et al.*, 2001). This receptor is not present on male urethral tissue Since the neisserial outer membrane proteins porin and pilin have both been shown to bind CR3 (Edwards *et al.*, 2002), it suggests that women should be more susceptible to gonococcal infections.

The complement system is a vital component of innate immunity, playing a crucial part in host defence (See Chapter 8 for details). Complement is also an important effector for adaptive immunity because complement activation modulates specific responses to foreign antigens. The female genital tract is a source of alternative pathway (but not classical) complement components (Price and Boettcher, 1979; Vanderpuye *et al.*, 1992). This suggests that women should be less susceptible to gonococcal infections. A molecular explanation of these two conflicting conclusions is lacking.

After exposure to *N. gonorrhoeae*, the urethras of most men resist infection (Holmes *et al.*, 1970), and in patients with gonococcal urethritis the bladder and kidney are never infected. This natural resistance of the male urethra to gonococci has not been explained by classical immune mechanisms but could be explained by antibacterial properties of urine. Normal urine contains only trivial amounts of immunoglobulins and complement (Turner *et al.*, 1970), but it contains solutes, which might kill gonococci and thereby explain natural resistance. McCutchan and coworkers (McCutchan *et al.*, 1977) demonstrated that the major killing factor in urine was its acidic pH, but that high concentrations of urea and NaCl were also bactericidal. Several studies have also indicated that antagonistic interactions between components of the normal flora or other potential pathogens may play a role in host resistance to infection. Hipp and associates reported the antagonism of colonization with *N. gonorrhoeae* by *Candida albicans* (Hipp *et al.*, 1974). Inhibitory lactobacilli are more prevalent not only in all uninfected women but also in women who did not become infected following known exposure to infected consorts. These data suggest that lactobacilli may play an active role in resistance to gonococcal infection of the endocervix (Saigh *et al.*, 1978). Subsequent research has shown that this protective effect is due to the lowering of vaginal pH by commensal organisms (Redondo-Lopez *et al.*, 1990).

With respect to gender and meningococcal disease, it is interesting to note that meningococcal carriage seems to be higher in men (Block *et al.*, 1999; Caugant *et al.*, 1994), but men do not suffer from an increased incidence of disease. An explanation of this observation at the molecular level is lacking and merits further study.

Immunostimulatory molecules

LOS

Lipopolysaccharide is a glycolipid component of the outer membrane of Gram-negative bacteria and exhibits the most potent immunostimulating activity (Ulevitch and Tobias, 1995). Trace amounts of LPS are able to activate the innate immune system, leading to the production of an array of proinflammatory mediators. Therefore, LPS is generally acknowledged to play a central role not only in eliciting inflammatory responses but also in causing septic shock during Gram-negative bacterial infection (Cohen, 2002). *Neisseria* LOS is the structural analogue to LPS, differing from LPS in that it lacks an oligosaccharide-repeat unit (Griffiss *et al.*, 1987).

LOS is composed of three major components: oligosaccharide chain extensions, a carbohydrate core region, and lipid A. Significant variability in the oligosaccharide and lipid A regions have been reported (Kahler and Stephens, 1998; Lee *et al.*, 1995; Rahman *et al.*, 2000; Tong *et al.*, 2002; Yamasaki *et al.*, 1994; Yamasaki *et al.*, 1991). Discovery of the genetic basis that underlies this heterogeneity has allowed for the construction of strains with invariant structures (Arking *et al.*, 2001; Burch *et al.*, 1997; Gotschlich, 1994; Kahler *et al.*, 1996; Petricoin III *et al.*, 1991; Stephens *et al.*, 1994). The oligosaccharide extensions of the LOS contain determinants which resemble human glycosphingolipid antigens that play a role in molecular mimicry (Mandrell and Apicella, 1993). Studies have also shown that the oligosaccharide region of the LOS can be involved in receptor–mediated interactions (Edwards and Apicella, 2002; Harvey *et al.*, 2000; John *et al.*, 2002; Song *et al.*, 2000). The importance of an intact LOS was clearly demonstrated by Post and coworkers (Post *et al.*, 2002), who demonstrated that the lipid A modification in the *N. gonorrhoeae msbB* mutant renders it more susceptible to killing when internalized by the urethral epithelial cells. Hence LOS is a critical factor in gonococcal survival in non-phagocytic cells.

The endotoxin structures of different Gram-negative bacteria influence the potency of these molecules during TLR4-mediated activation of human macrophages for pro-inflammatory cytokine release. The specific acylation pattern of the lipid A part largely determines the biological activity of LPS. The potency of meningococcal lipid A for activation of human macrophages and dendritic cells depends on the structure of lipid A, particularly the number of acyl chains and lipid A glycosylation (Zughaier *et al.*, 2006; Zughaier *et al.*, 2004). While TLR4 is the cell-signalling molecule responding to LOS, the accessory protein MD-2 is essential for TLR4 activation (Shimazu *et al.*, 1999). Zimmer and coworkers (Zimmer *et al.*, 2007), using a set of structurally defined variants on meningococcal lipid A, showed that the hierarchy in the binding affinity of different lipid A structures for human MD-2 directly correlated with differences in TLR4 pathway activation and cytokine production by human macrophages. Since Patrone and Stein (2007) showed that variation in the oligosaccharide portion of the molecule contributed very little to the overall signalling to monocytes, this clearly indicates that the biologically active portion of LOS, as it relates to signalling to phagocytic cells, resides in the lipid A region. However, it should not be construed that oligosaccharide variation is not important in innate immunity. For example, Gulati and coworkers (Gulati *et al.*, 2005) showed that enhanced factor H binding to sialylated gonococci is restricted to the sialylated lacto-*N*-neotetraose LOS, and that different oligosaccharide structures interact via a different mechanism with factor H.

The landmark construction of a strain of *N. meningitidis* that is completely devoid of LOS (Steeghs *et al.*, 1998) has made it possible to parse out the contributions of neisserial LOS from other neisserial surface components in stimulating innate immune responses. A completely LOS-deficient but viable mutant was obtained by insertional inactivation of the *lpxA* gene, encoding UDP-GlcNAc acyltransferase required for the first step of lipid A biosynthesis (Steeghs *et al.*, 1998). The expression and assembly of integral outer membrane proteins in the absence of LOS was largely unaffected. However, the expression of an iron limitation-inducible, cell surface-exposed lipoprotein is greatly reduced. Major changes were seen in the phospholipid composition, with a shift towards phosphatidylethanolamine and phosphatidylglycerol species containing mostly shorter chain, saturated fatty acids. The presence of the capsular polysaccharide turned out to be essential for viability without LOS. Human monocytes produce reduced amounts of pro-inflammatory cytokines in response to LPS-deficient meningococci as compared to wild-type organisms. However, even without LPS, a significant cytokine response was still observed, indicating that other neisserial components such as peptidoglycan, Por, lipoproteins, or CpG DNA could be acting in non-TLR4-mediated pathways (Van Der Ley and Steeghs, 2003).

Neisserial LOS initiates the production of host-derived inflammatory mediators, initiated by the interaction of LOS with TLR4. Immediately distal to TLR4 activation are two intracellular cascades that regulate signal transduction processes, gene expression, and production of proinflammatory mediators (Akira *et al.*, 2001). One pathway uses a MyD88-dependent cascade while the other utilizes TRIF. The MyD88-dependent pathway leads to up-regulation of cytokines and chemokines such as TNF-α, IL-1, IL-6, and monocyte chemoattractant protein 1 (MCP-1), whereas the TRIF-dependent pathway leads to the production of IFN-γ, which in turn activates the STAT-1 pathway resulting in the production of mediators such as interferon-inducible protein 10 (IP-10) and nitric oxide (Karaghiosoff *et al.*, 2003). It has been shown that at low concentrations meningococcal LOS is a potent inducer of MyD88- and TRIF-dependent cytokines, whereas at the same picomole concentrations *E. coli* LPS induced comparable levels of TNF-α, IL-1β, and MIP-3α but significantly less IFN-β, nitric oxide, and IP-10 (Zughaier *et al.*, 2005). Since the number of the fatty acids of *N. meningitidis* are shorter compared with those of *E. coli*, it is possible that the differences in the nature and pattern of fatty acid substitution of enteric and meningococcal lipid A affects their biological properties.

Meningococcal strains that possess defects in KDO biosynthesis express a lipid A with significantly reduced bioactivities compared to strains expressing wild-type LOS (Zughaier *et al.*, 2004). Removal of the KDO moieties of wild type LOS by mild acetic acid treatment also attenuated cellular responses. Dendritic cells stimulated with KDO_2-lipid A from meningococci but not lipid A alone stimulated naive allogeneic $CD4^+$ cells to secrete enhanced levels of IFN-γ relative to T-cells primed with immature dendritic cells (Zughaier *et al.*, 2006).

Gonococci isolated from urethral exudates are resistant to complement mediated killing by human serum but in most cases the resistance is lost on one subculture *in vitro*. Investigation of this unstable type of resistance showed that it was due to sialylation of LOS by host cytidine 5′-monophospho-N-acetyl neuraminic acid (CMP-NANA) (Smith *et al.*, 1995). Opsonophagocytosis by human neutrophils indicated that LOS sialylation results in resistance to killing by human sera (Estabrook *et al.*, 1992). It is clear that LOS sialylation occurs in both gonococci and meningococci and affects pathogenicity.

Porins

Porins (PorA and PorB) are the most abundant neisserial outer membrane proteins (Judd *et al.*, 1987). Neisserial porins are essential for bacterial survival as they modulate the exchange of ions between the bacteria and the surrounding environment. They form aqueous transmembrane channels (Haines *et al.*, 1988), and the insertion process leads to the formation of a functional channel. Because they can insert into mammalian cell membranes (Weel *et al.*, 1991), this insertion causes a transient change in membrane potential and interference with host-cell signalling.

Porins have a variety of non-specific effects on human neutrophils, such as the inhibition of actin polymerization, degranulation, opsonin receptor expression, phagocytosis of opsonized neisserial organisms and neutrophil–phagolysosome fusion (Bjerknes *et al.*, 1995). Another important effect of neisserial porins is activation of B-cells. One of its best described activities is its effect as an adjuvant, eliciting B lymphocyte proliferation *in vitro* (Wetzler *et al.*, 1996), enhancing Ab secretion and amplifying the immune response (Snapper *et al.*, 1997). The mechanism of PorB adjuvanticity correlates with its ability to up-regulate the expression of the costimulatory molecule CD86 on the surface of phagocytes (Wetzler *et al.*, 1996). Among the variety of intracellular signal transduction events elicited by PorB, NF-κB nuclear translocation occurs but is abrogated in TLR2 and MyD88 knockout mice (Massari *et al.*, 2002). While TLR2 is the primary mediator for PorB binding, this interaction appears to be maximized in the presence of TLR1 (Massari *et al.*, 2006).

Porins also influence the development of serum resistance, enabling the bacteria to evade complement-mediated killing. The gonococcal porin PorB1A is critical in modulating stable serum resistance, which is not mediated by LOS sialylation or by binding to factor H (the main regulatory protein of the alternative pathway of complement) (Ram *et al.*, 1998). The effect of neisserial porins on complement activity could be another effect of these proteins on neisserial pathogenesis. Preventing interaction between factor H or C4b with gonococcal porins results in enhanced bactericidal killing. The ability of certain gonococcal strains to down-regulate complement early in the cascade (prior to and at the level of C3 deposition) might be partly responsible for milder symptoms seen in individuals infected with such strains (Rice and Goldenberg, 1981).

Gonococcal PorB elicits a variety of effects on human cell physiology and signalling. These include disruption of PMN degranulation (Haines *et al.*, 1991) and phagosome maturation (Mosleh *et al.*, 1998), and induction of apoptotic death in epithelial cells and phagocytes (See Chapter 9) (Muller *et al.,* 1999). In addition, the PorB encoded by the PIA allele, has been implicated in a unique invasion mechanism exhibited in human Chang epithelial cells under conditions of low phosphate (Van Putten *et al.,* 1998a). These *in vitro* effects observed would presumably have important consequences *in vivo.*

An analogous study using *N. meningitidis* PorB came to the conclusion that PorB protects cells from apoptosis (Massari *et al.,* 2000). There are several explanations for the contrasting effects of *N. gonorrhoeae* PorB1B and *N. meningitidis* PorB on susceptibility to apoptosis. These include actual differences in the two porins, differences in the purification of the porins, differences in the cell culture conditions used, or intrinsic differences in the cell lines used. However, given that subtle variations in experimental protocols can lead to such dramatic differences in the conclusions of the work are cause for pause as we attempt to ascribe functions to neisserial surface molecules.

Pili

Pili are composed of a major structural subunit, pilin (PilE protein), that assembles into a helical pilus fibre (Parge *et al.,* 1990). Pioneering studies by Kellogg and coworkers demonstrated that a loss of virulence was observed when gonococci lost their ability to express pili (Kellogg *et al.,* 1963). Subsequent work has shown that pili are necessary for initial attachment of these pathogens to epithelial and endothelial cell monolayers (Punsalang and Sawyer, 1973; Stephens and McGee, 1981; Swanson, 1973), twitching motility (Merz *et al.,* 2000), and natural competence (Sparling, 1966). The pilin subunit undergoes antigenic variation (Meyer *et al.,* 1982) and can be post-translationally modified (Stimson *et al.,* 1995). The protein PilC was localized to the pilus tip and is the major pilus adhesion molecule (Rudel *et al.,* 1995). PilC resides in the bacterial outer membrane (Rahman *et al.,* 1997) and has additional functions in pilus biogenesis and natural transformation competence (Jonsson *et al.,* 1991). In addition, purified PilC protein binds to host cells and can block pilus-mediated adherence of both *N. gonorrhoeae* and *N. meningitidis* (Scheuerpflug *et al.,* 1999).

Infection by pathogenic *Neisseria* spp. is initiated by interaction between the type IV pili and cellular receptors including CD46 (Kallstrom *et al.,* 2001), the I-domain containing integrins (Edwards and Apicella, 2005), and the complement regulatory protein 3 (Edwards *et al.,* 2001). Since CD46 is involved in the induction the T-regulatory 1 (Tr1) phenotype of human CD4$^+$ T-cells (Kemper *et al.,* 2003), Plant and Jonsson examined the capacity of gonococcal strains with mutations in pilus components to modulate the activity of primary human CD4$^+$ T lymphocytes (Plant and Jonsson, 2006). Their data indicate that bacterial interaction induce the activation and proliferation of cells and that proliferation can be observed with purified pili alone. Gonococcal infection of T-cells also stimulated the production of IL-10. These findings provide further evidence that gonococcal pili is an important cellular component modulating T-cell activity and show that the gonococcal pilus is involved in the induction of signalling events leading to modulation of the CD4$^+$ T-cell response.

Opa proteins

The pathogenic *Neisseria* express a family of phase-variable and antigenically distinct Opa proteins (Swanson, 1978) that allow bacterial attachment to virtually every cell type encountered during infection. Opa proteins rapidly change their structures, due to frequent interchromosomal recombination between the conserved regions of *opa* genes (Connell *et al.,* 1988). Some Opa variants bind cell surface-expressed heparan sulphate proteoglycans (HSPG), including members of the syndecan family of receptors, and extracellular matrix proteins such as fibronectin and vitronectin. Other variants bind members of the family of carcinoembryonic antigen cellular adhesion molecules (CEACAMs). Despite such surface sequence variation, distinct Opa proteins can bind to a common site on human CEACAMs allowing for intimate adhesion to and invasion of host cells. Both the gonococcus and the meningococcus are efficiently engulfed by human neutrophils via an opsonin independent process that depends on Opa binding to neutrophil-expressed CEACAMs (Belland *et al.,* 1992). While this interaction leads to an oxidative burst that kills the engulfed bacteria, the organism must derive some advantage to compensate for this detrimental effect (Gray-Owen *et al.*, 1997a; Gray-Owen *et al.*, 1997b), as it continues to express these proteins.

The Opa phenotype of gonococci isolated from urine and urethral swab cultures from experimentally infected subjects has been determined and an increase in the proportion of isolates expressing multiple Opa proteins occurred over time, demonstrating that there is strong selection for expression of one or more Opa proteins during infection in the male urethra (Jerse *et al.*, 1994). The observed selection for and/or induction of gonococcal Opa protein expression during female murine infection model appears to parallel events that occur during experimental urethritis (Jerse, 1999), indicating that Opa-expressing gonococci have a distinct evolutionary advantage, despite their ability to facilitate interactions with innate defence molecules.

CEACAMs occur on a variety of human cell types with CEACAM1 having the widest distribution. The receptors are often present at low density on cellular surfaces but their expression can be augmented under the influence of proinflammatory cytokines (Dansky-Ullmann *et al.*, 1995). Given these observations, and because Opa proteins are commonly expressed in disease isolates, consequences of Opa–CEACAM interactions in post-inflammation state of target cells may be crucial in determining host susceptibility to infection. High receptor densities also encourages a more rapid association of bacteria, especially if synergized by pili as observed by Rowe and coworkers (Rowe *et al.*, 2007).

CEACAM contains an immunoreceptor tyrosine-based inhibitory motif (ITIM). Ligation of this ITIM-containing receptor causes a Src kinase-dependent phosphorylation of the ITIM and recruitment of phosphotyrosine-specific and/or phosphoinositol- specific phosphatases down regulating signals derived from the T-cell receptor (TCR) (Taylor *et al.*, 2000). Since gonococcal expression of CEACAM1-specific Opas suppresses the activation of $CD4^+$ lymphocytes, such a phenomenon would help to explain why cytokine and antibody responses to gonococcal infection are unexpectedly low and lack signs of developing immunological memory (Hedges *et al.*, 1999; Hedges *et al.*, 1998). When gonococci interact with CEACAM1 on human B-cells, they inhibit antibody production and this interaction leads to the induction of cell death (Pantelic *et al.*, 2005).

Meningococcal Opa and Opc become effective adhesins only in acapsulate bacteria. As acapsulate meningococci are highly susceptible to serum bactericidal activity following blood infiltration, capsule up-regulation or selection of capsulate phase variants would be required for survival (Virji *et al.*, 1996). These observations demonstrate the significant interplay between Opa and the other neisserial surface antigens, and compound our understanding of the immunological contributions of Opa in disease.

IgA protease

IgA1 protease cleaves human IgA1 both in serum and at mucosal surfaces of the respiratory and urogenital tracts (Plaut *et al.*, 1975). An intact IgA1 protects against microorganisms infecting mucosal tissues by prevention of microbial adherence, toxin neutralization, and activation of Fc-g receptor-dependent phagocytosis. Hence the cleavage of IgA1 should impair IgA1-mediated effector functions (Kilian *et al.*, 1996). Lorenzen and coworkers demonstrated that IgA1 protease exhibits important immunostimulatory properties and may contribute substantially to the pathogenesis of neisserial infections by inducing large amounts of TNF-α and other proinflammatory cytokines. The stimulation of cytokine synthesis by IgA1 protease occurred in the nanomolar range. The rapid cytokine induction within 24 h indicates a role of monocytes/macrophages in this process. However, compared with peripheral blood mononuclear cells, significantly lower amounts of TNF-α and IL-8 and practically no IL-6 and IL-1β were elicited by purified monocytes in response to IgA1 protease, whereas LPS induced a strong cytokine response with secretions that included IL-6 and IL-1β. These results suggest that the presence of lymphocytes is required for full cytokine induction by IgA1 protease (Lorenzen *et al.*, 1999). Besides the cleavage of IgA1, the protease also cleaves LAMP-1, a lysosomal protein associated with the endosomal maturation (Hauck and Meyer, 1997). IgA1 protease cleavage of LAMP-1 promotes intracellular survival of pathogenic *Neisseria* spp. (Lin *et al.*, 1997) and their transcytosis across polarized epithelia (Hopper *et al.*, 2000), though contrasting results also exist (Hopper *et al.*, 2000).

Other outer membrane proteins

The H.8 antigen is named after a monoclonal antibody (mAb) that recognizes a surface epitope that is common to pathogenic *Neisseria* species including *N. gonorrhoeae* and *N. meningitidis* but absent from other commensal *Neisseria* species (Cannon *et al.*, 1984). This mAb is capable of recognizing several neisserial lipoproteins, including Lip and Azu (Cannon, 1989). Although the size and amino acid sequence of Lip can differ between *Neisseria* species and strains (Trees *et al.*, 2000), it primarily consists of 11–20 pentameric AAEAP amino acid repeats and a lipid-modified N-terminal cysteine residue and lacks any aromatic residues. Cannon and coworkers (Kawula *et al.*, 1987) reported the cloning and sequencing of a meningococcal gene (*laz*) encoding an H.8 mAb-binding protein with a consensus lipoprotein processing site, an N-terminal domain containing the epitope for H.8 mAb binding, and a C-terminal domain with extensive similarity to the sequences of azurins from other organisms. This protein was present as a single copy in the chromosome, was highly conserved in gonococci and meningococci, and was less conserved in commensal *Neisseria* species. The lipid-modified azurin-like protein contains imperfect AAEAP motifs at the N-terminal domain of the mature protein.

Fisette and coworkers (Fisette *et al.*, 2003) purified Lip from the outer membrane of *N. gonorrhoeae* and demonstrated its ability to stimulate the production of proinflammatory mediators from epithelial cells in a TLR2-dependent manner. Although the nature of the lipid modification of Lip is not precisely known, it can be modified with palmitic acid or oleic acid (Hoehn and Clark, 1992) demonstrating its lipoprotein nature. Lip is able to activate epithelial cells and trigger the release of cytokines and chemokines in a TLR2-dependent manner. The contributions of Lip to immunostimulation are unclear, because inactivation of Lip leads to changes in the expression of other outer membrane proteins. Although the role of H.8/Lip in the pathogenesis of *Neisseria* infections has not been fully elucidated, sera from patients with gonococcal infection contain antibodies that react with the H.8 antigen (Cannon, 1989). Furthermore, the distribution of the H.8 antigen in primarily pathogenic species of *Neisseria* hints at a possible role in infection, survival, and/or disease progression.

Capsules

The meningococcal capsule plays contrasting roles in disease pathogenesis. Its antiphagocytic properties are essential for bacterial growth in the bloodstream, a prerequisite for meningococcal septicaemia and meningitis (Jarvis and Vedros, 1987). However, the expression of the capsular polysaccharide inhibits the colonization and invasion of the nasopharyngeal barrier by masking the meningococcal adhesins/invasins (Hammerschmidt *et al.*, 1996a). Capsular gene expression is regulated both by phase variation via slipped-strand mispairing or reversible insertion of mobile elements (Hammerschmidt *et al.*, 1996a; Hammerschmidt *et al.*, 1996b) and at the transcriptional level. Capsule biosynthesis and assembly appear to be down-regulated during the early stages of the infectious cycle, and this facilitates the adhesion to and invasion of the host cells (Dietrich *et al.*, 2003). Expression of capsule is required for the intracellular survival of this microorganism in both human phagocytic and non-phagocytic cells. Capsule may enhance the resistance to cationic antimicrobial peptides (CAMPs). Spinosa and coworkers (Spinosa *et al.*, 2007), using a serogroup B strain, and its isogenic derivatives, which lack either the capsule or both the capsule and the lipooligosaccharide outer core, infected human phagocytic and non-phagocytic cells and monitor invasion and intracellular growth. Their data indicate that the capsule is required for the intracellular survival of this microorganism. *In vitro* assays suggested that an increased resistance to CAMPs was a possible mechanism by which the capsule protects meningococci in the intracellular environment.

Efflux pumps

CAMPs are constitutively expressed in macrophages and neutrophils and inducibly produced by epithelial cells at mucosal surfaces. Certain CAMPs hold effector functions that modulate expression of certain host genes important in host defence, thus providing communication between innate and adaptive immunity (Biragyn *et al.*, 2002; Bowdish *et al.*, 2004). Because bacterial pathogens constantly encounter CAMPs during infection, they have efflux pumps (see Chapter 13) and transport systems that export CAMPs from the periplasmic and intracellular compartments (Shafer *et al.*, 1998). Tzeng and coworkers (Tzeng *et al.*, 2005) demonstrated a critical contribution of both PEA modification of meningococcal lipid A and the Mtr efflux pump in the ability of meningococci to resist the action of CAMPs. Changes in outer membrane proteins such as PorB and outer membrane lipoprotein and alteration in pilin secretion or biogenesis also influence meningococcal susceptibility to CAMPs. A functional gonococcal efflux pump enhanced bacterial survival in the mouse model of genital tract infection (Jerse *et al.*, 2003), providing *in vivo* evidence of the importance of efflux pumps in neisserial disease.

Neisseria interactions with cells of the innate immune response

Neisserial interactions with PMNs

Researchers studying the fate of *N. gonorrhoeae* within PMNs have generated conflicting results. Differences in strains and experimental systems probably account for the discrepancies in conclusions regarding the relative importance of oxidants in killing the organism and in the ability of the organism to replicate within these cells. However, it is important to remember that in natural infections, a variety of clinical outcomes are observed, and the differences described in the research laboratory may in fact represent real differences, based on differential surface expression of gonococcal virulence determinants. In a typical gonococcal infection in men, PMNs are rapidly recruited to the site of infection. However, in the absence of treatment, disease persists for weeks to months. Given the number of PMNs recruited, and their failure to clear infection, these observations suggest that *N. gonorrhoeae* possesses mechanisms to evade killing by PMNs. Casey *et al.* (Casey *et al.*, 1986) demonstrated that piliated gonococci resisted intra-leucocytic killing by PMNs in both aerobic and anaerobic PMNs monolayers for as long as 165 min. Since anaerobic PMNs were as effective as aerobic PMNs in the intracellular killing of gonococci over extended periods of time, they concluded that an O_2-independent antimicrobial system played a significant role in the intra-leucocytic killing of gonococci. Since Virgi and Heckels (1986) demonstrated that Opa-negative bacteria showed better survival within PMNs, killing by PMNs may be associated with a specific interaction between Opa and a phagolysosome component.

Immunity to meningococcal infections depends on components of both the innate and adaptive branches of the immune system. Periods of nasopharyngeal colonization by commensal *Neisseria* strains (perhaps non-encapsulated variants of meningococci) elicit antibody responses, which may be protective against infections by different meningococcal strains due to cross-reactivity (Goldschneider *et al.*, 1969a, b). Meningococcal disease is associated with a severe systemic inflammatory response. Progression to multiple organ failure is characterized by shock, haemorrhagic skin lesions, and coagulopathy. Platelets and PMNs are involved in the evolution of this process, and symptoms occur via a variety of mechanisms: the acceleration of thrombin production; cytokine and chemokine production; and the release of reactive oxygen metabolites and proteases by phagocytes (Klein *et al.*, 1993; Zimmerman *et al.*, 1995).

Neisseria interactions with monocytes

Knepper *et al.* (1997) demonstrated that the monocyte response towards gonococci is highly dependent on the bacterial phenotype and differs from the neutrophil response. Both pili and Opa were able to direct interactions between the gonococcus and human monocytes and give rise to diverse biological responses. Pili and Opa are variably expressed and undergo structural variation during natural infections (Ghosh *et al.*, 2004; Jonsson *et al.*, 1994; Makepeace *et al.*, 2001; Schmidt *et al.*, 2001) generating a heterogeneous population of bacteria that differ in antigenic structures and functions. The fact that the gonococcus has evolved to variably express its surface constituents, and that these variations influence its interactions with monocytes indicates that this cell type is a potential target during gonococcal infection. During natural infections, gonococci appear to be primarily associated with PMNs (King *et al.*, 1978; Veale *et al.*, 1979). However, this is during the acute stage of infection. During prolonged gonococcal infection, cells of the monocyte/macrophage lineages are likely to be involved in the persistence and spread of the microorganisms. Study of gonococcal interactions with these cells has been largely ignored.

Neisseria interactions with lymphocytes

Boulton and Gray-Owen (2002) have demonstrated an interaction between a specific Opa (Opa_{52}) and CEACAM1 on $CD4^+$ T lymphocytes. This receptor contains an ITIM and provides a possible mechanism by which gonococci may suppress T-cell activation, thus evading the host immune response via Opa variation. However, CEACAM binding does not appear necessary for the selection of Opa^+ variants during the early stages of infection (Simms and Jerse, 2006). In addition, Song and coworkers (Song *et al.*, 2000) have demonstrated the Opa-independent invasion of polarized epithelial cells, indicating only a partial role for this molecule in gonococcal pathogenesis.

Though some vaccine studies indicate the potential for the production of antibodies against gonococcal surface molecules, individuals who have had gonococcal infections can be re-infected by the very same strain (Black *et al.*, 1985; Boslego *et al.*, 1991; Zhu *et al.*, 2005). This occurs in spite of the fact that low gonococcal-specific antibody titres can sometimes be measured in previously infected individuals (See Chapter 10) (Hedges *et al.*, 1999; Plummer *et al.*, 1993; Plummer *et al.*, 1994). This phenomenon indicates the lack of a protective memory response to gonococci. These immunological responses against gonococci can be contrasted with the immune response generated against meningococcal colonization, where protective immunity occurs after disease.

Neisseria interactions with dendritic cells

Dendritic cells (DCs) are key regulators of the immune response (Banchereau and Steinman, 1998). Immature DCs in the periphery and submucosa sample their environment and capture antigen. DCs can internalize microbes and microbial products, which are then processed for presentation to T-cells (Palucka and Banchereau, 1999). This process also induces surface co-stimulatory molecules and cytokine production by DCs (Rescigno *et al.*, 1999). The outcome of the interaction between DCs and lymphocyte is regulated by cytokines and chemokines and by the expression and function of co-stimulatory molecules. Hence, DCs provide a critical link between the innate and adaptive immune response with proper cross-talk between APC and T-cells needed to achieve an optimal adaptive immune response. Perturbations in how the signals are sent or received can have important and lasting consequences for the nature of adaptive immune response.

The immunoregulatory role of DCs is determined by the ligation of pattern recognition receptors (PRRs). This process results in the development of different DCs subsets that induce differentiation of naive T-cells into T helper 1 (T_H1) cells or T_H2 cells. The best studied PRR families on DCs are TLRs and C-type lectin receptors (CLR) that are specific for pathogen-derived carbohydrate structures. Triggering of TLR leads to the production of regulatory cytokines and up-regulation of major histocompatibility complex (MHC) and co-stimulatory molecules required for T-cell signalling (Schnare *et al.*, 2001), while ligation of C-type lectins has been demonstrated to result in binding and internalization of pathogens for antigen processing and presentation to T-cells (Engering *et al.*, 2002). *N. meningitidis* expressing a GlcNAc(β1–3)-Gal(β1–4)-Glc-R outer core targets the C-type lectin receptor DC-SIGN thereby mediating highly efficient bacterial uptake, and skewing of the immune response into a T_H1 direction (Steeghs *et al.*, 2006).

Wild-type killed meningococci induce both DC maturation and production of cytokines, particularly IL-10 and bioactive IL-12, whereas neither purified LOS nor LOS deficient meningococci, even with exogenously added LOS, are able to induce high levels of these cytokines (Dixon *et al.*, 2001). Optimal IL-12 and TNF-α production by human monocyte-derived DC in response to killed serogroup B *N. meningitidis* depends on physical contact and internalization of the bacteria by DCs (Uronen-Hansson *et al.*, 2004). Internalization of *N. meningitidis* is dependent on LOS, with LOS deficient bacteria internalized poorly compared to wild-type bacteria. Restoration of LOS biosynthesis in this LOS deficient strain by using an inducible *lpxA* gene restored both internalization and cytokine production and was enhanced in the presence of LPS binding protein (LBP). These results suggest that DC phagocytosis depends on expression of LOS within the bacteria and that optimal cytokine production, particularly IL-12, requires internalization of the bacteria. These findings have important implications for designing vaccines that will induce protective immune responses to group B *N. meningitidis*. Hence it is necessary to understand how the *Neisseria* interact with DCs in terms of induction of co-stimulation molecules and cytokines. The importance of DC function is exemplified by T_H1 differentiation in response to IL-12 production by DCs activated with bacterial products, such as LPS (Macatonia *et al.*, 1995). The majority of DC that produced cytokines had internalized *N. meningitidis* while inhibition of bacterial internalization markedly impaired IL-12 and TNF-α, but not IL-6 production.

The reason why wild-type meningococci are capable of inducing high levels of IL-10 and IL-12 appears to stem from the requirement of LPS within the outer membrane for efficient internalization by human DCs (Uronen-Hansson *et al.*, 2004). Live meningococci were incapable of inducing full DC maturation as compared with their killed counterparts (Jones *et al.*, 2007). Despite the lack of a typical maturation response, live meningococci could induce both IL-12 production and IL-10 by DCs, although the ratios of these two cytokines were quite different compared with the response to killed organisms. The results suggest that live meningococci may specifically modulate DC function which could have important influences on the nature of T-cell and B-cell responses that occur during natural infection/carriage and may have implications in the design of novel vaccines against this pathogen.

Gonococcal interactions at the mucosal surface

Our knowledge of how gonococci infect human cells is derived from tissue culture cell models that support gonococcal interactions *in vitro*, including primary human organ or cell cultures and transformed cell lines (Binnicker *et al.*, 2003; Edwards *et al.*, 2000; Fichorova *et al.*, 2001; Harvey *et al.*, 2002; Jerse *et al.*, 2002; Maisey *et al.*, 2003), the estradiol-treated mouse model (Jerse, 1999; Jerse *et al.*, 2002), and the clinical human challenge model (Jerse *et al.*, 1994; Schmidt *et al.*, 2001). From these studies, research has demonstrated that gonococcal infection of epithelial cells can occur in several stages – initial contact, intimate adhesion to the plasma membrane, invasion into the cell, intracellular survival and persistence, and transcytosis to the subepithelium or escape back into the extracellular environment. However, it is not clear which neisserial surface molecules are needed for each step in this process, and how each step relates to the clinical outcome of disease. Fig. 7.2 is a simplified representation of what is known about gonococcal-activated signalling in epithelial cells, and how this signalling is linked to disease. It is important to note that this signalling is mediated by specific receptor–ligand interactions. This differs from the non-specific interactions that occur between bacteria and cells of the innate immune system. Depending on the nature of the receptor–ligand interaction, different signalling cascades are activated. We have hypothesized that this differential activation modulates the dynamics and organization of the

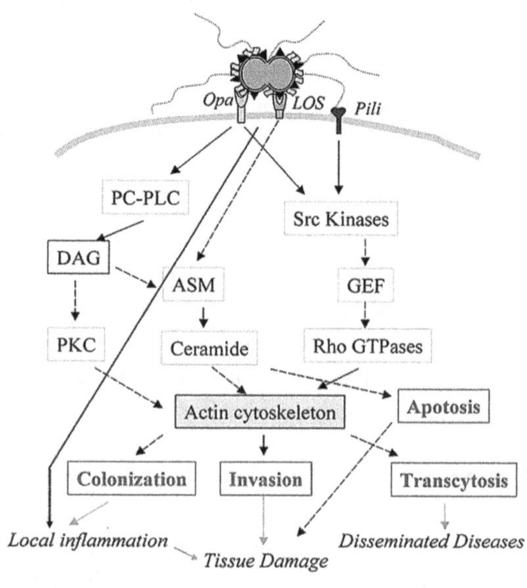

Figure 7.2 Potential signalling pathways and events that lead to disease.

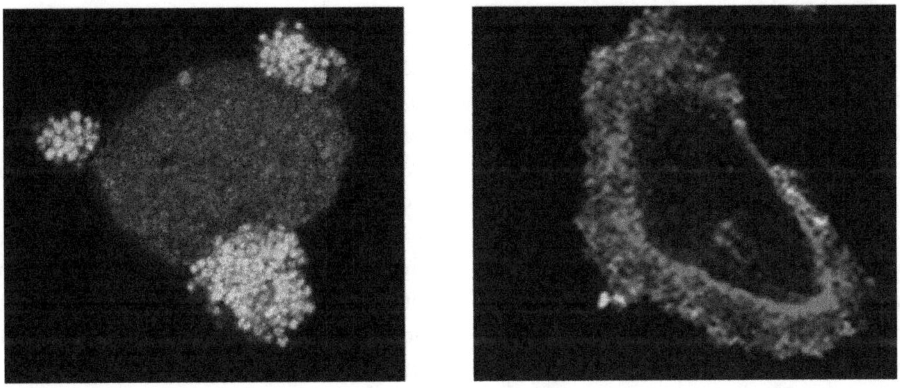

Figure 7.3 Cellular distribution of MS 11 variants on ME180 cells. Two variants of MS 11 differing in their Opa expression were allowed to interact with ME180 cells. In panel on the left, preparations were stained with SYTO green (a nucleic acid specific dye and Alexa fluor 546-labeled phalloidin. Cells appear as punctuate white spots on a grey background representing host cells.

actin cytoskeleton, with the outcome of infection determined by the sum-total of all the activation processes. Gonococci must overcome the innate immune responses of the host tissues and initiate signalling pathways in epithelial cells to orchestrate continued progression of the infection. A successful gonococcal infection occurs when gonococci persist within the host at a level where they do not elicit a profound proinflammatory response, yet are still transmitted readily within the population. This is somewhat different for meningococcal infections where disease production is a rare event, relative to the frequency of colonization.

Several aspects of host cell signalling in response to gonococcal contact have been examined. Gonococcal pili have been shown to bind CD46, inducing Ca^{2+} release from the intracellular stores of epithelial cells (Kallstrom *et al.*, 1998). Specific Opa proteins can bind HSPG and CEACAM, initiating internalization into epithelial cells. HSPG binding and uptake of gonococci has been shown to involve either phospholipase C and acid sphingomyelinase or the serum factors vitronectin or fibronectin (which can bind Opa_{CEA} and mediate bacterial internalization via integrin receptors and protein kinase C) (Duensing and van Putten, 1997; Gomez-Duarte *et al.*, 1997; Grassme *et al.*, 1997; van Putten *et al.*, 1998b). Gonococcal Opa-CEACAM binding in phagocytes leads to activation of Src family kinases and downstream signalling through the stress response kinases: p21-activated protein kinase (PAK) and Jun N-terminal kinase (JNK) (Hauck *et al.*, 1998). Neisserial PorB has been shown to interfere with phagocytosis by neutrophils and induce apoptotic signalling cascades in epithelial cells (Bjerknes *et al.*, 1995; Muller *et al.*, 1999). Several signalling pathways activated by *N. gonorrhoeae* lead to activation of the transcription factors NF-κB (p50/p65 heterodimer) and activator protein 1 (AP-1) (Muenzner *et al.*, 2001; Naumann *et al.*, 1997, 1998). Though TLR4/MD2 is important in recognition of some serotypes of neisserial LOS, activation of NF-κB during gonococcal challenge also occurs via TLR2 through other gonococcal surface structures (Fichorova *et al.*, 2002; Fisette *et al.*, 2003; Muenzner *et al.*, 2002; Pridmore *et al.*, 2003).

Epithelial cells connected by tight junctions provide an integral barrier function against invading microbes. These cells can produce a variety of antimicrobials [for review see Wira *et al.* (2005)]. The mucosal tissue of the urogenital tract is the first line of defence against *N. gonorrhoeae*. Upon transmission to a new host, gonococci encounter mucosal epithelial cells, which provide protection for underlying tissues and allow secretion and absorption exchanges with the external environment. They can be ciliated or non-ciliated, simple or stratified (multiple layers), and are classified into three general shape categories – columnar, cuboidal, and squamous. Gonococci seem to preferentially colonize non-ciliated, columnar or cuboidal mucosal epithelial cells, such as those found in the female cervix, endometrium, and fallopian tubes (Martini, 2001). This diversity in cell type at the site of infection provides further signalling diversity, as some cells will contain one set of receptors, capable of interacting with one set of neisserial components, and adjacent cells will be resistant to colonization, or will be engaged by different receptor–ligand interactions. In fact, this cellular tropism was demonstrated very elegantly many years ago by Stephens and coworkers (Stephens *et al.*, 1982), where they showed that attachment of gonococci is important, not only to secure the organism in the host, but also to initiate the disease process. They observed gonococci attaching to microvilli of non-ciliated cells of the fallopian tube, production of mucosal damage, engulfment by non-ciliated cells, transport of gonococci-containing phagocytic vacuoles to the base of the non-ciliated cells, and exocytosis of gonococci from phagocytic vacuoles into the subepithelial tissues. They were also able to replicate these observations using meningococci. However, at the time of these studies, little was known about the surface of the *Neisseria*, or the host cell targets, so molecular explanations of this phenomenon were not possible.

While many models have been employed to understand how gonococci and meningococci interact with host cells, little consideration has been given to how bacterial cell-bacterial cell interactions (microcolony formation versus biofilm formation) can affect the outcome of the interaction (see Chapter 4). Gonococci form microcolonies of tightly associated bacteria on the surface of most non-phagocytic cells. Gonococcal microcolonies are formed through Opa–LOS interactions on neighbouring gonococci and bacterial growth over time (Blake *et al.*, 1995). Gonococci also recruit host cell factors such as kinases, adaptor molecules, and actin-associated factors to the adherence site to promote intimate adhesion, microcolony forming interactions, host cell signalling and invasion (Duensing *et al.*, 1999; Higashi *et al.*, 2007; Lee *et al.*, 2005; Merz *et al.*, 1999). While it is believed that microcolonies may provide protection for gonococci sequestered in the core of the formation from antimicrobial molecules and phagocytes, this has not been clearly demonstrated. The data in Fig. 7.3 show two variants of gonococcal strain MS11 interacting with ME180 cells. While each of the ME180 cells shown in each panel is interacting with approximately the same number of gonococci, the distribution of gonococci is quite different on the two cells. The signalling between the gonococcus and the host cell in these two scenarios is likely to differ significantly. Since epithelial cells are the first cell type that the gonococcus encounters during colonization, the nature of the initial adhesion event may be important in dictating the disease outcome following colonization. In order to understand gonococcal pathogenesis, new methodologies that allow for study of the interaction at the single cell level are needed.

How host cells respond to the adhesion event and how they activate a variety of signalling cascades may provide differential signals to the innate immune system. Given the differences in host cell receptors on epithelial cells, relative to cells of the innate immune system, and the proclivity of immunological cells to be activated by non-specific signals, the signalling initiated at the epithelial surface is likely to differ from signalling initiated by tissue macrophages and neutrophils. Furthermore, neutrophils and macrophages can be activated indirectly in response to epithelial cell-elicited signals via cytokine responses, making it difficult to separate primary signalling events from secondary signals using currently available experimental methodology.

Stimulating epithelial cells to express immunomodulating compounds such as cytokines is the result of an innate immune response, which occurs due to recognition by host cells of several highly conserved structures found on the gonococcus. The cells of the innate immune system have evolved to recognize PAMPs through the PPRs such as the TLRs. Mononuclear phagocytes, including circulating monocytes and tissue macrophages, other phagocytic cells, and many epithelial cells, express these receptors.

As PAMPs are made only by microbial pathogens, detection of these via highly specific receptors allows the host to differentiate between its own surface molecules and those of an invading microorganism. In addition, PAMPs are usually invariant and necessary for the survival of the pathogen, increasing the likelihood that host PRRs will remain effective over time (Medzhitov and Janeway, 1997). There are three main functional classes of mammalian PRRs: secreted PRRs (e.g. mannan binding lectin), endocytic PRRs (e.g. the macrophage mannose receptor), and signalling PRRs (e.g. TLRs). These receptors play a major role in opsonization and phagocytosis of invading microbes, and induction of the inflammatory response. Binding by these receptors trigger signalling pathways within epithelial cells that lead to the activation of transcription factors, such as NF-κB and AP-1, and the expression of genes aimed at creating further defence responses (Akira *et al.*, 2001). Part of this response includes the increased production of proinflammatory cytokines and chemokines. Gonococci are known to elicit different levels of TNFα, IL-1, IL-6, and IL-8 secretion from a variety of epithelial cell types, which results in tissue damage and the recruitment of phagocytes (Fichorova *et al.*, 2001; Harvey *et al.*, 2002; Naumann *et al.*, 1997). Neutrophils and macrophages can engulf and destroy would-be adherent gonococci, but some studies show that a subset of gonococci may survive inside these cell types (for review see Rest and Shafer, 1989).

Several groups have investigated the production of cytokines during gonococcal challenge. One male challenge study using gonococcal strain MS11 found an increase in IL-8, IL-6 and TNFα levels in the urine samples of volunteers prior to the onset of symptoms. These cytokines were shown to increase concomitantly with the number of leucocytes found in urine sediment. At the onset of symptoms, IL-1β was detected (Ramsey *et al.*, 1995). Fichorova and coworkers measured IL-6, IL-8, and IL-1β production from challenged immortalized human cervical and vaginal epithelial cells at an MOI of 100:1, and also found that IL-1β was produced later than IL-6 and IL-8. However, TNFα was not measurable in these samples (Fichorova *et al.*, 2001). Using a cervical epithelial cell line (ME180) as a model of infection, and a challenge dose of 50:1 (MOI), Naumann and coworkers demonstrated production of TNFα and IL-8 by 6 hours post infection (Naumann *et al.*, 1997). Though these experiments allow some prediction regarding cytokine production during gonorrhoea, there is wide variety in the models, gonococcal strains, and challenge doses used to generate these cytokine profiles.

Interestingly, the lipid A portion of LOS likely serves as a site for C3b deposition on the bacterial cell surface (Edwards and Apicella, 2002). In addition, epithelial cells derived from vaginal, ectocervical and endocervical tissues have been shown to lack TLR4 and MD2, which are critical during host detection of LPS (Fichorova *et al.*, 2002). This suggests that these cells will be less efficient at activating the innate immune response.

Meningococcal interactions with host cells

N. meningitidis causes a diverse set of symptoms in infected individuals. Some patients develop meningitis, while others present with sepsis or even septic shock. The relationship between carriage of *N. meningitidis* and progression to invasive disease is not understood. Sim and coworkers (Sim *et al.*, 2000) assessed the rate of meningococcal carriage by swabbing the nasopharynx of volunteers and performing immunohistochemistry of tissue extracts. They found that meningococci were more widespread in the population than previously thought. More importantly, they found meningococci internal to the epithelial cell layer in the nasopharynx in asymptomatic individuals. This indicates that meningococci were invading into human tissues but were not eliciting symptoms. The duration of this asymptomatic carrier state varies but may be chronic, may last for several months, may be intermittent, or may be transient (Broome, 1986). Since carriage of commensal *Neisseria*, especially *N. lactamica*, is associated with a high titre of antibodies against *N. meningitidis*, and since *N. lactamica* shares several outer membrane antigens with *N. meningitidis*, it has been hypothesized that natural immunity against meningitis is partially acquired through colonization by commensal species. The precise nature of natural immunity to meningococci remains unknown, although a complex interaction between the organism and nasopharyngeal mucosal barrier, innate immune mechanisms, and acquired immunity is involved. Understanding the mechanisms of natural immunity may provide the key to development of better vaccines against meningococcal disease (See Chapter 11).

Colonization of the nasopharynx is a crucial step in initiating meningococcal disease. Colonization is a complex series of events that requires efficient scavenging of vital nutrients from human tissues, evasion of innate immune killing, and tropism to specific host cells. Studies using naturally occurring P^+ and P^- variants showed that pili play an important role in MC adhesiveness (Nassif *et al.*, 1997), however other surface proteins, including Opas and Opc, contribute to the initial adhesion of strains, especially those lacking pili and capsule (Aho *et al.*, 1991; Virji *et al.*, 1993).

Since the nasopharyngeal mucosa is rich in nitric oxide (NO)-producing cells, including macrophages and epithelial cells, successful colonization requires that the meningococcus overcome the local concentrations of NO. These local conditions have been shown to produce NO at concentrations that are likely to greatly exceed levels that are required to produce toxic effects in most bacteria (Lundberg *et al.*, 1995). Anjum and coworkers (Anjum *et al.*, 2002) demonstrated that *N. meningitidis* has two NO detoxification systems that contribute to meningococcal survival during colonization and immune evasion. Mutants lacking these genes retained their ability to be internalized by human macrophages and these mutants were more readily killed. This data indicates one mechanism by which *N. meningitidis* can evade the initial killing by macrophages.

After invasion of the bloodstream by the bacteria, three main immunological pathways are activated: complement cascade; the inflammatory response; and the coagulation and fibrinolysis pathway. These pathways do not act independently but are able to interact with each other (Stephens *et al.*, 2004). The main mechanism of complement evasion by group B meningococci is inhibition of membrane attack complex (MAC) insertion by their polysaccharide capsule. LOS structure may act in concert with capsule to prevent MAC insertion (Ram *et al.*, 1999).

Macrophage migration inhibitory factor (MIF) is a mediator of innate immunity and important in the pathogenesis of septic shock. Sprong and coworkers (Sprong *et al.*, 2007) studied MIF and cytokines *in vivo* in patients with meningococcal disease and found that MIF concentration in blood was increased during meningococcal disease and highest in the patients presenting with shock. Plasma concentration of MIF correlated with disease severity, the presence of shock and with the cytokines IL-1β, IL-10, IL-12, and vascular endothelial growth factor, but not with TNF-α. There have been several studies that indicate that an innate anti-inflammatory cytokine profile may contribute to fatal meningococcal disease.

Conclusions

Though the neisserial pathogens are genetically very similar, their capacity to cause disease differs dramatically. Furthermore, the same strain can elicit dramatically different symptoms in two closely related hosts. This suggests that differences in pathogenicity results, in part, from a highly variable expression of surface antigens. Constant alteration of the bacterial cell surface adds to the complexity of *Neisseria*-host cell interactions at each stage of infection, and obscures disease presentation.

While significant advances have been made in our understanding of how the pathogenic *Neisseria* evade the host innate immune response, the ability of these pathogens to antigenically vary their surfaces at high frequency has made it difficult to define molecular explanations for the observed results. Because many early studies were performed using strains with variable surfaces, we need to keep an open mind as to the relative contributions of surface antigens in evading the innate immune system. Given the diversity of diseases that the pathogenic *Neisseria* can cause, we need to recognize the limitations of the models that we are employing to explain disease pathogenesis.

Disease prevention will require the generation of vaccines. The identification of bacterial vaccine candidates has been accelerated by the availability of complete genome sequences. Proteomics and immunoproteomics have also allowed identification of a large number of novel proteins that elicit immune responses. Many of these antigens are conserved in sequence and are able to elicit bactericidal antibody responses in mice.

Future trends

With the panoply of variable surface molecules that interact with the innate immune system, it has been difficult to identify the role of individual molecules on the surface of the pathogenic *Neis-*

seria in interacting with this system. Since it is likely that different variants in the same inoculum can produce different host responses, we need to develop technologies that can analyse host responses in individual cells. Furthermore, we need to focus on generating a set of isogenic strains that lack the ability to change their surface, and that have well-defined surfaces, if we are to develop an understanding of how neisserial surface variation contributes to disease.

Paradoxically, genetic modification of the *Neisseria* to express defined surfaces by genetic deletion of one or more surface molecules may inadvertently change the bacterial surface in other unknown ways. Alterations may also facilitate/interfere with host cell interactions simply by changing the net surface charge on the bacterium.

Almost all work on neisserial biology has been performed with strains grown in nutrient replete media under aerobic conditions. These *in vitro* growth conditions suppress the expression of many important antigens. For example, anaerobically grown gonococci (see Chapter 2) both induce and repress the expression of several outer-membrane proteins, with AniA only being expressed under anaerobic conditions (Householder *et al.*, 1999). Antibody to AniA is found in sera from women with gonococcal infections, demonstrating that this protein is expressed *in vivo* and that anaerobic growth is a physiologically significant state for this organism (Clark *et al.*, 1987, 1988). NO is a critical component of the innate immune system and is toxic to bacterial cells (Davis *et al.*, 2001). NO is an important modulator of the cellular events that form the immune response, with low basal levels of NO being anti-inflammatory and high levels being pro-inflammatory. Since the *Neisseria* can detoxify NO, this has several implications in human health. The role of NO metabolism during gonococcal infection is largely unknown, but the reduction of host-produced NO to anti-inflammatory levels could contribute to asymptomatic infections by this organism in women (Cardinale and Clark, 2000).

Du and coworkers showed that the initial contact between gonococci and host epithelial cells signals the bacterium to modulate the expression of genes that will allow it to colonize and proliferate in the new host (Du *et al.*, 2005). Since the interaction between gonococci and the host results in changes in transcription of several genes which can lead to changes in bacterial cell metabolism and other basic functions, enabling the bacterium to quickly adapt to its new environment, we need to consider how the pathogenic *Neisseria* regulate changes in their surfaces in response to changes in their growth environment.

Most neisserial infections occur in non-sterile environments. Future research will need to focus more on the interplay between the pathogens and commensal organisms. This will become increasingly important with the development of vaccines that eliminate the carrier state of opportunistic pathogens.

Web resources

http://www.arthritis.org/innate-immunity.php
http://www.nlm.nih.gov/medlineplus/ency/article/000821.htm
http://www.innateimmunity-systemsbiology.org/
http://www3.niaid.nih.gov

Acknowledgement

This work was supported by grants AI024452 and AI068888 from the National Institutes of Health to D.C.S.

References

Aho, E.L., Dempsey, J.A., Hobbs, M.M., Klapper, D.G., and Cannon, J.G. (1991). Characterization of the opa (class 5) gene family of *Neisseria meningitidis*. Mol. Microbiol. *5*, 1429–1437.

Aho, E.L., Keating, A.M., and McGillivray, S.M. (2000). A comparative analysis of pilin genes from pathogenic and nonpathogenic *Neisseria* species. Microb. Pathog. *28*, 81–88.

Akira, S., Takeda, K., and Kaisho, T. (2001). Toll-like receptors: critical proteins linking innate and acquired immunity. Nat. Immunol. *2*, 675–680.

Alcala, B., Salcedo, C., Arreaza, L., Abad, R., Enriquez, R., De La Fuente, L., Uria, M.J., and Vazquez, J.A. (2004). Antigenic and/or phase variation of PorA protein in non-subtypable *Neisseria meningitidis* strains isolated in Spain. J. Med. Microbiol. *53*, 515–518.

Anjum, M.F., Stevanin, T.M., Read, R.C., and Moir, J.W. (2002). Nitric oxide metabolism in *Neisseria meningitidis*. J. Bacteriol. *184*, 2987–2993.

Apicella, M.A., Shero, M., Jarvis, G.A., Griffiss, J.M., Mandrell, R.E., and Schneider, H. (1987). Phenotypic variation in epitope expression of the *Neisseria gonorrhoeae* lipooligosaccharide. Infect. Immun. *55*, 1755–1761.

Arking, D., Tong, Y., and Stein, D.C. (2001). Analysis of lipooligosaccharide biosynthesis in the *Neisseriaceae*. J. Bacteriol. *183*, 934–941.

Banchereau, J., and Steinman, R.M. (1998). Dendritic cells and the control of immunity. Nature *392*, 245–252.

Banerjee, A., Wang, R., Supernavage, S.L., Ghosh, S.K., Parker, J., Ganesh, N.F., Wang, P.G., Gulati, S., and Rice, P.A. (2002). Implications of phase variation of a gene (*pgtA*) encoding a pilin galactosyl transferase in gonococcal pathogenesis. J. Exp. Med. *196*, 147–162.

Belland, R.J., Chen, T., Swanson, J., and Fischer, S.H. (1992). Human neutrophil response to recombinant neisserial Opa proteins. Mol. Microbiol. *6*, 1729–1737.

Binnicker, M.J., Williams, R.D., and Apicella, M.A. (2003). Infection of human urethral epithelium with *Neisseria gonorrhoeae* elicits an up-regulation of host anti-apoptotic factors and protects cells from staurosporine-induced apoptosis. Cell. Microbiol. *5*, 549–560.

Biragyn, A., Ruffini, P.A., Leifer, C.A., Klyushnenkova, E., Shakhov, A., Chertov, O., Shirakawa, A.K., Farber, J.M., Segal, D.M., Oppenheim, J.J., and Kwak, L.W. (2002). Toll-like receptor 4-dependent activation of dendritic cells by beta-defensin 2. Science *298*, 1025–1029.

Bish, S.E., Song, W., and Stein, D.C. (2008). Quantification of bacterial internalization by host cells using a b-lactamase reporter strain: *Neisseria gonorrhoeae* invasion into cervical epithelial cells requires bacterial viability. Microbes. Infect. *10*, 1182–1191.

Bjerknes, R., Guttormsen, H.K., Solberg, C.O., and Wetzler, L.M. (1995). Neisserial porins inhibit human neutrophil actin polymerization, degranulation, opsonin receptor expression, and phagocytosis but prime the neutrophils to increase their oxidative burst. Infect. Immun. *63*, 160–167.

Black, J.R., Black, W.J., and Cannon, J.G. (1985). Neisserial antigen H.8 is immunogenic in patients with disseminated gonococcal and meningococcal infections. J. Infect. Dis. *15*, 650–657.

Blake, M.S., Blake, C.M., Apicella, M.A., and Mandrell, R.E. (1995). Gonococcal opacity: lectin-like interactions between Opa proteins and lipooligosaccharide. Infect. Immun. *63*, 1434–1439.

Block, C., Gdalevich, M., Buber, R., Ashkenazi, I., Ashkenazi, S., and Keller, N. (1999). Factors associated with pharyngeal carriage of *Neisseria meningitidis* among Israel Defense Force personnel at the end of their compulsory service. Epidemiol. Infect. *122*, 51–57.

Boris, S., and Barbés, C. (2000). Role played by lactobacilli in controlling the population of vaginal pathogens. Microbes Infect. *2*, 543–546.

Boslego, J.W., Tramont, E.C., Chung, R.C., McChesney, D.G., Ciak, J., Sadoff, J.C., Piziak, M.V., Brown, J.D., Brinton Jr., C.C., and Wood, S.W. (1991). Efficacy trial of a parenteral gonococcal pilus vaccine in men. Vaccine *9*, 154–162.

Boulton, I.C., and Gray-Owen, S.D. (2002). Neisserial binding to CEACAM1 arrests the activation and proliferation of CD4+ T lymphocytes. Nat. Immunol. *3*, 229–236.

Bowdish, D.M., Davidson, D.J., Speert, D.P., and Hancock, R.E. (2004). The human cationic peptide LL-37 induces activation of the extracellular signal-regulated kinase and p38 kinase pathways in primary human monocytes. J. Immunol. *172*, 3758–3765.

Broome, C.V. (1986). The carrier state: *Neisseria meningitidis*. J. Antimicrob. Chemother. *18*, 25–34.

Burch, C.L., Danaher, R.J., and Stein, D.C. (1997). Antigenic variation in *Neisseria gonorrhoeae*: production of multiple lipooligosaccharides. J. Bacteriol. *179*, 982–986.

Cannon, J.G. (1989). Conserved lipoproteins of pathogenic *Neisseria* species bearing the H.8 epitope: lipid-modified azurin and H.8 outer membrane protein. Clin Microbiol. Rev. 2 Suppl., S1–4.

Cannon, J.G., Black, W.J., Nachamkin, I., and Stewart, P.W. (1984). Monoclonal antibody that recognizes an outer membrane antigen common to the pathogenic *Neisseria* species but not to most nonpathogenic *Neisseria* species. Infect. Immun. *43*, 994–999.

Cardinale, J.A., and Clark, V.L. (2000). Expression of AniA, the major anaerobically induced outer membrane protein of *Neisseria gonorrhoeae*, provides protection against killing by normal human sera. Infect. Immun. *68*, 4368–4369.

Carson, S.D., Stone, B., Beucher, M., Fu, J., and Sparling, P.F. (2000). Phase variation of the gonococcal siderophore receptor FetA. Mol. Microbiol. *36*, 585–593.

Casey, L.C. (2000). Immunologic response to infection and its role in septic shock. Crit. Care Clin. *16*, 193–213.

Casey, S.G., Shafer, W.M., and Spitznagel, J.K. (1986). *Neisseria gonorrhoeae* survive intraleukocytic oxygen-independent antimicrobial capacities of anaerobic and aerobic granulocytes in the presence of pyocin lethal for extracellular gonococci. Infect. Immun. *52*, 384–389.

Caugant, D.A., Hoiby, E.A., Magnus, P., Scheel, O., Hoel, T., Bjune, G., Wedege, E., Eng, J., and Froholm, L.O. (1994). Asymptomatic carriage of *Neisseria meningitidis* in a randomly sampled population. J. Clin. Microbiol. *32*, 323–330.

Chen, C.J., Elkins, C., and Sparling, P.F. (1998). Phase variation of hemoglobin utilization in *Neisseria gonorrhoeae*. Infect. Immun. *66*, 987–993.

Clark, V.L., Campbell, L.A., Palermo, D.A., Evans, T.M., and Klimpel, K.W. (1987). Induction and repression of outer membrane proteins by anaerobic growth of *Neisseria gonorrhoeae*. Infect. Immun. *55*, 1359–1364.

Clark, V.L., Knapp, J.S., Thompson, S., and Klimpel, K.W. (1988). Presence of antibodies to the major anaerobically induced gonococcal outer membrane protein in sera from patients with gonococcal infections. Microb Pathog. *5*, 381–390.

Cohen, J. (2002). The immunopathogenesis of sepsis. Nature *420*, 885–891.

Connell, T.D., Black, W.J., Kawula, T.H., Barritt, D.S., Dempsey, J.A., Kverneland, K., Jr., Stephenson, A., Schepart, B.S., Murphy, G.L., and Cannon, J.G. (1988). Recombination among protein II genes of *Neisseria gonorrhoeae* generates new coding sequences and increases structural variability in the protein II family. Mol. Microbiol. *2*, 227–236.

Cooper, M.D., McGee, Z.A., Mulks, M.H., Koomey, J.M., and Hindman, T.L. (1984). Attachment to and invasion of human fallopian tube mucosa by an IgA1 protease-deficient mutant of *Neisseria gonorrhoeae* and its wild-type parent. J. Infect. Dis. *150*, 737–744.

Danaher, R.J., Levin, J.C., Arking, D., Burch, C.L., Sandlin, R., and Stein, D.C. (1995). Genetic basis of *Neisseria gonorrhoeae* lipooligosaccharide antigenic variation. J. Bacteriol. *177*, 7275–7279.

Dansky-Ullmann, C., Salgaller, M., Adams, S., Schlom, J., and Greiner, J.W. (1995). Synergistic effects of IL-6 and IFN-gamma on carcinoembryonic antigen (CEA) and HLA expression by human colorectal carcinoma cells: role for endogenous IFN-beta. Cytokine *7*, 118–129.

Davis, K.L., Martin, E., Turko, I.V., and Murad, F. (2001). Novel effects of nitric oxide. Annu. Rev. Pharmacol. Toxicol. *41*, 203–236.

Dehio, C., Gray-Owen, S.D., and Meyer, T.F. (1998). The role of neisserial Opa proteins in interactions with host cells. Trends Microbiol. *6*, 489–495.

Dietrich, G., Kurz, S., Hubner, C., Aepinus, C., Theiss, S., Guckenberger, M., Panzner, U., Weber, J., and Frosch, M. (2003). Transcriptome analysis of *Neisseria meningitidis* during infection. J. Bacteriol. *185*, 155–164.

Dixon, G.L.J., Newton, P.J., Chain, B.M., Katz, D., Andersen, S.R., Wong, S., van der Ley, P., Klein, N., and Callard, R.E. (2001). Dendritic cell activation and cytokine production induced by group B *Neisseria meningitidis*: Interleukin-12 production depends on lipopolysaccharide expression in intact bacteria. Infect. Immun. *69*, 4351–4357.

Du, Y., Lenz, J., and Arvidson, C.G. (2005). Global gene expression and the role of sigma factors in *Neisseria gonorrhoeae* in interactions with epithelial cells. Infect. Immun. *73*, 4834–4845.

Duensing, T.D., and van Putten, J.P. (1997). Vitronectin mediates internalization of *Neisseria gonorrhoeae* by Chinese hamster ovary cells. Infect. Immun. *65*, 964–970.

Duensing, T.D., Wing, J.S., and van Putten, J.P. (1999). Sulfated polysaccharide-directed recruitment of mammalian host proteins: a novel strategy in microbial pathogenesis. Infect. Immun. *67*, 4463–4468.

Edwards, J.L., and Apicella, M.A. (2002). The role of lipooligosaccharide in *Neisseria gonorrhoeae* pathogenesis of cervical epithelia: lipid A serves as a C3 acceptor molecule. Cell. Microbiol. *4*, 585–598.

Edwards, J.L., and Apicella, M.A. (2005). I-domain-containing integrins serve as pilus receptors for *Neisseria gonorrhoeae* adherence to human epithelial cells. Cell. Microbiol. *7*, 1197–1211.

Edwards, J.L., Brown, E.J., Ault, K.A., and Apicella, M.A. (2001). The role of complement receptor 3 (CR3) in *Neisseria gonorrhoeae* infection of human cervical epithelia. Cell. Microbiol. *3*, 611–622.

Edwards, J.L., Brown, E.J., Uk-Nham, S., Cannon, J.G., Blake, M.S., and Apicella, M.A. (2002). A co-operative interaction between *Neisseria gonorrhoeae* and complement receptor 3 mediates infection of primary cervical epithelial cells. Cell. Microbiol. *4*, 571–584.

Edwards, J.L., Shao, J.Q., Ault, K.A., and Apicella, M.A. (2000). *Neisseria gonorrhoeae* elicits membrane ruffling and cytoskeletal rearrangements upon infection of primary human endocervical and ectocervical cells. Infect. Immun. *68*, 5354–5363.

Engering, A., Geijtenbeek, T.B., van Vliet, S.J., Wijers, M., van Liempt, E., Demaurex, N., Lanzavecchia, A., Fransen, J., Figdor, C.G., Piguet, V., and van Kooyk, Y. (2002). The dendritic cell-specific adhesion receptor DC-SIGN internalizes antigen for presentation to T-cells. J. Immunol. *168*, 2118–2126.

Estabrook, M.M., Christopher, N.C., Griffiss, J.M., Baker, C.J., and Mandrell, R.E. (1992). Sialylation and human neutrophil killing of group C *Neisseria meningitidis*. J. Infect. Dis. *166*, 1079–1088.

Fichorova, R.N., Cronin, A.O., Lien, E., Anderson, D.J., and Ingalls, R.R. (2002). Response to *Neisseria gonorrhoeae* by cervicovaginal epithelial cells occurs in the absence of toll-like receptor 4-mediated signaling. J. Immunol. *168*, 2424–2432.

Fichorova, R.N., Desai, P.J., Gibson, F.C., 3rd, and Genco, C.A. (2001). Distinct proinflammatory host responses to *Neisseria gonorrhoeae* infection in immortalized human cervical and vaginal epithelial cells. Infect. Immun. *69*, 5840–5848.

Fisette, P.L., Ram, S., Andersen, J.M., Guo, W., and Ingalls, R.R. (2003). The *lip* lipoprotein from *Neisseria gonorrhoeae* stimulates cytokine release and NF-kB activation in epithelial cells in a TLR2-dependent manner. J. Biol. Chem. *278*, 46252–46260.

Fitzgerald, K.A., Rowe, D.C., and Golenbock, D.T. (2004). Endotoxin recognition and signal transduction by the TLR4/MD2-complex. Microbes Infect. *6*, 1361–1367.

Ghosh, S.K., Zhao, J., Philogene, M.C., Alzaharani, A., Rane, S., and Banerjee, A. (2004). Pathogenic consequences of *Neisseria gonorrhoeae* pilin glycan variation. Microbes Infect. *6*, 693–701.

Goldschneider, I., Gotschlich, E.C., and Artenstein, M.S. (1969a). Human immunity to the meningococcus. I. The role of humoral antibodies. J. Exp. Med. *129*, 1307–1326.

Goldschneider, I., Gotschlich, E.C., and Artenstein, M.S. (1969b). Human immunity to the meningococcus. II. Development of natural immunity. J. Exp. Med. *129*, 1327–1348.

Gomez-Duarte, O.G., Dehio, M., Guzman, C.A., Chhatwal, G.S., Dehio, C., and Meyer, T.F. (1997). Binding of vitronectin to *opa*-expressing *Neisseria gonorrhoeae* mediates invasion of HeLa cells. Infect. Immun. *65*, 3857–3866.

Gotschlich, E.C. (1994). Genetic locus for the biosynthesis of the variable portion of *Neisseria gonorrhoeae* lipooligosaccharide. J. Exp. Med. *180*, 2181–2190.

Grassme, H., Gulbins, E., Brenner, B., Ferlinz, K., Sandhoff, K., Harzer, K., Lang, F., and Meyer, T.F. (1997). Acidic sphingomylenase mediates entry of *Neisseria gonorrhoeae* into nonphagocytic cells. Cell *91*, 605–615.

Gray-Owen, S.D., Dehio, C., Haude, A., Grunert, F., and Meyer, T.F. (1997a). CD66 carcinoembryonic antigens mediate interactions between Opa-expressing *Neisseria gonorrhoeae* and human polymorphonuclear phagocytes. EMBO J. *16*, 3435–3445.

Gray-Owen, S.D., Lorenzen, D.R., Haude, A., Meyer, T.F., and Dehio, C. (1997b). Differential Opa specificities for CD66 receptors influence tissue interactions and cellular response to *Neisseria gonorrhoeae*. Mol. Microbiol. *26*, 971–980.

Griffiss, J.M., Schneider, H., Mandrell, R.E., Jarvis, G.A., Kim, J.J., Gibson, B., and Apicella, M.A. (1987). The immunochemistry of neisserial LOS. Antonie Van Leeuwenhoek *53*, 501–507.

Guibourdenche, M., Popoff, M.Y., and Riou, J.Y. (1986). Deoxyribonucleic acid relatedness among *Neisseria gonorrhoeae, N. meningitidis, N. lactamica, N. cinerea* and *'Neisseria polysaccharea'.* Ann. Inst. Pasteur Microbiol. *137B*, 177–185.

Gulati, S., Cox, A., Lewis, L.A., Michael, F.S., Li, J., Boden, R., Ram, S., and Rice, P.A. (2005). Enhanced factor H binding to sialylated gonococci is restricted to the sialylated lacto-N-neotetraose lipooligosaccharide species: implications for serum resistance and evidence for a bifunctional lipooligosaccharide sialyltransferase in gonococci. Infect. Immun. *73*, 7390–7397.

Haines, K.A., Reibman, J., Tang, X.Y., Blake, M., and Weissmann, G. (1991). Effects of protein I of *Neisseria gonorrhoeae* on neutrophil activation: generation of diacylglycerol from phosphatidylcholine via a specific phospholipase C is associated with exocytosis. J. Cell Biol. *114*, 433–442.

Haines, K.A., Yeh, L., Blake, M.S., Cristello, P., Korchak, H., and Weissmann, G. (1988). Protein I, a translocatable ion channel from *Neisseria gonorrhoeae*, selectively inhibits exocytosis from human neutrophils without inhibiting O_2- generation. J. Biol. Chem. *263*, 945–951.

Hammerschmidt, S., Hilse, R., van Putten, J.P., Gerardy-Schahn, R., Unkmeir, A., and Frosch, M. (1996a). Modulation of cell surface sialic acid expression in *Neisseria meningitidis* via a transposable genetic element. EMBO J. *15*, 192–198.

Hammerschmidt, S., Muller, A., Sillmann, H., Muhlenhoff, M., Borrow, R., Fox, A., van Putten, J., Zollinger, W.D., Gerardy-Schahn, R., and Frosch, M. (1996b). Capsule phase variation in *Neisseria meningitidis* serogroup B by slipped-strand mispairing in the polysialyltransferase gene (*siaD*): correlation with bacterial invasion and the outbreak of meningococcal disease. Mol. Microbiol. *20*, 1211–1220.

Harvey, H.A., Porat, N., Campbell, C.A., Jennings, M., Gibson, B.W., Phillips, N.J., Apicella, M.A., and Blake, M.S. (2000). Gonococcal lipooligosaccharide is a ligand for the asialoglycoprotein receptor on human sperm. Mol. Microbiol. *36*, 1059–1070.

Harvey, H.A., Post, D.M., and Apicella, M.A. (2002). Immortalization of human urethral epithelial cells: a model for the study of the pathogenesis of and the inflammatory cytokine response to *Neisseria gonorrhoeae* infection. Infect. Immun. *70*, 5808–5815.

Hauck, C.R., and Meyer, T.F. (1997). The lysosomal/phagosomal membrane protein h-lamp-1 is a target of the IgA1 protease of *Neisseria gonorrhoeae*. FEBS Lett. *405*, 86–90.

Hauck, C.R., Meyer, T.F., Lang, F., and Gulbins, E. (1998). CD66-mediated phagocytosis of Opa52 *Neisseria gonorrhoeae* requires a Src-like tyrosine kinase- and Rac1-dependent signaling pathway. EMBO J. *17*, 443–454.

Hayashi, F., Smith, K.D., Ozinsky, A., Hawn, T.R., Yi, E.C., Goodlett, D.R., Eng, J.K., Akira, S., Underhill, D.M., and Aderem, A. (2001). The innate immune response to bacterial flagellin is mediated by Toll-like receptor 5. Nature *410*, 1099–1103.

Hedges, S.R., Mayo, M.S., Mestecky, J., Hook, E.W., 3rd, and Russell, M.W. (1999). Limited local and systemic antibody responses to *Neisseria gonorrhoeae* during uncomplicated genital infections. Infect. Immun. *67*, 3937–3946.

Hedges, S.R., Sibley, D.A., Mayo, M.S., Hook, E.W., 3rd, and Russell, M.W. (1998). Cytokine and antibody responses in women infected with *Neisseria gonorrhoeae*: effects of concomitant infections. J. Infect. Dis. *178*, 742–751.

Hemmi, H., Takeuchi, O., Kawai, T., Kaisho, T., Sato, S., Sanjo, H., Matsumoto, M., Hoshino, K., Wagner, H., Takeda, K., and Akira, S. (2000). A Toll-like receptor recognizes bacterial DNA. Nature *408*, 740–745.

Higashi, D.L., Lee, S.W., Snyder, A., Weyand, N.J., Bakke, A., and So, M. (2007). Dynamics of *Neisseria gonorrhoeae* attachment: Microcolony development, cortical plaque formation, and cytoprotection. Infect. Immun. *75*, 4743–4753.

Hipp, S.S., Lawton, W.D., Chen, N.C., and Gaafar, H.A. (1974). Inhibition of *Neisseria gonorrhoeae* by a factor produced by *Candida albicans*. Appl. Microbiol. *27*, 192–196.

Hoehn, G.T., and Clark, V.L. (1992). The major anaerobically induced outer membrane protein of *Neisseria gonorrhoeae*, Pan 1, is a lipoprotein. Infect. Immun. *60*, 4704–4708.

Holmes, K.K., Johnson, D.W., and Trostle, H.J. (1970). An estimate of the risk of men acquiring gonorrhea by sexual contact with infected females.. Am. J. Epidemiol. *91*, 170–174.

Hopper, S., Vasquez, B., Merz, A., Clary, S., Wilbur, J.S., and So, M. (2000). Effects of the immunoglobulin A1 protease on *Neisseria gonorrhoeae* trafficking across polarized T84 epithelial monolayers. Infect. Immun. *68*, 906–911.

Householder, T.C., Belli, W.A., Lissenden, S., Cole, J.A., and Clark, V.L. (1999). cis- and trans-acting elements involved in regulation of *aniA*, the gene encoding the major anaerobically induced outer membrane protein in *Neisseria gonorrhoeae*. J. Bacteriol. *181*, 541–551.

Jarva, H., Ram, S., Vogel, U., Blom, A.M., and Meri, S. (2005). Binding of the complement inhibitor C4bp to serogroup B *Neisseria meningitidis*. J. Immunol. *174*, 6299–6307.

Jarvis, G., and Vedros, N. (1987). Sialic acid of group B *Neisseria meningitidis* regulates alternative complement pathway activation. Infect. Immun. *55*, 174–180.

Jerse, A.E. (1999). Experimental gonococcal genital tract infection and opacity protein expression in estradiol-treated mice. Infect. Immun. *67*, 5699–5708.

Jerse, A.E., Cohen, M.S., Drown, P.M., Whicker, L.G., Isbey, S.F., Seifert, H.S., and Cannon, J.G. (1994). Multiple gonococcal opacity proteins are expressed during experimental urethral infection in the male. J. Exp. Med. *179*, 911–920.

Jerse, A.E., Crow, E.T., Bordner, A.N., Rahman, I., Cornelissen, C.N., Moench, T.R., and Mehrazar, K.

(2002). Growth of *Neisseria gonorrhoeae* in the female mouse genital tract does not require the gonococcal transferrin or hemoglobin receptors and may be enhanced by commensal lactobacilli. Infect. Immun. *70*, 2549–2558.

Jerse, A.E., Sharma, N.D., Simms, A.N., Crow, E.T., Snyder, L.A., and Shafer, W.M. (2003). A gonococcal efflux pump system enhances bacterial survival in a female mouse model of genital tract infection. Infect. Immun. *71*, 5576–5582.

John, C.M., Jarvis, G.A., Swanson, K.V., Leffler, H., Cooper, M.D., Huflejt, M.E., and Griffiss, J.M. (2002). Galectin-3 binds lactosaminylated lipooligosaccharides from *Neisseria gonorrhoeae* and is selectively expressed by mucosal epithelial cells that are infected. Cell. Microbiol. *4*, 649–662.

Jones, H.E., Uronen-Hansson, H., Callard, R.E., Klein, N., and Dixon, G.L.J. (2007). The differential response of human dendritic cells to live and killed *Neisseria meningitidis*. Cell. Microbiol. *9*, 2856–2869.

Jonsson, A.B., Ilver, D., Falk, P., Pepose, J., and Normark, S. (1994). Sequence changes in the pilus subunit lead to tropism variation of *Neisseria gonorrhoeae* to human tissue. Mol. Microbiol. *13*, 403–416.

Jonsson, A.B., Nyberg, G., and Normark, S. (1991). Phase variation of gonococcal pili by frameshift mutation in *pilC*, a novel gene for pilus assembly. EMBO J. *10*, 477–488.

Jose, J., Otto, G.W., and Meyer, T.F. (2003). The integration site of the *iga* gene in commensal *Neisseria sp*. Mol Genet Genomics *269*, 197–204.

Judd, R.C., Tam, M., and Joiner, K. (1987). Characterization of protein I from serum-sensitive and serum-resistant transformants of *Neisseria gonorrhoeae*. Infect. Immun. *55*, 273–276.

Kahler, C.M., Carlson, R.W., Rahman, M.M., Martin, L.E., and Stephens, D.S. (1996). Two glycosyltransferase genes, *lgt*F and *rfa*K, constitute the lipooligosaccharide *ice* (inner core extension) biosynthesis operon of *Neisseria meningitidis*. J. Bacteriol. *178*, 6677–6684.

Kahler, C.M., and Stephens, D.S. (1998). Genetic basis for biosynthesis, structure, and function of meningococcal lipooligosaccharide (endotoxin). Crit. Rev. Microbiol. *24*, 281–334.

Kallstrom, H., Blackmer Gill, D., Albiger, B., Liszewski, M.K., Atkinson, J.P., and Jonsson, A.B. (2001). Attachment of *Neisseria gonorrhoeae* to the cellular pilus receptor CD46: identification of domains important for bacterial adherence. Cel.l Microbio.l 3, 133–143.

Kallstrom, H., Islam, M.S., Berggren, P.O., and Jonsson, A.B. (1998). Cell signaling by the type IV pili of pathogenic *Neisseria*. J. Biol. Chem. *273*, 21777–21782.

Karaghiosoff, M., Steinborn, R., Kovarik, P., Kriegshäuser, G., Baccarini, M., Donabauer, B., Reichart, U., Kolbe, T., Bogdan, C., Leanderson, T., Levy, D., Decker, T., and Müller, M. (2003). Central role for type I interferons and Tyk2 in lipopolysaccharide-induced endotoxin shock. Nat. Immunol. *4*, 471–477.

Kawula, T.H., Spinola, S.M., Klapper, D.G., and Cannon, J.G. (1987). Localization of a conserved epitope and an azurin-like domain in the H.8 protein of pathogenic *Neisseria*. Mol. Microbiol. *1*, 179–185.

Kellogg, D.S., Peacock, W.L., Deacon, W.E., Brown, L., and Pirkle, C.I. (1963). *Neisseria gonorrhoeae* I. Virulence genetically linked to clonal variation. J. Bacteriol. *85*, 1274–1279.

Kemper, C., Chan, A.C., Green, J.M., Brett, K.A., Murphy, K.M., and Atkinson, J.P. (2003). Activation of human CD4+ cells with CD3 and CD46 induces a T-regulatory cell 1 phenotype. Nature *421*, 388–392.

Kilian, M., Reinholdt, J., Lomholt, H., Poulsen, K., and Frandsen, E.V. (1996). Biological significance of IgA1 proteases in bacterial colonization and pathogenesis: critical evaluation of experimental evidence. Acta Pathol. Microbiol. Immunol. Scand. *104*, 321–338.

King, G., James, J.F., and Swanson, J. (1978). Studies on gonococcus infection. XI. Comparison of *in vivo* and *vitro* association of *Neisseria gonorrhoeae* with human neutrophils. J. Infect. Dis. *137*, 38–43.

Klein, N.J., Levin, M., Strobel, S., and Finn, A. (1993). Degradation of glycosaminoglycans and fibronectin on endotoxin-stimulated endothelium by adherent neutrophils: relationship to CD11b/CD18 and L-selectin expression. J. Infect. Dis. *167*, 890–898.

Knepper, B., Heuer, I., Meyer, T.F., and van Putten, J.P. (1997). Differential response of human monocytes to *Neisseria gonorrhoeae* variants expressing pili and opacity proteins. Infect. Immun. *65*, 4122–4129.

Lambden, P.R., Heckels, J.E., Jones, L.T., and Watt, P.J. (1979). Variations in surface protein composition associated with virulence properties in opacity types of *Neisseria gonorrhoeae*. J. Gen. Microbiol. *114*, 305–312.

Latz, E., Franko, J., Golenbock, D.T., and Schreiber, J.R. (2004). *Haemophilus influenzae* type B-outer membrane protein complex glycoconjugate vaccine induces cytokine production by engaging human Toll-like receptor 2 (TLR2) and requires the presence of TLR2 for optimal immunogenicity. J. Immunol. *172*, 2431–2438.

Lavitola, A., Aoun, L., Ohayon, H., Crémieux, A.C., Ancelle, R., and Martin, P.M. (1987). The 70-Kd neisserial common antigen is a surface-exposed, antigenically stable peptidic structure. Ann. Inst. Pasteur Microbiol. *138*, 333–342.

Lee, F.K., Stephens, D.S., Gibson, B.W., Engstrom, J.J., Zhou, D., and Apicella, M.A. (1995). Microheterogeneity of *Neisseria* lipooligosaccharide: analysis of a UDP- glucose 4-epimerase mutant of *Neisseria meningitidis* NMB. Infect. Immun. *63*, 2508–2515.

Lee, S.W., Higashi, D.L., Snyder, A., Merz, A.J., Potter, L., and So, M. (2005). PilT is required for PI(3,4,5)P3-mediated crosstalk between *Neisseria gonorrhoeae* and epithelial cells. Cell. Microbiol. 7, 1271–1284.

Lin, L., Ayala, P., and So, M. (1997). The *Neisseria* type 2 IgA1 protease cleaves LAMP1 and promotes survival of bacteria within epithelial cells. Mol. Microbiol. *24*, 1083–1094.

Lorenzen, D.R., Düx, F., Wölk, U., Tsirpouchtsidis, A., Haas, G., and Meyer, T.F. (1999). Immunoglobulin A1 protease, an exoenzyme of pathogenic neisseriae, is a potent inducer of proinflammatory cytokines. J. Exp. Med. *190*, 1049–1058.

Lundberg, J.O., Farkas, S.T., Witzberb, E., Rinder, J., Lidholm, J., Anggaard, A., Hokfelt, T., Lundberg, J.M., and Alvin, K. (1995). High nitric oxide production in human sinuses. Nat. Med. *1*, 370–373.

Macatonia, S.E., Hosken, N.A., Litton, M., Vieira, P., Hsieh, C.S., Culpepper, J.A., Wysocka, M., Trinchieri, G., Murphy, K.M., and O'Garra, A. (1995). Dendritic cells produce IL12 and direct the development of Th1 cells from naive CD4+ T-cells. J. Immunol. *154*, 5071–5079.

Mackinnon, F.G., Ho, Y., Blake, M.S., Michon, F., Chandraker, A., Sayegh, M.H., and Wetzler, L.M. (1999). The role of B/T costimulatory signals in the immunopotentiating activity of neisserial porin. J. Infect. Dis. *180*, 755–761.

Maisey, K., Nardocci, G., Imarai, M., Cardenas, H., Rios, M., Croxatto, H.B., Heckels, J.E., Christodoulides, M., and Velasquez, L.A. (2003). Expression of proinflammatory cytokines and receptors by human fallopian tubes in organ culture following challenge with *Neisseria gonorrhoeae*. Infect. Immun. *71*, 527–532.

Makepeace, B.L., Watt, P.J., Heckels, J.E., and Christodoulides, M. (2001). Interactions of *Neisseria gonorrhoeae* with mature human macrophage opacity proteins influence production of proinflammatory cytokines. Infect. Immun. *69*, 1909–1913.

Mandrell, R.E., and Apicella, M.A. (1993). Lipooligosaccharides (LOS) of mucosal pathogens: molecular mimicry and host-modification of LOS. Immunobiol *187*, 382–402.

Mandrell, R.E., Griffiss, J.M., and Macher, B.A. (1988). Lipooligosaccharides (LOS) of *Neisseria gonorrhoeae* and *Neisseria meningitidis* have components that are immunochemically similar to precursors of human blood group antigens. J. Exp. Med. *168*, 107–126.

Martini, F.H. (2001). The Tissue Level of Organization. In Fundamentals of Anatomy and Physiology (Benjamin Cummings), pp. 106–117.

Massari, P., Henneke, P., Ho, Y., Latz, E., Golenbock, D.T., and Wetzler, L.M. (2002). Cutting edge: Immune stimulation by neisserial porins is toll-like receptor 2 and MyD88 dependent. J. Immunol. *168*, 1533–1537.

Massari, P., Ho, Y., and Wetzler, L.M. (2000). *Neisseria meningitidis* porin PorB interacts with mitochondria and protects cells from apoptosis. Proc. Natl. Acad. Sci. USA *97*, 9070–9075.

Massari, P., Visintin, A., Gunawardana, J., Halmen, K.A., King, C.A., Golenbock, D.T., and Wetzler, L.M. (2006). Meningococcal porin PorB binds to TLR2 and requires TLR1 for signaling. J. Immunol. *176*, 2373–2380.

McCutchan, J.A., Wunderlich, A., and Braude, A.I. (1977). Role of urinary solutes in natural immunity to gonorrhea. Infect. Immun. *15*, 149–155.

McKenna, W.R., Mickelsen, P.A., Sparling, P.F., and Dyer, D.W. (1988). Iron uptake from lactoferrin and transferrin by *Neisseria gonorrhoeae*. Infect. Immun. *56*, 785–791.

Medzhitov, R. (2007). Recognition of microorganisms and activation of the immune response. Nature *449*, 819–826.

Medzhitov, R., and Janeway, C.A., Jr. (1997). Innate immunity: The virtues of a nonclonal system of recognition. Cell *91*, 295–298.

Merz, A.J., Enns, C.A., and So, M. (1999). Type IV pili of pathogenic neisseriae elicit cortical plaque formation in epithelial cells. Mol. Microbiol. *32*, 1316–1332.

Merz, A.J., So, M., and Sheetz, M.P. (2000). Pilus retraction powers bacterial twitching motility. Nature *407*, 98–102.

Meyer, T.F., Mlawer, N., and So, M. (1982). Pilus expression in *Neisseria gonorrhoeae* involves chromosomal rearrangements. Cell *30*, 45–52.

Mosleh, I.M., Huber, L.A., Steinlein, P., Pasquali, C., Gunther, D., and Meyer, T.F. (1998). *Neisseria gonorrhoeae* porin modulates phagosome maturation. J. Biol. Chem. *273*, 35332–35338.

Muenzner, P., Billker, O., Meyer, T.F., and Naumann, M. (2002). Nuclear factor-kappa B directs carcinoembryonic antigen-related cellular adhesion molecule 1 receptor expression in *Neisseria gonorrhoeae*-infected epithelial cells. J. Biol. Chem. *277*, 7438–7446.

Muenzner, P., Naumann, M., Meyer, T.F., and Gray-Owen, S.D. (2001). Pathogenic *Neisseria* trigger expression of their carcinoembryonic antigen-related cellular adhesion molecule 1 (CEACAM1; previously CD66a) receptor on primary endothelial cells by activating the immediate early response transcription factor, nuclear factor-kappaB.J. Biol. Chem. *276*, 24331–24340.

Muller, A., Gunther, D., Dux, F., Naumann, M., Meyer, T.F., and Rudel, T. (1999). Neisserial porin (PorB) causes rapid calcium influx in target cells and induces apoptosis by the activation of cysteine proteases. EMBO J. *18*, 339–352.

Murphy, G.L., Connell, T.D., Barritt, D.S., Koomey, M., and Cannon, J.G. (1989). Phase variation of gonococcal protein II: regulation of gene expression by slipped-strand mispairing of a repetitive DNA sequence. Cell *56*, 539–547.

Nassif, X., Marceau, M., Pujol, C., Pron, B., Beretti, J.L., and Taha, M.K. (1997). Type-4 pili and meningococcal adhesiveness. Gene *192*, 149–153.

Naumann, M., Rudel, T., Wieland, B., Bartsch, C., and Meyer, T.F. (1998). Coordinate activation of activator protein 1 and inflammatory cytokines in response to *Neisseria gonorrhoeae* epithelial cell contact involves stress response kinases. J. Exp. Med. *188*, 1277–1286.

Naumann, M., Wessler, S., and Meyer, T.F. (1997). *Neisseria gonorrhoeae* epithelial cell interaction leads to the activation of the transcription factors nuclear factor kappaB and activator protein 1 and the induction of inflammatory cytokines. J. Exp. Med. *186*, 247–258.

Palucka, K., and Banchereau, J. (1999). Dendritic cells: a link between innate and adaptive immunity. J. Clin. Immunol. *19*, 12–25.

Pantelic, M., Kim, Y.-J., Bolland, S., Chen, I., Shively, J., and Chen, T. (2005). *Neisseria gonorrhoeae* kills carcinoembryonic antigen-related cellular adhesion molecule 1 (CD66a)-expressing human B-cells and inhibits antibody production. Infect. Immun. *73*, 4171–4179.

Parge, H.E., Bernstein, S.L., Deal, C.D., McRee, D.E., Christensen, D., Capozza, M.A., Kays, B.W., Fieser,

T.M., Draper, D., So, M., Getzoff, E.D., and Tainer, J. (1990). Biochemical purification and crystallographic characterization of the fiber-forming protein pilin from *Neisseria gonorrhoeae*. J. Biol. Chem. *265*, 2278–2285.

Patrone, J.B., Bish, S.E., and Stein, D.C. (2006). TNF-alpha-independent IL-8 expression: alterations in bacterial challenge dose cause differential human monocytic cytokine response. J. Immuno.l *177*, 1314–1322.

Patrone, J.B., and Stein, D.C. (2007). Effect of gonococcal lipooligosaccharide variation on human monocytic cytokine profile. BMC Microbiol. *7*, 7.

Petricoin III, E.F., Danaher, R., and Stein, D.C. (1991). Analysis of the *lsi* region involved in lipooligosaccharide biosynthesis in *Neisseria gonorrhoeae*. J. Bacteriol. *173*, 7896–7902.

Plant, L.J., and Jonsson, A.B. (2006). Type IV pili of *Neisseria gonorrhoeae* influence the activation of human CD4(+) T-cells. Infect. Immun. *74*, 442–448.

Plante, M., Cadieux, N., Rioux, C.R., Hamel, J., Brodeur, B.R., and Martin, D. (1999). Antigenic and molecular conservation of the gonococcal NspA protein. Infect. Immun. *67*, 2855–2861.

Plaut, A.G., Gilbert, J.V., Artenstein, M.S., and Capra, J.D. (1975). *Neisseria gonorrhoeae* and *Neisseria meningitidis*: extracellular enzyme cleaves human immunoglobulin A. Science *190*, 1103–1105.

Plummer, F.A., Chubb, H., Simonsen, J.N., Bosire, M., Slaney, L., Maclean, I., Ndinya-Achola, J.O., Waiyaki, P., and Brunham, R.C. (1993). Antibody to Rmp (outer membrane protein 3) increases susceptibility to gonococcal infection. J. Clin. Invest. *91*, 339–343.

Plummer, F.A., Chubb, H., Simonsen, J.N., Bosire, M., Slaney, L., Nagelkerke, N.J., Maclean, I., Ndinya-Achola, J.O., Waiyaki, P., and Brunham, R.C. (1994). Antibodies to opacity proteins (Opa) correlate with a reduced risk of gonococcal salpingitis. J. Clin. Invest. *93*, 1748–1755.

Porter, E., Yang, H., Yavagal, S., Preza, G.C., Murillo, O., Lima, H., Greene, S., Mahoozi, L., Klein-Patel, M., Diamond, G., Gulati, S., Ganz, T., Rice, P.A., and Quayle, A.J. (2005). Distinct defensin profiles in *Neisseria gonorrhoeae* and *Chlamydia trachomatis* urethritis reveal novel epithelial cell–neutrophil interactions. Infect. Immun. *73*, 4823–4833.

Post, D.M., Phillips, N.J., Shao, J.Q., Entz, D.D., Gibson, B.W., and Apicella, M.A. (2002). Intracellular survival of *Neisseria gonorrhoeae* in male urethral epithelial cells: importance of a hexaacyl lipid A. Infect. Immun. *70*, 909–920.

Price, R.J., and Boettcher, B. (1979). The presence of complement in human cervical mucus and its possible relevance to infertility in women with complement-dependent sperm- immobilizing antibodies. Fertil. Steril. *32*, 61–66.

Pridmore, A.C., Jarvis, G.A., John, C.M., Jack, D.L., Dower, S.K., and Read, R.C. (2003). Activation of toll-like receptor 2 (TLR2) and TLR4/MD2 by *Neisseria* is independent of capsule and lipooligosaccharide (LOS) sialylation but varies widely among LOS from different strains. Infect. Immun. *71*, 3901–3908.

Punsalang, A.P.J., and Sawyer, W.D. (1973). Role of pili in the virulence of *Neisseria gonorrhoeae*. Infect. Immun. *8*, 255–263.

Rahman, M., Kallstrom, H., Normark, S., and Jonsson, A.B. (1997). PilC of pathogenic *Neisseria* is associated with the bacterial cell surface. Mol. Microbiol. *25*, 11–25.

Rahman, M.M., Kolli, V.S., Kahler, C.M., Shih, G., Stephens, D.S., and Carlson, R.W. (2000). The membrane phospholipids of *Neisseria meningitidis* and *Neisseria gonorrhoeae* as characterized by fast atom bombardment mass spectrometry. Microbiology *146*, 1901–1911.

Ram, S., Cox, A.D., Wright, J.C., Vogel, U., Getzlaff, S., Boden, R., Li, J., Plested, J.S., Meri, S., Gulati, S., Stein, D.C., Richards, J.C., Moxon, E.R., and Rice, P.A. (2003). Neisserial lipooligosaccharide is a target for complement component C4b: Inner core phosphoethanolamine residues define C4b linkage specificity. J. Biol. Chem. *278*, 50851–50862.

Ram, S., Mackinnon, F.G., Gulati, S., McQuillen, D.P., Vogel, U., Frosch, M., Elkins, C., Guttormsen, H.K., Wetzler, L.M., Oppermann, M., Pangburn, M.K., and Rice, P.A. (1999). The contrasting mechanisms of serum resistance of *Neisseria gonorrhoeae* and group B *Neisseria meningitidis*. Mol. Immunol. *36*, 915–928.

Ram, S., McQuillen, D.P., Gulati, S., Elkins, C., Pangburn, M.K., and Rice, P.A. (1998). Binding of complement factor H to loop 5 of porin protein 1A: a molecular mechanism of serum resistance of nonsialylated *Neisseria gonorrhoeae*. J. Exp. Med. *188*, 671–680.

Ramsey, K.H., Schneider, H., Cross, A.S., Boslego, J.W., Hoover, D.L., Staley, T.L., Kuschner, R.A., and Deal, C.D. (1995). Inflammatory cytokines produced in response to experimental human gonorrhea. J. Infect. Dis. *172*, 186–191.

Redondo-Lopez, V., Cook, R.L., and Sobel, J.D. (1990). Emerging role of lactobacilli in the control and maintenance of the vaginal microflora.. Rev Infect Dis *12*, 856–872.

Rescigno, M., Granucci, F., Citterio, S., Foti, M., and Ricciardi-Castagnoli, P. (1999). Coordinated events during bacteria-induced DC maturation. Immunol. Today *20*, 200–203.

Rest, R.F., and Shafer, W.M. (1989). Interactions of *Neisseria gonorrhoeae* with human neutrophils. Clin. Microbiol. Rev. 2 Suppl., S83–91.

Rice, P.A., and Goldenberg, D.L. (1981). Clinical manifestations of disseminated infection caused by *Neisseria gonorrhoeae* are linked to differences in bactericidal reactivity of infecting strains. Ann Intern, Med *95*, 175–178.

Riordan, F.A., Marzouk, O., Thomson, A.P., Sills, J.A., and Hart, C.A. (1996). Proinflammatory and anti-inflammatory cytokines in meningococcal disease. Arch Dis Child *75*, 453–454.

Ross, S.C., and Densen, P. (1984). Complement deficiency states and infection: epidemiology, pathogenesis and consequences of neisserial and other infections in an immune deficiency. Medicine (Baltimore) *63*, 243–273.

Rowe, H.A., Griffiths, N.J., Hill, D.J., and Virji, M. (2007). Co-ordinate action of bacterial adhesins and human carcinoembryonic antigen receptors in enhanced cellular invasion by capsulate serum resistant *Neisseria meningitidis*. Cell Microbiol 9, 154–168.

Roy, K., Wang, S.A., and Meltzer, M.I. (2005). Optimizing treatment of antimicrobial-resistant *Neisseria gonorrhoeae*. Emerg. Infect. Dis. *11*, 1265–1273.

Rudel, T., Scheurerpflug, I., and Meyer, T.F. (1995). *Neisseria* PilC protein identified as type-4 pilus tip-located adhesin. Nature *373*, 357–359.

Russell, M.W., Hedges, S.R., Wu, H.Y., Hook, E.W., 3rd, and Mestecky, J. (1999). Mucosal immunity in the genital tract: prospects for vaccines against sexually transmitted diseases – a review. Am. J. Reprod. Immunol. *42*, 58–63.

Saigh, J.H., Sanders, C.C., and Sanders, W.E., J.r (1978). Inhibition of *Neisseria gonorrhoeae* by aerobic and facultatively anaerobic components of the endocervical flora: evidence for a protective effect against infection. Infect. Immun. *19*, 704–710.

Sarkari, J., Pandit, N., Moxon, E.R., and Achtman, M. (1994). Variable expression of the Opc outer membrane protein in *Neisseria meningitidis* is caused by size variation of a promoter containing poly- cytidine. Mol. Microbiol. *13*, 207–217.

Scheuerpflug, I., Rudel, T., Ryll, R., Pandit, J., and Meyer, T.F. (1999). Roles of PilC and PilE proteins in pilus-mediated adherence of *Neisseria gonorrhoeae* and *Neisseria meningitidis* to human erythrocytes and endothelial and epithelial cells. Infect. Immun. *67*, 834–843.

Schmidt, K.A., Schneider, H., Lindstrom, J.A., Boslego, J.W., Warren, R.A., Van de Verg, L., Deal, C.D., McClain, J.B., and Griffiss, J.M. (2001). Experimental gonococcal urethritis and reinfection with homologous gonococci in male volunteers. Sex. Transm. Dis. *28*, 555–564.

Schnare, M., Barton, G.M., Holt, A.C., Takeda, K., Akira, S., and Medzhitov, R. (2001). Toll-like receptors control activation of adaptive immune responses. Nat. Immunol. *2*, 947- 950.

Schneider, M.C., Exley, R.M., Ram, S., Sim, R.B., and Tang, C.M. (2007). Interactions between *Neisseria meningitidis* and the complement system. Trends Microbio.l *15*, 233–240.

Schryvers, A.B., and Morris, L.J. (1988). Identification and characterization of the human lactoferrin-binding protein from *Neisseria meningitidis*. Infect. Immun. *56*, 1144–1149.

Segal, E., Billyard, E., So, M., Storzbach, S., and Meyer, T.F. (1985). Role of chromosomal rearrangement in *Neisseria gonorrhoeae* pilus phase variation. Cell *40*, 293–302.

Shafer, W.M., Qu, X., Waring, A.J., and Lehrer, R.I. (1998). Modulation of *Neisseria gonorrhoeae* susceptibility to vertebrate antibacterial peptides due to a member of the resistance/nodulation/division efflux pump family. Proc. Natl. Acad. Sci. USA *95*, 1829–1833.

Shimazu, R., Akashi, S., Ogata, H., Nagai, Y., Fukudome, K., Miyake, K., and Kimoto, M. (1999). MD-2, a molecule that confers lipopolysaccharide responsiveness on Toll-like receptor 4. J. Exp. Med. *189*, 1777–1782.

Sim, R.J., Harrison, M.M., Moxon, E.R., and Tang, C.M. (2000). Underestimation of meningococci in tonsillar tissue by nasopharyngeal swabbing. Lancet *356*, 1653–1654.

Simms, A.N., and Jerse, A.E. (2006). In vivo selection for *Neisseria gonorrhoeae* opacity protein expression in the absence of human carcinoembryonic antigen cell adhesion molecules. Infect. Immun. *74*, 2965–2974.

Simons, M.P., Nauseef, W.M., and Apicella, M.A. (2005). Interactions of *Neisseria gonorrhoeae* with adherent polymorphonuclear leukocytes. Infect. Immun. *73*, 1971–1977.

Singleton, T.E., Massari, P., and Wetzler, L.M. (2005). Neisserial porin-induced dendritic cell activation is MyD88 and TLR2 dependent. J. Immunol *174*, 3545–3550.

Smith, H., Parsons, N.J., and Cole, J.A. (1995). Sialylation of neisserial lipopolysaccharide: a major influence on pathogenicity. Microb. Pathog. *19*, 365–377.

Snapper, C.M., Rosas, F.R., and Wetzler, L.M. (1997). Neisserial porins may provide critical second signals to polysaccharide-activated murine B-cells for induction of immunoglobulin secretion. Infect. Immun. *65*, 3203–3208.

Song, W., Ma, L., Chen, W., and Stein, D.C. (2000). Role of lipooligosaccharide in *opa*-independent invasion of *Neisseria gonorrhoeae* into human epithelial cells. J. Exp. Med. *191*, 949–959.

Sparling, P.F. (1966). Genetic transformation of *Neisseria gonorrhoeae* to streptomycin resistance. J. Bacteriol. *92*, 1364–1371.

Spinosa, M.R., Progida, C., Talà, A., Cogli, L., Alifano, P., and Bucci, C. (2007). The *Neisseria meningitidis* capsule is important for intracellular survival in human cells. Infect. Immun. *75*, 3594–3603.

Sprong, T., Pickkers, P., Geurts-Moespot, A., van der Ven-Jongekrijg, J., Noeleman, C., Knaup, M., Leroy, D., Calandra, T., van der Meer, J.W.M., Sweep, F., van Kooyk, Y., and van de Winkel, J.G.J. (2007). Macrophage migration inhibitory factor (MIF) in meningococcal septic shock and experimental human endotoxemia. Shock *27*, 482–487.

Steeghs, L., den Hartog, R., den Boer, A., Zomer, B., Roholl, P., and van der Ley, P. (1998). Meningitis bacterium without endotoxin. Nature *392*, 449–450.

Steeghs, L., van Vliet, S.J., Uronen-Hansson, H., van Mourik, A., Engering, A., Sanchez-Hernandez, M., Klein, N., Callard, R., van Putten, J.P.M., van der Ley, P., and van Deuren, M. (2006). *Neisseria meningitidis* expressing *lgtB* lipopolysaccharide targets DC-SIGN and modulates dendritic cell function. Cell. Microbiol. *8*, 316–325.

Stephens, D.S., McAllister, C.F., Zhou, D., Lee, F.K., and Apicella, M.A. (1994). Tn916-generated, lipooligosaccharide mutants of *Neisseria meningitidis* and *Neisseria gonorrhoeae*. Infect. Immun. *62*, 2947–2952.

Stephens, D.S., and McGee, Z.A. (1981). Attachment of *Neisseria meningitidis* to human mucosal surfaces: influence of pili and type of receptor cell. J. Infect. Dis. *143*, 525–532.

Stephens, D.S., McGee, Z.A., Melly, M.A., Hoffman, L.H., and Gregg, C.R. (1982). Attachment of pathogenic *Neisseria* to human mucosal surfaces: role in pathogenesis. Infection *10*, 192–195.

Stephens, D.S., Munford, R.S., and Wetzler, L.M. (2004) Meningococcal infections. In Harrison's Principles of Internal Medicine, D.L. Kasper, E. Braunwald, A.S. Fauci, S.L. Hauser, D.L. Longo, J.L. Jameson, and K.J. Isselbacher, eds. (McGraw-Hill, New York).

Stimson, E., Virji, M., Makepeace, K., Dell, A., Morris, H.R., Payne, G., Saunders, J.R., Jennings, M.P., Barker, S., Panico, M., Blench, I., and Moxon, E.R. (1995). Meningococcal pilin: a glycoprotein substituted with digalactosyl 2,4-diacetamido-2,4,6-trideoxyhexose. Mol. Microbiol. *17*, 1201–1214.

Strober, W., Murray, P.J., Kitani, A., and Watanabe, T. (2006). Signalling pathways and molecular interactions of NOD1 and NOD2. Nat Rev Immunol *6*, 9–20.

Swanson, J. (1973). Studies on gonococcus infection. IV. Pili: their role in attachment of gonococci to tissue culture cells. J. Exp. Med. *137*, 571–589.

Swanson, J. (1978). Studies on gonococcus infection. XIV. Cell wall protein differences among color/opacity colony variants of *Neisseria gonorrhoeae*. Infect. Immun. *21*, 292–302.

Takeda, K., Kaisho, T., and Akira, S. (2003). Toll-like receptors. Annu. Rev. Immuno.l *21*, 335–376.

Taylor, L.S., Paul, S.P., and McVicar, D.W. (2000). Paired inhibitory and activating receptor signals. Rev. Immunogenet. 2.

Tong, Y., Arking, D., Ye, S., Reinhold, B., Reinhold, V., and Stein, D.C. (2002). *Neisseria gonorrhoeae* strain PID2 simultaneously expresses six chemically related lipooligosaccharide structures. Glycobiology *12*, 523–533.

Tosi, M.F. (2005). Innate immune responses to infection. J. Allergy Clin. Immunol *116*, 241–249.

Trees, D.L., Schultz, A.J., and Knapp, J.S. (2000). Use of the neisserial lipoprotein (Lip) for subtyping *Neisseria gonorrhoeae*. J. Clin. Microbiol. *38*, 2914–2916.

Turner, M.W., Johansson, S.G.O., Barratt, T.M., and Bennich, H. (1970). Studies on the levels of immunoglobulins in normal human urine with particular reference to IgE. Int Arch Allergy *37*, 409–417.

Turner, P.C., Thomas, C.E., Stojiljkovic, I., Elkins, C., Kizel, G., Ala'Aldeen, D.A., and Sparling, P.F. (2001). Neisserial TonB-dependent outer-membrane proteins: detection, regulation and distribution of three putative candidates identified from the genome sequences. Microbiology *147*, 1277–1290.

Tzeng, Y.L., Ambrose, K.D., Zughaier, S., Zhou, X., Miller, Y.K., Shafer, W.M., and Stephens, D.S. (2005). Cationic antimicrobial peptide resistance in *Neisseria meningitidis*. J. Bacteriol. *187*, 5387–5396

Ulevitch, R.J., and Tobias, P.S. (1995). Receptor-dependent mechanisms of cell stimulation by bacterial endotoxin. Annu. Rev. Immunol. *13*, 437–457.

Uronen-Hansson, H., Steeghs, L., Allen, J., Dixon, G.L., Osman, M., van der Ley, P., Wong, S.Y., Callard, R., and Klein, N. (2004). Human dendritic cell activation by *Neisseria meningitidis*: phagocytosis depends on expression of lipooligosaccharide (LOS) by the bacteria and is required for optimal cytokine production. Cell. Microbiol. *6*, 625–637.

Valore, E.V., Park, C.H., Igreti, S.L., and Ganz, T. (2002). Antimicrobial components of vaginal fluid. Am. J. Obstet. Gynecol. *187*, 561–568.

Van Der Ley, P., and Steeghs, L. (2003). Lessons from an LPS-deficient *Neisseria meningitidis* mutant. J. Endotoxin Res. *9*, 124–128.

Van Putten, J.M., Duensing, T.D., and Carlson, J. (1998a). Gonococcal invasion of epithelial cells driven by PIA: a bacterial ion channel with GTP binding properties. J. Exp. Med. *188*, 941–952.

van Putten, J.P., Duensing, T.D., and Cole, R.L. (1998b). Entry of OpaA+ gonococci into HEp-2 cells requires concerted action of glycosaminoglycans, fibronectin and integrin receptors. Mol. Microbiol. *29*, 369–379.

Vanderpuye, O.A., Labarrere, C.A., and McIntyre, J.A. (1992). The complement system in human reproduction. Am J Reprod Immunol *27*, 145–155.

Veale, D.R., Goldner, M., Penn, C.W., Ward, J., and Smith, H. (1979). Intracellular survival and growth of gonococci in human phagocytes. J Gen Microbiol *113*, 383–393.

Virji, M., Evans, D., Hadfield, A., Grunert, F., Teixeira, A.M., and Watt, S.M. (1999). Critical determinants of host receptor targeting by *Neisseria meningitidis* and *Neisseria gonorrhoeae*: identification of Opa adhesiotopes on the N-domain of CD66 molecules. Mol. Microbiol. *34*, 538–551.

Virji, M., and Heckels, J.E. (1986). The effect of protein II and pili on the interaction of *Neisseria gonorrhoeae* with human polymorphonuclear leucocytes. J Gen Microbiol *132 (Pt 2)*, 503–512.

Virji, M., Makepeace, K., Ferguson, D.J., Achtman, M., and Moxon, E.R. (1993). Meningococcal Opa and Opc proteins: their role in colonization and invasion of human epithelial and endothelial cells. Mol. Microbiol. *10*, 499–510.

Virji, M., Makepeace, K., Peak, I.R., Ferguson, D.J., and Moxon, E.R. (1996). Pathogenic mechanisms of *Neisseria meningitidis*. Ann N Y Acad Sci *797*, 273–276.

Vitovski, S., Read, R.C., and Sayers, J.R. (1999). Invasive isolates of *Neisseria meningitidis* possess enhanced immunoglobulin A1 protease activity compared to colonizing strains. Faseb J *13*, 331–337.

Walport, M.J. (2001). Complement. N Engl J Med *344*, 1058–1066.

Weel, J.F., Hopman, C.T., and van Putten, J.P. (1991). Bacterial entry and intracellular processing of *Neisseria gonorrhoeae* in epithelial cells: immunomorphological evidence for alterations in the major outer membrane protein P.IB.J. Exp. Med. *174*, 705–715.

Wetzler, L.M., Ho, Y., and Reiser, H. (1996). Neisserial porins induce B lymphocytes to express costimulatory B7-2 molecules and to proliferate. J. Exp. Med. *183*, 1151–1159.

Wira, C.R., Grant-Tschudy, K.S., and Crane-Godreau, M.A. (2005). Epithelial cells in the female reproductive tract: a central role as sentinels of immune protection. Am J Reprod Immunol *53*, 65–76.

Yamasaki, R., Kerwood, D.E., Schneider, H., Quinn, K.P., Griffiss, J.M., and Mandrell, R.E. (1994). The structure of lipooligosaccharide produced by *Neisseria gonorrhoeae*, strain 15253, isolated from a patient with disseminated infection: Evidence for a new glycosylation pathway of the gonococcal lipooligosaccharide. J. Biol. Chem. *269*, 30345–30351.

Yamasaki, R., Nasholds, W., Schneider, H., and Apicella, M.A. (1991). Epitope expression and partial structural characterization of F62 lipooligosaccharide (LOS) of *Neisseria gonorrhoeae*: IgM monoclonal antibodies (3F11 and 1-1-M) recognize non-reducing termini of the LOS components. Mol Immunol *28*, 1233–1242.

Zhu, W., Thomas, C.E., Chen, C.J., Van Dam, C.N., Johnston, R.E., Davis, N.L., and Sparling, P.F. (2005). Comparison of immune responses to gonococcal PorB delivered as outer membrane vesicles, recombinant protein, or Venezuelan equine encephalitis virus replicon particles. Infect. Immun. *73*, 7558–7568.

Zimmer, S.M., Zughaier, S.M., Tzeng, Y.L., and Stephens, D.S. (2007). Human MD-2 discrimination of meningococcal lipid A structures and activation of TLR4. Glycobiol *17*, 847–856.

Zimmerman, G.A., Prescott, S.M., and McIntyre, T.M. (1995). Endothelial cell interactions with granulocytes: tethering and signalling molecules Immunol Today *13*, 93–100.

Zughaier, S., Agrawal, S., Stephens, D.S., and Pulendran, B. (2006). Hexa-acylation and KDO(2)-glycosylation determine the specific immunostimulatory activity of *Neisseria meningitidis* lipid A for human monocyte derived dendritic cells. Vaccine *24*, 1291–1297.

Zughaier, S.M., Tzeng, Y.L., Zimmer, S.M., Datta, A., Carlson, R.W., and Stephens, D.S. (2004). *Neisseria meningitidis* lipooligosaccharide structure-dependent activation of the macrophage CD14/Toll-like receptor 4 pathway. Infect. Immun. *72*, 371–380.

Zughaier, S.M., Zimmer, S.M., Datta, A., Carlson, R.W., and Stephens, D.S. (2005). Differential induction of the toll-like receptor 4-MyD88-dependent and -independent signaling pathways by endotoxins. Infect. Immun. *73*, 2940–2950.

Interactions of *Neisseria* with Complement

8

Lisa A. Lewis, Elizabeth Burrowes, Peter A. Rice and Sanjay Ram

Abstract

Neisseria gonorrhoeae and *Neisseria meningitidis* have evolved intricate mechanisms to evade killing by the complement system. Binding of complement inhibitors, LOS sialylation and expression of capsular polysaccharide in the case of *N. meningitidis* all play key roles in enabling these bacteria to evade complement. The multiplicity of complement evasion strategies reflects the importance of overcoming this immune barrier. Neisserial resistance to complement mediated killing appears to be restricted to humans and may provide an explanation for the host specificity of neisserial disease. Recent advances in our understanding of neisserial complement evasion strategies should aid in developing better animal models and vaccine design.

Introduction

The complement system is a key arm of innate immune defences against invading pathogens, both in the blood and on mucosal surfaces. The importance of complement in host defences against neisserial infections has been well established. *Neisseria gonorrhoeae* and *Neisseria meningitidis* have developed several sophisticated mechanisms to evade complement. Although both pathogens use similar mechanisms to evade complement, such as lipooligosaccharide (LOS) sialylation and the ability to bind to complement inhibitory proteins, recent comparative studies of complement evasion mechanisms across the two species have shed light on subtle mechanistic differences of these common strategies used by gonococci and meningococci. This chapter aims to provide a basic understanding of the fluid-phase of the complement system as it relates to neisserial complement evasion strategies. The role of membrane bound complement receptors and regulatory proteins in promoting bacterial adherence and invasion is beyond the scope of this chapter and is not reviewed here. While the literature on complement interactions with neisseriae is extensive, this chapter aims to provide a cohesive review to bridge several observations that have been made over the past three decades with more recent data and also to provide possible avenues for future investigation.

The complement system and its regulation

Introduction

The complement system is a key component of innate immunity, initially described as a heat-labile bactericidal substance that 'complemented' antibodies in destroying bacteria. Complement is a biochemical cascade of approximately 35 fluid-phase and membrane-bound proteins found in the blood, normally circulating as inactive zymogens, which protect the host against invading microorganisms by promoting the rapid recognition and clearance of pathogens, infected cells and immune complexes; virtually all invading pathogens are confronted by complement directly following entry into the human host. Three different pathways can activate the complement cascade, each is a series of proteolytic reactions: (1) the classical pathway, (2) the alternative pathway and (3) the mannose-binding lectin

(MBL) pathway (Fig. 8.1). The end result of this activation cascade is massive amplification of the response and activation of the cell-killing membrane attack complex, as may occur with Gram-negative bacteria, or enhanced deposition of C3 fragments that promote opsonophagocytic killing of Gram-positive and -negative bacteria or fungal pathogens. For a detailed review of the complement cascade refer to Walport (2001a,b).

The classical pathway

The classical pathway, the first to be discovered, is initiated by activation of the C1-complex. C1 is a multi-subunit protein composed of a C1q molecule and two molecules each of C1r and C1s. C1q can bind directly to many bacteria and also to IgM and IgG antibodies that have interacted with an antigen, resulting in activation of C1qrs (Cooper, 1985). Activated C1s cleaves the serum

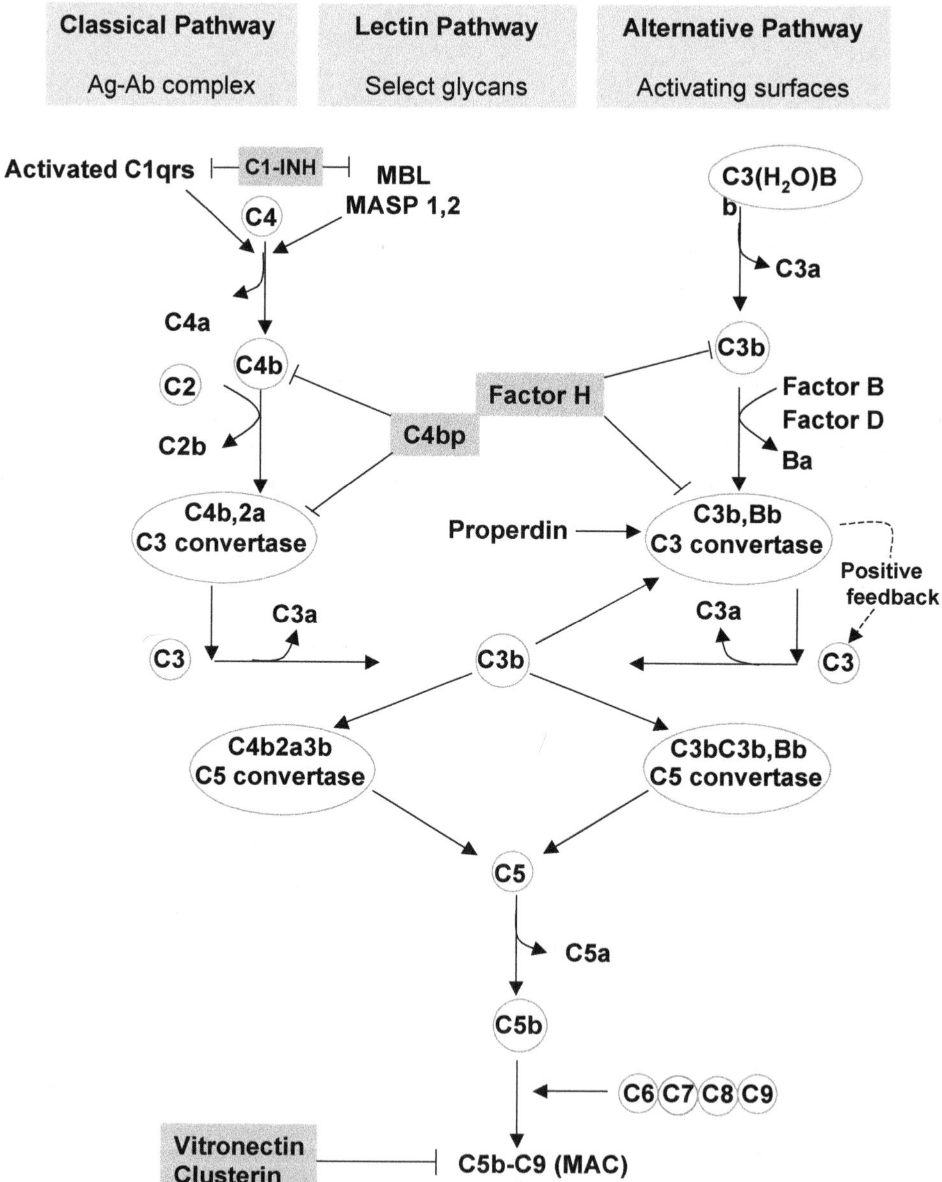

Figure 8.1 Schematic of the complement cascade. Negative regulators are indicated by grey boxes. Mannose binding lectin binds to a range of select glycans including *N*-acetyl-D-glucosamine, mannose, *N*-acetyl-mannosamine, fucose and glucose.

protein C4, separating 77 amino acids (C4a) from the *N*-terminus of the α-chain, and creating the metastable C4b molecule. This results in activation of the exposed, highly labile, internal thioester bond of C4b such that the carbonyl group linked to the native thioester sulphur becomes more electrophilic. Nucleophilic attack of this bond by either hydroxyl or amino groups exposed on nearby molecules leads to the formation of ester or amide linkages, respectively, that covalently anchor the nascently activated C4 to the target surface (Fearon, 1978; Lachmann and Hughes-Jones, 1984; Muller-Eberhard, 1988). Alternatively, the carbonyl group can react with H_2O and become hydrolysed. C1s also cleaves C2 into C2b, which binds non-covalently to a site on C4b to form C4b2a, the C3 convertase of the classical pathway. Production of C3 convertase leads to cleavage of C3 into C3a and C3b; C3b binds to the C3 convertase to make C5 convertase.

There are two isoforms of C4 in human serum, called C4A and C4B, each with distinct properties. C4A forms amide linkages with its targets, while C4B preferentially forms ester linkages. C4A is believed to engage CR1 more efficiently than C4B, while C4B may be haemolytically more active than C4A. The role of the isoforms of C4 in protecting against *N. meningitidis* infection remains controversial. Two studies suggested an increased risk of meningitis associated with homozygous C4B deficiency (Rowe *et al.*, 1989; Bishof *et al.*, 1990), while a larger study by Cates *et al.* *(1992)* did not confirm these findings.

The alternative pathway

The alternative pathway is initiated by the spontaneous and continuous low-grade hydrolysis of the thioester in circulating serum C3 to produce $C3(H_2O)$; $C3(H_2O)$ then binds Factor B, which forms the substrate for Factor D, whose action results in the formation of a C3 convertase, $C3(H_2O)Bb$, that initiates the alternative pathway. This complex is an unstable fluid-phase C3 convertase that cleaves C3 to yield C3a and C3b (Pangburn and Muller-Eberhard, 1984; Pangburn, 1986). Analogous to C4 cleavage, this reaction exposes the internal thioester bond in C3b and thereby promotes the formation of covalent ester or amide bonds with accessible hydroxyl or amino groups on target surfaces (Gordon and Hostetter, 1986). The C3b fragment covalently binds to cellular and microbial surfaces; if the cell or microbe surface is an activator, alternative pathway activation proceeds. If the C3b fragment binds to a non-activator cellular or microbial surface, alternative pathway activation is muted. Surface-bound C3b binds circulating factor B, which is cleaved by factor D to produce the membrane-bound, alternative-pathway C3 convertase, C3bBb. Cleavage of C3 by this enzymatic complex initiates an amplification loop and results in the rapid generation of additional C3b (Pangburn, 1986). This positive feedback loop is a salient characteristic of the alternative pathway. Addition of a second C3b to the C3 convertase leads to the formation of a C5 convertase (C3bC3bBb). Factor P (properdin) stabilizes both the C3 and C5 alternative pathway convertases from decay dissociation. The functional role of properdin in stabilizing the alternative pathway convertase has long been established; however, new evidence demonstrates that properdin can also bind to certain microbial surfaces (including a strain of *N. gonorrhoeae*), and provide a platform for *de novo* convertase assembly (Spitzer *et al.*, 2007). Properdin deficiency in humans, as discussed below, is associated with severe fulminant neisserial infections with a high mortality rate (Braconier *et al.*, 1983).

The mannose-binding lectin pathway

The lectin pathway is analogous to the classical pathway; both pathways share C2, C3 and C4. However, initiation of the mannose-binding lectin pathway is antibody independent. The lectin pathway is activated when mannose-binding-lectin (MBL) binds to mannose and other sugar residues of polysaccharide structures on microbial surfaces; MBL is thought to mediate killing via complement activation or by enhancing phagocytosis by acting as an opsonin. MBL is a member of the collectin family of proteins and is associated with three novel serine proteases; MASP-1, MASP-2 and MASP-3. MASP-2 mediates most of the protease activity and cleaves C4. MASP-1 has minimal C3 and C2 cleaving activity and its role in complement activation is uncertain. Although the function of MASP-3 is not resolved, recombinant MASP-3 can inhibit C4-cleaving activity of recombinant as well as naturally occurring MBL–MASP-2 complexes. Like C1, once activated, the MBL–MASP-2 complex is able to cleave C4 and C2 to form C4b2a, the classical pathway C3 convertase that is able to enzymatically split hundreds of molecules of C3 into C3a and C3b.

Assembly of the membrane attack complex

Complement activation by all three pathways results in the formation of the membrane attack complex (MAC), also known as C5b-9. After binding to C5 convertase, C5 is cleaved into C5a and C5b. C5a diffuses away and acts as an inflammatory mediator, activating a number of cells through G-protein-linked receptors. The C5b fragment binds directly to the target microorganism or cell wall and serves as the binding site for other terminal complement components (Muller-Eberhard, 1986). C5b binds to C6, forming a stable bimolecular complex that then binds C7. The addition of C7 changes the conformation of the complex so that it becomes hydrophobic and able to fully insert into membranes. The binding of one molecule of C8 to each C5b67 complex gives rise to small transmembrane channels that disrupt the target membranes. Each membrane-bound C5b678 complex acts as a receptor for multiple C9 molecules; binding of the first C9 molecule initiates a process of C9 oligomerization at the site of membrane attack. Once assembled in the target cell membrane, the MAC creates transmembrane channels that allows free passage of ions and small molecules that leads to osmotic lysis of a cell and death in the case of Gram-negative bacteria.

Fluid-phase regulation of the complement cascade

Complement activation occurs at sites of tissue damage and on foreign surfaces; under normal physiological conditions activation is kept to a minimum. Because uncontrolled activation of the complement cascade is detrimental to the host, it must be tightly regulated, which is achieved by fluid-phase and membrane-bound regulator proteins. A soluble-phase regulator of the early part of the classical pathway is C1-inhibitor, a serine protease inhibitor (SERPIN) that binds to and inactivates C1r, C1s and the MASPs (Arlaud *et al.*, 2001). C1-inhibitor can also inhibit the alternative pathway because it binds to C3b and prevents factor B binding to C3b (Jiang *et al.*, 2001). The MBL pathway is regulated by α_2-macroglobulin.

The multimeric C4b-binding protein (C4BP) is an important regulator of the classical pathway. C4BP binds to C4b, accelerates the decay of the classical pathway C3 and C5 convertases (C4b2a and C4bC3bC2a) and acts as a co-factor in the factor I-mediated cleavage of C4b to the haemolytically inactive C4d. Factor H (fH) is the analogous regulator of the alternative pathway. Factor H controls the activity of the alternative pathway C3 and C5 convertases (C3bBb and C3bC3bBb) by competing with factor B for C3b binding, by displacing the Bb subunit from the convertase, and by serving as a co-factor in the factor I-mediated cleavage and inactivation of C3b to iC3b. Several bacteria, including pathogenic neisseriae, bind to human fH and C4BP and regulate complement activation on their surfaces.

In serum, vitronectin (S protein) binds to the metastable C5b-9 complex and prevents the formation of MAC; vitronectin acts through its heparin binding site and prevents the polymerization of C9. Because MAC is capable of inserting into any cell membrane, vitronectin protects cells that are adjacent to sites of complement activation from accidental attack.

Membrane associated complement receptors and regulators

Several host cell surface molecules act as ligands for complement components or their breakdown products. These include the C1q receptor (C1qR), the complement receptors (CRs) 1, 2, 3 and 4 and receptors for complement activation products such as the C5a receptor. These complement receptors play an integral part in the immune response; for example, the CR3 receptor binds iC3b and is subsequently involved in the opsonophagocytic clearance of bacteria.

Membrane-bound regulators play a key role in controlling complement activity and prevent inappropriate activation against self: (a) they interfere with the assembly and function of convertase activity on cell membranes (decay-accelerating protein DAF or CD55), by serving as a cofactor for factor I mediated cleavage of C3b and C4b (membrane co-factor protein MCP or CD46) or in the case of CR1, perform both these functions, protecting the cells of the body from complement attack; (b) they transport and clear immune complexes (CR1) and (c) they mediate transmembrane and intracellular trafficking (CR1) and transmembrane signalling (CR2).

Neisserial disease in complement deficient individuals

Complement deficiencies and neisserial infections

The complement system is an essential component of the innate immune defence against infection by *N. meningitidis* and *N. gonorrhoeae*. People who lack or have a deficiency in antibody-dependent, complement-mediated immune bactericidal activity are most susceptible to neisserial disease; terminal complement component (C5 through C9) deficiencies and deficiencies of the alternative pathway have a strong effect on susceptibility to, as well as severity of, neisserial infections. Inherited defects affecting individual complement proteins are rare in the general population but the prevalence depends largely upon the ethnic origin of the population that is examined. C9 deficiency appears to be uncommon in Caucasians, but its reported prevalence among normal Japanese blood donors is 0.095% (Fukumori *et al.*, 1989); in the south-eastern United States, C6 deficiency is reported to be more prevalent in African Americans (1 in 1600) than in Caucasians (Zhu *et al.*, 2000).

Deficiencies in components of the alternative pathway, namely properdin and factor D, have been associated with increased susceptibility, almost exclusively, to meningococcal infections (Sjoholm *et al.*, 1982; Braconier *et al.*, 1983; Soderstrom *et al.*, 1989; Sprong *et al.*, 2006). Properdin stabilizes the alternative pathway C3 convertase (C3bBb) and persons with properdin deficiency have a higher mortality rate than normal individuals (Sjoholm *et al.*, 1982). Factor D is necessary for cleavage of factor B to yield Bb, a component of the alternative pathway C3 convertase. Deficiencies in regulatory proteins of the alternative pathway, such as fH and factor I, are also associated with meningococcal disease (Vyse *et al.*, 1994; Zipfel *et al.*, 1999). These deficiencies result in unimpeded consumption (and reduction) of C3, a major neisserial opsonin, and a decrease in the ability to form C3 convertases that are necessary to activate the terminal portion of the complement pathway. Conversely, it was reported recently that a single nucleotide polymorphism (C496T), which enhances the NF-κB-responsive element of the promoter of the gene encoding fH, is associated with higher fH levels, and results in a 2.9-fold increase in susceptibility to serogroup C meningococcal disease (Haralambous *et al.*, 2006). Binding of fH to meningococci constitutes an important mechanism of complement evasion (discussed below). Deficiency in the mannose binding lectin (MBL) pathway as evidenced by reduced serum mannan-binding-lectin levels has also been shown to predispose to meningococcal disease (Bax *et al.*, 1999; Kuipers *et al.*, 2003).

The prevalence of complement deficiency, particularly deficiency of one of the terminal components, in patients with disseminated gonococcal infection exceeds that of the general population and is an important host factor predisposing to systemic infection with *N. gonorrhoeae* (Ellison *et al.*, 1987).

The majority of persons with meningococcal disease do not have complement deficiency. Infants are protected from meningococcal disease for the first few months of life by transferred maternal antibodies. Susceptibility to meningococcal infection peaks at age 6–12 months when maternal antibodies wane but decreases again after colonization of closely related non-pathogenic bacteria such as *Neisseria lactamica* that have surface antigens in common with virulent strains of *N. meningitidis* (Gold *et al.*, 1978). Colonization with *N. meningitidis* itself gradually replaces non-pathogenic *Neisseria* and induces antibodies to infecting strains, thus reinforcing natural immunity (Bakir *et al.*, 2001; Yazdankhah and Caugant, 2004). Invasive disease occurs if no protective bactericidal antibodies are mounted against the infecting strain.

Complement activation by *Neisseria*

Classical pathway activation

Extensive complement consumption, resulting from activation of both the classical and alternative pathways, is a hallmark of fulminant meningococcal disease (Hoffman and Edwards, 1972; Greenwood *et al.*, 1976; Zwahlen and Waldvogel, 1984). Ingwer *et al.* demonstrated that efficient complement-dependent killing of neisseriae requires a functional classical pathway; Mg-EGTA-treated serum (inactive classical pathway) was unable to support killing of gonococci (Ingwer *et al.*, 1978). This finding was corroborated by showing that the kinetics of gonococcal killing is identical in normal and properdin deficient serum (containing an intact classical pathway) while gonococcal killing is suppressed in C2-deficient serum (containing an intact alternative pathway) (Densen *et al.*, 1988). Furthermore, absorption of bacteria-specific antibodies from bactericidal sera abrogates complement-dependent killing of *Neisseria* (Jarvis and Vedros, 1987; McQuillen *et al.*, 1994). Together, these data solidify the importance of the antibody-dependent classical pathway in initiating complement activation on *Neisseria*.

Alternative pathway activation

Efficient killing of neisseriae also requires a fully functional alternative pathway. Relatively normal C4 levels coupled with the dramatic reduction in factor B concentration that occurs during meningococcemia indicates that the alternative pathway plays a key role in complement activation (Hoffman and Edwards, 1972). In individuals with classical pathway defects, specific IgG antibody can act synergistically with properdin to enhance alternative pathway mediated killing of *N. meningitidis* (Soderstrom *et al.*, 1987). Monomeric IgA1 antibody directed against certain somatic antigens also initiates bacterial killing of certain serogroup C strains (Jarvis and Griffiss, 1989). Similarly, classical pathway initiated killing of gonococci can be augmented by activation of the alternative pathway; the ability of a given gonococcal strain to bind to properdin can be correlated with the titre required to kill the strain (Griffiss *et al.*, 1991).

Role of the MBL pathway

The role of the MBL pathway in the defence against infection is still debated (Cartwright *et al.*, 1987). MBL deficiency is common and this condition is believed to predispose individuals to a variety of infections (Eisen and Minchinton, 2003). Microbial surfaces with a high content of repetitive (and terminal) mannose and/or *N*-acetylglucosamine (GlcNAc) residues are efficiently recognized by MBL. A number of bacterial pathogens that cause meningitis have been examined for MBL binding; high binding occurs to unencapsulated variants of *N. meningitidis, Listeria monocytogenes, Haemophilus influenzae,* intermediate binding to serogroup A *N. meningitidis, Escherichia coli K1* and *Streptococcus pneumoniae* and weak binding to non-serogroup A *N. meningitidis* and *H. influenzae* type B (van Emmerik *et al.*, 1994). *N. gonorrhoeae* also binds MBL (Devyatyarova-Johnson *et al.*, 2000).

Mammalian carbohydrates often terminate in sialic acid residues, which shield the relevant neutral sugars and hence block recognition by MBL. Capsular polysaccharides, LOS structure and sialylation each play a role in modulating binding of MBL to *N. meningitidis* B1940 (Jack *et al.*, 1998). MBL binds to *galE* mutants with truncated LOS but not to the encapsulated wild type and only weakly to an un-encapsulated *siaD* mutant, suggesting that LOS structure modulates MBL binding. In addition, bound MBL–MASP complexes activate both isoforms of C4 (C4A and C4B). The addition of exogenous CMP-NANA (the donor molecule for sialic acid) to growth medium enhances sialylation of lacto-*N*-neotetraose LOS and results in decreased binding of MBL and decreased complement killing of MBL pre-opsonized *N. meningitidis* (Jack *et al.*, 2001). MBL also binds to gonococcal porin (Por) B3 and the opacity-associated protein (Opa), but does not bind directly to purified LOS. Binding of MBL to PorB3 and Opa occurs in the absence of calcium and is not inhibited by mannose or *N*-acetylglucosamine (natural carbohydrate ligands for MBL) (Estabrook *et al.*, 2004). Sialylation of gonococcal LOS decreases MBL binding (Devyatyarova-Johnson *et al.*, 2000). Taken together, these data suggest that binding of MBL to PorB3 and Opa may be modulated by LOS structure and sialylation.

Of note, studies demonstrating MBL pathway activation on *Neisseria* were conducted using bacteria that were 'pre-incubated' with purified MBL (Jack *et al.*, 1998; Jack *et al.*, 2001); when MBL is added to complement and bacteria simultaneously, no killing results. Killing occurs only when bacteria are pre-incubated with MBL followed by the addition of complement (Gulati *et al.*, 2002). MBL function on the gonococcal surface is regulated by C1-inhibitor and α_2-macroglobulin. In addition, MBL may play a role in enhancing opsonophagocytosis via CR1 (CD35) binding (Ghiran *et al.*, 2000); it may also enhance complement activation in individuals who lack adequate levels of protective antibody.

Complement evasion by *Neisseria*

Activation of the complement system can result in rapid death of an invading organism and therefore it is not surprising that pathogenic microbes have evolved mechanisms to evade complement activation at nearly every step (Rooijakkers and van Strijp, 2007). Common strategies involve hijacking (or mimicking) host complement regulatory molecules and inactivating complement proteins by complement inhibitory proteins or proteases. In addition, many pathogens subvert the complement system using more general mechanisms; for example the Gram-positive cell wall inhibits MAC mediated lysis. Several recent comprehensive overviews of immune evasion by human pathogens are available (Rooijakkers and van Strijp, 2007; Zipfel *et al.*, 2007; Lambris *et al.*, 2008). The ability of pathogenic *Neisseria* to infect humans depends on effective evasion of complement and these uniquely human pathogens have evolved an array of strategies to limit activation of human complement on their surfaces (summarized in Table 8.1).

Deposition of C4 and C3

The complement system recognizes invading pathogens such as pathogenic neisseriae and facilitates killing either directly or by enhancing opsonophagocytosis using a process of targeted complement activation on the microbial surface. Complement proteins C3 and C4 play an important role in this activation. Both proteins possess highly reactive internal thioester bonds, which bind to proximate hydroxyl or amide groups on bacterial surface targets that serve as electron donors (Hostetter *et al.*, 1982; Dodds *et al.*, 1996). Complement mediated killing versus resistance to complement is determined, in large part, by the balance between covalent attachment of C3 and C4 to the surface targets (followed by activation) on the one hand vs. the binding of complement regulators (e.g. C4BP and fH) on the other. In *Klebsiella pneumoniae* the amount of C3b bound to porin dictates serum sensitivity verses resistance; smooth lipopolysaccharide inhibits binding of C3b to porin and increases serum resistance (Alberti *et al.*, 1996). Identification of neisserial target structures where complement deposits and characterization of the balance between activation and regulation has enhanced our understanding of susceptibility to neisserial disease.

Classical pathway activation is initiated by binding of the activated C1 complex to surface-bound IgG or IgM molecules. Activated C1s cleaves the α-chain of C4, which results in the formation of metastable C4b. The labile thioester bond in C4b covalently attaches via an ester or an amide linkage to bacterial surface targets and leads to the formation of the C3 convertase (C4b2a) of the classical pathway. Binding of a C3b molecule to or close to the classical pathway C3 convertase imparts C5 convertase (C4b2aC3b) activity to the enzyme complex (reviewed in Walport, 2001a). Similarly, C3b covalently bound to the bacterium can bind factor B to form the C3bB complex. Cleavage of factor B in this complex by the enzyme, factor D, leads to formation of the alternative pathway C3 convertase (C3bBb), which, in turn, cleaves additional C3 molecules to C3b. C3b binding to the bacterial surface is the central event in all three of the complement activation pathways and leads to activation of the terminal pathway of complement. C3b is part of both the classical and alternative pathway C5 convertases (C4b2aC3b and C3bC3bBb, respectively), which cleave C5 to initiate membrane attack complex (C5b-9) formation. In addition, iC3b, an inactive degradation product of C3b that results from the action of the proteolytic enzyme, Factor I, and its cofactor, fH, promotes opsonophagocytosis primarily via engagement of CR1 and CR3, respectively. Most of the C3b that binds to *N. meningitidis* and *N. gonorrhoeae* is converted to iC3b

Table 8.1 Neisserial complement interactions and complement evasion strategies.

Strategy	*N. gonorrhoeae*	*N. meningitidis*
C3 deposition	1 LOS lipid A and glycose residues (ester) 2 PorB1.1B (ester) 3 Opa proteins (ester)	1 LOS glycose residues (ester)
C4 deposition	1 LOS (amide and ester) • HepII 3- and 6- PEA serve equally as acceptors for C4b • HepII 3-lactose decreases C4 deposition and blocks 6-PEA 2 PorB1.1B (amide) • loop 4 and 5 3 Opa (amide)	1 LOS (amide and ester) • HepII 6-PEA preferentially binds C4b • 6-PEA bearing strains are more serum sensitive 2 Por1A and PorB2 do NOT bind C4b 3 Opa
C4BP binding	Porin from stably serum resistant strains[a] PorB1.1A • Loop 1 of PorB1.1A (strain FA19) required but not sufficient • CCP1 of C4BP required • Hydrophobic • Not blocked by C4b, NaCl or heparin PorB1.1B • Loops 5 and 7 of PorB1.1B (strain MS11) • CCP1 of C4BP • Ionic • Blocked by C4b, NaCl and heparin	PorA • Loops 1 and 4 of PorA • CCP 2 and CCP3 of C4BP • Ionic • NaCl and heparin block binding • Observed only in hypotonic conditions
fH binding	PorB.1A and PorB.1B with LOS sialylation • Loop 5 of PorB.1A (strain FA19) binds fH • PorB.1A (strains 252, 15253) interacts with CCP6 and CCP18–20 of fH • Sialylation of LNT (not P^k/L1) LOS enhances fH binding The gonococcal homolog of fHbp, NGO033, is not surface exposed in FA1090 and is not involved in serum resistance	fHbp (GNA1870) • Level of fH binding varies with the level of FHbp expression • LOS sialylation does not enhance fH binding • fH does not bind meningococcal porins
FHL-1	PorB.1A	
FHR-1	PorB.1A and sialylated PorB.1B	
Blocking Ab	Anti-RMP Ab	Anti-Class 4 Ab (maybe) Anti-group C capsule IgA1 Anti-serogroup W IgG
LOS sialic acid.	LOS sialylation confers unstable serum resistance and enhances binding of fH to gonococcal Por	Not essential for serum resistance; may play a minor role Does not enhance fH binding
Capsule	NA	Most important determinant of meningococcal serum resistance • Serogroup A capsules do not contain sialic acid but regulate complement. • Unencapsulated mutants (all serogroups) bind more C3 and C4

a Binding modulated by HepI chain length and lipid A PEA content.

(Jarvis, 1994; McQuillen *et al.*, 1999). Recently, LOS, gonococcal PorB.1B, and the opacity proteins have been identified as targets for activated C3 and C4 (Edwards and Apicella, 2002; Ram *et al.*, 2003; Lewis *et al.*, 2008).

Neisserial LOS serves as a site for deposition of both C3b and C4b (Edwards and Apicella, 2002; Ram *et al.*, 2003; Lewis *et al.*, 2008). C3 binds to lipid A (Edwards and Apicella, 2002) and together with C4 also binds to carbohydrate moieties that extend distally from core heptose residues on gonococcal LOS (Ram *et al.*, 2003; Lewis *et al.*, 2008). Phosphoethanolamine (PEA) residues on the second heptose (HepII) and glycose extensions on the first heptose (HepI) play a pivotal role in determining the type of linkage that C4b forms with LOS (Ram *et al.*, 2003). In *N. meningitidis*, ester C4b-LOS bonds occur with ≥4 glycose extensions on HepI and C4A forms amide linkages with PEA. PEA at the 6-position of HepII binds C4b over PEA at the 3-position preferentially. LOS species with more truncated chain extensions from HepI bind C4b via amide linkages to 3-PEA on HepII. Meningococcal strains that express 6-PEA are more serum sensitive than those possessing a 3-PEA. These findings may explain why 6-PEA-bearing strains constitute fewer than 30% of all clinical meningococcal isolates (Plested *et al.*, 1999). In gonococci, 3- and 6- PEA serve equally as targets for amide linked C4, even in the presence of an extended HepI LOS. Interestingly, 3-lactose on HepII decreases C4 deposition and blocks accessibility of 6-PEA. Greater than 90% of *N. gonorrhoeae* clinical isolates possess a lactose substitution at the 3-position of HepII; expression of HepII lactose may, via modifications in C4b binding to LOS, impart a survival advantage to gonococci.

Gonococcal PorB.1A (previously designated Por1A) strains are associated with stable serum resistance and disseminated disease while PorB.1B (previously called Por1B) strains are usually serum sensitive in the unsialylated state and are associated with local genital infection, particularly pelvic inflammatory disease (Rice *et al.*, 1980). PorB.1A strains preferentially bind to C4BP and fH as described below. Gonococcal PorB.1B forms amide links with C4b and ester links with C3b; the site of C4b binding has been localized to the region spanned by loops 4 and 5 of PorB.1B. Preferential binding of C4b and C3b to PorB.1B may contribute further to enhanced sensitivity of PorB.1B strains relative to PorA.1A strains (Lewis *et al.*, 2008).

The opacity-associated proteins (Opa) from both *N. gonorrhoeae* and *N. meningitidis* are acceptors for C4b and C3b. An earlier report indicated that gonococcal colonies identified in women around menses (when menstrual blood contributes to a higher level of complement locally) and examined after a single agar subpassage are mostly of the transparent phenotype (i.e. more likely to be Opa-negative), while colonies identified at other times in the cycle are usually opaque (Swanson, 1978). Women are uniquely susceptible to disseminated gonococcal infection (DGI) during menses (Rice, 2005) and serum-resistant strains isolated from blood are also more likely to produce transparent colonies (McCutchan *et al.*, 1976). Most *N. meningitidis* isolates from blood or cerebrospinal fluid commonly grow as transparent colonies, whereas meningococci isolated from the nasopharynxes of asymptomatic carriers usually form opaque or a mixture of opaque and transparent colonies (Stephens and McGee, 1983; McGee *et al.*, 1990). Thus, it appears that *Neisseria* that encounter high levels of complement (as occurs during menses or during bloodstream infection) often turn Opa expression 'off'. It is possible that loss of a target for C4b and C3b may confer an advantage in combating complement.

Binding of *Neisseria* to complement regulators

The most widely described complement evasion strategy employed by bacteria, yeast, fungi and viruses is to bind host complement inhibitory proteins such as fH and C4BP. Factor H and C4BP possess decay accelerating activity which results in irreversible dissociation of the C3-convertases of the alternative and classical pathways respectively. The net effect of decay acceleration is to limit the total amount of C3b deposited on an organism. *N. gonorrhoeae* and *N. meningitidis* have evolved independent strategies to evade complement by binding to complement regulators.

C4 binding protein

C4BP inhibits both the classical and MBL pathways of complement by acting as a cofactor for factor I mediated cleavage of C4b and as indicated above by accelerating decay of the classical pathway C3 convertase (C4b2a) (Gigli *et al.*, 1979). Diverse microbes, such as *Moraxella catarrhalis* (Nordstrom *et al.*, 2004), *Bordetella pertussis* (Berggard *et al.*, 1997), *E. coli* (Prasadarao *et al.*, 2002), *Borrelia recurrentis* (Meri *et al.*, 2006), *Candida albicans* (Meri et al., 2002) and *Haemophilus influenzae* (Hallstrom *et al.*, 2007), bind C4BP as a mechanism to enhance their pathogenic potential. In group A streptococci, binding of functionally active C4BP to the M protein family proteins Arp and Sir results in resistance to opsonophagocytosis (Berggard *et al.*, 2001). Conversely, C4BP bound to *B. pertussis* by filamentous haemagglutinin does not protect against complement mediated lysis.

Binding of *Neisseria gonorrhoeae* to C4BP

Serum resistance of gonococcal strains is strongly correlated with the ability to bind C4BP; in one study 16 of 21 stable serum resistant strains were found to bind C4BP while 0 of 8 serum sensitive strains (unsialylated) bound C4BP. When C4BP binding to serum resistant gonococci is prevented (i.e. by anti-C4BP antibody) the strains become serum sensitive. C4BP that binds to gonococci is functional as evidenced by the cleavage of C4b to the haemolytically inactive C4d fragment (Ram *et al.*, 2001).

Gonococcal porin is the ligand for C4BP on serum-resistant strains. Wild-type *N. gonorrhoeae* F62, isolated from a case of uncomplicated gonorrhoea (Schneider *et al.*, 1982), is serum sensitive, when unsialylated, and does not bind C4BP. Allelic replacement of the F62 porin with PorB.1A from either serum resistant *N. gonorrhoeae* FA19 or PorB.1B from serum resistant *N. gonorrhoeae* MS11 enables F62 to bind C4BP. FA19 was isolated from the bloodstream of an individual with disseminated gonococcal disease (Cannon *et al.*, 1981) while MS11, like F62, was isolated from a case of uncomplicated gonococcal infection (Swanson, 1978). Hybrid porin constructs show that loop 1 of PorB.1A and loops 5 and 7 of PorB.1B are involved in binding to C4BP (Ram *et al.*, 2001). C4BP binding to PorB.1A and PorB.1B is also affected by LOS composition; both the HepI chain length of LOS (Ram *et al.*, 2007) and levels of phosphoethanolamine (PEA) content in lipid A (L.A. Lewis, unpublished observations) modulate binding of C4BP to gonococci. PorB.1A and PorB.1B interact with C4BP in distinct ways. The PorB.1A–C4BP interaction is hydrophobic and involves sites on C4BP distinct from the binding sites for C4b and heparin. C4BP cannot be dissociated from PorB.1A bearing gonococci with high concentrations of NaCl, C4b or heparin. In contrast, PorB.1B–C4BP interactions are ionic and likely involve the binding sites for C4b and heparin; NaCl, C4b and heparin are all able to interfere with binding of PorB.1B to C4BP (Ram *et al.*, 2001).

C4BP is a large plasma protein that consists of seven identical α-chains and a unique β-chain linked together by short amphipathic helices that are stabilized by disulphide bridges (Kask *et al.*, 2002). The α- and β-chains contain eight and three complement control protein (CCP) modules respectively. Each CCP module consist of ~60 amino acids that form a compact hydrophobic core surrounded by five or more β strands organized into β-sheets (Kirkitadze and Barlow, 2001). Recombinant C4BP molecules that lack individual CCP domains have been used to demonstrate that human C4BP binds to gonococci exclusively via the N-terminal CCP1 of the α-chain (Ram *et al.*, 2001).

Although human C4BP binds to both PorB.1A and PorB.1B of serum resistant *N. gonorrhoeae*, chimpanzee C4BP binds only to PorB.1B (not PorB.1A) and C4BP from other primates does not bind to gonococcal porins (Ngampasutadol *et al.*, 2005). By exploiting the amino acid differences between CCP1 of humans, chimpanzees and rhesus macaque C4BPs, the key amino acids of human C4BP that are required for binding to the gonococcal porin molecules have been identified. Seven amino acids in CCP1 (K24, T43, T45, A12, M14, R22 and L34) are important for the C4BP–PorB.1A interaction (Jarva *et al.*, 2007). Of note, this interaction is hydrophobic and all of these amino acids are either hydrophobic or capable of making hydrophobic contacts. The ionic interaction between PorB.1B and C4BP relies on a single critically charged residue in CCP1, K24 (Jarva *et al.*, 2007).

Binding of *Neisseria meningitidis* to C4BP

PorA of *N. meningitidis* also binds C4BP, however deletion of meningococcal PorB, which is homologous to gonococcal PorB (PorB.1A and porB.1B), does not affect binding of C4BP to meningococci. C4BP that binds to the meningococcal surface is functionally active and promotes inactivation of C4b by factor I. Analogous to the interaction with gonococcal PorB.1B, the C4BP–PorA interaction can be inhibited with NaCl (ionic) and heparin. Peptide spot analysis of PorA has indicated preferential binding of loops 1 and 4 to C4BP. CCP2 and CCP3 are the domains of C4BP that are most likely involved in binding to PorA (Jarva *et al.*, 2005). There are salient differences in the binding of C4BP to *N. gonorrhoeae* and *N. meningitidis*; binding of C4BP to meningococci is primarily observed under hypotonic conditions and binding is reduced by 90% when physiological buffers are used, while gonococcal C4BP interactions occur under iso-osmolar conditions.

Meningococci that bind C4BP withstand killing better in serum than meningococci that do not bind C4BP; PorA bearing strains consistently survive better than PorA negative strains. Furthermore, as occurs with gonococci, prevention of C4BP binding to meningococci also facilitates their survival in serum (Jarva *et al.*, 2005). Of note, meningococcal strains that cause invasive disease are encapsulated; capsular polysaccharide obstructs binding of C4BP to meningococci (Jarva *et al.*, 2005) and capsule imparts serum resistance independent of C4BP binding. C4BP binding to meningococci may be an important immune evasion mechanism when organisms are unencapsulated, as may occur when they traverse mucosal epithelium.

Factor H

Factor H, a soluble negative regulator of the alternative pathway of complement functions as a cofactor for factor I-mediated cleavage of C3b to iC3b and as indicated above also dissociates the alternative pathway C3 convertase (C3bBb). Numerous human pathogens recruit functionally active fH to circumvent the AP. As an example, in *Yersinia enterocolitica* YadA binds fH and mediates serum resistance; YadA expressing strains have reduced C3b deposition.

Factor H is a 150-kDa glycoprotein composed of 20 complement control protein (CCP) (also called short consensus repeat or SCR) domains. Each CCP contains about 60 amino acids including four invariant cysteine residues that direct folding into compact units (Ripoche *et al.*, 1988). In addition, fH binds to polyanions such as sialic acid, heparin and sulphated glycosaminoglycans (Meri and Pangburn, 1994). Both *N. gonorrhoeae* and *N. meningitidis* recruit fH to their surfaces to promote survival *in vivo*.

Binding of *Neisseria gonorrhoeae* to factor H

In 1970, Ward observed that *N. gonorrhoeae* recovered directly from human urethral exudates were resistant to killing by normal human serum but that this property was lost upon culture in laboratory medium (Ward *et al.*, 1970). This phenomenon was termed 'unstable' serum resistance. Harry Smith and his colleagues subsequently discovered that gonococci scavenge 5′-cytidine-monophospho-N-acetylneuraminic acid (CMP-NANA) from the host and sialylated their LOS, which leads to a serum resistant phenotype (Tan *et al.*, 1986; Nairn *et al.*, 1988; Parsons *et al.*, 1988). In eukaryotic cells sialic acid plays an important role in preventing destruction of host tissues by the alternative pathway of complement. Sialylation enhances the interaction between C3b and fH, which results in efficient cleavage of C3b to iC3b and inactivation of the alternative pathway. Removing sialic acid decreases binding of fH and increases activation of the alternative pathway. Sialylation of gonococcal LOS enhances binding of the complement regulator fH to gonococci and provides one mechanism whereby sialylation enhances serum resistance (Ram *et al.*, 1998b). Neuraminidase enzymatically removes sialic acid from gonococcal LOS and abrogates binding of fH to gonococci. Factor H bound to gonococci functions to increase conversion of C3b to iC3b and decrease total C3 deposition by inactivating the alternative pathway C3 convertase (C3bBb). Interestingly, only sialylation of the lactoneotetraose (LNT) epitope of LOS enhances fH binding and serum resistance; sialylation of the alternative P^k structure of LOS (called L1 in *N. meningitidis*) is not associated with increased serum resistance or fH binding (Gulati *et al.*, 2005). Factor H binding to gonococci with porin PorB.1B requires sialylated LOS. In the case of meningococcal strain Y2220, sialylation of LNT LOS does not enhance fH binding. However, Y2220 engineered to express gonococcal PorB.1B binds to fH in a sialylation dependent manner. Consistent with what was seen in gonococci, sialylated meningococci expressing gonococcal PorB.1B are more resistant to killing by normal human serum (Madico *et al.*, 2007).

Some gonococcal strains, particularly those associated with disseminated gonococcal infection (DGI), remain resistant to serum even after they are cultured on laboratory media. This serum resistant phenotype, termed stable serum resistance, is not dependent on CMP-NANA or LOS sialylation. Stable serum-resistant *N. gonorrhoeae* are often of the PorB.1A serotype and bind fH in an LOS sialic acid independent manner (Ram *et al.*, 1998a). For example, allelic replacement of the PorB.1B of MS11 with the PorB.1A of FA19 enables MS11 to bind fH in a sialylation independent manner (Ram *et al.*, 1998a). *N. meningitidis* Y2220 expressing gonococcal PorB.1A binds fH in a sialylation independent manner further confirming the role of PorB.1A as the ligand for fH in the context of another pathogenic *Neisseria* species (Madico *et al.*, 2007). Hybrid porin molecules where loop 5 from PorB.1A has been substituted for loop 5 in PorB.1B bind fH in the absence of LOS sialylation suggesting that loop 5 of PorB.1A contains the fH binding site. Synthetic peptides that correspond to loop 5 of FA19 PorB.1A inhibit fH binding and decrease serum resistance (Ram *et al.*, 1998a). A PorB.1A strain has been shown to interact with 2 distinct regions of fH; CCP 6 and CCP 18–20. A sialylated PorB.1B strain has been shown to bind to fH CCP18–20 (Ram *et al.*, 1998a; Ngampasutadol *et al.*, 2008).

In addition to binding to full-length fH, gonococci also bind Factor H-like protein (FHL-1) and Factor H-related protein (FHR-1); these molecules are related to fH, having homology with CCP1–7 and CCP 6,7,18–20 respectively (Ngampasutadol *et al.*, 2008). FHR-1 binds to both PorB.1A and sialylated PorB.1B while FHL-1 binds weakly to PorB.1A. Because FHL-1 has complement regulatory and cell attachment properties (Hellwage *et al.*, 1997), it is possible that binding of this alternatively spliced variant of fH may also contribute to serum resistance and adherence to host cells.

The binding of gonococci to complement regulators helps to explain stable and unstable serum resistant phenotypes. Strains that possess the unstable serum resistant phenotype do not bind C4BP and require sialylation to bind fH. Stable serum resistant strains bind C4BP and, sometimes, fH, in the absence of LOS sialylation.

Binding of *Neisseria meningitidis* to factor H

Factor H binds to the meningococcal surface, retains its activity and contributes to the ability of *N. meningitidis* to avoid complement-mediated killing by human serum (Madico *et al.*, 2006; Schneider *et al.*, 2006). Meningococcal porins do not mediate binding to fH and, unlike *N. gonorrhoeae*, binding of fH to *N. meningitidis* is independent of sialylation of LOS and sialic acid content of the meningococcal capsules. Binding of fH to meningococci is mediated by a 29-kDa protein designated fH binding protein (fHbp) (Madico *et al.*, 2006; Schneider *et al.*, 2006). FHbp, also known as GNA1870 or LP.2086, is a leading meningococcal vaccine candidate (Masignani *et al.*, 2003; Fletcher *et al.*, 2004). Strains that lack fHbp do not bind fH and anti-fHbp antibody blocks binding of fH to bacteria (Madico *et al.*, 2006). The expression level of fHbp varies widely among different strains and correlates with the level of fH binding; strains that express higher amounts of fHbp bind more fH (Masignani *et al.*, 2003; Madico *et al.*, 2006). FHbp is subdivided into 3 variant families (variants 1, 2 and 3), based on variation in amino acid sequence. All three fHbp variant proteins bind fH (Madico *et al.*, 2006). The solution structure of the 18-kDa C-terminal portion of fHbp shows an eight-stranded antiparallel β-barrel overlaid by a short α-helix with an unstructured N-terminus (Cantini *et al.*, 2006). *N. gonorrhoeae* strain FA1090 has a homologue of fHbp (called NGO033), however this protein is not predicted to contain a signal peptide and is not expressed on the gonococcal cell surface. Mutational analysis indicates that GNA033 is not associated with serum resistance or fH binding in *N. gonorrhoeae* (S. Ram, unpublished observations).

Species specificity of regulator binding

N. gonorrhoeae and *N. meningitidis* are exclusively human pathogens. Explanations for species restriction of neisserial diseases are multifactorial. Pathogenic neisseriae are only able to acquire iron from human transferrin, not transferrin from other mammalian species (Schryvers and Gonzalez, 1990; Zarantonelli *et al.*, 2007). Growth of meningococci in the mouse bloodstream is dependent on the presence of human transferrin; human transferrin enhances meningococcal infection in mice (Holbein, 1981) and transgenic mice expressing human transferrin, but not wild-type mice, support growth of meningococci (Zarantonelli *et al.*, 2007). Similarly, meningococci bind selectively to human CD46 and transgenic mice that express human CD46 are not only susceptible to meningococcal disease but allow bacteria to cross the blood–brain barrier (Johansson *et al.*, 2003). The ability of pathogenic neisseriae to evade complement-mediated killing in a species-specific fashion may also contribute to the narrow host range of infection caused by pathogenic neisseriae. Gonococcal strains that bind human C4BP are resistant to human serum but are readily killed by serum from other primate species and rodents. *N. gonorrhoeae* do not bind to non-human C4BP; resistance to rodent serum can be achieved by supplementing rodent sera with human C4BP (Ngampasutadol *et al.*, 2005). Likewise, gonococci also bind to human fH selectively and can be rescued from killing by primate sera by human fH (Ngampasutadol *et al.*, 2008). By comparison, *Yersinia pestis*, which infects essentially all mammals, binds mammalian C4BP generally and resists killing by serum from multiple species of mammal (Ngampasutadol *et al.*, 2005). Knowledge of specific factors that enhance survival of neisserial species in non-human hosts may lead to the development of superior animal models to better simulate human disease. Species restriction of binding of complement inhibitory proteins to *N. meningitidis* has not yet been reported.

Blocking antibodies

Blocking antibodies are antibodies that inhibit complement-mediated killing of microbial organisms by bactericidal antibodies in human serum. Antibodies (IgG and $F(ab')_2$) directed against the reduction-modifiable protein of *N. gonorrhoeae* (Rmp or PIII protein; the homologue of Class 4 protein in *N. meningitidis*) block killing mediated by bactericidal antibodies directed against other outer membrane molecules (Rice *et al.*, 1986). The presence of blocking antibodies in serum prior to infection with *N. gonorrhoeae* has been shown to increase susceptibility to gonococcal infections (Plummer *et al.*, 1993) and together with (bactericidal) antibodies directed against porin and LOS influence the outcome of human challenge with *N. gonorrhoeae* (Rice *et al.*, 1994). The blocking mechanism has not been fully elucidated, but it is clear that blocking antibodies activate complement on the bacterial surface and enhance binding of C3 and C9 to bacteria. Blocking antibodies compete for binding with bactericidal antibodies and may (also) function by diverting C3 away from bactericidal targets onto non-bactericidal targets (Joiner *et al.*, 1985a). Complement activation at these alternative site(s) could result in insertion of 'non-bactericidal' C5b-9 (Joiner *et al.*, 1985b). Consistent with this idea, several studies have suggested that proper MAC insertion, rather than quantity of MAC, dictates efficient bacterial killing (Harriman *et al.*, 1982; Joiner *et al.*, 1983; Joiner *et al.*, 1989; Drogari-Apiranthitou *et al.*, 2002).

Monoclonal antibodies against Rmp block the activity of bactericidal monoclonal antibodies that are directed against other meningococcal structures (Munkley *et al.*, 1991). Anti-Class 4 antibodies induced by vaccination with a meningococcal outer-membrane vesicle vaccine containing the Class 4 protein did not exhibit blocking properties (Rosenqvist *et al.*, 1999). The Class 4 antibodies elicited by this vaccine were shown to react strongly with the Class 4 protein by immunoblot, but failed to bind to whole bacteria, suggesting that these antibodies were directed against regions in the Class 4 molecule that were not surface exposed. Nevertheless, natural anti-Class 4 antibodies that blocked the bactericidal effect of an anti-PorA (Class 1) monoclonal Ab were detected in pre-immune serum from a single individual.

Griffiss discovered that IgA antibodies purified from convalescent sera of individuals recovering from infection with *N. meningitidis* (serogroups B, C and Y) could block the bactericidal activity of IgG and IgM isolated from the same sera. Blocking was dependent on the ratio of bactericidal to blocking antibody, was strain specific and was more effective against IgG-mediated killing than IgM (Griffiss, 1975). More recently, IgA1 directed against serogroup C meningococcal capsular polysaccharide has been shown to block the bactericidal activity of IgG antibody directed against the same serogroup C capsule. IgA1 was shown to block IgG-mediated killing via either the classical pathway (as seen in factor B-depleted and properdin-deficient serum), the alternative pathway (using Mg-EGTA-chelated serum) or the two pathways combined (intact normal serum). The mechanism of IgA1-mediated blocking did not involve competition with IgG and F(ab')2 and Fab derivatives of the blocking IgA1 blocked as well as the intact IgA1 (Jarvis and Griffiss, 1991). The blocking effect is not restricted to IgA; Selander *et al.* (2000) described an anti-serogroup W-135 capsular IgG (isolated from a vaccinee) that blocked killing by C2-deficient serum.

IgA1 can also mediate classical pathway killing of meningococci. IgA1 directed against meningococcal outer membrane proteins has been shown to mediate complement-dependent killing of meningococci. Killing is dependent on the Fc portion of the antibody and is not associated with increased C3 binding to bacteria (Jarvis and Griffiss, 1989). The specificity of IgA binding may determine whether it acts as a blocking or a bactericidal antibody against *N. meningitidis* (Jarvis and Griffiss, 1991).

Lipooligosaccharide sialylation and serum resistance

N. gonorrhoeae and *N. meningitidis* both decorate their LOSs with sialic acid but there are important differences in the mechanisms of complement regulation by sialylated LOS in these two pathogenic species.

Harry Smith and his colleagues discovered that *N. gonorrhoeae* scavenged CMP-NANA from infected human hosts to sialylate their LOS; this characteristic was associated with serum resistance (Tan *et al.*, 1986; Nairn *et al.*, 1988; Parsons *et al.*, 1988). Sialylation of LOS unequivocally promotes serum resistance in *N. gonorrhoeae*. As described in detail above, sialylation of gonococcal LNT LOS specifically enhances binding of the alternative complement pathway regulatory molecule, fH, which directly enhances gonococcal resistance to serum killing (Ram *et al.*, 1998b). Consistent with these observations, interruption of the gonococcal lactate permease gene (*lctP*) significantly decreases the ability of gonococci to colonize and survive in the murine model of lower genital tract infection. LctP mutants are more sensitive to complement than their parent strains; owing to reduced sialylation in the LctP mutants (Exley *et al.*, 2007). In addition, a cooperative interaction between LOS sialic acid and the gonococcal porin (Por) molecule is necessary for fH binding to *N. gonorrhoeae* (Madico *et al.*, 2004).

The role of lacto-*N*-neotetraose LOS sialylation in conferring serum resistance to *N. meningitidis* is less well defined. Shortly after reports on gonococcal LOS sialylation were published, it was observed that *N. meningitidis* could also sialylate their LOS but without the need for exogenous CMP-NANA added to growth media (Mandrell *et al.*, 1991). CMP-NANA is also the donor molecule necessary for the biosynthesis of sialic acid containing capsular polysaccharides; serogroup B, C, W-135 and Y (but not A). Although all non-serogroup A strains of *N. meningitidis* are capable of synthesizing CMP-NANA endogenously, sialylation of meningococcal LOS, in contrast to gonococcal LOS, does not enhance fH binding.

The role that LOS sialylation plays in serum resistance of meningococci remains uncertain. Fox *et al* reported that LOS sialylation does not enhance serum resistance of *N. meningitidis* (Fox *et al.*, 1989) however, the amount of unsialylated lacto-*N*-neotetraose LOS that is expressed by epidemiologically related serogroup C strains can be directly correlated to the degree of killing by serum (Estabrook *et al.*, 1997). Interruption of the meningococcal lactate permease gene (*lctP*) in unencapsulated (*siaD*) *N. meningitidis* results in a slight decrease in serum resistance when compared to the unencapsulated parent (Exley *et al.*, 2005). In addition, *lctP* mutants of wild-type encapsulated meningococci have been shown to be attenuated during bloodstream infection in the infant rat model. Restoration of virulence of the *lctP* mutant in the C3-deficient mouse is consistent with the role that C3 plays in this system. *LctP* mutants have defects in the sialic acid biosynthetic pathway and the importance of this molecule in both capsule biosynthesis and LOS sialylation must be considered when evaluating the results observed in the infant rat and C3-deficient mouse. Protection against serum killing conferred by LOS sialylation, when it occurs, may also be strain or serogroup specific. Isogenic *lst* (LOS sialyltransferase) mutants of serogroup B strain B1940 and serogroup C strain 2120 that lacked the ability to sialylate LOS were reportedly slightly more sensitive to serum killing (≥25%) than their respective parent strains. However, the *lst* mutation was not reported to impact survival in a third, serogroup B strain, MC58 (Vogel *et al.*, 1999).

An *lst* mutant, lacking the ability to sialylate LOS, was able to cause disseminated disease in the infant rat model to the same extent as the wild-type strain (Vogel *et al.*, 1999). In humans, meningococci isolated from blood and CSF often express LOS species that lack the ability to sialylate (e.g. the L8 immunotype (Gal→Glc→HepI)), suggesting that LOS sialylation may not be essential for pathogenesis or serum resistance. Overall, we suggest that meningococcal LOS sialylation may play only a minor role in enhancing serum resistance, is probably not critical for pathogenesis in experimental animal models and is not required absolutely for growth in the human bloodstream. Unencapsulated meningococci are often observed during nasopharyngeal carriage. In the nasopharynx where concentrations of complement are low (Andoh *et al.*, 1997), sialylation of LOS might be important for survival (Claus *et al.*, 2005). Nevertheless, a systematic analysis of this question has not yet been reported.

Meningococcal capsular polysaccharides and serum resistance

Meningococcal capsular polysaccharide is undoubtedly the most important determinant of meningococcal virulence and serum resistance. Essentially all meningococci recovered from the blood or cerebrospinal fluid of persons with invasive disease possess a polysaccharide capsule. Capsular polysaccharides are the basis for serogrouping; serogroup A and B capsules are homopolymers of α(2,8)- and α(2,9)-linked sialic acid respectively, capsular polysaccharides of serogroups W-135 and Y are repeating units of 6-D-gal α(1,4)-NANA α(2,6)- and 6-D-glc α(1,4)-NANA α(2,6)-, respectively and the serogroup A capsule is a homopolymer of *N*-acetylmannosamine phosphate. Serogroups B, C, W-135 and Y capsules all contain sialic acid, which on host cells prevents complement-mediated killing. The polysaccharides of serogroups A, C, W-135, and Y are also O-acetylated to varying extents. De-O-acetylated polysaccharides are slightly more immunogenic than acetylated polysaccharide (Richmond *et al.*, 2001); however, O-acetyltransferase mutants are as serum resistant as the wild-type strains (Claus and Vogel, unpublished observations).

It is well established that encapsulated meningococci are more serum resistant than their unencapsulated counterparts (Jarvis and Vedros, 1987; Vogel *et al.*, 1997; Kahler *et al.*, 1998). Recently, invasive serogroup C strains of *N. meningitidis* that are able to resist killing by bactericidal Abs induced by the MenC conjugate vaccine have been isolated and their mechanism of serum resistance characterized. These highly serum resistant strains contain an IS1301 insertion sequence in the intergenic region between the operons for capsule biosynthesis (*sia* operon) and export (*ctr* operon). The insertion sequence apparently up-regulates both *siaA* and *ctrA* and leads to an increase in production of capsular polysaccharide. Strains with increased capsule production effectively down regulate of the AP of complement and bind reduced levels of C3 (Uria *et al.*, 2008). Increased capsule expression may also inhibit proper MAC insertion or mask subcapsular antigens (Uria *et al.*, 2008). The critical role of capsular polysaccharide and LOS in serum resistance is further highlighted by the studies of Geoffroy *et al.* (2003). Of 3000 meningococcal mutants analysed for their ability to withstand complement-mediated killing, the 18 mutants that were identified as highly sensitive to killing by 25% human serum had all lost function in genes already known or subsequently discovered to be required for the biosynthesis of capsule or LOS.

The precise mechanism of complement regulation by capsule is not well characterized. Similar amounts of C3b and iC3b deposit on wild-type serogroup B *N. meningitidis* strain B1940 and its unencapsulated (*siaD*) mutant. These data seem to indicate that the capsule does not affect steps leading up to and including cleavage of C3 and are consistent with the notion that capsule prevents insertion of the MAC. However, we have noted (unpublished data) that unencapsulated isogenic mutant strains derived from four of the five major meningococcal serogroups (A, B, C and W-135) bind more C3 and C4 than their wild-type parent strains. In 1987, Jarvis and Vedros (Jarvis and Vedros, 1987) demonstrated that a sialic acid-deficient mutant (chemically induced) of a serogroup B strain was unable to regulate the alternative pathway; in addition this strain was killed by serum that lacked bacteria-specific antibody. Enzymatic desialylation of the wild-type strain also augmented C3 deposition. Sialic acid is not likely to have decreased C3 binding by enhancing the interaction of fH with C3b because colominic acid (also a homopolymer of α(2,8)-linked sialic acid but derived from *E. coli* K1 capsule) does not enhance the interaction between C3b and fH (Meri and Pangburn, 1990; Meri and Pangburn, 1994). In any event, capsule-mediated complement regulation is not related exclusively to the sialic acid moiety; serogroup A capsule, which lacks sialic acid altogether, also regulates complement. Additional studies will be needed to elucidate the mechanism(s) by which capsule mediates resistance to complement.

Conclusion

Complement evasion is necessary for survival of the pathogenic neisseriae in humans, their only natural host. These bacteria achieve evasion using multiple, sometimes overlapping, strategies. The neisserial complement evasion mechanisms described are uniquely tailored to escape attack by human complement and may in part dictate the narrow host range of *Neisseria*. Pathogens that infect diverse hosts, such as *Borrelia species.* and *Y. pestis*, generally possess more broad spectrum complement evasion systems. Although quite similar at the genomic level, *N. gonorrhoeae* and *N. meningitidis* have developed alternative mechanisms that differ in subtle ways to combat complement mediated killing (Table 8.1). These subtle differences may enable these organisms to successfully occupy different ecological niches in the human host.

Over the past few decades, *N. gonorrhoeae* has developed remarkable capability to rapidly develop resistance against multiple antibiotics. The global spread of quinolone resistant *N. gonorrhoeae* has curtailed inexpensive options available for effective oral antibiotic therapy. Thus, there is an urgent need to develop an effective vaccine against *N. gonorrhoeae.* Vaccines that rely on antibody production to confer protection often require complement activation for efficacy. Recent advances in the understanding of how *N. gonorrhoeae* evade complement may aid in the design of effective vaccine candidates that elicit protective antibodies that can circumvent the down regulation of complement inhibitors that gonococci employ so effectively to thwart complement-dependent killing. In addition, molecules that block the binding of complement inhibitors to the gonococcal surface may also have therapeutic potential. Meningococcal polysaccharide-based vaccines have proven to be efficacious against four of the major meningococcal serogroups (A, C, W-135 and Y); however, a vaccine against serogroup B disease has remained out of reach. Furthermore, the emergence of epidemics caused by serogroups (such as serogroup X) not covered by the available polysaccharide vaccines necessitates development of vaccine candidates that will confer protection against diverse serogroups of *N. meningitidis.* 'Reverse vaccinology' has identified several promising candidates. Targeting an antigen such as fHbp that plays an important role in complement evasion represents an exciting advance in the field. In addition to activating complement by the 'conventional' mechanism, antibodies against fHbp have the added benefit of diverting fH away from the bacterial surface, crippling an important bacterial evasion strategy.

The precise mechanism of action of blocking antibodies remains an enigma and while the specificity and their role in gonococcal pathogenesis has been characterized and a possible role in enhancing susceptibility to disease suggested, the role of blocking antibodies in meningococcal pathogenesis remains uncertain, particularly their potential in undermining efficacious responses to vaccines.

Future trends

The mechanisms used by neisseriae to evade complement are gradually being unravelled. The past two decades have seen important advances in how neisseriae evade killing by serum. Structural studies that define how complement inhibitors such as factor H and C4BP bind to their bacterial ligands are needed because such information could help design new therapeutic agents and vaccines.

Recent work has shown that there exists 'cross-talk' between different arms of the innate immune systems, such as the toll-like receptors and the complement system, and a role for complement in bridging the innate and adaptive immune systems. An area that merits investigation is the role of complement receptors and membrane-bound inhibitors of complement in disease pathogenesis. Such studies will provide important perspectives in bridging the complement system with other arms of the immune system and with cell signalling pathways.

Although we have gained invaluable insights from the animal model currently used to study gonorrhoea, the observations have been limited to vaginal colonization. It is important to develop animal models that will permit ascending infection and dissemination to facilitate *in vivo* studies of host and bacterial factors in disease. The observation that gonococci selectively bind human complement down-regulators adds to the list of species-specific disease mechanisms (such as transferrin utilization and CD46 binding) and has set the stage to develop transgenic mice that express human complement inhibitors.

References

Alberti, S., Alvarez, D., Merino, S., Casado, M.T., Vivanco, F., Tomas, J.M., and Benedi, V.J. (1996). Analysis of complement C3 deposition and degradation on *Klebsiella pneumoniae*. Infect. Immun. *64*, 4726–4732.

Andoh, A., Fujiyama, Y., Kimura, T., Uchihara, H., Sakumoto, H., Okabe, H., and Bamba, T. (1997). Molecular characterization of complement components (C3, C4, and factor B) in human saliva. J. Clin. Immunol. *17*, 404–407.

Arlaud, G.J., Gaboriaud, C., Thielens, N.M., Rossi, V., Bersch, B., Hernandez, J.F., and Fontecilla-Camps, J.C. (2001). Structural biology of C1: dissection of a complex molecular machinery. Immunol. Rev. *180*, 136–145.

Bakir, M., Yagci, A., Ulger, N., Akbenlioglu, C., Ilki, A., and Soyletir, G. (2001). Asymtomatic carriage of *Neisseria meningitidis* and *Neisseria* lactamica in relation to Streptococcus pneumoniae and *Haemophilus influenzae* colonization in healthy children: apropos of 1400 children sampled. Eur. J. Epidemiol. *17*, 1015–1018.

Bax, W.A., Cluysenaer, O.J., Bartelink, A.K., Aerts, P.C., Ezekowitz, R.A., and van Dijk, H. (1999). Association of familial deficiency of mannose-binding lectin and meningococcal disease. Lancet *354*, 1094–1095.

Berggard, K., Johnsson, E., Mooi, F.R., and Lindahl, G. (1997). Bordetella pertussis binds the human complement regulator C4BP: role of filamentous hemagglutinin. Infect. Immun. *65*, 3638–3643.

Berggard, K., Johnsson, E., Morfeldt, E., Persson, J., Stalhammar-Carlemalm, M., and Lindahl, G. (2001). Binding of human C4BP to the hypervariable region of M protein: a molecular mechanism of phagocytosis resistance in *Streptococcus pyogenes*. Mol. Microbiol. *42*, 539–551.

Bishof, N.A., Welch, T.R., and Beischel, L.S. (1990). C4B deficiency: a risk factor for bacteremia with encapsulated organisms. J. Infect. Dis. *162*, 248–250.

Braconier, J.H., Sjoholm, A.G., and Soderstrom, C. (1983). Fulminant meningococcal infections in a family with inherited deficiency of properdin. Scand. J. Infect. Dis. *15*, 339–345.

Cannon, J.G., Lee, T.J., Guymon, L.F., and Sparling, P.F. (1981). Genetics of serum resistance in *Neisseria gonorrhoeae*: the sac-1 genetic locus. Infect. Immun. *32*, 547–552.

Cantini, F., Savino, S., Scarselli, M., Masignani, V., Pizza, M., Romagnoli, G., Swennen, E., Veggi, D., Banci, L., and Rappuoli, R. (2006). Solution structure of the immunodominant domain of protective antigen GNA1870 of *Neisseria meningitidis*. J. Biol. Chem. *281*, 7220–7227.

Cartwright, K.A., Stuart, J.M., Jones, D.M., and Noah, N.D. (1987). The Stonehouse survey: nasopharyngeal carriage of meningococci and *Neisseria* lactamica. Epidemiol. Infect. *99*, 591–601.

Cates, K.L., Densen, P., Lockman, J.C., and Levine, R.P. (1992). C4B deficiency is not associated with meningitis or bacteremia with encapsulated bacteria. J. Infect. Dis. *165*, 942–944.

Claus, H., Maiden, M.C., Wilson, D.J., McCarthy, N.D., Jolley, K.A., Urwin, R., Hessler, F., Frosch, M., and Vogel, U. (2005). Genetic analysis of meningococci carried by children and young adults. J. Infect. Di.s *191*, 1263–1271.

Cooper, N.R. (1985). The classical complement pathway: activation and regulation of the first complement component. Adv. Immunol. *37*, 151–216.

Densen, P., McRill, C.M., and Ross, S.C. (1988). Assembly of the membrane attack complex promotes decay of the alternative pathway C3 convertase on *Neisseria gonorrhoeae*. J. Immunol. *141*, 3902–3909.

Devyatyarova-Johnson, M., Rees, I.H., Robertson, B.D., Turner, M.W., Klein, N.J., and Jack, D.L. (2000). The lipopolysaccharide structures of *Salmonella enterica* serovar Typhimurium and *Neisseria gonorrhoeae* determine the attachment of human mannose-binding lectin to intact organisms. Infect. Immun. *68*, 3894–3899.

Dodds, A.W., Ren, X.D., Willis, A.C., and Law, S.K. (1996). The reaction mechanism of the internal thioester in the human complement component C4. Nature *379*, 177–179.

Drogari-Apiranthitou, M., Kuijper, E.J., Dekker, N., and Dankert, J. (2002). Complement activation and formation of the membrane attack complex on serogroup B *Neisseria meningitidis* in the presence or absence of serum bactericidal activity. Infect. Immun. *70*, 3752–3758.

Edwards, J.L., and Apicella, M.A. (2002). The role of lipooligosaccharide in *Neisseria gonorrhoeae* pathogenesis of cervical epithelia: lipid A serves as a C3 acceptor molecule. Cell. Microbiol. *4*, 585–598.

Eisen, D.P., and Minchinton, R.M. (2003). Impact of mannose-binding lectin on susceptibility to infectious diseases. Clin. Infect. Dis. *37*, 1496–1505.

Ellison, R.T., 3rd, Curd, J.G., Kohler, P.F., Reller, L.B., and Judson, F.N. (1987). Underlying complement deficiency in patients with disseminated gonococcal infection. Sex. Transm. Dis. *14*, 201–204.

Estabrook, M.M., Griffiss, J.M., and Jarvis, G.A. (1997). Sialylation of *Neisseria meningitidis* lipooligosaccharide inhibits serum bactericidal activity by masking lacto-N-neotetraose. Infect. Immun. *65*, 4436–4444.

Estabrook, M.M., Jack, D.L., Klein, N.J., and Jarvis, G.A. (2004). Mannose-binding lectin binds to two major outer membrane proteins, opacity protein and porin, of *Neisseria meningitidis*. J. Immunol. *172*, 3784–3792.

Exley, R.M., Shaw, J., Mowe, E., Sun, Y.H., West, N.P., Williamson, M., Botto, M., Smith, H., and Tang, C.M. (2005). Available carbon source influences the resistance of *Neisseria meningitidis* against complement. J. Exp. Med. *201*, 1637–1645.

Exley, R.M., Wu, H., Shaw, J., Schneider, M.C., Smith, H., Jerse, A.E., and Tang, C.M. (2007). Lactate acquisition promotes successful colonization of the murine genital tract by *Neisseria gonorrhoeae*. Infect. Immun. *75*, 1318–1324.

Fearon, D.T. (1978). Regulation by membrane sialic acid of β1H-dependent decay-dissociation of amplification C3 convertase of the alternative complement pathway. Proc. Natl. Acad. Sci. USA *75*, 1971–1975.

Fletcher, L.D., Bernfield, L., Barniak, V., Farley, J.E., Howell, A., Knauf, M., Ooi, P., Smith, R.P., Weise, P., Wetherell, M., Xie, X., Zagirsky, R., Zhang, Y., and Zlotnik, G.W. (2004). Vaccine potential of the *Neisseria meningitidis* 2086 lipoprotein. Infect. Immun. *72*, 2088–2100.

Fox, A.J., Jones, D.M., Scotland, S.M., Rowe, B., Smith, A., Brown, M.R., Fitzgeorge, R.G., Baskerville, A., Parsons, N.J., and Cole, J.A. (1989). Serum killing of meningococci and several other gram-negative bacterial species is not decreased by incubating them with cytidine 5′- monophospho-N-acetyl neuraminic acid [letter]. Microb. Pathog. *7*, 317–318.

Fukumori, Y., Yoshimura, K., Ohnoki, S., Yamaguchi, H., Akagaki, Y., and Inai, S. (1989). A high incidence of C9 deficiency among healthy blood donors in Osaka, Japan. Int. Immunol. *1*, 85–89.

Geoffroy, M.C., Floquet, S., Metais, A., Nassif, X., and Pelicic, V. (2003). Large-scale analysis of the meningococcus genome by gene disruption: resistance to complement-mediated lysis. Genome Res. *13*, 391–398.

Ghiran, I., Barbashov, S.F., Klickstein, L.B., Tas, S.W., Jensenius, J.C., and Nicholson-Weller, A. (2000). Complement receptor 1/CD35 is a receptor for mannan-binding lectin. J. Exp. Med. *192*, 1797–1808.

Gigli, I., Fujita, T., and Nussenzweig, V. (1979). Modulation of the classical pathway C3 convertase by plasma proteins C4 binding protein and C3b inactivator. Proc. Natl. Acad. Sci. USA *76*, 6596–6600.

Gold, R., Goldschneider, I., Lepow, M.L., Draper, T.F., and Randolph, M. (1978). Carriage of *Neisseria meningitidis* and *Neisseria* lactamica in infants and children. J. Infect. Dis. *137*, 112–121.

Gordon, D.L., and Hostetter, M.K. (1986). Complement and host defence against microorganisms. Pathology *18*, 365–375.

Greenwood, B.M., Onyewotu, II, and Whittle, H.C. (1976). Complement and meningococcal infection. Br. Med. J. *1*, 797–799.

Griffiss, J.M. (1975). Bactericidal activity of meningococcal antisera. Blocking by IgA of lytic antibody in human convalescent sera. J. Immunol. *114*, 1779–1784.

Griffiss, J.M., Jarvis, G.A., O'Brien, J.P., Eads, M.M., and Schneider, H. (1991). Lysis of *Neisseria gonorrhoeae* initiated by binding of normal human IgM to a hexosamine-containing lipooligosaccharide epitope(s) is augmented by strain-specific, properdin-binding-dependent alternative complement pathway activation. J. Immunol. *147*, 298–305.

Gulati, S., Cox, A., Lewis, L.A., Michael, F.S., Li, J., Boden, R., Ram, S., and Rice, P.A. (2005). Enhanced factor H binding to sialylated Gonococci is restricted to the sialylated lacto-N-neotetraose lipooligosaccharide species: implications for serum resistance and evidence for a bifunctional lipooligosaccharide sialyltransferase in gonococci. Infect. Immun. *73*, 7390–7397.

Gulati, S., Sastry, K., Jensenius, J.C., Rice, P.A., and Ram, S. (2002). Regulation of the mannan-binding lectin pathway of complement on *Neisseria gonorrhoeae* by C1-inhibitor and alpha 2-macroglobulin. J. Immunol. *168*, 4078–4086.

Hallstrom, T., Jarva, H., Riesbeck, K., and Blom, A.M. (2007). Interaction with C4b-binding protein contributes to nontypeable *Haemophilus influenzae* serum resistance. J. Immunol. *178*, 6359–6366.

Haralambous, E., Dolly, S.O., Hibberd, M.L., Litt, D.J., Udalova, I.A., O'Dwyer, C., Langford, P.R., Simon Kroll, J., and Levin, M. (2006). Factor H, a regulator of complement activity, is a major determinant of meningococcal disease susceptibility in UK Caucasian patients. Scand. J. Infect. Dis. *38*, 764–771.

Harriman, G.R., Podack, E.R., Braude, A.I., Corbeil, L.C., Esser, A.F., and Curd, J.G. (1982). Activation of complement by serum-resistant *Neisseria gonorrhoeae*. Assembly of the membrane attack complex without subsequent cell death. J. Exp. Med. *156*, 1235–1249.

Hellwage, J., Kuhn, S., and Zipfel, P.F. (1997). The human complement regulatory factor-H-like protein 1, which represents a truncated form of factor H, displays cell-attachment activity. Biochem. J. *326*, 321–327.

Hoffman, T.A., and Edwards, E.A. (1972). Group-specific polysaccharide antigen and humoral antibody response in disease due to *Neisseria meningitidis*. J. Infect. Dis. *126*, 636–644.

Holbein, B.E. (1981). Enhancement of *Neisseria meningitidis* infection in mice by addition of iron bound to transferrin. Infect. Immun. *34*, 120–125.

Hostetter, M.K., Thomas, M.L., Rosen, F.S., and Tack, B.F. (1982). Binding of C3b proceeds by a transesterification reaction at the thiolester site. Nature *298*, 72–75.

Ingwer, I., Petersen, B.H., and Brooks, G. (1978). Serum bactericidal action and activation of the classic and alternate complement pathways by *Neisseria gonorrhoeae*. J. Lab. Clin. Med. *92*, 211–220.

Jack, D.L., Dodds, A.W., Anwar, N., Ison, C.A., Law, A., Frosch, M., Turner, M.W., and Klein, N.J. (1998). Activation of complement by mannose-binding lectin on isogenic mutants of *Neisseria meningitidis* serogroup Br. J. Immunol. *160*, 1346–1353.

Jack, D.L., Jarvis, G.A., Booth, C.L., Turner, M.W., and Klein, N.J. (2001). Mannose-binding lectin accelerates complement activation and increases serum killing of *Neisseria meningitidis* serogroup C.J. Infect. Dis. *184*, 836–845.

Jarva, H., Ngampasutadol, J., Ram, S., Rice, P.A., Villoutreix, B.O., and Blom, A.M. (2007). Molecular characterization of the interaction between porins of *Neisseria gonorrhoeae* and C4b-binding protein. J. Immunol. *179*, 540–547.

Jarva, H., Ram, S., Vogel, U., Blom, A.M., and Meri, S. (2005). Binding of the complement inhibitor C4bp to serogroup B *Neisseria meningitidis*. J. Immunol *174*, 6299–6307.

Jarvis, G.A. (1994). Analysis of C3 deposition and degradation on *Neisseria meningitidis* and *Neisseria gonorrhoeae*. Infect. Immun. *62*, 1755–1760.

Jarvis, G.A., and Griffiss, J.M. (1989). Human IgA1 initiates complement-mediated killing of *Neisseria meningitidis*. J. Immunol. *143*, 1703–1709.

Jarvis, G.A., and Griffiss, J.M. (1991). Human IgA1 blockade of IgG-initiated lysis of *Neisseria meningitidis* is a function of antigen-binding fragment binding to the polysaccharide capsule. J. Immunol. *147*, 1962–1967.

Jarvis, G.A., and Vedros, N.A. (1987). Sialic acid of group B *Neisseria meningitidis* regulates alternative complement pathway activation. Infect. Immun. *55*, 174–180.

Jiang, H., Wagner, E., Zhang, H., and Frank, M.M. (2001). Complement 1 inhibitor is a regulator of the alternative complement pathway. J. Exp. Med. *194*, 1609–1616.

Johansson, L., Rytkonen, A., Bergman, P., Albiger, B., Kallstrom, H., Hokfelt, T., Agerberth, B., Cattaneo, R., and Jonsson, A.B. (2003). CD46 in meningococcal disease. Science *301*, 373–375.

Joiner, K.A., Scales, R., Warren, K.A., Frank, M.M., and Rice, P.A. (1985a). Mechanism of action of blocking immunoglobulin G for *Neisseria gonorrhoeae*. J. Clin. Invest. *76*, 1765–1772.

Joiner, K.A., Tartanian, A.B., Hammer, C.H., and Schweinle, J.E. (1989). Multimeric C9 within C5b-9 deposits in unique locations in the cell wall of Salmonella typhimurium. J. Immunol. *142*, 4450–4457.

Joiner, K.A., Warren, K.A., Brown, E.J., Swanson, J., and Frank, M.M. (1983). Studies on the mechanism of bacterial resistance to complement-mediated killing. IV. C5b-9 forms high molecular weight complexes with bacterial outer membrane constituents on serum-resistant but not on serum-sensitive *Neisseria gonorrhoeae*. J. Immunol. *131*, 1443–1451.

Joiner, K.A., Warren, K.A., Hammer, C., and Frank, M.M. (1985b). Bactericidal but not nonbactericidal C5b-9 is associated with distinctive outer membrane proteins in *Neisseria gonorrhoeae*. J. Immunol. *134*, 1920–1925.

Kahler, C.M., Martin, L.E., Shih, G.C., Rahman, M.M., Carlson, R.W., and Stephens, D.S. (1998). The (alpha2 ->8)-linked polysialic acid capsule and lipooligosaccharide structure both contribute to the ability of serogroup B *Neisseria meningitidis* to resist the bactericidal activity of normal human serum. Infect. Immun. *66*, 5939–5947.

Kask, L., Hillarp, A., Ramesh, B., Dahlback, B., and Blom, A.M. (2002). Structural requirements for the intracellular subunit polymerization of the complement inhibitor C4b-binding protein. Biochemistry *41*, 9349–9357.

Kirkitadze, M.D., and Barlow, P.N. (2001). Structure and flexibility of the multiple domain proteins that regulate complement activation. Immunol. Rev. *180*, 146–161.

Kuipers, S., Aerts, P.C., Cluysenaer, O.J., Bartelink, A.K., Ezekowitz, R.A., Bax, W.A., Salimans, M., and Vandyk, H. (2003). A case of familial meningococcal disease due to deficiency in mannose-binding lectin (MBL). Adv. Exp. Med. Biol. *531*, 351–355.

Lachmann, P.J., and Hughes-Jones, N.C. (1984). Initiation of complement activation. Springer Semin. Immunopathol. *7*, 143–162.

Lambris, J.D., Ricklin, D., and Geisbrecht, B.V. (2008). Complement evasion by human pathogens. Nat. Rev. Microbiol. *6*, 132–142.

Lewis, L.A., Ram, S., Prasad, A., Gulati, S., Getzlaff, S., Blom, A.M., Vogel, U., and Rice, P.A. (2008). Defining targets for complement components C4b and C3b on the pathogenic neisseriae. Infect. Immun. *76*, 339–350.

Madico, G., Ngampasutadol, J., Gulati, S., Vogel, U., Rice, P.A., and Ram, S. (2007). Factor H binding and function in sialylated pathogenic neisseriae is influenced by gonococcal, but not meningococcal, porin. J. Immunol. *178*, 4489–4497.

Madico, G., Ram, S., Getzlaff, S., Prasad, A., Gulati, S., Ngampasutadol, J., Vogel, U., and Rice, P.A. (2004). Sialylation of lacto-N-tetraose lipooligosaccharide in gonococci, but not meningococci, results in enhanced factor H binding: the modulatory role of gonococcal porin, Paper presented at: 14th International Pathogenic *Neisseria* Conference (Milwaukee, Wisconsin).

Madico, G., Welsch, J.A., Lewis, L.A., McNaughton, A., Perlman, D.H., Costello, C.E., Ngampasutadol, J., Vogel, U., Granoff, D.M., and Ram, S. (2006). The meningococcal vaccine candidate GNA1870 binds the complement regulatory protein factor H and enhances serum resistance. J. Immunol. *177*, 501–510.

Mandrell, R.E., Kim, J.J., John, C.M., Gibson, B.W., Sugai, J.V., Apicella, M.A., Griffiss, J.M., and Yamasaki, R. (1991). Endogenous sialylation of the lipooligosaccharides of *Neisseria meningitidis*. J. Bacteriol. *173*, 2823–2832.

Masignani, V., Comanducci, M., Giuliani, M.M., Bambini, S., Adu-Bobie, J., Arico, B., Brunelli, B., Pieri, A., Santini, L., Savino, S., Serruto, D., Litt, D., Kroll, S., Welsch, J.A., Granoff, D.M., Rappuoli, R., and Pizza, M. (2003). Vaccination against *Neisseria meningitidis* using three variants of the lipoprotein GNA1870. J. Exp. Med. *197*, 789–799.

McCutchan, J.A., Levine, S., and Braude, A.I. (1976). Influence of colony type on susceptibility of gonococci to killing by human serum. J. Immunol. *116*, 1652–1655.

McGee, Z., Clemens, C., Klein, J., Barley, L., and Gorby, G. (1990). Role of tumor necrosis factor in gonococcal damage to human fallopian tube mucosa, Paper presented at: Seventh International Pathogenic *Neisseria* Conference (Berlin, West Germany).

McQuillen, D.P., Gulati, S., Ram, S., Turner, A.K., Jani, D.B., Heeren, T.C., and Rice, P.A. (1999). Complement processing and immunoglobulin binding to *Neisseria gonorrhoeae* determined *in vitro* simulates *in vivo* effects. J. Infect. Dis. *179*, 124–135.

McQuillen, D.P., Gulati, S., and Rice, P.A. (1994). Complement-mediated bacterial killing assays. Methods Enzymol. *236*, 137–147.

Meri, S., and Pangburn, M.K. (1990). Discrimination between activators and nonactivators of the alternative pathway of complement: regulation via a sialic acid/polyanion binding site on factor H. Proc. Natl. Acad. Sci. USA *87*, 3982–3986.

Meri, S., and Pangburn, M.K. (1994). Regulation of alternative pathway complement activation by glycosaminoglycans: specificity of the polyanion binding site on factor H. Biochem. Biophys. Res. Commun. *198*, 52–59.

Meri, T., Cutler, S.J., Blom, A.M., Meri, S., and Jokiranta, T.S. (2006). Relapsing fever spirochetes Borrelia recurrentis and B. duttonii acquire complement regulators C4b-binding protein and factor H. Infect. Immun. *74*, 4157–4163.

Meri, T., Hartmann, A., Lenk, D., Eck, R., Wurzner, R., Hellwage, J., Meri, S., and Zipfel, P.F. (2002). The yeast *Candida albicans* binds complement regulators factor H and FHL-1. Infect. Immun. *70*, 5185–5192.

Muller-Eberhard, H.J. (1986). The membrane attack complex of complement. Annu. Rev. Immunol. *4*, 503–528.

Muller-Eberhard, H.J. (1988). Molecular organization and function of the complement system. Annu. Rev. Biochem. *57*, 321–347.

Munkley, A., Tinsley, C.R., Virji, M., and Heckels, J.E. (1991). Blocking of bactericidal killing of *Neisseria meningitidis* by antibodies directed against class 4 outer membrane protein. Microb Pathog *11*, 447–452.

Nairn, C.A., Cole, J.A., Patel, P.V., Parsons, N.J., Fox, J.E., and Smith, H. (1988). Cytidine 5′-monophospho-N-acetylneuraminic acid or a related compound is the low Mr factor from human red blood cells which induces gonococcal resistance to killing by human serum. J. Gen. Microbiol. *134*, 3295–3306.

Ngampasutadol, J., Ram, S., Blom, A.M., Jarva, H., Jerse, A.E., Lien, E., Goguen, J., Gulati, S., and Rice, P.A. (2005). Human C4b-binding protein selectively interacts with *Neisseria gonorrhoeae* and results in species-specific infection. Proc. Natl. Acad. Sci. USA *102*, 17142–17147.

Ngampasutadol, J., Ram, S., Gulati, S., Agarwal, S., Li, C., Visintin, A., Monks, B., Madico, G., and Rice, P.A. (2008). Human Factor H Interacts Selectively with *Neisseria gonorrhoeae* and Results in Species-Specific Complement Evasion. J. Immunol. *180*, 3426–3435.

Nordstrom, T., Blom, A.M., Forsgren, A., and Riesbeck, K. (2004). The emerging pathogen Moraxella catarrhalis interacts with complement inhibitor C4b binding protein through ubiquitous surface proteins A1 and A2. J. Immunol. *173*, 4598–4606.

Pangburn, M.K. (1986). Differences between the binding sites of the complement regulatory proteins DAF, CR1, and factor H on C3 convertases. J. Immunol. *136*, 2216–2221.

Pangburn, M.K., and Muller-Eberhard, H.J. (1984). The alternative pathway of complement. Springer Semin. Immunopathol. *7*, 163–192.

Parsons, N.J., Patel, P.V., Tan, E.L., Andrade, J.R., Nairn, C.A., Goldner, M., Cole, J.A., and Smith, H. (1988). Cytidine 5′-monophospho-N-acetyl neuraminic acid and a low molecular weight factor from human blood cells induce lipopolysaccharide alteration in gonococci when conferring resistance to killing by human serum. Microb. Pathog. *5*, 303–309.

Plested, J.S., Makepeace, K., Jennings, M.P., Gidney, M.A., Lacelle, S., Brisson, J., Cox, A.D., Martin, A., Bird, A.G., Tang, C.M., Mackinnon, F.M., Richards, J.C., and Moxon, E.R. (1999). Conservation and accessibility of an inner core lipopolysaccharide epitope of *Neisseria meningitidis*. Infect. Immun. *67*, 5417–5426.

Plummer, F.A., Chubb, H., Simonsen, J.N., Bosire, M., Slaney, L., Maclean, I., Ndinya-Achola, J.O., Waiyaki, P., and Brunham, R.C. (1993). Antibody to Rmp (outer membrane protein 3) increases susceptibility to gonococcal infection. J. Clin. Invest. *91*, 339–343.

Prasadarao, N.V., Blom, A.M., Villoutreix, B.O., and Linsangan, L.C. (2002). A novel interaction of outer membrane protein A with C4b binding protein mediates serum resistance of *Escherichia coli* K1. J. Immunol. *169*, 6352–6360.

Ram, S., Cox, A.D., Wright, J.C., Vogel, U., Getzlaff, S., Boden, R., Li, J., Plested, J.S., Meri, S., Gulati, S., Stein, D.C., Richards, J.C., Moxon, E.R., and Rice, P.A. (2003). Neisserial lipooligosaccharide is a target for complement component C4b. Inner core phosphoethanolamine residues define C4b linkage specificity. J. Biol. Chem. *278*, 50853–50862.

Ram, S., Cullinane, M., Blom, A.M., Gulati, S., McQuillen, D.P., Monks, B.G., O'Connell, C., Boden, R., Elkins, C., Pangburn, M.K., Dahlback, B., and Rice, P.A. (2001). Binding of C4b-binding protein to porin: a molecular mechanism of serum resistance of *Neisseria gonorrhoeae*. J. Exp. Med. *193*, 281–295.

Ram, S., McQuillen, D.P., Gulati, S., Elkins, C., Pangburn, M.K., and Rice, P.A. (1998a). Binding of complement factor H to loop 5 of porin protein 1A: a molecular mechanism of serum resistance of nonsialylated *Neisseria gonorrhoeae*. J. Exp. Med. *188*, 671–680.

Ram, S., Ngampasutadol, J., Cox, A.D., Blom, A.M., Lewis, L.A., St Michael, F., Stupak, J., Gulati, S., and Rice, P.A. (2007). Heptose I glycan substitutions on *Neisseria gonorrhoeae* lipooligosaccharide influence C4b-binding protein binding and serum resistance. Infect. Immun. *75*, 4071–4081.

Ram, S., Sharma, A.K., Simpson, S.D., Gulati, S., McQuillen, D.P., Pangburn, M.K., and Rice, P.A. (1998b). A novel sialic acid binding site on factor H mediates serum resistance of sialylated *Neisseria gonorrhoeae*. J. Exp. Med. *187*, 743–752.

Rice, P.A. (2005). Gonococcal arthritis (disseminated gonococcal infection). Infect. Dis. Clin. North Am. *19*, 853–861.

Rice, P.A., McCormack, W.M., and Kasper, D.L. (1980). Natural serum bactericidal activity against *Neisseria gonorrhoeae* isolates from disseminated, locally invasive, and uncomplicated disease. J. Immunol. *124*, 2105–2109.

Rice, P.A., McQuillen, D.P., Gulati, S., Jani, D.B., Wetzler, L.M., Blake, M.S., and Gotschlich, E.C. (1994). Serum resistance of *Neisseria gonorrhoeae*. Does it thwart the inflammatory response and facilitate the transmission of infection? Ann. NY Acad. Sci. *730*, 7–14.

Rice, P.A., Vayo, H., Tam, M., and Blake, M.S. (1986). Immunoglobulin G antibodies directed against protein III block killing of serum-resistant *Neisseria gonorrhoeae* by immune serum. J. Exp. Med. *164*, 1735–1748.

Richmond, P., Borrow, R., Goldblatt, D., Findlow, J., Martin, S., Morris, R., Cartwright, K., and Miller, E. (2001). Ability of 3 different meningococcal C conjugate vaccines to induce immunologic memory after a single dose in UK toddlers. J. Infect. Dis. *183*, 160–163.

Ripoche, J., Day, A.J., Harris, T.J., and Sim, R.B. (1988). The complete amino acid sequence of human complement factor H. Biochem. J. *249*, 593–602.

Rooijakkers, S.H., and van Strijp, J.A. (2007). Bacterial complement evasion. Mo.l Immunol. *44*, 23–32.

Rosenqvist, E., Musacchio, A., Aase, A., Hoiby, E.A., Namork, E., Kolberg, J., Wedege, E., Delvig, A., Dalseg, R., Michaelsen, T.E., and Tommassen, J. (1999). Functional activities and epitope specificity of human and murine antibodies against the class 4 outer membrane protein (Rmp) of *Neisseria meningitidis*. Infect. Immun. *67*, 1267–1276.

Rowe, P.C., McLean, R.H., Wood, R.A., Leggiadro, R.J., and Winkelstein, J.A. (1989). Association of homozygous C4B deficiency with bacterial meningitis. J. Infect. Dis. *160*, 448–451.

Schneider, H., Griffiss, J.M., Williams, G.D., and Pier, G.B. (1982). Immunological basis of serum resistance of *Neisseria gonorrhoeae*. J. Gen. Microbiol. *128*, 13–22.

Schneider, M.C., Exley, R.M., Chan, H., Feavers, I., Kang, Y.H., Sim, R.B., and Tang, C.M. (2006). Functional significance of factor H binding to *Neisseria meningitidis*. J. Immunol. *176*, 7566–7575.

Schryvers, A.B., and Gonzalez, G.C. (1990). Receptors for transferrin in pathogenic bacteria are specific for the host's protein. Can. J. Microbiol. *36*, 145–147.

Selander, B., Kayhty, H., Wedege, E., Holmstrom, E., Truedsson, L., Soderstrom, C., and Sjoholm, A.G. (2000). Vaccination responses to capsular polysaccharides of *Neisseria meningitidis* and *Haemophilus influenzae* type b in two C2-deficient sisters: alternative pathway-mediated bacterial killing and evidence for a novel type of blocking IgG.J. Clin. Immunol. *20*, 138–149.

Sjoholm, A.G., Braconier, J.H., and Soderstrom, C. (1982). Properdin deficiency in a family with fulminant meningococcal infections. Clin. Exp. Immunol. *50*, 291–297.

Soderstrom, C., Braconier, J.H., Danielsson, D., and Sjoholm, A.G. (1987). Bactericidal activity for *Neisseria meningitidis* in properdin-deficient sera. J. Infect. Dis. *156*, 107–112.

Soderstrom, C., Sjoholm, A.G., Svensson, R., and Ostenson, S. (1989). Another Swedish family with complete properdin deficiency: association with fulminant meningococcal disease in one male family member. Scand. J. Infect. Dis. *21*, 259–265.

Spitzer, D., Mitchell, L.M., Atkinson, J.P., and Hourcade, D.E. (2007). Properdin can initiate complement activation by binding specific target surfaces and providing a platform for de novo convertase assembly. J. Immunol. *179*, 2600–2608.

Sprong, T., Roos, D., Weemaes, C., Neeleman, C., Geesing, C.L., Mollnes, T.E., and van Deuren, M. (2006). Deficient alternative complement pathway activation due to factor D deficiency by 2 novel mutations in the complement factor D gene in a family with meningococcal infections. Blood *107*, 4865–4870.

Stephens, D.S., and McGee, Z.A. (1983). Association of virulence of *Neisseria meningitidis* with transparent colony type and low-molecular-weight outer membrane proteins. J. Infect. Dis. *147*, 282–292.

Swanson, J. (1978). Studies on gonococcus infection. XII. Colony color and opacity variants of gonococci. Infect. Immun. *19*, 320–331.

Tan, E.L., Patel, P.V., Parsons, N.J., Martin, P.M., and Smith, H. (1986). Lipopolysaccharide alteration is associated with induced resistance of *Neisseria gonorrhoeae* to killing by human serum. J. Gen. Microbiol. *132*, 1407–1413.

Uria, M.J., Zhang, Q., Li, Y., Chan, A., Exley, R.M., Gollan, B., Chan, H., Feavers, I., Yarwood, A., Abad, R., Borrow, R., Fleck, R.A., Mulloy, B., Vazquez, J.A., and Tang, C.M. (2008). A generic mechanism in *Neisseria meningitidis* for enhanced resistance against bactericidal antibodies. J. Exp. Med. *205*, 1423–1434.

van Emmerik, L.C., Kuijper, E.J., Fijen, C.A., Dankert, J., and Thiel, S. (1994). Binding of mannan-binding protein to various bacterial pathogens of meningitis. Clin. Exp. Immunol. *97*, 411–416.

Vogel, U., Claus, H., Heinze, G., and Frosch, M. (1999). Role of lipopolysaccharide sialylation in serum resistance of serogroup B and C meningococcal disease isolates. Infect. Immun. *67*, 954–957.

Vogel, U., Weinberger, A., Frank, R., Muller, A., Kohl, J., Atkinson, J.P., and Frosch, M. (1997). Complement factor C3 deposition and serum resistance in isogenic capsule and lipooligosaccharide sialic acid mutants of serogroup B *Neisseria meningitidis*. Infect. Immun. *65*, 4022–4029.

Vyse, T.J., Spath, P.J., Davies, K.A., Morley, B.J., Philippe, P., Athanassiou, P., Giles, C.M., and Walport, M.J. (1994). Hereditary complement factor I deficiency. Q.J. Med. *87*, 385–401.

Walport, M.J. (2001a). Complement. First of two parts. N. Engl. J. Med. *344*, 1058–1066.

Walport, M.J. (2001b). Complement. Second of two parts. N. Engl. J Med. *344*, 1140–1144.

Ward, M.E., Watt, P.J., and Glynn, A.A. (1970). Gonococci in urethral exudates possess a virulence factor lost on subculture. Nature *227*, 382–384.

Yazdankhah, S.P., and Caugant, D.A. (2004). *Neisseria meningitidis*: an overview of the carriage state. J. Med. Microbiol. *53*, 821–832.

Zarantonelli, M.L., Szatanik, M., Giorgini, D., Hong, E., Huerre, M., Guillou, F., Alonso, J.M., and Taha, M.K. (2007). Transgenic mice expressing human transferrin as a model for meningococcal infection. Infect. Immun. *75*, 5609–5614.

Zhu, Z., Atkinson, T.P., Hovanky, K.T., Boppana, S.B., Dai, Y.L., Densen, P., Go, R.C., Jablecki, J.S., and Volanakis, J.E. (2000). High prevalence of complement component C6 deficiency among African-Americans in the south-eastern USA. Clin. Exp. Immunol. *119*, 305–310.

Zipfel, P.F., Hellwage, J., Friese, M.A., Hegasy, G., Jokiranta, S.T., and Meri, S. (1999). Factor H and disease: a complement regulator affects vital body functions. Mol. Immunol. *36*, 241–248.

Zipfel, P.F., Wurzner, R., and Skerka, C. (2007). Complement evasion of pathogens: common strategies are shared by diverse organisms. Mol. Immunol. *44*, 3850–3857.

Zwahlen, A., and Waldvogel, F.A. (1984). Magnitude of bacteremia and complement activation during *Neisseria meningitidis* infection: study of two coprimary cases with different clinical presentations. Eur. J. Clin. Microbiol. *3*, 439–441.

Consequences of Pathogenic *Neisseria* Infection on Cellular Apoptosis

9

Sarah A. Follows and Paola Massari

Abstract

A number of pathogens, including viruses, bacteria, and parasites, have evolved mechanisms to subvert apoptosis by either positively or negatively modulating host defences. In particular, the inhibition of the apoptotic process by microbial pathogens has previously demonstrated importance in securing intracellular niches, which may be an important mechanism for microbial survival, replication or immune evasion. Similarly, the induction of apoptosis may be beneficial to pathogens by promoting escape from host cells or diminishing their effector functions, again promoting immune evasion. Conflicting studies have reported that infection with the facultative intracellular pathogens *Neisseria gonorrhoeae* and *Neisseria meningitidis* can either inhibit or induce apoptosis. Reports on *neisserial* influence on the apoptotic response of host cells have been published consistently for almost ten years and will be the focus of this chapter. Overall, it seems that *Neisseria* species more often inhibit apoptosis, and this inhibition may allow time for adaptation to a new environment, intracellular replication, or immune evasion presumably leading to the spread of infection.

Introduction

Apoptosis is an active process of cell death involved in normal cell turnover and removal of abnormal or infected cells (Cohen, 1997; Krammer *et al.*, 1994; Thatte and Dahanukar, 1997). Apoptosis protects intracellular contents from being released into the surrounding environment and leading to inflammation, such as in the case of necrosis (Proskuryakov *et al.*, 2002; Savill *et al.*, 1993). Apoptosis may also be induced during inflammatory responses via death receptor ligation caused by either TNF-α or FasL expressed on immune effector cells (Weinrauch and Zychlinsky, 1999). This can trigger death in the immune effectors themselves, resulting in resolution of inflammation (Boise *et al.*, 1995; Kabelitz and Janssen, 1997), as well as non-inflammatory death of target cells. Apoptosis can also be initiated during infection of host cells by pathogenic microorganisms. For example, microbial infection of non-phagocytic cells, such as epithelial cells, can cause production of reactive oxygen species, intracellular stress, and DNA damage (Cuff and Ruby, 1996; Weinrauch and Zychlinsky, 1999) and apoptosis (Le Bras *et al.*, 2005; Storz and Imlay, 1999). In this context, apoptosis acts as a host defence against infection, which will eventually eliminate the infected host cell via clearance by phagocytosis (Platt *et al.*, 1998; Savill *et al.*, 1993). In contrast to apoptosis as an anti-inflammatory event, apoptotic cell death can also induce inflammation when it occurs following infection (Restifo, 2000; Zychlinsky and Sansonetti, 1997). Inflammatory signals may be a secondary result of the caspase cascade, for example the production of the pro-inflammatory cytokines IL-1β and IL-18 by active capase-1 (Restifo, 2000; Zychlinsky and Sansonetti, 1997). Therefore, apoptosis appears to be important for both the initiation and resolution of inflammation.

Various pathogens have demonstrated an ability to exploit apoptosis in both phagocytic and non-phagocytic cell types. The induction of apoptosis in phagocytes may be beneficial to pathogens by promoting escape from these cells or diminishing their effector functions (Kobayashi *et al.*, 2003b). Induction of apoptosis in non-phagocytic cells may prevent the tissue destruction and inflammation associated with cell necrosis, thereby avoiding recognition by the immune system (Zychlinsky and Sansonetti, 1997). The inhibition of apoptosis in non-phagocytic cells may also be beneficial to pathogens, by securing an intracellular niche, inhibiting any inflammatory processes that may be initiated via apoptosis, and potentially shielding organisms from recognition by phagocytes (Hacker and Fischer, 2002). Furthermore, inhibition of apoptosis by obligate intracellular pathogens has proven to be an important mechanism which allows for bacterial replication. *Neisseria* species are among those pathogens which have demonstrated an ability to interfere with the apoptotic cellular machinery.

The majority of work characterizing the apoptotic response to *Neisseria* supports the finding that infection promotes resistance to apoptosis. There are several reasons why this organisms may have evolved mechanisms to inhibit apoptosis. First, upon infection of epithelial cells at the mucosal surface, *Neisseria* may need to prevent the host defence of epithelial exfoliation. Second, prohibiting cell death may provide an intracellular niche for the organisms to replicate, while evading recognition by the immune system. Inhibition of epithelial cell death by apoptosis also prevents them from being detected by circulating phagocytes, which function to pick up and destroy apoptotic cells; therefore, preventing epithelial cell apoptosis also prevents killing by phagocytes. Third, infected epithelial cells promote inflammation via cytokine and chemokine production, including the apoptosis-inducing cytokine TNF-α; the anti-apoptotic response may have evolved to prevent this event. Last, prevention of cell death in innate immune cells (i.e. neutrophils) clearly benefits the organisms by preventing neutrophil function, as well as providing an intracellular niche for replication. Many components of *Neisseria* have been implicated in the ability to promote resistance to apoptosis, and likely a combination of several factors involved in establishing and maintaining infection leads to the anti-apoptotic effect. Furthermore, different components of the bacteria may play a role in resistance to apoptosis during different stages of infection. Depending on the cell type and location in the body, host cells may also be more susceptible or more resistant to apoptosis and immune responses. As facultative intracellular pathogens, it is conceivable that the interaction of *Neisseria* with the host cell would be improved if apoptosis were inhibited, allowing to the bacteria to reach their specific niche *in vivo* where optimal bacterial multiplication can occur in the absence of inflammation.

The process of apoptosis

Apoptosis is a predetermined homeostatic mechanism, first observed by Kerr *et al.* in hepatocytes, that describes a type of cell death different from necrotic cell death. (Kerr *et al.*, 1972). Small round bodies surrounded by membranes and containing intact cytosolic organelles or parts of the nucleus are characteristics of apoptotic cells, and result from cellular condensation and budding. Apoptotic cells package their contents into apoptotic bodies and expose surface molecules, such as phosphatidylserine, to target phagocytic uptake and removal that results in a non-inflammatory process *in vivo*. Together with chromatin condensation and nuclear fragmentation (Kroemer and Martin, 2005), these are the main morphological changes that define apoptosis. As apoptosis is used to maintain tissue homeostasis, it is a very controlled mechanism with multiple, crucial steps. Activation of a distinct subset of caspases (cysteine-dependent aspartate-specific proteases) and involvement of mitochondria have been described as the central intracellular checkpoint of apoptosis (Green and Kroemer, 2004).

Extrinsic and intrinsic apoptosis pathways

Apoptosis can be triggered by two main distinct pathways, the extrinsic or receptor-activated pathway and the intrinsic or mitochondrial pathway. Both pathways have an initiation phase, a commitment phase and an execution phase (Kroemer, 1997) (Fig. 9.1). The various death-inducing stimuli (i.e. extrinsic or intrinsic) determine the initiation phase, the commitment phase is mediated by both caspases activation and involvement of mitochondria, and the last phase leads to cell shrinkage, chromatin condensation, nuclear

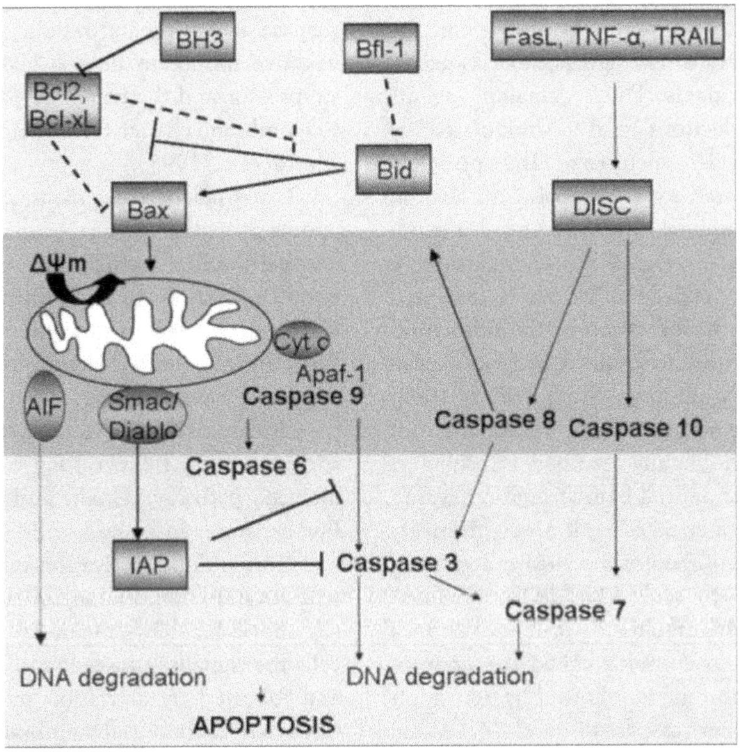

Figure 9.1 Schematic representation of the two major apoptosis pathways. In the intrinsic pathway, various cellular stress signals induce recruitment of pro-apoptotic members of the Bcl-2 family (initiation phase), followed by involvement of mitochondria and activation of initiator caspase-9 (commitment phase) and subsequent activation of effector caspase-3 and -7 (execution phase) leading to final DNA degradation and cell death. In the extrinsic pathway, death-receptor mediated signals induce formation of the DISC complex (initiation phase), followed by activation of initiator caspase-8 and -10 (commitment phase) and subsequent activation of effector caspase-3 and -7 (execution phase), merging the two pathways downstream of the commitment phase. Pro-apoptotic member of the Bcl-2 family, Bid, provides further cross-talk between the two pathways upstream of mitochondrial events.

fragmentation, blebbing, and phosphatidylserine exposure on the surface of the plasma membrane. Among the key proteins involved in apoptosis, initiator caspases (i.e. caspase-8, -9, and -10) and effector caspases (i.e. caspase-3, -6, and -7) are activated in response to both pathways, as well as members of the Bcl-2 family (Danial, 2007; Letai *et al.*, 2002) (Fig. 9.1). This latter protein family is divided into three groups which contain both anti-apoptotic and pro-apoptotic members. The first group, which has four regions of shared Bcl-2 homology (BH4), includes Bcl-2, Bcl-XL, Mcl-1, Bcl-w, Bcl-b and Bfl-1. The second group includes proteins with three BH-domains, (Bax, Bak and Bok), and a further group with one BH-domain, the BH3-only proteins such as Bid, Bad, Bim, Bik, PUMA and more (Letai *et al.*, 2002).

In the extrinsic pathway, apoptosis is triggered by activation of death receptors on the cell surface, including the tumour necrosis factor (TNF) receptor-1, CD95/Fas, the TNF-related apoptosis inducing ligand (TRAIL) receptors-1 and -2 and the death receptor 3 (TRAMP) (Schulze-Osthoff *et al.*, 1998). Activation of death receptors induces recruitment and oligomerization of the adapter molecule FADD (Fas-associating death domain-containing protein) and the formation of a complex called DISC (death-inducing signalling complex) (Budihardjo *et al.*, 1999). This event is followed by activation of initiator caspase-8 and -10 (Debatin and Krammer, 2004), which then proceed to activate executioner caspase-3 and 7 (Green and Kroemer, 2004) (Fig. 9.1).

The intrinsic apoptotic pathway can be triggered by several stress signals, such as excessive calcium, oxidants, DNA damaging agents and bacterial infections, and is modulated by pro-apoptotic Bcl-2 members. The principal event is represented by release of a number of mitochondrial proteins into the cytosol, including cytochrome c (Fig. 9.1). Cytochrome c release has been explained via three potential mechanisms: permeabilization of the mitochondrial membrane, due to changes in its potential (Kroemer, 1997), ion flow (Scaffidi *et al.*, 1998; Vander Heiden *et al.*, 1997), and involvement of pro-apoptotic Bcl-2 family members (Cosulich *et al.*, 1997). Mitochondrial cytochrome c released into the cytosol associates with a constitutively expressed protein, apoptosis protease activating factor 1 (APAF-1) and ATP/dATP (Cain *et al.*, 2002), and the initiator caspase-9 to form a multimeric protein complex called the apoptosome. The apoptosome is followed by activation of the executioner caspases-3 and -6. Along with cytochrome c, other mitochondrial factors such as activator of caspase/direct inhibitor of apoptosis (IAP) binding protein (Deveraux *et al.*, 1997), second mitochondria-derived activator of caspase/direct IAP binding protein with a low pI (Smac/DIABLO) (Verhagen *et al.*, 2000), caspase-independent apoptogenic death effectors including apoptosis inducing factor (AIF) (Susin *et al.*, 1999) and endonuclease G (EndoG) (Li *et al.*, 2001), as well as the accumulation of reactive oxygen species (ROS) (Fleury *et al.*, 2002), can be dependent on changes in the mitochondrial membrane potential. Mitochondrial membrane potential is controlled by mitochondrial proteins such as ANT, CypD and VDAC (Colombini, 2004; Colombini *et al.*, 1996) and by members of the Bcl-2 family (Green and Kroemer, 2004; Kroemer *et al.*, 2007). Loss of mitochondrial potential ($\Delta\Psi m$) has been considered as an early event in apoptosis (Zamzami *et al.*, 1996), but other reports indicate that this event may instead follow the triggering of apoptosis (Ly *et al.*, 2003). Moreover, the requirement of $\Delta\Psi m$ for cytochrome c release is controversial and the role of VDAC and mitochondrial depolarization might have a less significant impact on mitochondrial mediated cell death (Baines *et al.*, 2007; Galluzzi and Kroemer, 2007). Other pathways leading to caspase activation and cellular apoptosis include negative induction due to loss of activity of a suppressor and the ceramide pathway secondary to membrane change (Hofmann and Dixit, 1998; Martin *et al.*, 1995).

Cross-talk between the intrinsic and extrinsic apoptosis pathways has been described at the caspase level, as cytochrome c release and caspase-9 activation are also promoted by caspase-8 via the truncated form of the cytosolic protein Bid (tBid), thereby bridging the two signalling apoptotic cascades (Fig. 9.1). In addition, caspase-3 can also activate Bid, amplifying the extrinsic pathway by recruiting components of the intrinsic pathway (Green and Kroemer, 2004). Furthermore, in type 1 cells, ligation of death receptors induces activation of effector caspases without mitochondrial membrane depolarization (Scaffidi *et al.*, 1998), On the other hand, in type 2 cells the signalling cascade initiated by caspase-8 and followed by activation of Bid and of Bax, leads to mitochondrial membrane depolarization, cytochome c release and caspase-3 activation (Krammer, 2000; Scaffidi *et al.*, 1998).

Intracellular components involved in modulation of apoptosis

Members of the Bcl-2 protein family have a fundamental role in modulation of apoptosis. While some of these proteins are cytosolic and others reside in the mitochondria, most function by interacting with the mitochondria (Kroemer and Reed, 2000). Anti-apoptotic Bcl-2 family members inhibit apoptosis by modulating mitochondrial potential, either via a direct interaction with pro-apoptotic pore-forming Bcl-2 family members (Bax, Bak) or by neutralizing BH3-only proteins (Kroemer and Reed, 2000). Additionally, Bcl-2 and Bcl-X_L also interact with proteins that are part of the mitochondrial permeability transition pore ANT (Marzo *et al.*, 1998) and VDAC (Brenner and Kroemer, 2000; Shimizu *et al.*, 2000). Anti-apoptotic Bcl-2 family members might then have two functions on mitochondria: 1) inhibition of pore formation by Bax and Bak and 2) modulation of pores formed by mitochondrial resident proteins such as VDAC and ANT.

Pro-apoptotic Bcl-2 family members promote apoptosis in various ways. Bak is associated with the mitochondrial outer membrane, while Bax is a cytosolic protein (Wei *et al.*, 2001). Upon induction of apoptosis, Bax undergoes a conformational change and inserts into the mitochondrial outer membrane (Wolter *et al.*, 1997), where it can associate with other pro-apoptotic members (i.e. Bak or tBid) (Kuwana *et al.*, 2002), inducing mitochondrial membrane depolarization (Sharpe *et al.*, 2004). Bax and Bak have been described as ligands for isoforms of the mitochondrial proteins VDAC and TOM22 (Bellot *et al.*, 2007; Cheng *et al.*, 2003; Shimizu *et al.*, 1999)as well as Bid (Rostovtseva *et al.*, 2004). The formation of these complexes is specifically inhibited by Bcl-X_L and has been reported to occur in both a caspase-dependent and -independent manner (Nechushtan *et al.*, 2001). However, Bax also appears to be able to permeabilize the mitochondrial outer membrane in the absence of mitochondrial pore components (Eskes *et al.*, 1998; Kuwana *et al.*, 2002) or direct interaction with VDAC (Baines *et al.*, 2007). Pro-apoptotic Bad, a BH3-only protein, appears to act by two different mechanisms, by interacting with either Bcl-2 or Bcl-xL, dissociating them from other BH3-only proteins or from Bax/Bak, and promoting mitochondrial depolarization. Pro-apoptotic tBid can directly activate Bax by inducing its translocation into the mitochondrial membrane or can exert a direct effect on Bak (Letai *et al.*, 2002). However, the molecular mechanism of release of cytochrome c and other mitochondrial components still remains controversial.

Mitochondrial factor Smac/DIABLO promotes apoptosis indirectly by inactivating members of the IAP (inhibitor of apoptosis protein) family, a caspase-inhibitor protein family (Verhagen *et al.*, 2000). Moreover, the caspase-independent death effectors apoptosis-inducing factor (AIF) (Susin *et al.*, 1999) and EndoG (Li *et al.*, 2001) promote apoptosis by translocating from the cytosol to the nucleus and mediate DNA degradation and chromatin condensation.

Modulation of apoptosis by bacteria

Bacteria use different strategies to interfere with programmed cell death, mostly to avoid normal host defence responses. Most intracellular pathogens inhibit apoptosis during early phases of their intracellular life cycle, while they can induce apoptosis in a later phase, to minimize the immune and inflammatory response of the host (Gao and Kwaik, 2000). For example, inhibition of apoptosis in the early life cycle to promote intracellular survival has been shown for the obligate intracellular bacteria, *Chlamydia* (Fan *et al.*, 1998). However, at later times, *Chlamydia* can interfere with the host cell cycle by inducing apoptosis. Inhibition of apoptosis by *Chlamydia* is via an NFkB-independent inhibition of Bax and Bak activation (Xiao *et al.*, 2004; Zhong *et al.*, 2006). Furthermore, infection with *Chlamydia* renders infected cells resistant to apoptosis induced by different stimuli (Airenne *et al.*, 2002; Fan *et al.*, 1998; Fischer *et al.*, 2001). Other bacteria which can modulate cellular apoptotic responses include *Rickettsia rickettsii*, which can prevent programmed cell death of infected endothelial cells (Clifton *et al.*, 1998), *Brucella*, which inhibits spontaneous apoptosis in human monocytes (Gross *et al.*, 2000), *Mycobacterium tuberculosis*, which prevents apoptosis in macrophages (Balcewicz-Sablinska *et al.*, 1998), the periodontal pathogen *Porphyromonas gingivalis*, which induces activation of anti-apoptotic cell signalling pathways (Murray and Wilton, 2003; Nakhjiri *et al.*, 2001) and the facultative intracellular *Neisseria meningitidis* and *Neisseria gonorrhoeae*, which prevent induction of apoptosis mediated by both the intrinsic and the extrinsic pathway (Beck and Meyer, 2000; Binnicker *et al.*, 2003; Higashi *et al.*, 2007; Massari *et al.*, 2000; Massari *et al.*, 2003a; Morales *et al.*, 2006; Tunbridge *et al.*, 2006). However, in the case of both *P. gingivalis* and *Neisseria* species, contrasting reports exist on their ability to modulate apoptosis (Brozovic *et al.*, 2006; Desta and Graves, 2007; Muller *et al.*, 2000; Muller *et al.*, 1999) (Schubert-Unkmeir 2007). Regardless, delayed apoptosis may allow enough time for the pathogen to adapt to a new environment, multiply, and promote further infection.

Conversely, induction of apoptosis has been reported in response to other bacterial species (Monack and Falkow, 2000; Weinrauch and Zychlinsky, 1999). Intracellular organisms such as *Listeria monocytogenes* (Rogers *et al.*, 1996), *Shigella flexneri* (Zychlinsky and Sansonetti, 1997), *M. tuberculosis* (Rojas *et al.*, 1997) and *Salmonella* (Monack *et al.*, 1996) are some examples. Extracellular bacteria can modulate the apoptotic pathways by translocating bacterial toxins into the host cell membranes and cytoplasm, such as *Helicobacter pylori* (Kuck *et al.*, 2001), which induces apoptosis by up-regulating Bax expression and down-regulating Bcl-2 expression (Konturek *et al.*, 1999). The bovine reproductive tract pathogen *Tritrichomonas foetus* can induce caspase-3 activation, chromatin condensation and DNA fragmentation in bovine vaginal epithelial cells (Singh *et al.*, 2004). Therefore, while induction of apoptosis seems to be potentially beneficial to the host, allowing initiation of immune responses to a given pathogen, Induction of apoptosis by the pathogen may benefit the pathogen by dampening the inflammatory response that is evident with cell necrosis.

Neisseria species and apoptosis

Multiple cell types have been examined in several studies on the effects of *Neisseria* species on both the intrinsic and extrinsic pathways of apoptosis. Many *in vitro* systems for studying *Neisseria* infections have been developed, and have provided the basis for many discoveries related to pathogenesis. Importantly, each *in vitro* model provides information based on the specificity of the interaction between the bacterium and that particular host cell type. Epithelial cells are the initial sites of host invasion by pathogenic *Neisseria* species and represent the first line of defence against microbial pathogens (Merz and So, 2000; Stephens, 1989). For example, gonococcal infections begin at mucosal surfaces (Tramont, 1981), typically the urethra in males and the cervix in females (Edwards and Apicella, 2004). However other anatomical locations, such as the fallopian tubes, the rectum, and the pharynx, may also become infected (Cohen and Sparling, 1992; Givan and Keyl, 1974; Heinonen *et al.*, 1985; Klein *et al.*, 1977). Once inside host epithelial cells, bacteria may reside in vacuoles, although free organisms in the cytoplasm have been observed (Griffiss *et al.*, 1999; Mosleh *et al.*, 1997; Shaw and Falkow, 1988). *N. gonorrhoeae* has also demonstrated the ability to replicate intracellularly in epithelial cells (Hagen and Cornelissen, 2006; Lin *et al.*, 1997; McGee *et al.*, 1981; Merz and So, 2000; Shaw and Falkow, 1988). In addition to epithelial cells, the response of immune cell types to these bacteria is also important. For example, polymorphonuclear leucocytes (PMNs) may become infected with *N. gonorrhoeae.* Although PMNs are a central component of the gonococcal immune response, some gonococcal strains can resist killing and actually replicate inside PMNs (Simons *et al.*, 2005; Veale *et al.*, 1979), presumably contributing to persistence of infection (Veale *et al.*, 1976).

With regard to meningococcal infections, the bacteria also initially colonize epithelial cells. Although the respiratory tract is the preferred site of colonization, during meningococcal infection, the bacteria can disseminate from infected sites and penetrate the blood–brain barrier via the bloodstream, where they can cause sepsis (Cartwright and Ala'Aldeen, 1997). The mechanism of meningococcal invasion has not been fully characterized (Merz *et al.*, 1996). Studies on the effect of whole intact meningococci on epithelial cells and endothelial cells are of the outmost importance, and *N. meningitidis* infection has been investigated in various cell systems, including HeLa cells (Massari *et al.*, 2003a), human macrophages (Tunbridge *et al.*, 2006), endothelial cells (Linhartova *et al.*, 2006; Taha, 2000), meningeal cells (Christodoulides *et al.*, 2002; Fowler *et al.*, 2006; Wells *et al.*, 2001), and mouse cerebrovascular cells (Constantin *et al.*, 2002).

Both *N. gonorrhoeae* and *N. meningitidis* have been implicated in their abilities to modulate apoptosis. Several strains of *N. gonorrhoeae* have demonstrated protective effects against induction of apoptosis, while other studies indicate that *N. gonorrhoeae* promotes apoptosis during infection. However, the majority of data regarding *N. gonorrhoeae* infection and apoptotic responses of host cells suggest that infection promotes resistance to apoptosis. Similarly, *N. meningitidis* has also been shown to both induce and prevent apoptosis in various models. For example, N. *meningitidis* infection of HeLa cells did not induce apoptosis even when the bacteria were used at high MOI (Massari 2003). Induction of minimal cell death by necrosis following prolonged meningococcal infection in meningeal cells (Fowler *et al.*, 2006) and NO-induced cell death in mouse cerebrovascular endothelium have also been reported (Constantin *et al.*, 2002), although with no significant apoptosis. Results of the studies related to the role of apoptosis in *Neisseria* infections are summarized in Table 9.1, and will be reviewed below. In addition, many of these studies have suggested a role for bacterial outer membrane proteins in contributing to apoptosis. Examples are pili, an important initial attachment factor (Punsalang and Sawyer, 1973; Swanson *et al.*, 1971); opacity-associated proteins (Opa), which lend themselves to an opaque colony phenotype and are responsible for intimate attachment (Blake and Gotschlich, 1984); porins, trimeric proteins that form pores in the outer membrane and function in the exchange of ions (Massari *et al.*, 2003b); and lipooligosaccharides (LOS), outer membrane glycolipids (Griffiss *et al.*, 1988).

Induction of apoptosis by *Neisseria gonorrhoeae*

Infection with *N. gonorrhoeae* has been reported to induce apoptosis in epithelial cells, neutrophils, and B-cells. Early evidence suggested that the invasive *N. gonorrhoeae* strain N242 was capable of inducing apoptosis in HeLa cells and monocytic cell lines following prolonged infections, as opposed to non-invasive gonococcal strains or the commensal strain *N. sicca* (Muller *et al.*, 1999). This study also demonstrated that cell death was modulated by the bacterial outer membrane protein, porin (type PIA), which induced calcium flux, calpain activation (a calcium-dependent

Table 9.1 List of studies on *N. gonorrhoeae* and *N. meningitidis* and their effect on the host cell apoptotic response

Citation (listed in chronological order)	Host cell type examined	Effect on apoptosis
N. gonorrhoeae studies		
Muller *et al.* (1999)	HeLa cervical epithelial, monocyte-like cell lines	Induces
Muller *et al.* (2000)	HeLa cervical epithelial	Induces
Beck *et al.* (2000)	Monocyte-like cell line	Inhibits
Chen *et al.* (2001)	B-cell line, primary PMN	Induces
Binnicker *et al.* (2003)	Urethral epithelial	Inhibits
Binnicker *et al.* (2004)	Urethral epithelial	Inhibits
Howie *et al.* (2005)	Colonic epidermoid	Inhibits
Pantelic *et al.* (2005)	Primary B-cells	Induces
Muenzner *et al.* (2005)	Cervical epithelial	Inhibits
Morales *et al.* (2006)	Fallopian tube epithelial	Inhibits
Simons *et al.* (2006)	Primary PMN	Inhibits
Kepp *et al.* (2007)	HeLa cervical epithelial	Induces
Higashi *et al.* (2007)	Epidermoid	Inhibits
Reyes *et al.* (2007)	Fallopian tube epithelial	Induces
Howie *et al.* (2008)	Colonic epidermoid	Inhibits
Follows *et al.* (2009)	Cervical epithelial	Inhibits
N. meningitidis studies		
Massari *et al.* (2000)	B-cell line, Jurkat cells	Inhibits
Wells *et al.* (2001)	Meningeal cells	Inhibits
Massari *et al.* (2003)	HeLa cervical cells	Inhibits
Robinson *et al.* (2004)	Meningeal cells	Inhibits
Schubert-Unkmeir *et al.* (2007)	Brain endothelial cells	Induces

protease) and caspase activation. Porin was subsequently demonstrated to translocate into host cell mitochondria, resulting in mitochondrial activation of apoptosis (Muller *et al.*, 2000). However, this induction of apoptosis in HeLa cells might have been influenced by the method of cell culture in the absence of serum. More recently, another study showed that *N. gonorrhoeae*-induced apoptosis in HeLa cells cultured without serum was dependent on both Bak and Bax, pro-apoptotic Bcl-2 proteins that have similar or redundant functions (Kepp *et al.*, 2007).

Opa engagement of the host cell receptors, CEACAM molecules, was also examined in relation to cell death. CEACAM molecules are known receptors for Opas and are expressed on various human host cell types, including epithelial and immune cells (Billker *et al.*, 2000, 2002). These receptors promote bacterial uptake and also contain immunoglobulin tyrosine-based activating motifs (ITAMs) in their cytoplasmic tails (Naumann *et al.*, 1999). The activation of signalling cascades, mediated through ITAM motifs on the CEACAM molecules, was responsible for induction of cell death in DT40 B-cells expressing CEACAM molecules, which phagocytose gonococci. However, death did not occur during infection with *N. gonorrhoeae* lacking Opa or in DT40 B-cells lacking CEACAM expression. Increased cell death was also demonstrated in primary neutrophils incubated with *N. gonorrhoeae* expressing OpaC and OpaI molecules, suggesting that Opa-CEACAM mediated phagocytosis of gonococci results in cell death. A similar effect of Opa–CEACAM interactions has been reported in B-cells (Pantelic *et al.*, 2005). Human B-cells expressing CEACAM1 *in vitro* were capable of interacting with OpaI-expressing *N. gonorrhoeae* strain MS11, and this interaction appeared to prevent antibody secretion and induce cell death, suggesting that *N. gonorrhoeae* can inhibit immune cell functions. Furthermore, cell death mediated by CEACAM1 was inhibited by a caspase-3 inhibitor (Pantelic *et al.*, 2005), which implicates apoptosis as the cause of cell death. As a whole, the results from the above studies suggest that, given the right environmental conditions, gonococcal infection may cause death of its host cell. In some cases those environmental conditions may involve a lack of nutrients, while is other cases the particular Opa variants or host cell receptors that are expressed may contribute to induction of apoptosis. However, the culture conditions may also promote resistance to apoptosis, which has been demonstrated many times and is detailed below.

Inhibition of apoptosis by *Neisseria gonorrhoeae*

Early studies on *N. gonorrhoeae* modulation of apoptosis demonstrated that a secreted protein from *N. gonorrhoeae*, IgA1 protease, blocked TNF-α mediated apoptosis in the U937 monocyte cell line by cleaving the TNF receptor II (Beck and Meyer, 2000). Future studies went on to demonstrate that bacterial infection of other host cell types was also capable of inhibiting apoptosis via alternative mechanisms. Gonococcal invasion of epithelial cells specific to both the human male and female genital tracts has been widely studied (Merz and So, 2000; Stephens, 1989). In an *in vitro* model system of infection using both primary and transformed urethral epithelial cells (UECs) from male donors, microarray studies identified the up-regulation of several anti-apoptotic genes, including *bfl-1*, *cox-2*, and *c-IAP-2* (Binnicker *et al.*, 2003). Additional experiments proved that the anti-apoptotic gene *mcl-1* was also up-regulated, while pro-apoptotic genes were not up-regulated in response to infection. This result appears cell type-specific, as expression of anti-apoptotic genes in response to gonococcal infection was not detected in Chang human conjunctival epithelial cells (Binnicker *et al.*, 2003).

To study resistance to apoptosis, staurosporine (STS), a protein kinase C inhibitor, was used to induce apoptosis through the intrinsic pathway. STS induces apoptosis by promoting the translocation of pro-apoptotic Bax into the mitochondria, where it interacts with Bak to induce mitochondrial pathway activation (Wei *et al.*, 2001). Early apoptotic events that occur following STS treatment include acidification of the cytosol, followed by release of cytochrome c into the cytosol, which can be inhibited by overexpression of anti-apoptotic Bcl-2 (Matsuyama *et al.*, 2000). STS treatment is a common method for induction of apoptosis, and had been used in previous studies of *Neisseria* and apoptosis

(Binnicker *et al.*, 2003; Massari *et al.*, 2000; Massari *et al.*, 2003a). Specifically, infection with *N. gonorrhoeae* strain 1291 for 4 h was sufficient to inhibit STS-induced mitochondrial release of cytochrome c and DNA fragmentation (Binnicker *et al.*, 2003). The anti-apoptotic gene up-regulation was presumably the basis of resistance to staurosporine (STS)-induced apoptosis in these cells.

Further evidence supporting an anti-apoptotic effect of *N. gonorrhoeae* derive from the studies of Morales *et al.*, which demonstrated that primary fallopian tube (FT) epithelial cells were protected from TNF-α induced apoptosis following infection with *N. gonorrhoeae* strain P9, variant -17, (a Pil$^+$ and Opa$^+$ strain) (Morales *et al.*, 2006). This study was the first to describe *N. gonorrhoeae*-mediated protection against apoptosis in a relevant female model system, as fallopian tubes may become infected with gonococci *in vivo* if the infection ascends into the upper female genital tract. Induction of apoptosis was demonstrated when low a multiplicity of infection (MOI of 1) was used, as a consequence of TNF-α production by the FT epithelial cells. A follow-up study demonstrated that pro-apoptotic genes were not induced by infection with this low MOI, suggesting that apoptosis was regulated by post-transcriptional mechanisms in these cells. Interestingly, apoptosis was not observed following infection of FT-cells with *N. gonorrhoeae* at high MOI, demonstrated by lack of caspase-3 activation and DNA degradation. This study highlights the potential importance of innate immune mechanisms (i.e. TNF-α mediated apoptosis) as a defence against gonococcal infection, and suggests that *N. gonorrhoeae* may have evolved a mechanism to overcome host defences upon colonization at high levels.

The ability of *N. gonorrhoeae* to protect cells from apoptosis has also been established in another female cell model, using transformed endocervical and ectocervical epithelial cell lines (Fichorova *et al.*, 2001; Fichorova *et al.*, 1997; Follows, 2009). Infection of these cells with *N. gonorrhoeae* strain FA1090B resulted in resistance to STS-induced apoptosis, as determined by decreased mitochondrial depolarization, caspase-3 activation and DNA degradation (Follows, 2009). This effect was shown to be dose-dependent, as higher MOI of infection led to greater protection from apoptosis. The anti-apoptotic effect of *N. gonorrhoeae* in this system was attributed to the presence of live bacteria, as such response was absent in cells incubated with heat-killed or antibiotic-killed organisms. Furthermore, *N. gonorrhoeae* infection in the endocervical epithelial cell model also induced significant up-regulation of numerous genes, including the anti-apoptotic genes *bfl-1*, *cIAP-2*, and *c-flip*. Collectively, these results suggest that *N. gonorrhoeae* may establish infection in the cervix by inhibiting cellular apoptotic responses and thus potentially avoid host defences and allow time for intracellular replication. This is an important observation, as the cervix is the primary site of infection in female hosts, from which infections can ascend to other tissues causing more complicated disease.

Evidence of *N. gonorrhoeae*-mediated protection against apoptosis also exists for non-epithelial cell types, including polymorphonuclear leucocytes (PMNs) (Simons *et al.*, 2006), short-lived cells of the immune system that activate programmed cell death following phagocytosis (Kobayashi *et al.*, 2003a; Luo and Loison, 2007). As mentioned previously, PMNs are a central component of the gonococcal immune response, and although PMNs are capable of phagocytosing gonococci, some bacteria are resistant to killing and have been shown to replicate inside PMNs (see Chapter 7). Upon *N. gonorrhoeae* phagocytosis, PMNs showed delayed nuclear condensation and DNA degradation compared to phagocytosis of opsonized zymosan particles (OPZ) (Simons *et al.*, 2006). Additionally, mitochondrial integrity was maintained and caspase activation was reduced in PMN that phagocytosed either live or dead gonococci, suggesting that *N. gonorrhoeae* induced resistance to apoptosis independent of bacterial viability in this system. Expression of genes involved in apoptosis was also examined in PMN, and upon phagocytosis of *N. gonorrhoeae*, expression of *cIAP2* increased and levels of *XIAP* were unchanged (Simons *et al.*, 2006). However, expression of the same genes decreased in response to OPZ phagocytosis (Simons *et al.*, 2006), suggesting that anti-apoptotic gene up-regulation was a specific event elicited by infection with *N. gonorrhoeae* and not simply related to phagocytosis-mediated PMN activation.

Gonococcal components involved in modulation of apoptosis

One of the most studied gonococcal component involved in modulation of apoptosis is the outer membrane protein, porin. Gonococcal porins, PIA and PIB, (and their meningococcal homologues PorA and PorB, discussed in the next chapter), are the major outer membrane gonococcal proteins and function to form pores for the passage of solutes in the bacteria. Neisserial porins have been shown to translocate from the bacterial outer membrane into artificial membranes as well as into target cell membranes (Lynch, 1984; Weel *et al.*, 1991; Weel and van Putten, 1991), forming a functional channel regulated by ATP and GTP by the host cell (Rudel *et al.*, 1996). Induction of apoptosis by gonococcal porin PIB has been first demonstrated by Muller *et al.* in HeLa cells and monocytic cell lines (Muller *et al.*, 1999) cultured in the absence of serum supplement. Incubation of cells with PIB was consistent with colocalization of the porin within the mitochondrial compartment, followed by phosphatidylserine externalization and DNA degradation (Muller *et al.*, 1999). As a further evidence of induction of apoptosis, specific substrates indicative of caspases and calpain activation, α-fodrin and PARP (a downstream marker of apoptosis) were induced (Muller *et al.*, 1999), suggesting that the mechanism for induction of apoptosis by *N. gonorrhoeae* porin might be mediated by Ca^{2+} influx. Moreover, ATP and specific caspase and calpain inhibitors abolished the apoptotic effect of gonococcal porin. Interestingly, structural and functional parallels have been drawn between neisserial porins and mitochondrial porins VDAC (Massari *et al.*, 2000; Muller *et al.*, 2002), and there is evidence of a specific effect of gonococcal porin on mitochondrial homeostasis, including membrane permeabilization and induction of cytochrome c release in isolated mitochondria from Jurkat T-cells (Muller *et al.*, 2000). These effects were inhibited by over-expression of anti-apoptotic members of the Bcl-2 family, Bcl-2 and Bcl-xL, suggesting modulation of apoptosis via the mitochondrial pathway.

However, more recent reports have demonstrated that purified gonococcal porin plays a role in the anti-apoptotic effect of *N. gonorrhoeae* infection described in UEC cells. Porin purified from *N. gonorrhoeae* strain MS11 was shown to direct the up-regulation of the genes encoding *bfl-1*, *cox-2* and *cIAP-2*, in agreement with the anti-apoptotic effect of live bacteria (Binnicker *et al.*, 2004). As these genes are under the control of the transcription factor, NF-κB, it was not surprising to find that NF-κB was also activated by purified PIB. These studies suggest that the innate immune response generated in UECs during infection of male hosts contributes to resistance to apoptosis in these cells. Studies in UECs aimed at identifying other bacterial components that modulate apoptosis revealed that Opa and pili mutant gonococcal strains were also capable of inducing *bfl-1* (Binnicker *et al.*, 2004), suggesting that that these outer membrane components are not involved in modulation of apoptosis. Similarly, LOS has not been implicated for an ability to modulate apoptosis, as purified gonococcal LOS did not affect *bfl-1* gene expression (Binnicker *et al.*, 2004).

The involvement of Opa outer membrane proteins in modulation of apoptosis has been described in a study examining the effect of *N. gonorrhoeae* strain N309 on ME-180 cervical carcinoma cells, which led to the conclusion that bacterial interactions with CEACAM receptors were responsible for inhibiting exfoliation of infected epithelial cells (Muenzner *et al.*, 2005). Exfoliation of epithelial cells, or anoikis, is an important host defence against infection that lowers bacterial load and is controlled by apoptosis (Gilmore, 2005; Muenzner *et al.*, 2005; Mulvey *et al.*, 2000; Schilling *et al.*, 2001). The Opa+ non-piliated *N. gonorrhoeae* strain N309 was capable of increasing epithelial cell adhesion to a collagen matrix, which was dependent on the expression of Opa as well as on host cell gene expression, in particular CD105 (Muenzner *et al.*, 2005). CD105, a member of the TGF-β1 receptor family, has been linked to cell migration and organization of cell adhesion sites (Muenzner *et al.*, 2005). Along with β1 integrins, CD105 is critical for prevention of bacteria-induced cell detachment *in vitro* (Muenzner *et al.*, 2005), suggesting that *N. gonorrhoeae* may be able to inhibit the innate host defense of epithelial exfoliation *in vivo*. This is important to the hypothesis that inhibition of apoptosis may help in the establishment of infection in the cervix, the first tissue encountered in the female reproductive tract which is susceptible to infection upon exposure. Furthermore, Opa interaction with CEACAM receptors was responsible for the gene and protein up-regulation that was demonstrated by Muenzner *et al.* (2005). Gonococcal engagement of CEACAM receptors, which are present on many *N. gonorrhoeae* host cell types, results in signalling that activates PI3-K pathways (Booth *et al.*, 2003). The possibility that this influences the observed anti-apoptotic response in other cell types is strong, as CEACAM signalling in human monocytes has been shown to block mitochondrial apoptosis in a PI3-K- and Akt-dependent manner (Yu *et al.*, 2006).

Lastly, pili outer membrane proteins have also been implicated to modulate apoptosis. *N. gonorrhoeae* type IV pilus (Tfp) has been demonstrated to activate mitogen-activated protein kinase (MAPK) pathways and enhance the expression of genes involved in cell stress response and survival (Howie *et al.*, 2005). Pili mediate attachment to host cells in a dynamic fashion through pilus retraction which contributes to uptake of the bacteria by host cells (Kallstrom *et al.*, 2001; Merz and So, 1997; Merz and So, 2000; Merz *et al.*, 2000; Punsalang and Sawyer, 1973). Interactions of host cells with pili result in the activation of several MAPKs, and has been demonstrated to activate cytoprotective signalling in the T84 human colonic epithelial cell line (Howie *et al.*, 2005). Infection of cells with *N. gonorrhoeae* strain N400 led to a reduction of both active caspase-8 as and cleaved PARP, while a pili retraction mutant, N400*pilT*, appeared to increase caspase-8 and PARP activation (Howie *et al.*, 2005). Interestingly, both WT and *pilT* mutant gonococcal strains used in these studies induced resistance to STS-mediated caspase-8 and PARP cleavage, suggesting that pilus retraction might enhance protection from apoptosis (Howie *et al.*, 2005). More recent work has demonstrated that induction of the *pilT* gene during infection of A431 human epidermoid carcinoma cells rescues these cells from infection-induced apoptosis (Higashi *et al.*, 2007). In these studies, infection with *N. gonorrhoeae* induced apoptosis in approximately 6% of cells, whereas apoptosis increased to approximately 70% of cells with the *pilT* mutant strain. However, restoring expression of *pilT* in this strain reduced the percentage of apoptotic cells to approximately 36% (Higashi *et al.*, 2007). Further, the same group has shown that Tfp-mediated ERK activation is responsible for cytoprotective effects on T84 cells. Specifically, ERK activation caused the down-regulation of the pro-apoptotic proteins Bad and Bim (Howie *et al.*, 2008). Overall, these studies suggest that pili is involved in modulation of apoptosis in epithelial cells, supporting the hypothesis that multiple gonococcal strains, as well as several individual components, are capable of promoting resistance to apoptosis in various cell systems.

Inhibition of apoptosis by *Neisseria meningitidis*

Modulation of host cell survival mechanisms by *N. meningitidis* has been demonstrated in various cell models. The first studies have described improved cell survival following infection of cells with *N. meningitidis* also improves cell survival when apoptosis is induced via the mitochondrial pathway by STS in both epithelial and meningeal cells (Massari *et al.*, 2003a; Robinson *et al.*, 2004). In the case of epithelial cells, HeLa cells have been infected for 18h with live *N. meningitidis* and several cellular markers of apoptosis have been examined, including mitochondrial membrane depolarization, cytochrome *c* release, caspase activation, and DNA degradation (Massari *et al.*, 2003a). Incubation of HeLa cells with meningococci reduced the loss of mitochondrial potential and release of cytochrome *c* in the cytosol of cells treated with staurosporine. Furthermore, STS-induced caspase-3 and -9 activation and DNA degradation were reduced, indicative of prevention from apoptosis (Massari *et al.*, 2003a). As previously mentioned, mitochondrial functions including outer membrane integrity and permeability can be influenced by Bcl-2 family member expression and intracellular localization. During STS-induced apoptosis, translocation of Bax from the cytosol into the mitochondrial membrane has been shown, followed by release of mitochondrial components into the cytosol, which may or may not require mitochondrial membrane depolarization (Tsujimoto and Shimizu, 2000). However, translocation of Bax was not inhibited by infection of HeLa cells with *N. meningitidis* (Massari *et al.*, 2003a), suggesting that the meningococcal anti-apoptotic effect would not modulate such pro-apoptotic protein. Furthermore, the expression levels of anti-apoptotic Bcl-2 in both the cell cytosol and in association with mitochondria did not appear to be affected by *N. meningitidis* (Massari *et al.*, 2003a). However, this does not exclude the possibility that other pro-apoptotic members or anti-apoptotic members of the Bcl-2 family might be modulated in response to meningococcal infection.

Obviously, although meningococci interact with epithelial cells in the very first stage of infection, other cell models including endothelial, meningeal and cerebrovascular cells might also provide relevant insights for drawing parallels between meningococcal disease, induction of inflammation and apoptosis.

In meningeal cells, for example, infection with *N. meningitidis* induces up-regulation of several anti-apoptotic genes (Robinson *et al.*, 2004; Wells *et al.*, 2001), including chemokine genes and genes specifically involved in prevention from cellular apoptosis (Robinson *et al.*, 2004). Up-regulation of anti-apoptotic gene expression, such as IEX-1L anti-death protein (Wu *et al.*, 1998), BB2-bombesin receptor (Kresch *et al.*, 1999), Opn (O'Regan and Berman, 2000) and down-regulation of pro-apoptotic genes such as caspase-10 (Wang *et al.*, 1999) is consistent with an increased anti-apoptotic cellular response (Wells *et al.*, 2001), although up-regulation of interferon regulatory factor 1 (IRF-1), a pro-apoptotic regulator, has also been shown (Iadecola *et al.*, 1999). Furthermore, expression of Cox-2 (Hinz and Brune, 2002) has been shown in these cells in response to prolonged meningococcal infection (Robinson *et al.*, 2004). In support of the evidence of the anti-apoptotic properties of N. *meningitidis*, prevention from STS-induced apoptosis has also been demonstrated in infected meningeal cells (Robinson *et al.*, 2004).

In contrast to the anti-apoptotic properties of *N. meningitidis*, Schubert-Unkmeir *et al.* have recently shown that meningococcal infection of human brain endothelial cells induced up-regulation of pro-apoptotic genes as soon as 4h after infection (Schubert-Unkmeir *et al.*, 2007). Increased gene expression of Bak, Bad, calpain, and proteins involved in Fas-mediated apoptosis were shown by both microarrays and real-time PCR. However, down-regulation of other pro-apoptotic genes, such as Bax and the death-associated protein 3, was also demonstrated, along with the down-regulation of several anti-apoptotic genes, such as Bcl-xL, dad1, adenosine A1 receptor, Akt1 and TRAF1 (which was subsequently up-regulated at 8h post-infection) was shown by both microarrays and real-time PCR. In the same report, activation of caspase-3 and AMPK-α (Blazquez *et al.*, 2001) was shown, and a small percentage of cells showed phosphatidylserine exposure after a 24-h incubation with live bacteria. Furthermore, PKB/Akt activity, which has a role in inhibition of pro-apoptotic Bad and in regulation of expression of genes involved in apoptosis, was decreased. This is the first contrasting report of induction of apoptosis by *N. meningitidis* and, as previously discussed for *N. gonorrhoeae*-mediated modulation of apoptosis, differences in the various cell types examined and in the experimental conditions might greatly influence the cellular response to a given pathogen.

Meningococcal components involved in modulation of apoptosis

The major neisserial component involved in modulation of apoptosis induced by both *N. meningitidis* and *N. gonorrhoeae* to date is the porin molecule (Binnicker *et al.*, 2004; Massari *et al.*, 2000, 2003a; Muller *et al.*, 1999). As previously mentioned (see previous paragraph on gonococcal porin), the structure and function of meningococal and gonococcal porins are very similar. Porins are native trimers composed of three meningococcal polypeptides (35 kDa) with a high content of β-pleated sheet (~ 36%) (Derrick *et al.*, 1999), and a predicted 16-strand β-barrel fold native structure with multiple surface-exposed, variable, hydrophilic loops. They function as pores for the exchange of ions in the bacteria (Minetti *et al.*, 1998), and can also directly interact with artificial membranes and eukaryotic target cell membranes (Lynch, 1984) to form functional channels at low membrane potential regulated by ATP and GTP by the eukaryotic host cell (Rudel *et al.*, 1996). It has been shown that purified *N. meningitidis* PorB can associate with mitochondria within the intracellular compartment and contributes to the modulation of mitochondrial membrane potential (Massari *et al.*, 2000). A direct interaction with the mitochondrial protein VDAC has also been shown

(Massari *et al.*, 2000), which might explain the ability of meningococcal PorB to modulate mitochondrial potential upon induction of apoptosis, contributing to reduced cytochrome *c* release and DNA degradation (Massari *et al.*, 2000). Purified meningococcal porin has been shown to modulate mitochondrial potential in a variety of cell systems, including human epithelial and T-cell lines (i.e. HeLa, Jurkat) and murine cells (i.e. primary B-cells and B-cell lines) in response to induction of apoptosis by STS (Massari *et al.*, 2000). Porin has also been described to translocate from the bacteria into the mitochondrial compartment upon infection of cells with live meningococci (Massari *et al.*, 2003a) with a porin-specific interaction type, which is not reproduced by other outer membrane proteins (i.e. RmpM; Klugman *et al.*, 1989).

Meningococcal secreted proteins (MSPs) have also been investigated for their ability to induce an anti-apoptotic response in meningeal cells (Robinson *et al.*, 2004). Similar to live meningococci, MSPs induced positive regulation of genes involved in cellular apoptosis, in particular up-regulation of anti-apoptotic genes *cox-2* and *mcl-1*. Although induction of low levels of cell death was observed after 48h incubation with MSPs, this was independent from apoptosis, shown by a lack of apoptosis-specific markers such as caspase-3 activation and DNA degradation (Robinson *et al.*, 2004). Furthermore, as described for purified meningococcal porin, cells incubated with MSPs were protected from staurosporine-induced apoptosis (Robinson *et al.*, 2004), demonstrated by trypan blue exclusion from live cells.

Conclusions

Similar to various other bacterial pathogens, *Neisseria* species have developed the ability to interfere with the apoptotic cellular machinery. There is a large body of literature that supports the anti-apoptotic role of both *N. meningitidis* and *N. gonorrhoeae*. However, the process of apoptosis is based on a delicate balance of intracellular events that are regulated by the host cell, and are often host-cell type dependent. The ability of neisserial species to modulate apoptosis has been characterized in several different cell types and correlates with various physiological responses, such as modulation of mitochondrial permeability as well as changes in host cell gene expression, which has been investigated by various research groups. The effect of live bacteria and of bacterial components, such as the porins, has been shown to mostly down-regulate apoptosis. The studies discussed in this chapter indicate that, by interfering with host cell death processes, neisserial species might be 'buying time' to allow bacterial replication and further spreading of infection. Prevention of apoptosis would then be an immune evasion mechanism optimized not only by obligate intracellular bacteria but also by facultative intracellular bacteria such as neisseriae.

References

Airenne, S., Surcel, H.M., Tuukkanen, J., Leinonen, M., and Saikku, P. (2002). *Chlamydia pneumoniae* inhibits apoptosis in human epithelial and monocyte cell lines. Scand. J. Immunol. *55*, 390–398.

Baines, C.P., Kaiser, R.A., Sheiko, T., Craigen, W.J., and Molkentin, J.D. (2007). Voltage-dependent anion channels are dispensable for mitochondrial-dependent cell death. Nat. Cell. Biol. *9*, 550–555.

Balcewicz-Sablinska, M.K., Keane, J., Kornfeld, H., and Remold, H.G. (1998). Pathogenic Mycobacterium tuberculosis evades apoptosis of host macrophages by release of TNF-R2, resulting in inactivation of TNF-alpha. J. Immunol. *161*, 2636–2641.

Beck, S.C., and Meyer, T.F. (2000). IgA1 protease from *Neisseria gonorrhoeae* inhibits TNFalpha-mediated apoptosis of human monocytic cells. FEBS Lett. *472*, 287–292.

Bellot, G., Cartron, P.F., Er, E., Oliver, L., Juin, P., Armstrong, L.C., Bornstein, P., Mihara, K., Manon, S., and Vallette, F.M. (2007). TOM22, a core component of the mitochondria outer membrane protein translocation pore, is a mitochondrial receptor for the proapoptotic protein Bax. Cell Death Differ. *14*, 785–794.

Billker, O., Popp, A., Brinkmann, V., Wenig, G., Schneider, J., Caron, E., and Meyer, T.F. (2002). Distinct mechanisms of internalization of *Neisseria gonorrhoeae* by members of the CEACAM receptor family involving Rac1- and Cdc42-dependent and -independent pathways. EMBO J. *21*, 560–571.

Billker, O., Popp, A., Gray-Owen, S.D., and Meyer, T.F. (2000). The structural basis of CEACAM-receptor targeting by neisserial Opa proteins. Trends Microbiol *8*, 258–260; discussion 260–251.

Binnicker, M.J., Williams, R.D., and Apicella, M.A. (2003). Infection of human urethral epithelium with *Neisseria gonorrhoeae* elicits an up-regulation of host anti-apoptotic factors and protects cells from staurosporine-induced apoptosis. Cell. Microbiol. *5*, 549–560.

Binnicker, M.J., Williams, R.D., and Apicella, M.A. (2004). Gonococcal porin IB activates NF-kappaB in human urethral epithelium and increases the expression of host antiapoptotic factors. Infect. Immun. *72*, 6408–6417.

Blake, M.S., and Gotschlich, E.C. (1984). Purification and partial characterization of the opacity-associated proteins of *Neisseria gonorrhoeae*. J. Exp. Med. *159*, 452–462.

Blazquez, C., Geelen, M.J., Velasco, G., and Guzman, M. (2001). The AMP-activated protein kinase prevents ceramide synthesis de novo and apoptosis in astrocytes. FEBS Lett. *489*, 149–153.

Boise, L.H., Minn, A.J., and Thompson, C.B. (1995). Receptors that regulate T-cell susceptibility to apoptotic cell death. Ann N Y Acad Sci *766*, 70–80.

Booth, J.W., Telio, D., Liao, E.H., McCaw, S.E., Matsuo, T., Grinstein, S., and Gray-Owen, S.D. (2003). Phosphatidylinositol 3-kinases in carcinoembryonic antigen-related cellular adhesion molecule-mediated internalization of *Neisseria gonorrhoeae*. J. Biol. Chem. *278*, 14037–14045.

Brenner, C., and Kroemer, G. (2000). Apoptosis. Mitochondria – the death signal integrators. Science *289*, 1150–1151.

Brozovic, S., Sahoo, R., Barve, S., Shiba, H., Uriarte, S., Blumberg, R.S., and Kinane, D.F. (2006). Porphyromonas gingivalis enhances FasL expression via up-regulation of NFkappaB-mediated gene transcription and induces apoptotic cell death in human gingival epithelial cells. Microbiology *152*, 797–806.

Budihardjo, I., Oliver, H., Lutter, M., Luo, X., and Wang, X. (1999). Biochemical pathways of caspase activation during apoptosis. Annu. Rev. Cell. Dev. Biol. *15*, 269–290.

Cain, K., Bratton, S.B., and Cohen, G.M. (2002). The Apaf-1 apoptosome: a large caspase-activating complex. Biochimie *84*, 203–214.

Cartwright, K.A., and Ala'Aldeen, D.A. (1997). *Neisseria meningitidis*: clinical aspects. J. Infect. *34*, 15–19.

Chen, T., Bolland, S., Chen, I., Parker, J., Pantelic, M., Grunert, F., and Zimmermann, W. (2001). The CGM1a (CEACAM3/CD66d)-mediated phagocytic pathway of *Neisseria gonorrhoeae* expressing opacity proteins is also the pathway to cell death. J. Biol. Chem. *276*, 17413–17419.

Cheng, E.H., Sheiko, T.V., Fisher, J.K., Craigen, W.J., and Korsmeyer, S.J. (2003). VDAC2 inhibits BAK activation and mitochondrial apoptosis. Science *301*, 513–517.

Christodoulides, M., Makepeace, B.L., Partridge, K.A., Kaur, D., Fowler, M.I., Weller, R.O., and Heckels, J.E. (2002). Interaction of *Neisseria meningitidis* with human meningeal cells induces the secretion of a distinct group of chemotactic, proinflammatory, and growth-factor cytokines. Infect. Immun. *70*, 4035–4044.

Clifton, D.R., Goss, R.A., Sahni, S.K., van Antwerp, D., Baggs, R.B., Marder, V.J., Silverman, D.J., and Sporn, L.A. (1998). NF-kappa B-dependent inhibition of apoptosis is essential for host cellsurvival during Rickettsia rickettsii infection. Proc. Natl. Acad. Sci. USA *95*, 4646–4651.

Cohen, G.M. (1997). Caspases: the executioners of apoptosis. Biochem. J. *326*, 1–16.

Cohen, M.S., and Sparling, P.F. (1992). Mucosal infection with *Neisseria gonorrhoeae*. Bacterial adaptation and mucosal defenses. J. Clin. Invest. *89*, 1699–1705.

Colombini, M. (2004). VDAC: the channel at the interface between mitochondria and the cytosol. Mol. Cell. Biochem. *256–257*, 107–115.

Colombini, M., Blachly-Dyson, E., and Forte, M. (1996). VDAC, a channel in the outer mitochondrial membrane. Ion Channels *4*, 169–202.

Constantin, D., Ala'Aldeent, D., and Murphy, S. (2002). Transcriptional activation of nitric oxide synthase-2, and NO-induced cell death, in mouse cerebrovascular endothelium exposed to *Neisseria meningitidis*. J. Neurochem. *81*, 270–276.

Cosulich, S.C., Worrall, V., Hedge, P.J., Green, S., and Clarke, P.R. (1997). Regulation of apoptosis by BH3 domains in a cell-free system. Curr. Biol. *7*, 913–920.

Cuff, S., and Ruby, J. (1996). Evasion of apoptosis by DNA viruses. Immunol. Cell. Bio.l *74*, 527–537.

Danial, N.N. (2007). BCL-2 Family proteins: critical checkpoints of apoptotic cell death. Clin. Cancer Res. *13*, 7254–7263.

Debatin, K.M., and Krammer, P.H. (2004). Death receptors in chemotherapy and cancer. Oncogene *23*, 2950–2966.

Derrick, J.P., Urwin, R., Suker, J., Feavers, I.M., and Maiden, M.C. (1999). Structural and evolutionary inference from molecular variation in *Neisseria* porins. Infect. Immun. *67*, 2406–2413.

Desta, T., and Graves, D.T. (2007). Fibroblast apoptosis induced by Porphyromonas gingivalis is stimulated by a gingipain and caspase-independent pathway that involves apoptosis-inducing factor. Cell. Microbiol. *9*, 2667–2675.

Deveraux, Q.L., Takahashi, R., Salvesen, G.S., and Reed, J.C. (1997). X-linked IAP is a direct inhibitor of cell-death proteases. Nature *388*, 300–304.

Edwards, J.L., and Apicella, M.A. (2004). The molecular mechanisms used by *Neisseria gonorrhoeae* to initiate infection differ between men and women. Clin. Microbiol. Rev. *17*, 965–981.

Eskes, R., Antonsson, B., Osen-Sand, A., Montessuit, S., Richter, C., Sadoul, R., Mazzei, G., Nichols, A., and Martinou, J.C. (1998). Bax-induced cytochrome C release from mitochondria is independent of the permeability transition pore but highly dependent on Mg2+ ions. J. Cell. Biol. *143*, 217–224.

Fan, T., Lu, H., Hu, H., Shi, L., McClarty, G.A., Nance, D.M., Greenberg, A.H., and Zhong, G. (1998). Inhibition of apoptosis in chlamydia-infected cells: blockade of mitochondrial cytochrome c release and caspase activation. J. Exp. Med. *187*, 487–496.

Fichorova, R.N., Desai, P.J., Gibson, F.C., 3rd, and Genco, C.A. (2001). Distinct proinflammatory host responses to *Neisseria gonorrhoeae* infection in immortalized human cervical and vaginal epithelial cells. Infect. Immun. *69*, 5840–5848.

Fichorova, R.N., Rheinwald, J.G., and Anderson, D.J. (1997). Generation of papillomavirus-immortalized cell lines from normal human ectocervical, endocervical, and vaginal epithelium that maintain expression of tissue-specific differentiation proteins. Biol Reprod *57*, 847–855.

Fischer, S.F., Schwarz, C., Vier, J., and Hacker, G. (2001). Characterization of antiapoptotic activities of Chlamydia pneumoniae in human cells. Infect. Immun. *69*, 7121–7129.

Fleury, C., Mignotte, B., and Vayssiere, J.L. (2002). Mitochondrial reactive oxygen species in cell death signaling. Biochimie *84*, 131–141.

Fowler, M.I., Yin, K.Y., Humphries, H.E., Heckels, J.E., and Christodoulides, M. (2006). Comparison of the inflammatory responses of human meningeal cells following challenge with *Neisseria* lactamica and with *Neisseria meningitidis*. Infect. Immun. *74*, 6467–6478.

Follows, S., Murlidharan, J., Massari, P., Wetzler, L., and Genco, C.A. (2009). *Neisseria gonorrhoeae* infection protects human epithelial cells from apoptosis via expression of host antiapoptotic proteins. Infect. Immun. *77*, 3602–3610.

Galluzzi, L., and Kroemer, G. (2007). Mitochondrial apoptosis without VDAC. Nat. Cell Biol. *9*, 487–489.

Gao, L.Y., and Kwaik, Y.A. (2000). The modulation of host cell apoptosis by intracellular bacterial pathogens. Trends Microbiol. *8*, 306–313.

Gilmore, A.P. (2005). Anoikis. Cell Death Differ *12 Suppl 2*, 1473–1477.

Givan, K.F., and Keyl, A. (1974). The isolation of *Neisseria* species from unusual sites. Can. Med. Ass. J. *111*, 1077–1079.

Green, D.R., and Kroemer, G. (2004). The pathophysiology of mitochondrial cell death. Science *305*, 626–629.

Griffiss, J.M., Lammel, C.J., Wang, J., Dekker, N.P., and Brooks, G.F. (1999). *Neisseria gonorrhoeae* coordinately uses Pili and Opa to activate HEC-1-B-cell microvilli, which causes engulfment of the gonococci. Infect. Immun. *67*, 3469–3480.

Griffiss, J.M., Schneider, H., Mandrell, R.E., Yamasaki, R., Jarvis, G.A., Kim, J.J., Gibson, B.W., Hamadeh, R., and Apicella, M.A. (1988). Lipooligosaccharides: the principal glycolipids of the neisserial outer membrane. Rev Infect Dis *10 Suppl 2*, S287–295.

Gross, A., Terraza, A., Ouahrani-Bettache, S., Liautard, J.P., and Dornand, J. (2000). In vitro Brucella suis infection prevents the programmed cell death of human monocytic cells. Infect. Immun. *68*, 342–351.

Hacker, G., and Fischer, S.F. (2002). Bacterial antiapoptotic activities. FEMS Microbiol Lett *211*, 1–6.

Hagen, T.A., and Cornelissen, C.N. (2006). *Neisseria gonorrhoeae* requires expression of TonB and the putative transporter TdfF to replicate within cervical epithelial cells. Mol. Microbiol. *62*, 1144–1157.

Heinonen, P.K., Teisala, K., Punnonen, R., Miettinen, A., Lehtinen, M., and Paavonen, J. (1985). Anatomic sites of upper genital tract infection. Obstet Gynecol *66*, 384–390.

Higashi, D.L., Lee, S.W., Snyder, A., Weyand, N.J., Bakke, A., and So, M. (2007). Dynamics of *Neisseria gonorrhoeae* attachment: microcolony development, cortical plaque formation, and cytoprotection. Infect. Immun. *75*, 4743–4753.

Hinz, B., and Brune, K. (2002). Cyclooxygenase-2 – 10 years later. J. Pharmacol. Exp. Ther. *300*, 367–375.

Hofmann, K., and Dixit, V.M. (1998). Ceramide in apoptosis – does it really matter? Trends Biochem. Sci. *23*, 374–377.

Howie, H.L., Glogauer, M., and So, M. (2005). The N. *gonorrhoeae* type IV pilus stimulates mechanosensitive pathways and cytoprotection through a pilT-dependent mechanism. PLoS Biol 3, e100.

Iadecola, C., Salkowski, C.A., Zhang, F., Aber, T., Nagayama, M., Vogel, S.N., and Ross, M.E. (1999). The transcription factor interferon regulatory factor 1 is expressed after cerebral ischemia and contributes to ischemic brain injury. J. Exp. Med. *189*, 719–727.

Kabelitz, D., and Janssen, O. (1997). Antigen-induced death of T-lymphocytes. Front Biosci 2, d61–77.

Kallstrom, H., Blackmer Gill, D., Albiger, B., Liszewski, M.K., Atkinson, J.P., and Jonsson, A.B. (2001). Attachment of *Neisseria gonorrhoeae* to the cellular pilus receptor CD46: identification of domains important for bacterial adherence. Cell Microbiol 3, 133–143.

Kerr, J.F., Wyllie, A.H., and Currie, A.R. (1972). Apoptosis: a basic biological phenomenon with wide-ranging implications in tissue kinetics. Br J Cancer *26*, 239–257.

Klein, E.J., Fisher, L.S., Chow, A.W., and Guze, L B. (1977). Anorectal gonococcal infection. Ann Intern Med 86, 340–346.

Klugman, K.P., Gotschlich, E.C., and Blake, M.S. (1989). Sequence of the structural gene (rmpM) for the class 4 outer membrane protein of *Neisseria meningitidis*, homology of the protein to gonococcal protein III and *Escherichia coli* OmpA, and construction of meningococcal strains that lack class 4 protein. Infect. Immun. *57*, 2066–2071.

Kobayashi, S.D., Braughton, K.R., Whitney, A.R., Voyich, J.M., Schwan, T.G., Musser, J.M., and DeLeo, F.R. (2003a). Bacterial pathogens modulate an apoptosis differentiation program in human neutrophils. Proc. Natl. Acad. Sci. USA *100*, 10948–10953.

Kobayashi, S.D., Voyich, J.M., and DeLeo, F.R. (2003b). Regulation of the neutrophil-mediated inflammatory response to infection. Microbes Infect 5, 1337–1344.

Konturek, P.C., Pierzchalski, P., Konturek, S.J., Meixner, H., Faller, G., Kirchner, T., and Hahn, E.G. (1999). Helicobacter pylori induces apoptosis in gastric mucosa through an up-regulation of Bax expression in humans. Scand J Gastroenterol *34*, 375–383.

Krammer, P.H. (2000). CD95's deadly mission in the immune system. Nature *407*, 789–795.

Krammer, P.H., Behrmann, I., Daniel, P., Dhein, J., and Debatin, K.M. (1994). Regulation of apoptosis in the immune system. Curr Opin Immunol *6*, 279–289.

Kresch, M., Christian, C., Zhu, L., Obe, M., Sanders, M.M., and Hussain, N. (1999). Bombesin inhibits apoptosis in developing fetal rat lung. Lung *177*, 241–251.

Kroemer, G. (1997). The proto-oncogene Bcl-2 and its role in regulating apoptosis. Nat Med 3, 614–620.

Kroemer, G., Galluzzi, L., and Brenner, C. (2007). Mitochondrial membrane permeabilization in cell death. Physiol Rev *87*, 99–163.

Kroemer, G., and Martin, S.J. (2005). Caspase-independent cell death. Nat Med *11*, 725–730.

Kroemer, G., and Reed, J.C. (2000). Mitochondrial control of cell death. Nat Med *6*, 513–519.

Kuck, D., Kolmerer, B., Iking-Konert, C., Krammer, P.H., Stremmel, W., and Rudi, J. (2001). Vacuolating cytotoxin of Helicobacter pylori induces apoptosis in the human gastric epithelial cell line AGS. Infect. Immun. *69*, 5080–5087.

Kuwana, T., Mackey, M.R., Perkins, G., Ellisman, M.H., Latterich, M., Schneiter, R., Green, D.R., and Newmeyer, D.D. (2002). Bid, Bax, and lipids cooperate to form supramolecular openings in the outer mitochondrial membrane. Cell *111*, 331–342.

Le Bras, M., Clement, M.V., Pervaiz, S., and Brenner, C. (2005). Reactive oxygen species and the mitochondrial signaling pathway of cell death. Histol Histopathol *20*, 205–219.

Letai, A., Bassik, M.C., Walensky, L.D., Sorcinelli, M.D., Weiler, S., and Korsmeyer, S.J. (2002). Distinct BH3 domains either sensitize or activate mitochondrial apoptosis, serving as prototype cancer therapeutics. Cancer Cell *2*, 183–192.

Li, L.Y., Luo, X., and Wang, X. (2001). Endonuclease G is an apoptotic DNase when released from mitochondria. Nature *412*, 95–99.

Lin, L., Ayala, P., Larson, J., Mulks, M., Fukuda, M., Carlsson, S.R., Enns, C., and So, M. (1997). The *Neisseria* type 2 IgA1 protease cleaves LAMP1 and promotes survival of bacteria within epithelial cells. Mol. Microbiol. *24*, 1083–1094.

Linhartova, I., Basler, M., Ichikawa, J., Pelicic, V., Osicka, R., Lory, S., Nassif, X., and Sebo, P. (2006). Meningococcal adhesion suppresses proapoptotic gene expression and promotes expression of genes supporting early embryonic and cytoprotective signaling of human endothelial cells. FEMS Microbiol. Lett. *263*, 109–118.

Luo, H.R., and Loison, F. (2007). Constitutive neutrophil apoptosis: Mechanisms and regulation. Am J Hematol.

Ly, J.D., Grubb, D.R., and Lawen, A. (2003). The mitochondrial membrane potential (deltapsi(m)) in apoptosis; an update. Apoptosis *8*, 115–128.

Lynch, E.C., Blake, M.S., Gotschlich, E.C., and Mauro, A. (1984). Studies of porins: Spontaneously transferred from whole cells and reconstituted from purified proteins of *Neisseria gonorrhoeae* and *Neisseria meningitidis*. Biophys. J. *45*, 104–107.

Martin, S.J., Takayama, S., McGahon, A.J., Miyashita, T., Corbeil, J., Kolesnick, R.N., Reed, J.C., and Green, D.R. (1995). Inhibition of ceramide-induced apoptosis by Bcl-2. Cell Death Differ. *2*, 253–257.

Marzo, I., Brenner, C., Zamzami, N., Susin, S.A., Beutner, G., Brdiczka, D., Remy, R., Xie, Z.H., Reed, J.C., and Kroemer, G. (1998). The permeability transition pore complex: a target for apoptosis regulation by caspases and bcl-2-related proteins. J. Exp. Med. *187*, 1261–1271.

Massari, P., Ho, Y., and Wetzler, L.M. (2000). *Neisseria meningitidis* porin PorB interacts with mitochondria and protects cells from apoptosis. Proc. Natl. Acad. Sci. USA *97*, 9070–9075.

Massari, P., King, C.A., Ho, A.Y., and Wetzler, L.M. (2003a). Neisserial PorB is translocated to the mitochondria of HeLa cells infected with *Neisseria meningitidis* and protects cells from apoptosis. Cell Microbiol *5*, 99–109.

Massari, P., Ram, S., Macleod, H., and Wetzler, L.M. (2003b). The role of porins in neisserial pathogenesis and immunity. Trends Microbiol *11*, 87–93.

Matsuyama, S., Llopis, J., Deveraux, Q.L., Tsien, R.Y., and Reed, J.C. (2000). Changes in intramitochondrial and cytosolic pH: early events that modulate caspase activation during apoptosis. Nat Cell Biol *2*, 318–325.

McGee, Z.A., Johnson, A.P., and Taylor-Robinson, D. (1981). Pathogenic mechanisms of *Neisseria gonorrhoeae*: observations on damage to human fallopian tubes in organ culture by gonococci of colony type 1 or type 4. J. Infect. Dis. *143*, 413–422.

Merz, A.J., Rifenbery, D.B., Arvidson, C.G., and So, M. (1996). Traversal of a polarized epithelium by pathogenic neisseriae: facilitation by type IV pili and maintenance of epithelial barrier function. Mol. Med. *2*, 745–754.

Merz, A.J., and So, M. (1997). Attachment of piliated, Opa- and Opc- gonococci and meningococci to epithelial cells elicits cortical actin rearrangements and clustering of tyrosine-phosphorylated proteins. Infect. Immun. *65*, 4341–4349.

Merz, A.J., and So, M. (2000). Interactions of pathogenic neisseriae with epithelial cell membranes. Annu. Rev. Cell Dev. Biol. *16*, 423–457.

Merz, A.J., So, M., and Sheetz, M.P. (2000). Pilus retraction powers bacterial twitching motility. Nature *407*, 98–102.

Minetti, C.A., Blake, M.S., and Remeta, D.P. (1998). Characterization of the structure, function, and conformational stability of PorB class 3 protein from *Neisseria meningitidis*. A porin with unusual physicochemical properties. J. Biol. Chem. *273*, 25329–25338.

Monack, D., and Falkow, S. (2000). Apoptosis as a common bacterial virulence strategy. Int. J. Med. Microbiol. *290*, 7–13.

Monack, D.M., Raupach, B., Hromockyj, A.E., and Falkow, S. (1996). *Salmonella typhimurium* invasion induces apoptosis in infected macrophages. Proc. Natl. Acad. Sci. USA *93*, 9833–9838.

Morales, P., Reyes, P., Vargas, M., Rios, M., Imarai, M., Cardenas, H., Croxatto, H., Orihuela, P., Vargas, R., Fuhrer, J., Heckels, J.E., Christodoulides, M., Velasquez, L. (2006). Infection of human fallopian tube epithelial cells with *Neisseria gonorrhoeae* protects cells from tumor necrosis factor alpha-induced apoptosis. Infect. Immun. *74*, 3643–3650.

Mosleh, I.M., Boxberger, H.J., Sessler, M.J., and Meyer, T.F. (1997). Experimental infection of native human ureteral tissue with *Neisseria gonorrhoeae*: adhesion, invasion, intracellular fate, exocytosis, and passage through a stratified epithelium. Infect. Immun. *65*, 3391–3398.

Muenzner, P., Rohde, M., Kneitz, S., and Hauck, C.R. (2005). CEACAM engagement by human pathogens enhances cell adhesion and counteracts bacteria-induced detachment of epithelial cells. J. Cell. Biol. *170*, 825–836.

Muller, A., Gunther, D., Brinkmann, V., Hurwitz, R., Meyer, T.F., and Rudel, T. (2000). Targeting of the pro-apoptotic VDAC-like porin (PorB) of *Neisseria gonorrhoeae* to mitochondria of infected cells. EMBO J. *19*, 5332–5343.

Muller, A., Gunther, D., Dux, F., Naumann, M., Meyer, T.F., and Rudel, T. (1999). Neisserial porin (PorB) causes rapid calcium influx in target cells and induces apoptosis by the activation of cysteine proteases. EMBO J. *18*, 339–352.

Muller, A., Rassow, J., Grimm, J., Machuy, N., Meyer, T.F., and Rudel, T. (2002). VDAC and the bacterial porin PorB of *Neisseria gonorrhoeae* share mitochondrial import pathways. EMBO J. *21*, 1916–1929.

Mulvey, M.A., Schilling, J.D., Martinez, J.J., and Hultgren, S.J. (2000). Bad bugs and beleaguered bladders: interplay between uropathogenic *Escherichia coli* and innate host defenses. Proc. Natl. Acad. Sci. USA *97*, 8829–8835.

Murray, D.A., and Wilton, J.M. (2003). Lipopolysaccharide from the periodontal pathogen Porphyromonas gingivalis prevents apoptosis of HL60-derived neutrophils *in vitro*. Infect. Immun. *71*, 7232–7235.

Nakhjiri, S.F., Park, Y., Yilmaz, O., Chung, W.O., Watanabe, K., El-Sabaeny, A., Park, K., and Lamont, R.J. (2001). Inhibition of epithelial cell apoptosis by Porphyromonas gingivalis. FEMS Microbiol. Lett. *200*, 145–149.

Naumann, M., Rudel, T., and Meyer, T.F. (1999). Host cell interactions and signalling with *Neisseria gonorrhoeae*. Curr. Opin. Microbiol. 2, 62–70.

Nechushtan, A., Smith, C.L., Lamensdorf, I., Yoon, S.H., and Youle, R.J. (2001). Bax and Bak coalesce into novel mitochondria-associated clusters during apoptosis. J. Cell. Biol. *153*, 1265–1276.

O'Regan, A., and Berman, J.S. (2000). Osteopontin: a key cytokine in cell-mediated and granulomatous inflammation. Int. J. Exp. Pathol. *81*, 373–390.

Pantelic, M., Kim, Y.J., Bolland, S., Chen, I., Shively, J., and Chen, T. (2005). *Neisseria gonorrhoeae* kills carcinoembryonic antigen-related cellular adhesion molecule 1 (CD66a)-expressing human B-cells and inhibits antibody production. Infect. Immun. 73, 4171–4179.

Platt, N., da Silva, R.P., and Gordon, S. (1998). Recognizing death: the phagocytosis of apoptotic cells. Trends Cell. Biol. *8*, 365–372.

Proskuryakov, S.Y., Gabai, V.L., and Konoplyannikov, A.G. (2002). Necrosis is an active and controlled form of programmed cell death. Biochemistry (Mosc) *67*, 387–408.

Punsalang, A.P., Jr., and Sawyer, W.D. (1973). Role of pili in the virulence of *Neisseria gonorrhoeae*. Infect. Immun. *8*, 255–263.

Restifo, N P. (2000). Building better vaccines: how apoptotic cell death can induce inflammation and activate innate and adaptive immunity. Curr Opin. Immunol. *12*, 597–603.

Robinson, K., Taraktsoglou, M., Rowe, K.S., Wooldridge, K.G., and Ala'Aldeen, D.A. (2004). Secreted proteins from *Neisseria meningitidis* mediate differential human gene expression and immune activation. Cell. Microbiol. *6*, 927–938.

Rogers, H.W., Callery, M.P., Deck, B., and Unanue, E.R. (1996). *Listeria monocytogenes* induces apoptosis of infected hepatocytes. J. Immunol. *156*, 679–684.

Rojas, M., Barrera, L.F., Puzo, G., and Garcia, L.F. (1997). Differential induction of apoptosis by virulent *Mycobacterium tuberculosis* in resistant and susceptible murine macrophages: role of nitric oxide and mycobacterial products. J. Immunol. *159*, 1352–1361.

Rostovtseva, T.K., Antonsson, B., Suzuki, M., Youle, R.J., Colombini, M., and Bezrukov, S.M. (2004). Bid, but not Bax, regulates VDAC channels. J. Biol. Chem. *279*, 13575–13583.

Rudel, T., Schmid, A., Benz, R., Kolb, H.A., Lang, F., and Meyer, T.F. (1996). Modulation of *Neisseria* porin (PorB) by cytosolic ATP/GTP of target cells: parallels between pathogen accommodation and mitochondrial endosymbiosis. Cell *85*, 391–402.

Savill, J., Fadok, V., Henson, P., and Haslett, C. (1993). Phagocyte recognition of cells undergoing apoptosis. Immunol. Today *14*, 131–136.

Scaffidi, C., Fulda, S., Srinivasan, A., Friesen, C., Li, F., Tomaselli, K.J., Debatin, K.M., Krammer, P.H., and Peter, M.E. (1998). Two CD95 (APO-1/Fas) signaling pathways. EMBO J. *17*, 1675–1687.

Schaumburg, F., Hippe, D., Vutova, P., and Luder, C.G. (2006). Pro- and anti-apoptotic activities of protozoan parasites. Parasitology *132* Suppl., S69–85.

Schilling, J.D., Mulvey, M.A., and Hultgren, S.J. (2001). Dynamic interactions between host and pathogen during acute urinary tract infections. Urology *57*, 56–61.

Schubert-Unkmeir, A., Sokolova, O., Panzner, U., Eigenthaler, M., and Frosch, M. (2007). Gene expression pattern in human brain endothelial cells in response to *Neisseria meningitidis*. Infect. Immun. *75*, 899–914.

Schulze-Osthoff, K., Ferrari, D., Los, M., Wesselborg, S., and Peter, M.E. (1998). Apoptosis signaling by death receptors. Eur. J. Biochem. *254*, 439–459.

Sharpe, J.C., Arnoult, D., and Youle, R.J. (2004). Control of mitochondrial permeability by Bcl-2 family members. Biochim. Biophys. Acta *1644*, 107–113.

Shaw, J.H., and Falkow, S. (1988). Model for invasion of human tissue culture cells by *Neisseria gonorrhoeae*. Infect. Immun. *56*, 1625–1632.

Shen, Y., and Shenk, T.E. (1995). Viruses and apoptosis. Curr. Opin. Genet. De.v *5*, 105–111.

Shimizu, S., Konishi, A., Kodama, T., and Tsujimoto, Y. (2000). BH4 domain of antiapoptotic Bcl-2 family members closes voltage-dependent anion channel and inhibits apoptotic mitochondrial changes and cell death. Proc. Natl. Acad. Sci. USA *97*, 3100–3105.

Shimizu, S., Narita, M., and Tsujimoto, Y. (1999). Bcl-2 family proteins regulate the release of apoptogenic cytochrome c by the mitochondrial channel VDAC. Nature *399*, 483–487.

Simons, M.P., Nauseef, W.M., and Apicella, M.A. (2005). Interactions of *Neisseria gonorrhoeae* with adherent polymorphonuclear leukocytes. Infect. Immun. *73*, 1971–1977.

Simons, M.P., Nauseef, W.M., Griffith, T.S., and Apicella, M.A. (2006). *Neisseria gonorrhoeae* delays the onset of apoptosis in polymorphonuclear leukocytes. Cell. Microbiol. *8*, 1780–1790.

Singh, B.N., Lucas, J.J., Hayes, G.R., Kumar, I., Beach, D.H., Frajblat, M., Gilbert, R.O., Sommer, U., and Costello, C.E. (2004). Tritrichomonas foetus induces apoptotic cell death in bovine vaginal epithelial cells. Infect. Immun. 72, 4151–4158.

Stephens, D.S. (1989). Gonococcal and meningococcal pathogenesis as defined by human cell, cell culture, and organ culture assays. Clin. Microbiol. Rev. 2 Suppl., S104–111.

Storz, G., and Imlay, J.A. (1999). Oxidative stress. Curr. Opin. Microbiol. 2, 188–194.

Susin, S.A., Lorenzo, H.K., Zamzami, N., Marzo, I., Snow, B.E., Brothers, G.M., Mangion, J., Jacotot, E., Costantini, P., Loeffler, M., Larochette, N., Goodlett, D.R., Aebersold, R., Siderovski, D.P., Penninger, J.M., Kroemer, G. (1999). Molecular characterization of mitochondrial apoptosis-inducing factor. Nature *397*, 441–446.

Swanson, J., Kraus, S.J., and Gotschlich, E.C. (1971). Studies on gonococcus infection. I. Pili and zones of adhesion: their relation to gonococcal growth patterns. J. Exp. Med. *134*, 886–906.

Taha, M.K. (2000). *Neisseria meningitidis* induces the expression of the TNF-alpha gene in endothelial cells. Cytokine *12*, 21–25.

Thatte, U., and Dahanukar, S. (1997). Apoptosis: clinical relevance and pharmacological manipulation. Drugs *54*, 511–532.

Tramont, E.C. (1981). Adhesion of *Neisseria gonorrhoeae* and disease. Ciba Found. Symp. *80*, 188–201.

Tsujimoto, Y., and Shimizu, S. (2000). Bcl-2 family: life-or-death switch. FEBS Lett. *466*, 6–10.

Tunbridge, A.J., Stevanin, T.M., Lee, M., Marriott, H.M., Moir, J.W., Read, R.C., and Dockrell, D.H. (2006). Inhibition of macrophage apoptosis by *Neisseria meningitidis* requires nitric oxide detoxification mechanisms. Infect. Immun. *74*, 729–733.

Vander Heiden, M.G., Chandel, N.S., Williamson, E.K., Schumacker, P.T., and Thompson, C.B. (1997). Bcl-xL regulates the membrane potential and volume homeostasis of mitochondria. Cell *91*, 627–637.

Veale, D.R., Finch, H., and Smith, H. (1976). Penetration of penicillin into human phagocytes containing *Neisseria gonorrhoeae*: intracellular survival and growth at optimum concentrations of antibiotic. J. Gen. Microbiol. *96*, 353–363.

Veale, D.R., Goldner, M., Penn, C.W., Ward, J., and Smith, H. (1979). The intracellular survival and growth of gonococci in human phagocytes. J. Gen. Microbiol. *113*, 383–393.

Verhagen, A.M., Ekert, P.G., Pakusch, M., Silke, J., Connolly, L.M., Reid, G.E., Moritz, R.L., Simpson, R.J., and Vaux, D.L. (2000). Identification of DIABLO, a mammalian protein that promotes apoptosis by binding to and antagonizing IAP proteins. Cell *102*, 43–53.

Wang, J., Zheng, L., Lobito, A., Chan, F.K., Dale, J., Sneller, M., Yao, X., Puck, J.M., Straus, S.E., and Lenardo, M.J. (1999). Inherited human Caspase 10 mutations underlie defective lymphocyte and dendritic cell apoptosis in autoimmune lymphoproliferative syndrome type II. Cell *98*, 47–58.

Weel, J.F., Hopman, C.T., and van Putten, J.P. (1991). Bacterial entry and intracellular processing of *Neisseria gonorrhoeae* in epithelial cells: immunomorphological evidence for alterations in the major outer membrane protein P.IB.J. Exp. Med. *174*, 705–715.

Weel, J.F., and van Putten, J.P. (1991). Fate of the major outer membrane protein P.IA in early and late events of gonococcal infection of epithelial cells. Res. Microbiol. *142*, 985–993.

Wei, M.C., Zong, W.X., Cheng, E.H., Lindsten, T., Panoutsakopoulou, V., Ross, A.J., Roth, K.A., MacGregor, G.R., Thompson, C.B., and Korsmeyer, S.J. (2001). Proapoptotic BAX and BAK: a requisite gateway to mitochondrial dysfunction and death. Science *292*, 727–730.

Weinrauch, Y., and Zychlinsky, A. (1999). The induction of apoptosis by bacterial pathogens. Annu. Rev. Microbiol. *53*, 155–187.

Wells, D.B., Tighe, P.J., Wooldridge, K.G., Robinson, K., and Ala' Aldeen, D.A. (2001). Differential gene expression during meningeal–meningococcal interaction: evidence for self-defense and early release of cytokines and chemokines. Infect. Immun. *69*, 2718–2722.

Wolter, K.G., Hsu, Y.T., Smith, C.L., Nechushtan, A., Xi, X.G., and Youle, R.J. (1997). Movement of Bax from the cytosol to mitochondria during apoptosis. J. Cell. Biol. *139*, 1281–1292.

Wu, M.X., Ao, Z., Prasad, K.V., Wu, R., and Schlossman, S.F. (1998). IEX-1L, an apoptosis inhibitor involved in NF-kappaB-mediated cell survival. Science *281*, 998–1001.

Xiao, Y., Zhong, Y., Greene, W., Dong, F., and Zhong, G. (2004). Chlamydia trachomatis infection inhibits both Bax and Bak activation induced by staurosporine. Infect. Immun. *72*, 5470–5474.

Yu, Q., Chow, E.M., Wong, H., Gu, J., Mandelboim, O., Gray-Owen, S.D., and Ostrowski, M.A. (2006). CEACAM1 (CD66a) promotes human monocyte survival via a phosphatidylinositol 3-kinase- and AKT-dependent pathway. J. Biol. Chem. *281*, 39179–39193.

Zamzami, N., Susin, S.A., Marchetti, P., Hirsch, T., Gomez-Monterrey, I., Castedo, M., and Kroemer, G. (1996). Mitochondrial control of nuclear apoptosis. J. Exp. Med. *183*, 1533–1544.

Zhong, Y., Weininger, M., Pirbhai, M., Dong, F., and Zhong, G. (2006). Inhibition of staurosporine-induced activation of the proapoptotic multidomain Bcl-2 proteins Bax and Bak by three invasive chlamydial species. J. Infect. *53*, 408–414.

Zychlinsky, A., and Sansonetti, P.J. (1997). Apoptosis as a proinflammatory event: what can we learn from bacteria-induced cell death? Trends Microbiol. *5*, 201–204.

Role of Acquired Immunity in *Neisseria* Infections

10

Manish Sadarangani, Matthew D. Snape, Dominic F. Kelly, Gunnstein Norheim, J. Claire Hoe, Susan Lewis, Lee Wetzler and Andrew J. Pollard

Abstract

A description of the clinical features of meningococcal disease first appears in the literature in 1805 in a thesis on an outbreak in Geneva, yet the global distribution of the organism and the diversity of its surface structures indicate a far longer history of association with man (perhaps as an acapsulate commensal), with diversity likely driven by immune-selection. Carriage studies indicate that almost all humans are likely to be exposed to *Neisseria meningitidis* at some stage and yet most do not develop disease. In early childhood, after loss of maternal antibody and before acquisition of adaptive immunity, the innate immune system effectively controls the meningococcus in most exposed individuals and prevents onset of invasive disease. However, the lower rates of disease in later childhood and amongst adults who have acquired specific antibody to the meningococcal surface indicates that the adaptive immune system is better still at preventing invasion into the bloodstream. In the first portion of this chapter we will consider both the natural acquisition of adaptive immunity to the meningococcus and the active induction of specific antibody through immunization. There will also be a discussion regarding gonococcal acquired immunity at the end of this chapter.

Natural acquisition of humoral immunity to meningococci

Despite widespread carriage of potentially pathogenic *N. meningitidis* in the normal population, an extremely low proportion of the carriers develop meningococcal disease. Protection is believed to be mediated by functional anti-meningococcal antibodies in sera of a high proportion of the healthy normal population (Goldschneider *et al.*, 1969a). Antibodies may exert their protective function by binding to meningococcal surface structures and serve as deposition sites for complement; leading to insertion of the membrane attack complex and lysis of the bacterium (bactericidal effect). This is considered the main protective mechanism against meningococcal disease (Goldschneider *et al.*, 1969a). Alternatively, the antibodies may bind to meningococci and facilitate uptake into neutrophils (opsonophagocytosis), where the conditions inside the neutrophil phagolysosome leads to lysis (DeVoe, 1976). Another protective mechanism is antibody-dependent cellular cytotoxicity, which may be observed following meningococcal infection (Lowell *et al.*, 1979; Lowell *et al.*, 1980) and plays a role in anatomical sites with low levels of complement.

A seroepidemiology study was performed in the USA in 1960s on sera from 282 children (age range newborn to 12 years) and 587 army recruits (aged 19–26 years) (Goldschneider *et al.*, 1969a). These sera were measured for serum bactericidal activity (SBA) against serogroup A, B and C meningococcal strains in the presence of human complement. The proportion of individuals with a putatively protective SBA titre (≥ 4) within each age group was imposed upon the age-related attack rate, and an inverse relation was observed (Fig. 10.1). A comparable recent study analysed 2706 residual diagnostic sera in the U.K. (age range 0–93 years) for SBA against a serogroup B isolate and in ELISA against outer

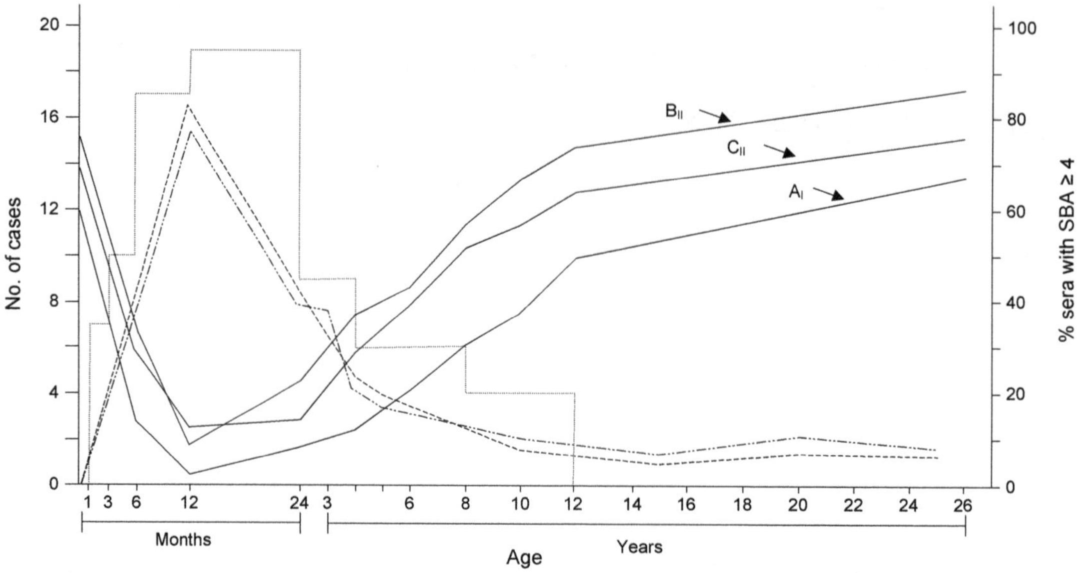

Figure 10.1 Inverse relation between disease incidence and putative protective serum bactericidal titres against serogroups A, B and C meningococcal disease in the USA in late 1960s (taken from Goldschneider *et al.*, 1969a). Key: - - - - - , no. of cases/100,000 age-specific population, 1965; —··—·· , no. of cases/100,000 age-specific population, 1966; ·············, age distribution of 72 cases admitted to Los Angeles Children's Hospital 1944–1953; ——— ,percent of sera in each group having a bactericidal titre of 1:4 or greater against meningococcal strains A_I, B_{II} and C_{II}. Sera from 282 children (at least 20 in each age group) and 657 army recruits (aged 19–26 years) were tested in the presence of exogenous complement. Each point in the figure represents the incidence of disease or prevalence of bactericidal activity among subjects in the age range encompassed by it and the previous point. (e.g. 2.5% of children 6–12 months of age had serum bactericidal activity against strain A_1).

membrane vesicles (OMVs) (Trotter *et al.*, 2007). However, the previously described relationship (Goldschneider *et al.*, 1969a) was not so clearly observed from their data, as incidence declined during childhood although no significant change in proportion of children with putatively protective titres was observed (Fig. 10.2) (Trotter *et al.*, 2007). This confirmed findings indicated in another study in Canada (Pollard *et al.*, 2004). The differences evident between recent seroepidemiological studies (Ceyhan *et al.*, 2007; Pollard *et al.*, 2004; Trotter *et al.*, 2003; Trotter *et al.*, 2007) and the widely recognized Goldschneider study (Goldschneider *et al.*, 1969a) may be caused by different epidemiological settings and antigenic exposure, but also questions the sensitivity of SBA alone as a correlate of protection (Pollard *et al.*, 2004).

When the observed population immunity is not generated by immunization with meningococcal vaccines, it is considered to be 'naturally acquired'. The four most described mechanisms inducing naturally acquired immunity are (1) exposure to non-neisserial organisms, (2) carriage of *Neisseria lactamica*, (3) asymptomatic carriage of *N. meningitidis* and (4) contracting meningococcal disease.

Exposure to non-neisserial organisms

Some of the anti-meningococcal antibodies observed in healthy human sera may have been induced by exposure to non-neisserial bacteria with cross-reactive antigenic structures. For example, most adults have significant levels of antibodies against serogroup B polysaccharide, primarily of the IgM class (Granoff *et al.*, 1995; Leinonen and Frasch, 1982). These may have been stimulated by asymptomatic carriage of either *Escherichia coli* K1 or *Moraxella nonliquefaciens*, which are part of the normal flora in the intestinal and respiratory tracts and have chemically identical capsule polysaccharides (Devi *et al.*, 1991). Cross reactivity is also observed between serogroup C and *E. coli* K92 polysaccharide (Glode *et al.*, 1977),

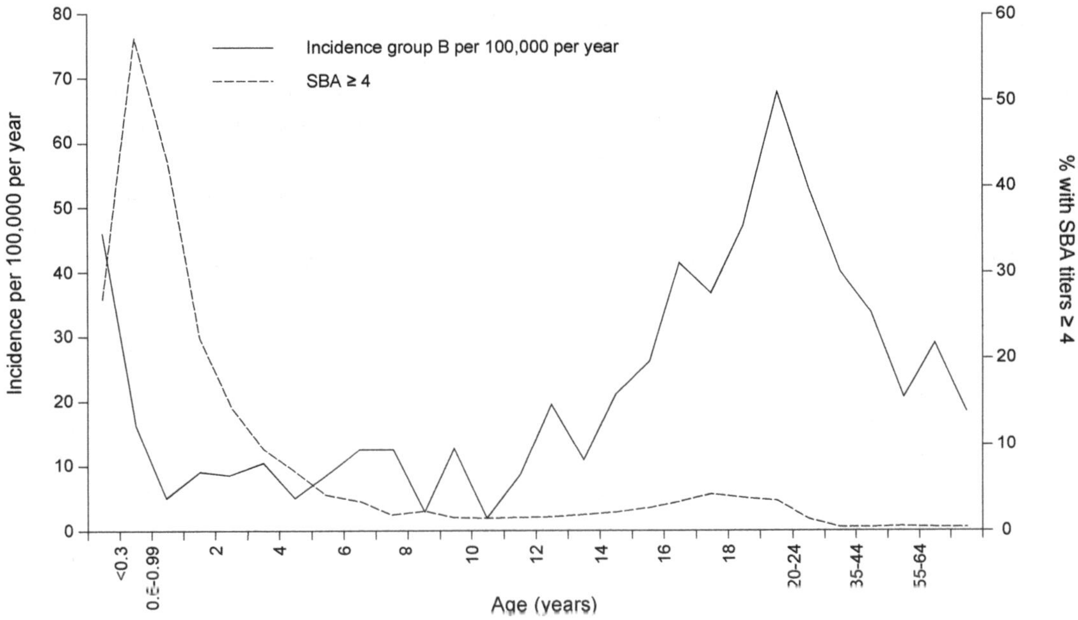

Figure 10.2 Age-related distribution of individuals with protective serum bactericidal titre against a serogroup B strain with human complement versus incidence of serogroup B meningococcal disease in the U.K. (taken from Trotter *et al.*, 2007).

and serogroup A and *Bacillus pumilis* Sh-17, *E. coli* K51 and K93 polysaccharides (Filice *et al.*, 1985; Robbins *et al.*, 1972). Significant levels of anti-A polysaccharide antibodies were also observed after immunization of laboratory animals with inactivated whole cells of these enteric organisms (Guirguis *et al.*, 1985; Robbins *et al.*, 1972). However, long-term exposure with non-neisserial organisms (e.g. *B. pumilus*) may be needed to induce such immunity (Jin *et al.*, 2003). The multitude of enteric organisms presents the possibility that bacteria other than *E. coli* and *B. pumilus* may be involved in inducing natural anti-meningococcal immunity.

Cross-reactive anti-meningococcal antibodies may also be induced by non-meningococcal bacterial meningitis, as observed with the induction of anti-serogroup A polysaccharide IgG induced in Finnish patients (Kayhty, 1982). This might be caused by polyclonal stimulation of serogroup A polysaccharide-specific plasma B-cells, a mechanism especially relevant in developing countries, where both bacterial and parasite infection prevalence rates are high.

The contribution of the antibodies induced by non-neisserial organisms to protection against meningococcal disease is however unclear. For example, IgG and IgM antibodies against serogroup B polysaccharide are of low avidity and show low bactericidal activity with human complement (Mandrell *et al.*, 1995; Mandrell and Zollinger, 1982). The importance of antibodies generated by cross-reactive organisms have also been downplayed as a factor in generating naturally acquired bactericidal antibodies against serogroup A (Amir *et al.*, 2005; Goldschneider *et al.*, 1969a) and against serogroup B meningococci in U.K. (Trotter *et al.*, 2007). The importance of antibodies attributed to protection may however be dependent on the assay in use. For example, human IgM against serogroup B polysaccharide demonstrated protection against infection in a study using an infant rat model, an ability that seemed to increase with age (Toropainen *et al.*, 2005), whereas another study showed that these were not responsible for the observed killing of meningococci in a whole blood model using non-bactericidal sera (Welsch and Granoff, 2007).

Carriage of *Neisseria lactamica*

Carriage of *N. lactamica* is high in infants and young adults but declines with age, as opposed to carriage of *N. meningitidis,* which is low in infancy and increases by adolescence and early adulthood (Bennett *et al.*, 2005; Blakebrough *et al.*, 1982; Cartwright *et al.*, 1987; Gold *et al.*, 1978; Olsen *et al.*, 1991). Since most meningococcal disease in Western countries occurs among children aged 0–4 years and the bactericidal titres against meningococci rise throughout infancy, it has been suggested that the carriage of *N. lactamica* is central to induction of the observed bactericidal activity against *N. meningitidis* (Gold *et al.*, 1978; Olsen *et al.*, 1991). However this was disputed by a recent seroepidemiological study in the UK showing that there was little change in SBA or anti-meningococcal IgG concentrations in young children, at a time when *N. lactamica* carriage is common (Trotter *et al.*, 2007).

Since *N. lactamica* is an acapsulate organism (Bennett *et al.*, 2005), it is assumed that the *N. lactamica* carriage induces immunity directed against relatively conserved subcapsular meningococcal antigens. Potential proteins of *N. lactamica* that could induce such cross-reactive immunity were recently identified by comparative proteomic analyses and immunization of mice (Finney *et al.*, 2007; Vaughan *et al.*, 2006). Another candidate is lipooligosaccharide (LOS) (Kim *et al.*, 1989). Direct evidence to support the hypothesis that exposure to *N. lactamica* carriage induces bactericidal anti-meningococcal antibodies is lacking as described above (Trotter *et al.*, 2007). However, it has been suggested that *N. lactamica* carriage may provide protection by generation of opsonophagocytic antibodies (Finney *et al.*, 2007; Trotter *et al.*, 2007).

Asymptomatic carriage of *Neisseria meningitidis*

Asymptomatic carriage of *N. meningitidis* is related to age, intimate personal contact, smoking, crowding and respiratory tract infections (Caugant *et al.*, 2007; MacLennan *et al.*, 2006; Stuart *et al.*, 1989). Carriage rates fluctuate with time, epidemic setting and geographical areas, and the antigenic diversity among isolates is high. Asymptomatic colonization with *N. meningitidis* at nasopharyngeal mucosal surfaces often stimulates an immune response (Gold *et al.*, 1978; Goldschneider *et al.*, 1969a). Since the carriage rates usually range from 10% to 35% in young adults and close to 100% in closed or semi-closed populations (Caugant *et al.*, 2007; Trotter *et al.*, 2007), carriage may be one of the most important contributors to induction of natural immunity against meningococci.

The antibody response following carriage is seen as increased IgG and/or IgA concentrations against meningococcal antigens both in the circulation and at mucosal surfaces, as recently confirmed by a number of studies in U.K. students (Horton *et al.*, 2005; Jones *et al.*, 1998; Jordens *et al.*, 2004; Robinson *et al.*, 2002). The specific IgG rise in serum induced by carriage may provide protection against disease once meningococci enter the circulation, and is considered the most reliable measure of protection. The IgG and IgA induced in nasopharyngeal secretions may play a major role in preventing colonization and invasion of submucosal tissues (Hamadeh *et al.*, 1995; Heyderman *et al.*, 2006). There is usually a good correlation between concentrations of specific IgG in saliva and in serum in individuals, and salivary IgG detected is therefore thought to be mainly passively diffused from serum (Zhang and Finn, 2004). On the contrary, the IgA detected in saliva is thought to be mainly derived from long-lived plasma cells resident in the mucosa (Horton *et al.*, 2005; Robinson *et al.*, 2002).

The specificity of immunity induced by carriage of a particular meningococcal strain may be strain-specific and directed against both capsular and non-capsular components (Goldschneider *et al.*, 1969a; Horton *et al.*, 2005; Jones *et al.*, 1998; Wedege *et al.*, 2003; Williams *et al.*, 2003). Rises in serum IgG against variants of PorA, PorB, Opa and OpcA outer membrane proteins homologous to the colonizing strain have been observed (Jones *et al.*, 1998; Rosenqvist *et al.*, 1995). The carriage-induced serum antibodies are able to exert bactericidal activity (Goldschneider *et al.*, 1969a; Jones *et al.*, 1998; Reller *et al.*, 1973; Robinson *et al.*, 2002; Wedege *et al.*, 2003). The SBA has been relatively strain-specific in some studies, whereas other studies have reported SBA also against heterologous strains (Goldschneider *et al.*, 1969b; Jones *et al.*, 1998; Jordens *et al.*, 2004). Studies in the African meningitis belt have revealed conflicting results concerning the relationship between carriage and protective immunity in serum for serogroup W-135 meningococci (Amadou Hamidou *et al.*, 2006; Boisier *et al.*, 2006; Mueller *et al.*, 2006). In the case of carriage of serogroup A meningococci, a higher level of anti-serogroup A protective immunity in serum was observed among the population in an area of Burkina Faso with a high carriage rate compared to in an area with a low carriage rate (Raghunathan *et al.*, 2006). However, study design and previous vaccination may affect the observed results.

Whereas the duration of carriage itself can vary from days to many months (Cartwright *et al.*, 1987; Stephens, 1999), the duration of the protection induced by carriage is estimated to be up to ~33 months after an assumed 6 months of carriage (Trotter *et al.*, 2007). However, a study from Gambia showed the duration of antibody levels to vary between different antigens (Thiesen *et al.*, 1997).

Immunity following meningococcal disease

As the incidence of meningococcal disease typically is very low (e.g. 1–3 cases per 100,000 inhabitants in Western countries), the contribution conferred by invasive disease to naturally acquired immunity at the population level is nearly absent. It is however relevant to study this induction mechanism for the purpose of serological confirmation of diagnosis in cases where cerebrospinal fluid (CSF) or blood samples are culture-negative or CSF is not available, and for the purpose of revealing the *in vivo* response to antigens considered for vaccine development.

Infection with *N. meningitidis* is often characterized by a significant increase in antigen-specific antibody levels: by the second week after onset of disease, seroconversion has been observed in 47–90% of the patients (Griffiss *et al.*, 1984; Guttormsen *et al.*, 1994; Harthug *et al.*, 1986; Milagres *et al.*, 2000; Norheim *et al.*, 2007; Pollard *et al.*, 1999b; Weis *et al.*, 2005). The seroconversion rate is dependent on the test in use and the level of pre-existing immunity. It is further complicated by the challenge in defining the exact date of the onset of disease, as onset of immune stimulation may not directly correspond with the onset of the clinically recognized symptoms during the transition from subclinical to systemic disease (Thompson *et al.*, 2006); thus defining a serum sample as 'acute phase' may not be straightforward (Milagres *et al.*, 2000; Norheim *et al.*, 2007). However, acute sera from septicaemia patients who have a more rapid onset of disease have significantly lower anti-meningococcal antibody levels than other meningococcal patients (Flaegstad *et al.*, 1990).

The specificity of the IgG and IgM response mounted in convalescent sera from meningococcal disease patients is often directed against both capsular and non-capsular antigens. A series of studies have shown that patients respond to capsular polysaccharides during infection, e.g. those of serogroups A, B, and C (Goldblatt *et al.*, 2002; Granoff *et al.*, 1995; Kayhty, 1982; Norheim *et al.*, 2007). The list of meningococcal surface proteins for which reaction with convalescent sera is seen includes among others: FetA, PorA, PorB, RmpM, Opa, OpcA, NadA, IgA1 protease, H.8 antigen, NspA, TbpB, AutA and LbpA (Ait-Tahar *et al.*, 2000; Ala'Aldeen *et al.*, 1994; Black *et al.*, 1985; Brooks *et al.*, 1992; Idanpaan-Heikkila *et al.*, 1995; Johnson *et al.*, 1999; Kizil *et al.*, 1999; Kortekaas *et al.*, 2006; Litt *et al.*, 2004; Mandrell and Zollinger, 1989; Norheim *et al.*, 2007; Pettersson *et al.*, 2006; Robinson *et al.*, 2002; Rosenqvist *et al.*, 1995; Sjursen *et al.*, 1990; Wedege *et al.*, 1998). Many of the surface exposed proteins recently revealed by whole-genome analyses and high-throughput screening further adds to this list (Litt *et al.*, 2004), however a reaction with convalescent serum should preferably be interpreted in comparison to an acute serum reaction. Infection can also induce antibodies against LOS (Estabrook *et al.*, 1993; Norheim *et al.*, 2008; Wedege *et al.*, 1998). The central question is what the relative contribution is of the antibodies stimulated by each of these antigens to naturally acquired protective immunity in humans.

Depending on age, between 21 and 100% of meningococcal patients may acquire putatively protective serum bactericidal titres following disease (Idanpaan-Heikkila *et al.*, 1995; Norheim *et al.*, 2007; Pollard *et al.*, 1999b). Serogroup B disease induces a bactericidal antibody response in adults but not infants, and there seems to be a maturational defect in the development of high-avidity antibodies in this age group (Pollard *et al.*, 1999b; Pollard and Levin, 2000). Since the serogroup B polysaccharide is poorly immunogenic in humans, the bactericidal activity mainly targets subcapsular antigens (Zollinger and Mandrell, 1983). Following serogroup A disease, a bactericidal antibody response was seen in nearly all patients during the second week after onset of disease, being directed both against the polysaccharide and the subcapsular antigens (Norheim *et al.*, 2007). However, complement source and target strain in the SBA assay may have a significant impact on results (Norheim *et al.*, 2007; Santos *et al.*, 2001; Zollinger and Mandrell, 1983). Opsonophagocytic activity is induced following meningococcal infection at least against PorA, PorB and TbpA+B complexes (Lehmann *et al.*, 1999a,b). When measuring such activity against whole meningococcal bacteria, the procedure for preparation of bacteria prior to antibody binding seems important (Findlow *et al.*, 2007). The duration of protection following disease is variable and exposure to intermittent meningococcal carriage complicates attempts to attribute the origin of the observed immunity (Wedege *et al.*, 1998).

The immunoglobulin class response stimulated by meningococcal infection is primarily IgM and IgG (Flaegstad *et al.*, 1990; Goldschneider *et al.*, 1969a; Harthug *et al.*, 1986; Pollard *et al.*, 1999b; Sjursen *et al.*, 1990). Antibody quantity and functional activity seems to increase with age (Flaegstad *et al.*, 1990; Sjursen *et al.*, 1990). The IgG antibodies are mainly of subclass IgG1 and IgG3 (Pollard *et al.*, 1999b; Sjursen *et al.*, 1990). In older children, there may also be a rise in IgG2 but no significant increase in IgG4. The relative ability of these to activate lysis in the presence of complement is suggested to be IgG1 = IgG3 > IgG2 >> IgG4, whereas the ability to trigger opsonophagocytosis seems to be IgG3 > IgG1 > IgG2 >> IgG4 (Aase and Michaelsen, 1994; Vidarsson *et al.*, 2001).

Summary

Naturally acquired antibody against meningococci plays a major role in defence against disease (Goldschneider *et al.*, 1969a). Vaccine development may benefit from studies on how the protective proportion of these antibodies is generated, in order to mimic the generation of systemic and mucosal immunity. The development of a conjugate vaccine based on the cross-reacting polysaccharide of *B. pumilus* and a vaccine based on *N. lactamica* OMVs are examples of this approach. Although the role of bactericidal activity of IgG in naturally acquired immunity is well established, there are important pieces missing in the puzzle to explain the recently described discrepancies in seroepidemiological studies (Goldschneider *et al.*, 1969a; Trotter *et al.*, 2007). Thus, further studies are warranted to describe the exact pattern and mechanism of naturally acquired protection versus age-related disease incidence, focusing on particular antigens to provide baseline data relevant for vaccine development.

Immune responses to plain polysaccharide and protein–polysaccharide conjugate meningococcal vaccines.

Introduction

Whilst vaccines based on meningococcal outer membrane vesicles and proteins have recently been developed, the most commonly employed meningococcal vaccines are based on the organism's polysaccharide capsule, either presented on its own (the plain polysaccharide vaccines) or covalently bound to a carrier protein (the protein–polysaccharide conjugate vaccines, also known as the glycoconjugate vaccines). Both plain polysaccharide and glycoconjugate vaccines stimulate the humoral immune system to produce antibodies against the polysaccharide capsule that can then facilitate lysis via the classical complement pathway, opsonophagocytosis or antibody-dependent cellular cytotoxicity (Pollard and Frasch, 2001). The primary role of the humoral system in protecting against meningococcal disease is reflected in the use of the SBA assay as the standard correlate of protection used to assess the immunogenicity of meningococcal vaccines. The SBA assay uses complement sourced from either humans (hSBA: protective titre ≥1:4; Goldschneider *et al.*, 1969a) or baby rabbits (rSBA: protective titre ≥1:8; Borrow *et al.*, 2001a).

Although both plain polysaccharide and protein-polysaccharide vaccines aim to induce humoral immunity, they differ in their ability to induce an immune response in infants and to generate an anamnestic response on repeat immunization. In order to understand these differences it is necessary to consider the different means by which the two vaccine types stimulate the humoral arm of the immune system.

Plain polysaccharide vaccines

Vaccines based on the polysaccharide capsules of meningococcal serogroups A, C, W-135 and/or Y have been available since the 1970s, and are currently available as bivalent (MenACP), trivalent (MenACWP) or tetravalent (MenACWYP) vaccines. These vaccines are used in a variety of contexts such as mass immunization for epidemic control (De Wals *et al.*, 2001; WHO, 2002) or as traveller vaccines (WHO, 2002), e.g. for Hajj pilgrims (Saudi Ministry of Health, 2007). The structural similarity of the serogroup B meningococcal polysaccharide capsule to polysaccharides on neural cell adhesion molecule (PSA-NCAM), expressed on the surface of human fetal neural tissue, makes the serogroup B capsule poorly immunogenic and a problematic vaccine candidate (Finne *et al.*, 1983).

Following immunization with a plain polysaccharide vaccine, the repeating epitopes of the polysaccharide antigen are thought to cross link antibodies on the surface of splenic marginal zone B-cells, thus stimulating them to proliferate into antibody secreting plasma cells (Zandvoort and Timens, 2002). However, unlike peptide antigens, it has conventionally been considered that polysaccharide antigens are unable to be presented to T helper cells in conjunction with major histocompatibility complex (MHC) class II molecules, and the B-cell response therefore receives no support from T helper cells. While this T-independent response results in the production of plasma cells that can secrete both IgM and IgG (Beuvery *et al.*, 1982), no antigen specific memory B-cells are produced (Zandvoort and Timens, 2002).

The T-independent response observed following immunization with plain polysaccharide vaccines explains some of the limitations of these vaccines. As the splenic marginal zone of children less than two years of age is immature, in this age group these vaccines are in general poorly immunogenic (Gold *et al.*, 1975; Twumasi *et al.*, 1995), and have been shown to be ineffective in the prevention of serogroup C disease (De Wals *et al.*, 2001).

The elevation of bactericidal antibodies induced by plain polysaccharide meningococcal vaccines wanes over the first few years after immunization in both children (Espin Rios *et al.*, 2000) and adults (Zangwill *et al.*, 1994). This is possibly related to the failure to induce antigen specific memory B-cells. Crucially, with the exception of serogroup A, administration of additional doses of plain polysaccharide vaccines in an attempt to boost waning antibody levels results in a relatively attenuated immune response. For example, serogroup C specific rSBA titres were significantly higher following a dose of MenACP in 10 – 29 year olds who were MenACP naïve than those who had previously received one or more doses of this vaccine (Jokhdar *et al.*, 2004). Similar results have been seen in young children (Al-Mazrou *et al.*, 2005; Leach *et al.*, 1997) and infants (MacDonald *et al.*, 1998). This immunological hyporesponsiveness is thought to be related to there being only a limited pool of splenic marginal zone B-cells specific to each antigen (Pollard and Granoff, 2007). In the absence of the generation of memory B-cells, immunization with a plain polysaccharide vaccine may deplete this pool, with the result that there are less antigen specific B-cells available to respond on subsequent exposure to the polysaccharide capsule. While this phenomenon clearly argues against the administration of repeated doses of polysaccharide vaccines, it is as yet unclear whether immune hyporesponsiveness also results in an attenuated response to natural infection following plain polysaccharide immunization.

An important exception to the phenomena described above is that the response to the serogroup A component of meningococcal plain polysaccharide vaccines appears to differ from that of the other meningococcal serogroups. A Cochrane review of the effectiveness of plain polysaccharide meningococcal serogroup A concluded that this vaccine effectively prevents disease for at least a year after immunization of Finnish children aged 3 months to 5 years and Nigerian and Sudanese children aged from 1 to 5 years (Patel and Lee, 2005). This review suggested that evidence of effectiveness of the serogroup A vaccine in children under the age of 2 years was inconclusive however. In addition, some studies demonstrate that higher serogroup A specific rSBA titres are seen following MenACP in individuals who have previously received this vaccine compared to those who are receiving it for the first time (Jokhdar *et al.*, 2004; Leach *et al.*, 1997). This implies effective immune boosting by repeated doses of the serogroup A polysaccharide, in contrast to the results seen for serogroup C. Of note, however, is that one study showed significantly higher rSBA geometric mean titres (GMTs) 1 month after MenACP in 5-year-olds receiving MenACP for the first time compared with those who had received three previous doses of MenACP (MacLennan *et al.*, 2001), suggesting immune hyporesponsiveness for serogroup A can be induced by multiple doses of the vaccine.

While the discussion above primarily relates to humoral immunity in circulating blood, given the usual portal of entry of the meningococcus is via the nasopharynx, it is important to consider the role of mucosal immunity. An increase in MenC-specific salivary IgA and IgG is seen after immunization of adolescents with MenACP (Zhang *et al.*, 2000). The evidence regarding the impact of polysaccharide vaccines on nasopharyngeal carriage is inconsistent (Dellicour and Greenwood, 2007), however it is likely that any reduction in nasopharyngeal carriage of the meningococcus is short lived, and is therefore unable to generate herd immunity.

Despite these shortcomings, plain polysaccharide vaccines do have a proven efficacy in outbreak settings. Population-based studies in Quebec following a mass immunization campaign with a plain polysaccharide vaccines in response to an increasing incidence of serogroup C meningococcal disease reported an effectiveness of 83% in 15- to 19-year-olds (De Wals *et al.*, 2001). The Cochrane review of serogroup A polysaccharide vaccines concluded that they are 95% effective for the first year after immunization, with a protective effect persisting through the second and third year after immunization (Patel and Lee, 2005). Therefore, although there are legitimate concerns regarding the administration of repeated doses of plain polysaccharide vaccines to individuals, there is still a useful role for these vaccines in outbreak settings.

Glycoconjugate meningococcal vaccines

In view of the limitations of a T-independent response, glycoconjugate vaccines have been developed in which the polysaccharide capsule is covalently bound to a carrier protein such as a tetanus toxoid, diphtheria toxoid or CRM_{197}, a mutant diphtheria toxoid. Although antigenically unrelated to the polysaccharide capsule, the carrier protein is processed by the polysaccharide specific B-cell and its peptides presented to T-cells in conjunction with MHC class II molecules. This in then able to initiate a T-cell-dependent response to the polysaccharide capsule in germinal centre B-cells. As a result, naive B-cells are stimulated to differentiate into both antibody secreting plasma cells and memory B-cells (Blanchard Rohner *et al.*, 2008).

Such technology was originally employed for the Hib glycoconjugate vaccine, and subsequently adapted for meningococcal vaccines. Three versions of a serogroup C meningococcal vaccine (MenCC) have been developed, 2 of which employ CRM_{197} as the carrier protein, while the third uses tetanus toxoid. These three vaccines have been successfully incorporated into routine childhood immunization schedules in Europe, Canada and Australia (Booy *et al.*, 2007; De Wals, 2004; Trotter and Ramsay, 2007). Two different versions of a tetravalent (serogroup A, C, W-135 and Y) meningococcal vaccine have also been developed (Rennels *et al.*, 2002; Snape *et al.*, 2008), one of which is routinely recommended for all children in the United States aged 10–11 years (CDC, 2007). A serogroup A glycoconjugate vaccine designed for use in Africa is also in development (LaForce *et al.*, 2007).

The expectation that the generation of a T-cell-dependent, germinal centre B-cell response would result in meningococcal vaccines that are immunogenic in infancy and able to generate sustained immunity has generally been met, with notable exceptions. For reasons that are unclear, the currently licensed tetravalent serogroup A, C, W-135 and Y glycoconjugate vaccine (with diphtheria as the carrier protein) is moderately immunogenic in infancy (Rennels *et al.*, 2004). This vaccine is therefore only licensed from the age of 2 years upwards (FDA, 2007), however licensure in younger age groups is being sought for a novel tetravalent meningococcal-CRM_{197} glycoconjugate vaccine (Snape *et al.*, 2008). In addition, bactericidal antibody levels induced by infant immunization with the MenC glyco-onjugate vaccine wane rapidly during the first year of life, and in the absence of a booster dose effectiveness of the vaccine at more than one year after immunization is negligible for a 2, 3, 4 month schedule (Trotter *et al.*, 2004). By contrast, immunization of teenagers with a single dose of the MenC glycoconjugate vaccine results in elevated antibody levels for at least three years (Snape *et al.*, 2006) and effectiveness is maintained at 90% or above more than 1 year after immunization of 3- to 4-year-olds and 11- to 16-year-olds (Trotter *et al.*, 2004). The reasons for MenC glycoconjugate vaccine becoming more immunogenic with increasing age through childhood and adolescence are uncertain, but may include maturation of lymphoid germinal centres or increased survival of plasma cells in the bone marrow (Pihlgren *et al.*, 2006). Alternatively, increased levels of natural priming by nasopharyngeal carriage of MenC or cross reacting antigens with increasing age may play a role.

The induction of immunological memory by MenC glycoconjugate vaccines has been directly demonstrated by the detection of specific memory B-cells in the blood following immunization (Blanchard Rohner *et al.*, 2008; Kelly *et al.*, 2006) and by evidence of priming with the glycoconjugate vaccine resulting in an augmented response to immunization with the plain polysaccharide vaccine compared to vaccine naive controls (MacDonald *et al.*, 1998; MacLennan *et al.*, 2000). Nevertheless, the association of declining bactericidal antibody levels with reduced effectiveness of the vaccine has emphasized that, in the absence of adequate levels of bactericidal antibody in the serum at the time of exposure to the organism, the secondary rise in bactericidal antibodies is not protective. This is supported by evidence that invasive meningococcal disease can occur within the first few days after nasopharyngeal acquisition (Edwards *et al.*, 1977) and that no rise in SBA titres is detectable in teenagers primed with MenC glycoconjugate vaccine until day 5 following exposure to the serogroup C polysaccharide capsule (Snape *et al.*, 2006). In order to maintain an ongoing reduction in the incidence of meningococcal disease, it is therefore essential that immunization schedules are designed to sustain SBA titres above the threshold of protection in susceptible populations.

In contrast to the plain polysaccharide vaccines, the glycoconjugate vaccines have been shown to reduce nasopharyngeal carriage of the meningococcus (Maiden and Stuart, 2002), and are therefore able to induce herd immunity (Ramsay *et al.*, 2003). This may be related to the fact that administration of MenCC generating higher serogroup C specific salivary IgG concentrations than MenACP (Zhang *et al.*, 2000), and has had a considerable impact on the cost effectiveness of mass immunization with this vaccine (Trotter and Edmunds, 2006).

Immunogenicity of alternating plain polysaccharide and protein–polysaccharide conjugate meningococcal vaccines

Studying the immunogenicity of sequential immunization with alternating plain polysaccharide or glycoconjugate meningococcal vaccines is important both from a public health perspective and as a tool to help understand the mechanisms underlying the immune response to these vaccines. With one exception (Lakshman *et al.*, 2002), most studies show that adults previously immunized with MenACP generate lower MenC SBA GMTs following MenCC immunization than MenACP naïve controls (Richmond *et al.*, 2000; Southern *et al.*, 2004). This contrasts with the concentrations of MenC IgG observed following MenCC, which are higher in those previously receiving MenACP (Richmond *et al.*, 2000). In children, those receiving MenACP at less than one year of age developed lower MenC SBA GMTs following subsequent MenCC than historical MenACP naïve controls. This was not the case for children receiving MenACP at more than 1 year of age, however no account was taken of the potential differences in pre-immunization SBA GMTs between the two groups (Borrow *et al.*, 2001b). It is clear, however, that following initial immunization with MenACP, the glycoconjugate vaccine is able to boost SBA titres more effectively than a further dose of MenACP (Richmond *et al.*, 2000), and that this can also 'reset' the immune system so that the response to a subsequent dose of MenACP is comparable to those receiving MenACP for the first time (MacLennan *et al.*, 1999). It is therefore suggested that the glycoconjugate vaccine is able to 'rescue' the immunological hyporesponsiveness induced by the plain polysaccharide vaccine. Conversely, it appears that the plain polysaccharide vaccine can induce hypo-responsiveness even after priming with the glycoconjugate vaccine (MacLennan *et al.*, 2001).

Of relevance to this last observation is a study demonstrating that, in teenagers previously immunized with MenCC, a small and transient rise in MenC specific memory B-cells is detectable in the circulation seven days after subsequent immunization with MenACP, compared with a relatively large and sustained increase in circulating memory B-cells following a second dose of MenCC (Kelly *et al.*, 2006). It is suggested that the dose of the plain polysaccharide vaccine stimulates the conversion of MenC specific memory B-cells to plasma cells without provoking further proliferation of memory B-cells, thus depleting the pool of MenC specific memory B-cells induced by the initial dose of the glycoconjugate vaccine. By contrast, the second dose of MenCC would appear to stimulate both the production of MenC specific plasma cells, and the production of new memory B-cells.

Summary

In view of their relatively poor immunogenicity in childhood, the lack of sustained effectiveness and the concerns regarding immunological hypo-responsiveness, it is likely that the plain polysaccharide meningococcal vaccines will soon be superseded by the glycoconjugate vaccines that are able to induce a T-dependent response in antigen specific B-cells. The development of relatively inexpensive serogroup A specific glycoconjugate vaccines designed for use in sub-Saharan Africa offers the hope that a sustained reduction in the global burden of meningococcal disease may be possible.

Antibody responses to serogroup B meningococcal vaccines

A summary of the responses to different serogroup B meningococcal OMV vaccines is given in Table 10.1. This is discussed in depth in Chapter 11.

T-cell responses to meningococcal proteins

Introduction

The humoral immune response is central to natural and vaccine-induced protection against *N. meningitidis*, since disease occurs in those who lack SBA (Hobbs *et al.*, 1967; Salit, 1981) or complement (D'Amelio *et al.*, 1992; Ross and Densen, 1984). In the context of induction of protective antibody following carriage or immunization, cellular immune responses to meningococcal antigens are of interest and importance. Furthermore, these responses are central to the inflammatory response that occurs during the pathophysiology of invasive meningococcal disease.

T-helper (T_H) cells are the major immune cells responsible for initiation, regulation, and memory of the immune response. T_H cells indirectly induce killing of bacteria by activating phagocytes (Paulnock, 1992), and are required for immunoglobulin class switching and affinity maturation (Ada, 1990). These cells recognize antigen peptides associated with MHC class II on antigen-presenting cells (APCs), such as dendritic cells (DCs). Depending on the nature of the T-cell help, contact with antigen can either lead to a predominantly pro-inflammatory T_H1-type response, an antibody-associated T_H2 response (Pollard *et al.*, 1999a) or, in the absence of a 'danger signal', tolerance (Groux *et al.*, 1997; Taams *et al.*, 2001).

T-cell responses to meningococcal outer membrane proteins

Early studies focussed on the outer membrane protein (OMP) PorA as the dominant meningococcal T-cell-stimulating protein because anti-meningococcal antibodies are frequently directed against it in immune individuals. In one study, 30 healthy adults with pre-existing antibodies to *N. meningitidis* and no previous invasive meningococcal disease, all had proliferative T-cell responses to PorA, which were greater than their responses to PorB, another OMP (Wiertz *et al.*, 1991). Conversely, 6 control cases who did not have anti-meningococcal antibodies did not have proliferative T-cell responses. Similar T-cell proliferative responses have been found with the Por protein of *N. gonorrhoeae* (Simpson *et al.*, 1999a), which is homologous to the meningococcal PorA and PorB proteins. Flow cytometry analysis revealed that T-cells showing strong proliferative responses to PorA peptides were of the $CD4^+$ T_H phenotype, rather than the $CD8^+$ cell type, as expected.

Detailed analysis of the potential T-cell epitopes of PorA by simultaneous multiple peptide synthesis revealed that most of the immunodominant T-cell epitopes were located outside the variable regions of PorA (Wiertz *et al.*, 1992) in surface loops 1, 3 and 5. Some B-cell and T-cell epitopes of PorA mapped identical regions of the protein (Wiertz *et al.*, 1991), the B-cell epitopes being located predominantly in surface loops 1, 4 and 5 (McGuinness *et al.*, 1990; van der Ley *et al.*, 1991). A number of T-cell epitopes from PorA show homology to constant regions of PorB, as well as the gonococcal Por protein. Some T-cell epitopes of PorA appear to be HLA-DR restricted (Wiertz *et al.*, 1991), which would have implications for vaccine design. These data are consistent with those found from work in mice, where immunization with a vaccine containing OMPs complexed to capsular polysaccharide revealed cross-reactive T-cell epitopes among PorA and PorB (Lifely *et al.*, 1992). T-cell epitope mapping of PorB in mice has revealed at least three murine T-cell epitopes, which are located in the highly conserved transmembrane regions, which share homology with PorA and the gonococcal Por (Delvig *et al.*, 1997). This may account for the cross-reactivity seen with these OMPs.

Further evidence of the T-cell immunogenicity of PorA has come from studies of OMV vaccines in humans. Different OMV vaccines manufactured in Norway and The Netherlands stimulated strong proliferative T-cell responses following immunization, reflecting antibody responses in vaccinees (Naess *et al.*, 1998; Oftung *et al.*, 1999; van der Voort *et al.*, 1997). PorA-negative OMVs derived from mutant versions of strains used to produce a hexavalent PorA OMV vaccine showed diminished responses compared to PorA-positive strains (van der Voort *et al.*, 1997), demonstrating that PorA was the immunodominant T-cell antigen, although PorB also elicited T-cell proliferation. The positive correlation observed between T-cell responses and anti-OMV antibody levels in these studies is consistent with the assumption that the antibody response is T-cell dependent.

Interestingly, the immunostimulatory effect of OMV vaccines was initially believed to be the result of their structure and the presence of LOS. However, DC-driven T-cell proliferation can also be enhanced by outer membrane complexes that are deficient in LOS (Al-Bader *et al.*, 2003). These findings confirm that components of the outer membrane other than LOS, such as OMPs or capsular polysaccharide, possess immunoadjuvant activity and can significantly modulate the antigen-presenting capacity of DCs to initiate and promote differentiation of primary T-cell responses.

The outcome of surface immunoglobulin-initiated B-cell activation is strongly dependent on the helper signals provided by $CD4^+$ T_H2 cells. Hence, the significant T-cell immunodominance of PorA presents an attractive explanation for the high antibody titres to these proteins observed following immunization. Further evidence for this is provided by a study in which recombinant PorA (rPorA) induced DC maturation, reflected by reduced receptor-mediated endocytosis, increased production of the chemokines interleukin-8 (IL-8), RANTES, macrophage inflammatory protein-1 alpha (MIP-1α) and MIP-1β, and augmented expression of the surface markers CD40, CD54, CD80, CD86 and MHC class II molecules, indicating T-cell activation (Al-Bader *et al.*, 2003). However, rPorA induced either low level or no significant secretion of pro-inflammatory cytokines from DCs, which would be the hallmark of a T_H1 type response. In addition, rPorA appeared to inhibit the production of the p70 subunit of IL-12 that results from the interaction between CD40 on the DC and CD40-ligand on T-cells, thereby directing T-cell differentiation towards a T_H2 type response. These indicate that PorA is important in determining the levels of DC activation and the nature of the T_H response.

Subsequent work has considered other meningococcal OMPs as potential T-cell immunogens. Antibodies directed against the opacity-associated adhesin (Opa) proteins have been detected in human serum following

Table 10.1 Summary of results of serogroup B meningococcal OMV vaccine trials

Vaccine and developer	Vaccine constituents	Population and number of doses		Immunogenicity: Percentage of vaccinees with a four fold increase in SBA titre	Immunogenicity: Percentage of participants with a bactericidal titre higher or equal to 1:4	Efficacy
Finlay Institute, Cuba	VA-MENGOC-BC Aluminium hydroxide adsorbed OMV from CU385 (B:4:P1.15) + C polysaccharide	Cuban teenagers[1]	2 doses			83%, 16 months post vaccination
		Cuban infants[2]	2 doses	27%		
		Chilean infants under 1 year [3]	2 doses	56% against the vaccine strain		
			3 doses	90% against the vaccine strain		
		Chilean toddlers 2–4 years [3]	2 doses	38% against the vaccine strain		
			3 doses	78% against the vaccine strain		
		Chilean adults 17–30 years [3]	2 doses	53% against the vaccine strain		
			3 doses	67% against the vaccine strain		
		Brazilian children 4–7 years[4]	2 doses			74%
		Brazilian children 2–4 years[4]	2 doses			47%
		Brazilian children under 2 years[4]	2 doses			-37%
		Icelandic teenagers 15–20 years[5]	2 doses	15% against the vaccine strain		
			3 doses	44% against the vaccine strain		

NIPH, Norway	MenBVac Aluminium hydroxide adsorbed OMV from 44/76 (B:15:P1.7,16)	Norwegian teenagers[6]	2 doses			57%, 29 months post vaccination
		Icelandic teenagers 15-20 years[5]	2 doses	47% against the vaccine strain		
			3 doses	84% against the vaccine strain		
		Chilean infants under 1 year [3]	2 doses	96% against the vaccine strain		
			3 doses	98% against the vaccine strain		
		Chilean toddlers 2–4 years[3]	2 doses	75% against the vaccine strain		
			3 doses	98% against the vaccine strain		
		Chilean adults 17–30 years[3]	2 doses	88% against the vaccine strain		
			3 doses	96% against the vaccine strain		
NVI, Netherlands	HexaMen Aluminium phosphate adsorbed combination of 2 different OMVs expressing 6 PorA proteins (P1.7,16; P1.5c,10; P1.5,2; P1.7h,4; P1.19,15; P1.12,13)	Infants[7]	4 doses	78–95% depending on PorA expressed on SBA target strain	98–100% depending on PorA expressed on SBA target strain	
		Toddlers[8]	3 doses	After 3rd dose for both 16–100% depending on PorA expressed on SBA target strain	29.5–100% low dose 27.5- 98% high dose Depending on strain	
		Children[8]	3 doses		19–100% low dose 16.1–95.2% high dose Depending on strain	
NVI, Netherlands	Aluminium hydroxide or aluminium phosphate adsorbed OMVs expressing one PorA protein (P1.7h,4)	Toddlers using vaccine with $AlPO_4$ adjuvant [9]	3 doses	96.4%	Immunogenic in Dutch toddlers (55–80% response). Response dependent on adjuvant used	
			4 doses	100%		

Table 10.1 Continued

Vaccine and developer	Vaccine constituents	Population and number of doses		Immunogenicity: Percentage of vaccinees with a four fold increase in SBA titre	Immunogenicity: Percentage of participants with a bactericidal titre higher or equal to 1:4	Efficacy
New Zealand government; NIPH, Norway; Novartis Vaccines	MeNZB Aluminium hydroxide adsorbed OMV from NZ98/254 B:4:P1.7,16	New Zealand adults[10]	3 doses	96%		
		New Zealand 8-12 years[10]	3 doses	76%		
		New Zealand 16-24 months[10-12]	3 doses	75%	16 months post 3rd dose 34%	
		New Zealand 6-8 months[10,11]	3 doses	74%	After 7 months post 3rd dose 28%	
		New Zealand 6-10 weeks[11]	3 doses	53%	After 4–5 months post 3rd dose 27%	
Finlay Institute, Cuba; GSK Belgium	VA-MENGOC-BC + Aluminium hydroxide absorbed to OMV from NZ228/98 B:4:P1.19.15	Teenagers[13]	3 doses	51–66% against homologous vaccine strains 28–62% against heterologous strains		

[1]Sierra *et al.* (1991), [2]Morley *et al.* (2001), [3]Tappero *et al.* (1999), [4]Milagres *et al.* (1998), [5]Perkins *et al.* (1998), [6]Bjune *et al.* (1991), [7]Cartwright *et al.* (1999) [8]de Kleijn *et al.* (2000), [9]de Kleijn *et al.* (2000), [10]Oster *et al.* (2005), [11]Oster *et al.* (2007), [12]Wong *et al.* (2007), [13]Boutriau *et al.* (2007).

meningococcal infection (Mandrell and Zollinger, 1989; Sjursen *et al.*, 1990) and after immunization with vaccines containing outer membrane complexes or vesicles (Mandrell and Zollinger, 1989; Milagres *et al.*, 1998), suggesting they may be involved in the development of protective immunity. A comparison of the OMPs PorA, PorB, Opa and Opc showed that Opa was more immunogenic for T-cells than the other OMPs, including PorA (Wiertz *et al.*, 1996). In most cases the proliferative responses elicited by Opc were also higher than responses observed for PorA, although not statistically different. As had been found for PorA, many peptides from the conserved regions of Opa showed positive T-cell responses (Wiertz *et al.*, 1996). Taken in combination with previous results, this appeared to establish a hierarchy of T cell immunogenicity within the meningococcal OMPs: Opa > Opc ≈ PorA > PorB.

Conversely, when T_H cells were exposed to OMVs from Opa-expressing bacteria, their activation and proliferation in response to a variety of stimuli were effectively halted via Opa binding to CEACAM1, a member of the carcinoembryonic antigen cell adhesion molecule family (Lee *et al.*, 2007). This potent immunosuppressive effect suggested that localized infection generates a 'zone of inhibition' resulting from the diffusion of outer membrane blebs into the surrounding tissues. There was also a suggestion in the early studies of PorA that prior immunization with an OMV vaccine reduced the proliferative T-cell response (Wiertz *et al.*, 1991). This may be due to a reduction in carriage, or a down-regulation of OMP-specific T-cells by the vaccine. Furthermore, infection by gonococci expressing CEACAM-binding Opa suppressed expression of the early T-cell activation marker CD69 and the subsequent proliferation of $CD4^+$ T-cells in response to various activating stimuli (Boulton and Gray-Owen, 2002). This effect was also mediated via binding to CEACAM1. CEACAM1-dependent effects were shown to result from a specific arrest in cell division, rather than from infection-induced cytotoxicity in this case.

A proposed mechanism by which T-cell proliferation can be suppressed by CEACAM1 ligation is based on findings that the cytoplasmic domain of CEACAM1 contains tyrosine residues that potentially interact with protein tyrosine kinases of the Src family, as well as an immunoreceptor tyrosine inhibition motif (ITIM) interacting with the protein tyrosine phosphatases SHP-1 and SHP-2 (Hauck *et al.*, 1998). Activation of the T-cell receptor (TCR) would then activate Src kinase, allowing tyrosine phosphorylation of CEACAM1. Ligation of CEACAM1 by Opa, however, would allow recruitment of SHP-1 and SHP-2 to ITIM, thereby increasing the activation threshold for T lymphocytes by dephosphorylating CEACAM1. CEACAM1 ligation to gonococcal Opa has been shown to induce SHP-1 and SHP-2 to bind this receptor, and this binding is dependent on TCR activation (Boulton and Gray-Owen, 2002).

Additional experiments are required to support the conclusion that the high antibody titres to the Opa proteins are indeed due to Opa-specific T_H cells. Furthermore, the analysis of cytokine profiles in cell culture supernatants of T-cells activated with the various meningococcal OMPs or peptides derived there from may provide evidence for the existence of immunomodulation of the B-cell response by Opa-specific T-cells.

T-cell responses to other meningococcal proteins

A major disadvantage of OMP antigens as meningococcal vaccine candidates is their substantial inter-strain variability (Callaghan *et al.*, 2006; Frasch *et al.*, 1985). Indeed, this variability is used for serological typing. OMP-based vaccination may result in only strain-specific protection. To circumvent this potential problem, multiple attempts have been made to try and identify T-cell epitopes within non-variable proteins that are conserved between genetically diverse strains.

IgA_1 protease is an exoenzyme of pathogenic neisseriae, which is secreted with an amphipathic α-protein, and cleaves human IgA_1. It has been found to stimulate the induction of the pro-inflammatory cytokines tumour necrosis factor-alpha (TNF-α), IL-1β, IL-6, and IL-8 from peripheral blood mononuclear cells (PBMCs) (Lorenzen *et al.*, 1999). Rapid cytokine induction indicated a role for monocytes and/or macrophages, but this response was enhanced in the presence of T-cells. In a parallel experiment, CD69 was slightly up-regulated, suggesting T-cell activation, but the low number of IL-2 receptor–positive cells suggested that no large T-cell population was activated. In a subsequent study, the majority of PBMCs obtained from 20 volunteers with high serum antibody titres against IgA_1 protease preparations exhibited a proliferative T-cell response to synthetic peptides with sequences derived from α-proteins (Jose *et al.*, 2000). Similar findings have been discovered with the gonococcal IgA_1 protease, which stimulates activation of $CD4^+$ and $CD8^+$ T-cells, with a bias towards a T_H1-based pro-inflammatory response (Tsirpouchtsidis *et al.*, 2002).

T-cell proliferation to meningococcal proteins, in both normal healthy adult volunteers and patients in the convalescent phase following meningococcal infection, is higher in iron-restricted conditions (Kizil *et al.*, 1999). The dominant T-cell immunogen in these circumstances is a protein named T-cell stimulating protein A (TspA). Sera from patients recovering from meningococcal disease were shown to contain antibodies specific for TspA, thus demonstrating that TspA is immunogenic for B-cells during natural infection. In a further study, T-cell proliferative responses to recombinant TspA were detected in PBMCs of convalescent patients and carriers, confirming that TspA-specific T-cell responses were stimulated by invasive disease and nasopharyngeal colonization (Robinson *et al.*, 2005). Following stimulation of PBMCs from convalescent patients with TspA, increased frequencies of both T_H1 and T_H2 cells were observed, indicating that invasive meningococcal disease can induce an unbiased T-helper subset response. The response of PBMCs from the carriers to TspA stimulation, however, was very weak, and the frequencies of cytokine-positive $CD4^+$ cells were not significantly greater than the frequencies in unstimulated control cultures.

A meningococcal genomic expression library has also been screened for potent $CD4^+$ T-cell antigens, using peripheral blood lymphocytes (PBLs) from four patients. One of the most promising positive clones was fully characterized and the gene was designated *autA* (auto-transporter A) as its peptide sequence shares molecular characteristics with the auto-transporter family of proteins (Ait-Tahar *et al.*, 2000). Purified recombinant AutA recalled strong secondary T-cell responses in PBLs of patients and some healthy donors, and induced strong primary T-cell responses in healthy donors. Immunoblots with rabbit polyclonal antibodies to recombinant AutA demonstrated the conserved nature, B-cell antigenicity and cross-reactivity of AutA amongst meningococci of different serogroups and strains representing different hyperinvasive lineages.

T-cell responses to *Neisseria meningitidis* during invasive disease

Elevated levels of TNF-α, IL-1, IL-6, and IL-8 have been found in the cerebrospinal fluid (CSF) and serum of patients with invasive meningococcal disease (Kornelisse *et al.*, 1995; Ohga *et al.*, 1994; van Deuren *et al.*, 1995; Waage *et al.*, 1989). These cytokines are produced by activated monocytes and macrophages, and promote an inflammatory immune response, which can ultimately lead to septic shock and death (Brouckaert *et al.*, 1993; Dinarello, 1997; Harada *et al.*, 1994; Kishimoto *et al.*, 1995; Kornelisse *et al.*, 1995; Ohga *et al.*, 1994; van Deuren *et al.*, 1995). This cytokine profile is associated with a T_H1-type response, and evidence for this is further supported by a study of 35 children with bacterial meningitis (of which 13 were caused by *N. meningitidis*), who had elevated levels of IL-12 and gamma-interferon (IFN-γ) (T_H1-associated cytokines) in both serum and CSF (Kornelisse *et al.*, 1997). In a different study, however, cytokine production by PBMCs isolated from 49 children convalescing from invasive meningococcal infection was skewed towards a T_H1 response (low IL-10/IFN-γ ratio) in the youngest children, while older children produced a more T_H2-like cytokine profile (higher IL-10/IFN-γ ratio). The exact cellular sources of cytokines were not identified in these studies, and it is therefore possible that the cytokines were being produced by cells other than T-cells. In particular, it has been shown that IL-10 is produced by cells other than T_H2 cells (Moore *et al.*, 2001; Trinchieri, 2001). It is likely that both a T_H1-type inflammatory response and a T_H2-type antibody-associated response are possible, as demonstrated in the studies of TspA.

The most important component of *N. meningitidis* in stimulating the release of pro-inflammatory cytokines is LOS. However, there is no consistent relationship between the levels of cytokines and endotoxin during invasive meningococcal disease (Prins *et al.*, 1998), suggesting that factors other than endotoxin may be involved in cytokine induction during acute meningococcal infection. This would support the data already described regarding the importance of OMPs in the interaction between the meningococcus and the cellular immune system.

T-cell responses to *Neisseria meningitidis* during carriage

Naturally acquired protective immunity against *N. meningitidis* is thought to partially explain the disparity between the high levels of carriage in the human nasopharynx and the rare incidence of disease (Ala'Aldeen *et al.*, 2000). In one study, proliferative T-cell responses to meningococcal OMVs were induced by tonsil mononuclear cells and were strongly associated with age, with high levels observed in teenagers, followed by a steady rise towards adulthood (Davenport *et al.*, 2003). Furthermore, stronger proliferative responses were observed with a LOS-deficient mutant, suggesting that it is the OMPs that are responsible for this proliferation. Responses were both PorA cross-specific and non-PorA dependent. The PorA cross-specificity may be explained by the location of T-cell epitopes within the conserved regions of the protein (Wiertz *et al.*, 1992), within homologous regions of other proteins, such as PorB, or may be induced by carriage of related commensal species, such as *N. lactamica*.

As with cytokine responses to invasive disease, meningococcal carriage seems to elicit an unbiased T-helper subset response. In a study of undergraduate students in the UK, PBMCs from 15 individuals were analysed for expression of CD69 (to indicate activation), IFN-γ (a T_H1-associated cytokine) and IL-5 (a T_H2-associated cytokine). Following a brief meningococcal stimulation both IFN-γ^+ and IL-5^+ events were detected among the CD69$^+$ CD4$^+$ population (Robinson *et al.*, 2002).

Meningococcal proteins as vaccine adjuvants

The polysaccharide capsule of meningococci is considered to be a T-independent antigen since T-helper (T_H) cells are not thought to be required for the induction of antibody, but the immunological basis for this is unknown (Basten and Feldmann, 1973; Mond *et al.*, 1995a; Mond *et al.*, 1995b; Snapper *et al.*, 1997). As a consequence of this, the immunogenicity of capsular polysaccharide vaccines in infants is poor, and the immunity is of short duration (Goldschneider *et al.*, 1973; Peltola *et al.*, 1977). This problem has been circumvented by conjugation of polysaccharide epitopes to protein or peptide carrier molecules that provide the T-cell epitopes, as in other conjugate vaccines (Schneerson *et al.*, 1980; Weinberg and Granoff, 1988). Using meningococcal proteins or peptides could enhance vaccine effectiveness by acting as protective immunogens in their own right, in addition to being carrier molecules.

Meningococcal outer membrane complexes, which include OMPs, LOS and lipid, have pronounced carrier activity, and have been used for *Haemophilus influenzae* type b (Hib) (Ambrosino *et al.*, 1992; Donnelly *et al.*, 1990; Jelonek *et al.*, 1993; Liu *et al.*, 1992; Perez-Melgosa *et al.*, 2001), meningococcal group C (Fukasawa *et al.*, 1999), and pneumococcal (Vella *et al.*, 1992) capsular polysaccharide conjugate vaccines, in addition to an experimental malaria peptide-based vaccine (Lowell *et al.*, 1988). These complexes have also been found to enhance the immunogenicity of vaccines for viruses (Slavik *et al.*, 1991), for human melanomas (Livingston *et al.*, 1993) and whole-cell tumours (Mesa *et al.*, 2006). Moreover, the carrier facilitated induction of significant antibody responses against Hib even in 2-month-old infants (Shapiro *et al.*, 1989), and to pneumococcal polysaccharide in young monkeys (Vella *et al.*, 1992), more potently than the commonly used protein and vaccine carrier diphtheria toxin. In addition, meningococcal OMPs have been found to have adjuvant activity, as described for the melanoma cancer vaccine (Livingston *et al.*, 1993), as well as mitogenic activity for lymphocytes in previous experimental Hib polysaccharide vaccines (Liu *et al.*, 1992). In the latter case, PorB was noted to be the immunodominant OMP. The precise mechanisms of these broad immunomodulatory effects remain largely unknown, but may involve up-regulation of the co-stimulatory molecule B7–2 (CD86) on APCs due to interaction with the pattern recognition receptor TLR2 (Massari *et al.*, 2002; Massari *et al.*, 2006; Wetzler *et al.*,

1996) or influence on mechanisms of antigen and presentation.

Summary

Immunodominant T-cell epitopes derived from immunogenic regions of OMPs appear to be the most promising in the development of a potential meningococcal vaccine, although T-cell responses to other meningococcal proteins are being increasingly recognized. Future vaccines should consist of well-characterized antigens that elicit long-lasting cross-protective immunity. T-cell determinants that could be recognized in the context of the vast majority of MHC class II haplotypes should be incorporated. It is hypothesized that there may be major benefit of using meningococcal T-cell epitopes for the induction of a specific memory immune response, compared to using tetanus or diphtheria-related carrier proteins, though direct evidence to support this is lacking. In addition, vaccination of adults who are at risk for meningococcal disease could easily lead to hyperimmunization with tetanus or diphtheria toxoid when conjugates of these proteins are injected. Thus, the identification and characterization of further meningococcal T- and B-cell protein antigens is a priority to enable the design of optimized vaccines against *N. meningitidis* infection. There is clear evidence of a T-cell response following disease, carriage and immunization, but the importance and phenotype of the optimal T-cell response for maximum and sustained immunity to *N. meningitidis* remains uncertain.

Acquired gonococcal immunity

N. gonorrhoeae is typically characterized by localized inflammation of the male urethra (urethritis) and the female endocervix in women (endocervicitis), which is often less symptomatic. Ascending infections in women can lead to complications such as pelvic inflammatory disease (PID), characterized most commonly by endometritis, and salpingitis, in 10–40% of infected women. Sequelae of PID include infertility, ectopic pregnancy, fetal wastage, neonatal blindness and disseminated gonococcal infections (DGI) (Hook *et al.*, 1990). Infection with *N. gonorrhoeae* also significantly increases the risk of infection with human immunodeficiency virus (HIV), with increased susceptibility to infection in HIV-negative patients and increased transmission by HIV-positive patients (Cohen *et al.*, 1997; Levine *et al.*, 1998). Innate immune response to gonococcal components are reviewed in Chapter 7 and will not be discussed here.

Does acquired gonococcal immunity exist?

The possibility that people develop natural acquired immunity to the gonococcus is controversial, as opposed to the solid data for the meningococcus discussed in the first part of this chapter. This is especially true for mucosal gonococcal disease, mainly characterized by urethritis. The nature of mucosal gonococcal disease likely accounts for the lack of easily measurable acquired immune response to the organism. Minimal inflammation is induced during disease, especially in women, where a majority of gonococcal cervicitis is asymptomatic (Sparling *et al.*, 1990). Moreover, as described in Chapter 8, structures on the gonococcus are known to down regulate complement activity, including LOS sialylation, and Factor H or C4bp binding to the porin (Massari *et al.*, 2003; Ram *et al.*, 1999), which probably decreases recognition of the bacteria by the immune system.

However, there are some studies that suggest that during mucosal infection causing urethritis, there is potential induction of protective immunity. Historical analysis of porin serotypes in Kenyan prostitutes performed by Brunham and Plummer in the early 1980s demonstrated that these individuals, though often re-infected with the gonococcus was rarely re-infected with organisms of the same porin (Protein I) serotype, suggesting some porin specific immunity was induced (Brunham *et al.*, 1985). Newer studies have shown that patients with gonococcal urethritis could develop specific antibodies towards gonococcal Protein I (Simpson *et al.*, 1999b), while another study found that a separate cohort of patients did not develop gonococcal specific antibodies (Hedges *et al.*, 1999). However, the latter study suffered from the lack of specificity of the immune response measured to the organism, as they found that even control patients had extremely high levels of antibodies recognizing the bacteria, and therefore could not detect any elevation in such responses as compared to baseline. Interestingly, a few studies found that patients with mucosal gonococcal disease did develop a humoral response towards gonococcal iron regulated proteins, which are known to be up-regulated *in vivo* (Agarwal *et al.*, 2005). Bactericidal ability of these antibodies, however, have never been demonstrated, so their protective ability is in doubt.

Is there a role for IgA?

One other area of controversy is the potential role of anti-gonococcal mucosal IgA antibodies in protection or lack of protection against mucosal disease. It is well known that patients with a complete lack of IgA are no more susceptible to mucosal infections as opposed to humans with normal levels of IgA. Interestingly, Apicella *et al.* (1986) found that if the gonococcus is covered in IgA, this may block the binding of more functionally active immunoglobulin isotypes, namely IgG, and may form a type of 'coat' preventing complement deposition . A similar phenomenon was also observed for the meningococcus (Jarvis and Griffiss, 1991). Though an intriguing *in vitro* concept, this has not been shown, thus far, to have any *in vivo* implications.

Immunity in complicated gonococcal infections

Despite the lack of data regarding induction of immunity for mucosal disease, there is much greater evidence that in both DGI and PID in women, anti-gonococcal humoral responses are induced. Similar to the studies by Brunham and Plummer mentioned above, there is early evidence that PID is less likely to occur with strains of the same PI serotype (Buchanan *et al.*, 1980). Moreover, Plummer has found that the pre-existence of antibodies to the opacity proteins present on the gonococcus (Opa or Protein 2 in older terminology) reduced the risk of gonococcal PID (Plummer *et al.*, 1994). It is likely, that due to the increased invasiveness of this disease and enhanced inflammatory response seen, as indicated by endometritis and salpingitis, a greater recruitment of immune cells and enhanced acquired immunity occurs.

Approximately 1–3% of patients, historically, with gonococcal disease develop DGI. Tenosynovitis and skin vesicles or suppurative joint infections characterize this disease. Unlike other infections that cause joint infections, only 50–60% have positive blood cultures (compared with >90%) (O'Brien *et al.*, 1983; Rice and Goldenberg, 1981). It has been shown that patients with disseminated disease develop antibodies that recognize outer membrane components of the organisms, including the porin, Protein I and most interestingly the reduction modifiable protein, Rmp, which was classically known as protein III (Blake *et al.*, 1989a; Brooks *et al.*, 1976; Christensen *et al.*, 1976; Hadfield and Glynn, 1982; Lammel *et al.*, 1985).

Rmp 'blocking' antibodies: implications in vaccine development

A very interesting phenomenon was elucidated by the analysis of the antibody response and bactericidal activity in sera from patients with DGI. Normally, the gonococcus is exquisitely sensitive to complement dependent killing (serum sensitive). Strains that cause DGI are normally not as sensitive to this type of killing and are termed serum resistant. What was found was that normal sera were more able to kill such strains, when gonococcal specific antibodies were added, but if sera from patients with DGI were then added, this bactericidal activity *diminished*. It was found that this was due to the presence of IgG that recognized Rmp on the surface of the gonococcus. If these Rmp specific IgG antibodies were removed, the remaining antibodies in the DGI sera were extremely bactericidal. If Rmp IgG was then added back, the bactericidal activity, again, decreased (Rice *et al.*, 1986; Virji and Heckels, 1988).

The mechanism of this blocking activity was not due to direct steric hindrance and prevention of binding of supposedly bactericidal antibodies, as various studies demonstrated that anti-Protein I antibodies still bound to the organisms even in the presence of the blocking Rmp antibodies. The mechanism of the blocking activity of anti-Rmp antibodies was eventually demonstrated to be due to an effect on complement deposition. When IgG bound to Rmp on the gonococcal surface, the complement cascade initiated by antibodies recognizing other gonococcal surface structures would not proceed to a point to allow the normal insertion of the terminal complement membrane attack complex into the bacterial membrane (Joiner *et al.*, 1985a; Joiner *et al.*, 1985b).

This finding had direct implications in anti-gonococcal vaccine development. Many investigators attempted to use the gonococcal porin, Protein I, as a potential vaccinogen. However, even when attempting to purify this component, significant contamination with Rmp always occurred, as it appears that Rmp associated closely with Protein I on the surface (Gotschlich *et al.*, 1988; Wetzler *et al.*, 1989a). An early vaccine using purified Protein I, contaminated with Rmp, demonstrated that patients immunized with this supposedly pure preparation had serum anti-gonococcal activity that diminished after immunization instead of increasing in activity (Arminjon *et al.*, 1987; Blake *et al.*, 1989b; Gulati *et al.*, 1991).

These results decreased the enthusiasm for the use of Protein I as an anti-gonococcal vaccinogen (Blake and Wetzler, 1995). However, mutant strains of gonococci were developed that lacked Rmp (Wetzler *et al.*, 1989b). Protein I purified from these strains, lacking Rmp or contaminating LOS, were able to induce bactericidal activity in sera from mice and rabbits (Wetzler *et al.*, 1992). However, no further human trials have been attempted. Interestingly, the meningococcal Rmp homolog, RmpM (classically termed class 4 protein), has similar properties to gonococcal Rmp. Anti-RmpM IgG can block the bactericidal activity of sera towards the meningococcus, but the potency of this blockade is less than the blockade induced on the gonococcus (Klugman *et al.*, 1989; Munkley *et al.*, 1991). As discussed in Chapter 11, this is still an issue in development of anti-meningococcal protein based subcapsular antigen based vaccines.

Gonococcal vaccines

There have been a number of failed attempts to develop an anti-gonococcal vaccine. Some of the reasons for these failures are discussed in detail in Chapter 11. One major reason that investigators felt this was a difficult endeavour, was the question of whether a proper mucosal response was induced by these vaccines, especially when delivered parenterally. Another is the lack of any adequate animal mode for the testing of such vaccines as described earlier in this chapter and in Chapter 11. One more issue was the contamination of early vaccine preparations with Rmp and subsequent induction of blocking antibodies as described above. Finally, unlike meningococci, gonococci lack capsular polysaccharides, which have been used successfully for many other bacteria as vaccines.

The earliest attempts to develop a vaccine to prevent gonococcal disease utilized heat-killed organisms. Braude immunized volunteers who had no previous history of exposure to *Neisseria*, and found that the whole organism was able to induce gonococcal specific adherence-blocking IgG in their sera, but the role of such antibodies in actual protection remained unclear(Braude *et al.*, 1985). Early on, Brinton, attempted to use purified pili, the main adherence factor on this organism, as a vaccinogen (Brinton *et al.*, 1980; Brinton *et al.*, 1978; Brinton *et al.*, 1981; Johnson *et al.*, 1991; McChesney *et al.*, 1982). A major issue with these preparations was their lack of significant cross reactivity to other gonococcal strains. This was likely due to the significant hyper-variable antigenic variation that occurs with the pilin subunits (Swanson *et al.*, 1987), but also due to the fact that when purified, the major immuno-dominant epitope of the pilin subunit, which a majority of the induced antibodies bound, was hidden on intact bacteria (Johnson *et al.*, 1991). This elucidated an extremely important caveat when utilizing isolated components of a pathogen as a vaccine candidate, which the induced immune response needs to recognize the intact living organism and not just the vaccinogen itself.

Despite these setbacks, there are substantial reasons to continue to develop an anti-gonococcal vaccine. These include (1) the high incidence of *N. gonorrhoeae* infection in the world (>60 million cases per year), (2) PID is still the major cause of acquired infertility, (3) the association of gonococcal disease and increased HIV infection and (4) the complications associated with asymptomatic cases and antibiotic resistance.

Since these early attempts at developing an anti-gonococcal vaccine, there has been other potential candidates, including recombinant porin, various iron binding proteins other outer membrane components, etc. However, none have proven promising as of yet, mainly because, there is still no agreed upon marker of immunity for gonococcal disease, unlike the meningococcus where bactericidal activity and/or anti-capsular antibody levels are felt to be adequate markers of immunity. There is no absolute evidence that serum bactericidal activity correlates with gonococcal immunity. Until there is, unless human trials are performed, it is difficult to ascertain what gonococcal vaccinogens may be protective.

References

Aase, A., and Michaelsen, T.E. (1994). Opsonophagocytic activity induced by chimeric antibodies of the four human IgG subclasses with or without help from complement. Scand. J. Immunol. *39*, 581–587.

Ada, G.L. (1990). The immunological principles of vaccination. Lancet 335, 523–526.

Agarwal, S., King, C.A., Klein, E.K., Soper, D.E., Rice, P.A., Wetzler, L.M., and Genco, C.A. (2005). The Gonococcal Fur-Regulated tbpA and tbpB Genes Are Expressed during Natural Mucosal Gonococcal Infection. Infect. Immun. 73, 4281–4287.

Ait-Tahar, K., Wooldridge, K.G., Turner, D.P., Atta, M., Todd, I., and Ala'Aldeen, D.A. (2000). Autotransporter A protein of *Neisseria meningitidis*: a potent CD4+ T-cell and B-cell stimulating antigen detected by expression cloning. Mol. Microbiol. *37*, 1094–1105.

Al-Bader, T., Christodoulides, M., Heckels, J.E., Holloway, J., Semper, A.E., and Friedmann, P.S. (2003). Activation of human dendritic cells is modulated by components of the outer membranes of *Neisseria meningitidis*. Infect. Immun. *71*, 5590–5597.

Al-Mazrou, Y., Khalil, M., Borrow, R., Balmer, P., Bramwell, J., Lal, G., Andrews, N., and Al-Jeffri, M. (2005). Serologic responses to ACYW135 polysaccharide meningococcal vaccine in Saudi children under 5 years of age. Infect. Immun. 73, 2932–2939.

Ala'Aldeen, D.A., Neal, K.R., Ait-Tahar, K., Nguyen-Van-Tam, J.S., English, A., Falla, T.J., Hawkey, P.M., and Slack, R.C. (2000). Dynamics of meningococcal long-term carriage among university students and their implications for mass vaccination. J. Clin. Microbiol. *38*, 2311–2316.

Ala'Aldeen, D.A., Stevenson, P., Griffiths, E., Gorringe, A.R., Irons, L.I., Robinson, A., Hyde, S., and Borriello, S.P. (1994). Immune responses in humans and animals to meningococcal transferrin-binding proteins: implications for vaccine design. Infect. Immun. *62*, 2984–2990.

Amadou Hamidou, A., Djibo, S., Elhaj Mahamane, A., Moussa, A., Findlow, H., Sidikou, F., Cisse, R., Garba, A., Borrow, R., Chanteau, S., and Boisier, P. (2006). Prospective survey on carriage of *Neisseria meningitidis* and protective immunity to meningococci in schoolchildren in Niamey (Niger): focus on serogroup W135. Microbes Infect. *8*, 2098–2104.

Ambrosino, D.M., Bolon, D., Collard, H., Van Etten, R., Kanchana, M.V., and Finberg, R.W. (1992). Effect of *Haemophilus influenzae* polysaccharide outer membrane protein complex conjugate vaccine on macrophages. J. Immunol. *149*, 3978–3983.

Amir, J., Louie, L., and Granoff, D.M. (2005). Naturally-acquired immunity to *Neisseria meningitidis* group A. Vaccine 23, 977–983.

Anonymous (2007). Saudi Ministry of Health Requirements.

Apicella, M.A., Westerink, M.A., Morse, S.A., Schneider, H., Rice, P.A., and Griffiss, J.M. (1986). Bactericidal antibody response of normal human serum to the lipooligosaccharide of *Neisseria gonorrhoeae*. J. Infect. Dis. *153*, 520–526.

Arminjon, P., Cadoz, M., Morse, S.A., Rock, J.P., and Sarafian, S.K. (1987). Bactericidal and opsonic activities of sera from individuals immunized with a gonococcal protein I vaccine. Abstracts Annual Meeting American Society of Microbiology, 118.

Basten, A., and Feldmann, M. (1973). The role of soluble factors in T-B-cell collaboration. Adv. Exp. Med. Biol. *29*, 171–178.

Bennett, J.S., Griffiths, D.T., McCarthy, N.D., Sleeman, K.L., Jolley, K.A., Crook, D.W., and Maiden, M.C. (2005). Genetic diversity and carriage dynamics of *Neisseria* lactamica in infants. Infect. Immun. 73, 2424–2432.

Beuvery, E.C., Leussink, A.B., Van Delft, R.W., Tiesjema, R.H., and Nagel, J. (1982). Immunoglobulin M and G antibody responses and persistence of these antibodies in adults after vaccination with a combined meningococcal group A and group C polysaccharide vaccine. Infect. Immun. 37, 579–585.

Black, J.R., Black, W.J., and Cannon, J.G. (1985). Neisserial antigen H.8 is immunogenic in patients with disseminated gonococcal and meningococcal infections. J. Infect. Dis. *151*, 650–657.

Blake, M.S., and Wetzler, L.M. (1995). Vaccines for gonorrhea: where are we on the curve? Trends Microbio.l 3, 469–474.

Blake, M.S., Wetzler, L.M., Gotschlich, E.C., and Rice, P.A. (1989a). Protein III: structure, function and genetics. Clin. Microbiol. Rev. *2*, S60-S63.

Blake, M.S., Wetzler, L.M., Gotschlich, E.C., Rice, P.A., and Atassi, M.Z. (1989b). Developing a gonococcal protein I vaccine. In The Immunobiology of Protein and Peptides V: Mechanisms, Designs and Applications (New York, Plenum Press), pp. 315–327.

Blakebrough, I.S., Greenwood, B.M., Whittle, H.C., Bradley, A.K., and Gilles, H.M. (1982). The epidemiology of infections due to *Neisseria meningitidis* and *Neisseria* lactamica in a northern Nigerian community. J. Infect. Dis. *146*, 626–637.

Blanchard Rohner, G., Snape, M.D., Kelly, D.F., John, T., Morant, A., Yu, L.M., Borkowski, A., Ceddia, F.C., Borrow, R., Siegrist, C.A., and Pollard, A.J. (2008). The magnitude of the antibody and memory B cell responses during priming with a protein-polysaccharide conjugate vaccine in human infants is associated with the persistence of antibody and the intensity of booster response. J. Immunol. *180*, 2165–2173.

Boisier, P., Nicolas, P., Djibo, S., Hamidou, A.A., Tenebray, B., Borrow, R., and Chanteau, S. (2006). Carriage of *Neisseria meningitidis* serogroup W135 ST-2881. Emerg. Infect. Dis. *12*, 1421–1423.

Booy, R., Jelfs, J., El Bashir, H., and Nissen, M.D. (2007). Impact of meningococcal C conjugate vaccine use in Australia. Med J Aust *186*, 108–109.

Borrow, R., Andrews, N., Goldblatt, D., and Miller, E. (2001a). Serological basis for use of meningococcal serogroup C conjugate vaccines in the United Kingdom: reevaluation of correlates of protection. Infect. Immun. *69*, 1568–1573.

Borrow, R., Goldblatt, D., Andrews, N., Richmond, P., Southern, J., and Miller, E. (2001b). Influence of prior meningococcal C polysaccharide vaccination on the response and generation of memory after meningococcal C conjugate vaccination in young children. J. Infect. Dis. *184*, 377–380.

Boulton, I.C., and Gray-Owen, S.D. (2002). Neisserial binding to CEACAM1 arrests the activation and proliferation of CD4+ T lymphocytes. Nat. Immunol. *3*, 229–236.

Braude, A., Wunderlich, A., Chikami, G., Douglas, H., McCutchan, J., and Schoolnik, G.K. (1985). Generation of adherence-blocking immunoglobin G (IgG) by vaccinating women with nonpiliated boiled gonococci: Observations on IgG binding to a high-molecular-weight protein in the outer membrane fraction. In The Pathogenic neisseriae (Washington D.C., American Society of Microbiology), pp. 323–328.

Brinton, C.C., Brown, A., Rogers, K., Guernia, N., Wood, S., Bryan, J., Labik, A., and Polen, S. (1980). Preparation, testing, safety, antigenicity, immunogenicity, and serological specificity of a purified gonococcal pilus vaccine for gonorrhea. In Current Chemotherapy and Infectious Disease Vol. II (Washington, American Society for Microbiology), pp. 1242–1245.

Brinton, C.C., Jr, Bryan, J., Dillon, J.A., Guernia, N., Jackobson, L., Labik, A., Lee, S., Levine, A., Lim, S., McMichael, J., Polen, S.A., Rogers, K., To, A.C.C., To, S.C.M., Brooks, G.F., Gotschlich, E.C., Holmes, K.K., Sawyer, W. D., and Young, F.E. (1978). Uses of pili in gonorrhea control; role of bacterial pili in disease, purification, and properties of gonococcal pili and progress in the development of a gonococcal pilus vaccine for gonorrhea. In Immunobiology of *Neisseria gonorrhoeae* (Washington, D.C., American Society of Microbiology), pp. 155–178.

Brinton, C.C., Wood, S.H., Brown, A., Labik, A.M., Bryan, J.R., Lee, S.W., Polen, S.E., Tramont, E.C., Fields, B., and Weinstein, L. (1981). The development of a neisserial pilus vaccine for gonorrhea and meningococcal meningitis. In Seminars in infectious diseases IV: Bacterial vaccines (New York, Thieme-Stratton), pp. 140–159.

Brooks, G.F., Israel, K.S., and Petersen, B.H. (1976). Bactericidal and opsonic activity against *Neisseria gonorrhoeae* in sera from patients with disseminated gonococcal infection. JInfectDis *134*, 450–462.

Brooks, G.F., Lammel, C.J., Blake, M.S., Kusecek, B., and Achtman, M. (1992). Antibodies against IgA1 protease are stimulated both by clinical disease and asymptomatic carriage of serogroup A *Neisseria meningitidis*. J. Infect. Dis. *166*, 1316–1321.

Brouckaert, P., Libert, C., Everaerdt, B., Takahashi, N., Cauwels, A., and Fiers, W. (1993). Tumor necrosis factor, its receptors and the connection with interleukin 1 and interleukin 6. Immunobiology *187*, 317–329.

Brunham, R.C., Plummer, F., Slaney, L., Rand, F., and DeWitt, W. (1985). Correlation of auxotype and protein I type with expression of disease due to *Neisseria gonorrhoeae*. JInfectDis *152*, 339–343.

Buchanan, T.M., Eschenbach, D.A., Knapp, J.S., and Holmes, K.K. (1980). Gonococcal salpingitis is less likely to recur with *Neisseria gonorrhoeae* of the same principal outer membrane antigen type. AmJObstetGynec *138*, 978–980.

Callaghan, M.J., Jolley, K.A., and Maiden, M.C. (2006). Opacity-associated adhesin repertoire in hyper-invasive *Neisseria meningitidis*. Infect. Immun. *74*, 5085–5094.

Cartwright, K.A., Stuart, J.M., Jones, D.M., and Noah, N.D. (1987). The Stonehouse survey: nasopharyngeal carriage of meningococci and *Neisseria* lactamica. Epidemiol Infect *99*, 591–601.

Caugant, D.A., Tzanakaki, G., and Kriz, P. (2007). Lessons from meningococcal carriage studies. FEMS Microbiol. Rev. *31*, 52–63.

CDC (2007). Revised Recommendations of the Advisory Committee on Immunization Practices to Vaccinate All Persons Aged 11 – 18 Years with Meningococcal Conjugate Vaccine. MMWR – Morbidity and Mortality Weekly Report *56*, 794–795.

Ceyhan, M., Yildirim, I., Balmer, P., Riley, C., Laher, G., Andrews, N., Borrow, R., Kurt, N., Turgut, M., Aydogan, A., Ecevit, C., Uysal, G., and Schultze, V. (2007). Age-specific seroprevalence of serogroup C meningococcal serum bactericidal antibody activity and serogroup A, C, W135 and Y-specific IgG concentrations in the Turkish population during 2005. Vaccine *25*, 7233–7237.

Christensen, K.K., Christensen, P., Mardh, P.A., and Westrom, L. (1976). Quantitation of serum antibodies to surface antigens of *Neisseria gonorrhoeae* with radiolabeled protein A of Staphylococcus aureus. JInfectDis *134*, 317–323.

Cohen, M.S., Hoffman, I.F., Royce, R.A., Kazembe, P., Dyer, J.R., Daly, C.C., Zimba, D., Vernazza, P.L., Maida, M., Fiscus, S.A, and Eron, J.J., Jr (1997). Reduction of concentration of HIV-1 in semen after treatment of urethritis: implications for prevention of sexual transmission of HIV-1. AIDSCAP Malawi Research Group. Lancet *349*, 1868–1873.

D'Amelio, R., Agostoni, A., Biselli, R., Brai, M., Caruso, G., Cicardi, M., Corvetta, A., Fontana, L., Misiano, G., Perricone, R., Quinti, I., Schena, F.P., Stroffolini, T., and Tedesco, F.(1992). Complement deficiency and antibody profile in survivors of meningococcal meningitis due to common serogroups in Italy. Scand. J. Immunol. *35*, 589–595.

Davenport, V., Guthrie, T., Findlow, J., Borrow, R., Williams, N.A., and Heyderman, R.S. (2003). Evidence for naturally acquired T-cell-mediated mucosal immunity to *Neisseria meningitidis*. J. Immunol. *171*, 4263–4270.

De Wals, P. (2004). Meningococcal C vaccines: the Canadian experience. Pediatr. Infect. Dis. J. *23*, S280–284.

De Wals, P., De Serres, G., and Niyonsenga, T. (2001). Effectiveness of a mass immunization campaign against serogroup C meningococcal disease in Quebec. JAMA *285*, 177–181.

Dellicour, S., and Greenwood, B. (2007). Systematic review: Impact of meningococcal vaccination on pharyngeal carriage of meningococci. Trop. Med. Int. Health *12*, 1409–1421.

Delvig, A.A., Rosenqvist, E., Oftung, F., and Robinson, J.H. (1997). T-Cell epitope mapping the PorB protein of serogroup B *Neisseria meningitidis* in B10 congenic strains of mice. Clin. Immunol. Immunopathol. *85*, 134–142.

Devi, S.J., Schneerson, R., Egan, W., Vann, W.F., Robbins, J.B., and Shiloach, J. (1991). Identity between polysaccharide antigens of Moraxella nonliquefaciens, group B *Neisseria meningitidis*, and *Escherichia coli* K1 (non-O acetylated). Infect. Immun. *59*, 732–736.

DeVoe, I.W. (1976). Egestion of degraded meningococci by polymorphonuclear leukocytes. J. Bacteriol. *125*, 258–266.

Dinarello, C.A. (1997). Interleukin-1. Cytokine Growth Factor Rev. *8*, 253–265.

Donnelly, J.J., Deck, R.R., and Liu, M.A. (1990). Immunogenicity of a *Haemophilus influenzae* polysaccharide-*Neisseria meningitidis* outer membrane protein complex conjugate vaccine. J. Immunol. *145*, 3071–3079.

Edwards, E.A., Devine, L.F., Sengbusch, G.H., and Ward, H.W. (1977). Immunological investigations of meningococcal disease. III. Brevity of group C acquisition prior to disease occurrence. Scand J. Infect. Dis. *9*, 105–110.

Espin Rios, I., Garcia-Fulgueiras, A., Navarro Alonso, J.A., Vazquez Moreno, J., Rodriguez Gonzalez, T., Navarro Sanchez, C., and Perez-Flores, D. (2000). Seroconversion and duration of immunity after vaccination against group C meningococcal infection in young children. Vaccine *18*, 2656–2660.

Estabrook, M.M., Baker, C.J., and Griffiss, J.M. (1993). The immune response of children to meningococcal lipooligosaccharides during disseminated disease is directed primarily against two monoclonal antibody-defined epitopes. J. Infect. Dis. *167*, 966–970.

FDA (2007). Product Approval Information, Meningococcal Polysaccharide (Serogroups A, C, Y and W-135) Diphtheria Conjugate Vaccine.

Filice, G.A., Hayes, P.S., Counts, G.W., Griffiss, J.M., and Fraser, D.W. (1985). Risk of group A meningococcal disease: bacterial interference and cross-reactive bacteria among mucosal flora. J. Clin. Microbiol. *22*, 152–156.

Findlow, J., Balmer, P., Yero, D., Niebla, O., Pajon, R., and Borrow, R. (2007). *Neisseria* vaccines 2007. Expert Rev Vaccines *6*, 485–489.

Finne, J., Leinonen, M., and Makela, P.H. (1983). Antigenic similarities between brain components and bacteria causing meningitis. Implications for vaccine development and pathogenesis. Lancet *2*, 355–357.

Finney, M., Vaughan, T., Taylor, S., Hudson, M.J., Pratt, C., Wheeler, J.X., Vipond, C., Feavers, I., Jones, C., Findlow, J., Borrow, R., and Gorringe, A. (2008). Characterization of the key antigenic components and pre-clinical immune responses to a meningococcal disease vaccine based on Neisseria lactamica outer membrane vesicles. Hum. Vaccin. *3*, 23–30.

Flaegstad, T., Gutteberg, T., and Kristiansen, B.E. (1990). Antibodies to meningococci in children with meningococcal disease. Scand. J. Infect. Dis. *22*, 547–551.

Frasch, C.E., Zollinger, W.D., and Poolman, J.T. (1985). Serotype antigens of *Neisseria meningitidis* and a proposed scheme for designation of serotypes. Rev. Infect. Dis. *7*, 504–510.

Fukasawa, L.O., Gorla, M.C., Schenkman, R.P., Garcia, L.R., Carneiro, S.M., Raw, I., and Tanizaki, M.M. (1999). *Neisseria meningitidis* serogroup C polysaccharide and serogroup B outer membrane vesicle conjugate as a bivalent meningococcus vaccine candidate. Vaccine *17*, 2951–2958.

Glode, M.P., Robbins, J.B., Liu, T.Y., Gotschlich, E.C., Orskov, I., and Orskov, F. (1977). Cross-antigenicity and immunogenicity between capsular polysaccharides of group C *Neisseria meningitidis* and of *Escherichia coli* K92. J. Infect. Dis. *135*, 94–104.

Gold, R., Goldschneider, I., Lepow, M.L., Draper, T.F., and Randolph, M. (1978). Carriage of *Neisseria meningitidis* and *Neisseria* lactamica in infants and children. J. Infect. Dis. *137*, 112–121.

Gold, R., Lepow, M.L., Goldschneider, I., Draper, T.L., and Gotschlich, E.C. (1975). Clinical evaluation of group A and group C meningococcal polysaccharide vaccines in infants. J. Clin. Invest. *56*, 1536–1547.

Goldblatt, D., Borrow, R., and Miller, E. (2002). Natural and vaccine-induced immunity and immunologic memory to *Neisseria meningitidis* serogroup C in young adults. J. Infect. Dis. *185*, 397–400.

Goldschneider, I., Gotschlich, E.C., and Artenstein, M.S. (1969a). Human immunity to the meningococcus. I. The role of humoral antibodies. J. Exp. Med. *129*, 1307–1326.

Goldschneider, I., Gotschlich, E.C., and Artenstein, M.S. (1969b). Human immunity to the meningococcus. II. Development of natural immunity. J. Exp. Med. *129*, 1327–1348.

Goldschneider, I., Lepow, M.L., Gotschlich, E.C., Mauck, F.T., Bachl, F., and Randolph, M. (1973). Immunogenicity of group A and group C meningococcal polysaccharides in human infants. J. Infect. Dis. *128*, 769–776.

Gotschlich, E.C., Seiff, M.E., Blake, M.S., and Horwitz, M.A. (1988). Studies on gonococcal protein I. In Bacteria-Host Cell Interaction (New York, Alan R. Liss, Inc.), pp. 63–73.

Granoff, D.M., Kelsey, S.K., Bijlmer, H.A., Van Alphen, L., Dankert, J., Mandrell, R.E., Azmi, F.H., and Scholten, R.J. (1995). Antibody responses to the capsular polysaccharide of *Neisseria meningitidis* serogroup B in patients with meningococcal disease. Clin. Diagn. Lab. Immunol *2*, 574–582.

Griffiss, J.M., Brandt, B.L., Broud, D.D., Goroff, D.K., and Baker, C.J. (1984). Immune response of infants and children to disseminated infections with *Neisseria meningitidis*. J. Infect. Dis. *150*, 71–79.

Groux, H., O'Garra, A., Bigler, M., Rouleau, M., Antonenko, S., de Vries, J.E., and Roncarolo, M.G. (1997). A CD4+ T-cell subset inhibits antigen-specific T-cell responses and prevents colitis. Nature *389*, 737–742.

Guirguis, N., Schneerson, R., Bax, A., Egan, W., Robbins, J.B., Shiloach, J., Orskov, I., Orskov, F., and el Kholy, A. (1985). *Escherichia coli* K51 and K93 capsular polysaccharides are crossreactive with the group A capsular polysaccharide of *Neisseria meningitidis*. Immunochemical, biological, and epidemiological studies. J. Exp. Med. *162*, 1837–1851.

Gulati, S., Rice, P.A., Blake, M.S., Sarafian, S.K., Morse, S.A., Quentin-Millet, M.J., Arminjon, F., Achtman, M., Kohl, P., Marchal, C., Morelli, G., Seiler, A., and Thiesen, B. (1991). Antibody responses in six volunteers immunized with a gonococcal Protein I vaccine. In neisseriae 1990 – Proceedings of the Seventh International Pathogenic *Neisseria* Conference, Berlin, Federal Republic of Germany, September 9–14, 1990 (Berlin, Walter de gruyter), pp. 229–234.

Guttormsen, H.K., Wetzler, L.M., and Solberg, C.O. (1994). Humoral immune response to class 1 outer membrane protein during the course of meningococcal disease. Infect. Immun. *62*, 1437–1443.

Hadfield, S.G., and Glynn, A.A. (1982). Analysis of antibodies in local and disseminated *Neisseria gonorrhoeae* infections by means of gel electrophoresis-derived ELISA. Immunology *47*, 283–288.

Hamadeh, R.M., Estabrook, M.M., Zhou, P., Jarvis, G.A., and Griffiss, J.M. (1995). Anti-Gal binds to pili of *Neisseria meningitidis*: the immunoglobulin A isotype blocks complement-mediated killing. Infect. Immun. *63*, 4900–4906.

Harada, A., Sekido, N., Akahoshi, T., Wada, T., Mukaida, N., and Matsushima, K. (1994). Essential involvement of interleukin-8 (IL-8) in acute inflammation. J. Leukoc. Biol. *56*, 559–564.

Harthug, S., Rosenqvist, E., Hoiby, E.A., Gedde-Dahl, T.W., and Froholm, L.O. (1986). Antibody response in group B meningococcal disease determined by enzyme-linked immunosorbent assay with serotype 15 outer membrane antigen. J. Clin. Microbiol. *24*, 947–953.

Hauck, C.R., Meyer, T.F., Lang, F., and Gulbins, E. (1998). CD66-mediated phagocytosis of Opa52 *Neisseria gonorrhoeae* requires a Src-like tyrosine kinase- and Rac1-dependent signalling pathway. EMBO J. *17*, 443–454.

Hedges, S.R., Mayo, M.S., Mestecky, J., Hook, E.W., III, and Russell, M.W. (1999). Limited local and systemic antibody responses to *Neisseria gonorrhoeae* during uncomplicated genital infections. Infect. Immun. *67*, 3937.

Heyderman, R.S., Davenport, V., and Williams, N.A. (2006). Mucosal immunity and optimizing protection with meningococcal serogroup B vaccines. Trends Microbiol. *14*, 120–124.

Hobbs, J.R., Milner, R.D., and Watt, P.J. (1967). Gamma-M deficiency predisposing to meningococcal septicaemia. Br, Med, J, *4*, 583–586.

Hook, J.E.W., Handsfield, H.H., Holmes, K.K., Mardh, P.A., Sparling, P.F., Wiesner, P.J., Cates, J.W., Lemon, S.M., and Stamm, W.E. (1990). Gonococcal infections in the adult. In Sexually Transmitted Diseases (New York, McGraw-Hill), pp. 149–160.

Horton, R.E., Stuart, J., Christensen, H., Borrow, R., Guthrie, T., Davenport, V., Finn, A., Williams, N.A., and Heyderman, R.S. (2005). Influence of age and carriage status on salivary IgA to *Neisseria meningitidis*. Epidemiol. Infect. *133*, 883–889.

Idanpaan-Heikkila, I., Hoiby, E.A., Chattopadhyay, P., Airaksinen, U., Michaelsen, T.M., and Wedege, E. (1995). Antibodies to meningococcal class 1 outer-membrane protein and its variable regions in patients with systemic meningococcal disease. J. Med. Microbiol. *43*, 335–343.

Jarvis, G.A., and Griffiss, J.M. (1991). Human IgA1 blockade of IgG-initiated lysis of *Neisseria meningitidis* is a function of antigen-binding fragment binding to the polysaccharide capsule. J. Immunol. *147*, 1962–1967.

Jelonek, M.T., Chang, S.J., Chiu, C.Y., Park, M.K., Nahm, M.H., and Ward, J.I. (1993). Comparison of naturally acquired and vaccine-induced antibodies to *Haemophilus influenzae* type b capsular polysaccharide. Infect. Immun. *61*, 5345–5350.

Jin, Z., Chu, C., Robbins, J.B., and Schneerson, R. (2003). Preparation and characterization of group A meningococcal capsular polysaccharide conjugates and evaluation of their immunogenicity in mice. Infect. Immun. *71*, 5115–5120.

Johnson, A.S., Gorringe, A.R., Mackinnon, F.G., Fox, A.J., Borrow, R., and Robinson, A. (1999). Analysis of the human Ig isotype response to lactoferrin binding protein A from *Neisseria meningitidis*. FEMS Immunol. Med. Microbiol. *25*, 349–354.

Johnson, S.C., Chung, R.C., Deal, C.D., Boslego, J.W., Sadoff, J.C., Wood, S.W., Brinton, C.C., Jr., and Tramont, E.C. (1991). Human immunization with Pgh 3–2 gonococcal pilus results in cross- reactive antibody to the cyanogen bromide fragment-2 of pilin. J. Infect. Dis. *163*, 128–134.

Joiner, K.A., Scales, R., Warren, K.A., Frank, M.M., and Rice, P.A. (1985a). Mechanism of action of blocking immunoglobin G for *Neisseria gonorrhoeae*. J. Clin. Invest. *76*, 1765–1772.

Joiner, K.A., Warren, K.A., Frank, M.M., Rice, P.A., and Schoolnik, G.K. (1985b). Blocking immunoglobulin G enhances complement consumption and deposition on *Neisseria gonorrhoeae* In The Pathogenic neisseriae (Washington, D.C., Amer. Soc. Microbiology), pp. 431–434.

Jokhdar, H., Borrow, R., Sultan, A., Adi, M., Riley, C., Fuller, E., and Baxter, D. (2004). Immunologic hyporesponsiveness to serogroup C but not serogroup A following repeated meningococcal A/C polysaccharide vaccination in Saudi Arabia. Clin. Diagn. Lab. Immunol. *11*, 83–88.

Jones, G.R., Christodoulides, M., Brooks, J.L., Miller, A.R., Cartwright, K.A., and Heckels, J.E. (1998). Dynamics of carriage of *Neisseria meningitidis* in a group of military recruits: subtype stability and specificity of the immune response following colonization. J. Infect. Dis. *178*, 451–459.

Jordens, J.Z., Williams, J.N., Jones, G.R., Christodoulides, M., and Heckels, J.E. (2004). Development of immunity to serogroup B meningococci during carriage of *Neisseria meningitidis* in a cohort of university students. Infect. Immun. *72*, 6503–6510.

Jose, J., Wolk, U., Lorenzen, D., Wenschuh, H., and Meyer, T.F. (2000). Human T-cell response to meningococcal immunoglobulin A1 protease associated alpha-proteins. Scand. J. Immunol. *51*, 176–185.

Kayhty, H. (1982). The unspecific antibody response to N. *meningitidis* group A capsular polysaccharide often seen in bacteraemic diseases. Parasite Immunol *4*, 157–170.

Kelly, D.F., Snape, M.D., Cutterbuck, E.A., Green, S., Snowden, C., Diggle, L., Yu, L.M., Borkowski, A., Moxon, E.R., and Pollard, A.J. (2006). CRM197-conjugated serogroup C meningococcal capsular polysaccharide, but not the native polysaccharide, induces persistent antigen-specific memory B-cells. Blood *108*, 2642–2647.

Kim, J.J., Mandrell, R.E., and Griffiss, J.M. (1989). *Neisseria lactamica* and *Neisseria meningitidis* share lipooligosaccharide epitopes but lack common capsular and class *1*, 2, and 3 protein epitopes. Infect. Immun. *57*, 602–608.

Kishimoto, T., Akira, S., Narazaki, M., and Taga, T. (1995). Interleukin-6 family of cytokines and gp130. Blood *86*, 1243–1254.

Kizil, G., Todd, I., Atta, M., Borriello, S.P., Ait-Tahar, K., and Ala'Aldeen, D.A. (1999). Identification and characterization of TspA, a major CD4(+) T-cell- and B-cell-stimulating *Neisseria*-specific antigen. Infect. Immun. *67*, 3533–3541.

Klugman, K.P., Gotschlich, E.C., and Blake, M.S. (1989). Sequence of the structural gene (rmpM) for the class 4 outer membrane protein of *Neisseria meningitidis*, homology to gonococcal protein III and *Escherichia coli* OmpA, and construction of meningococcal strains that lack class 4 protein. Infection and Immunity *57*, 2066–2071.

Kornelisse, R.F., de Groot, R., and Neijens, H.J. (1995). Bacterial meningitis: mechanisms of disease and therapy. Eur. J. Pediatr. *154*, 85–96.

Kornelisse, R.F., Hack, C.E., Savelkoul, H.F., van der Pouw Kraan, T.C., Hop, W.C., van Mierlo, G., Suur, M.H., Neijens, H.J., and de Groot, R. (1997). Intrathecal production of interleukin-12 and gamma interferon in patients with bacterial meningitis. Infect. Immun. *65*, 877–881.

Kortekaas, J., Muller, S.A., Ringler, P., Gregorini, M., Weynants, V.E., Rutten, L., Bos, M.P., and Tommassen, J. (2006). Immunogenicity and structural characterisation of an *in vitro* folded meningococcal siderophore receptor (FrpB, FetA). Microbes Infect. *8*, 2145–2153.

LaForce, F.M., Konde, K., Viviani, S., and Preziosi, M.P. (2007). The Meningitis Vaccine Project. Vaccine *25* Suppl. *1*, A97–100.

Lakshman, R., Burkinshaw, R., Choo, S., and Finn, A. (2002). Prior meningococcal A/C polysaccharide vaccine does not reduce immune responses to conjugate vaccine in young adults. Vaccine *20*, 3778–3782.

Lammel, C.J., Sweet, R.L., Rice, P.A., Knapp, J.S., Schoolnik, G.K., Heilbron, D.C., and Brooks, G.F. (1985). Antibody-antigen specificity in the immune response to infection with *Neisseria gonorrhoeae*. J. Infect. Dis. *152*, 990–1001.

Leach, A., Twumasi, P.A., Kumah, S., Banya, W.S., Jaffar, S., Forrest, B.D., Granoff, D.M., LiButti, D.E., Carlone, G.M., Pais, L.B., Broome, C.V., and Greenwood, B.M. (1997). Induction of immunologic memory in Gambian children by vaccination in infancy with a group A plus group C meningococcal polysaccharide–protein conjugate vaccine. J. Infect. Dis. *175*, 200–204.

Lee, H.S., Boulton, I.C., Reddin, K., Wong, H., Halliwell, D., Mandelboim, O., Gorringe, A.R., and Gray-Owen, S.D. (2007). Neisserial outer membrane vesicles bind the coinhibitory receptor carcinoembryonic antigen-related cellular adhesion molecule 1 and suppress CD4+ T lymphocyte function. Infect. Immun. *75*, 4449–4455.

Lehmann, A.K., Gorringe, A.R., Reddin, K.M., West, K., Smith, I., and Halstensen, A. (1999a). Human opsonins induced during meningococcal disease recognize transferrin binding protein complexes. Infect. Immun. *67*, 6526–6532.

Lehmann, A.K., Halstensen, A., Aaberge, I.S., Holst, J., Michaelsen, T.E., Sornes, S., Wetzler, L.M., and Guttormsen, H. (1999b). Human opsonins induced during meningococcal disease recognize outer membrane proteins PorA and PorB. Infect. Immun. *67*, 2552–2560.

Leinonen, M., and Frasch, C.E. (1982). Class-specific antibody response to group B *Neisseria meningitidis* capsular polysaccharide: use of polylysine precoating in an enzyme-linked immunosorbent assay. Infect. Immun. *38*, 1203–1207.

Levine, W.C., Pope, V., Bhoomkar, A., Tambe, P., Lewis, J.S., Zaidi, A.A., Farshy, C.E., Mitchell, S., and Talkington, D.F. (1998). Increase in endocervical CD4 lymphocytes among women with nonulcerative sexually transmitted diseases. J. Infect. Dis. *177*, 167–174.

Lifely, M.R., Rogers, M.V., Esdaile, J., Payne, M., and Tite, J.P. (1992). Murine cross-reactive T-cell epitopes of *Neisseria meningitidis* outer membrane proteins. Vaccine *10*, 159–163.

Litt, D.J., Savino, S., Beddek, A., Comanducci, M., Sandiford, C., Stevens, J., Levin, M., Ison, C., Pizza, M., Rappuoli, R., and Kroll, J.S. (2004). Putative vaccine antigens from *Neisseria meningitidis* recognized by serum antibodies of young children convalescing after meningococcal disease. J. Infect. Dis. *190*, 1488–1497.

Liu, M.A., Friedman, A., Oliff, A.I., Tai, J., Martinez, D., Deck, R.R., Shieh, J.T., Jenkins, T.D., Donnelly, J.J., and Hawe, L.A. (1992). A vaccine carrier derived from *Neisseria meningitidis* with mitogenic activity for lymphocytes. Proc. Natl. Acad. Sci. USA *89*, 4633–4637.

Livingston, P.O., Calves, M.J., Helling, F., Zollinger, W.D., Blake, M.S., and Lowell, G.H. (1993). GD3/proteosome vaccines induce consistent IgM antibodies against the ganglioside GD3. Vaccine *11*, 1199–1204.

Lorenzen, D.R., Dux, F., Wolk, U., Tsirpouchtsidis, A., Haas, G., and Meyer, T.F. (1999). Immunoglobulin A1 protease, an exoenzyme of pathogenic neisseriae, is a potent inducer of proinflammatory cytokines. J. Exp. Med. *190*, 1049–1058.

Lowell, G.H., Ballou, W.R., Smith, L.F., Wirtz, R.A., Zollinger, W.D., and Hockmeyer, W.T. (1988). Proteosome-lipopeptide vaccines: enhancement of immunogenicity for malaria CS peptides. Science *240*, 800–802.

Lowell, G.H., Smith, L.F., Artenstein, M.S., Nash, G.S., and MacDermott, R.P., Jr. (1979). Antibody-dependent cell-mediated antibacterial activity of human mononuclear cells. I. K lymphocytes and monocytes are effective against meningococi in cooperation with human imune sera. J. Exp. Med. *150*, 127–137.

Lowell, G.H., Smith, L.F., Griffiss, J.M., Brandt, B.L., and MacDermott, R.P. (1980). Antibody-dependent mononuclear cell-mediated antimeningococcal activity. Comparison of the effects of convalescent and postimmunization immunoglobulins G, M, and A.J. Clin. Invest. *66*, 260–267.

MacDonald, N.E., Halperin, S.A., Law, B.J., Forrest, B., Danzig, L.E., and Granoff, D.M. (1998). Induction of immunologic memory by conjugated vs plain meningococcal C polysaccharide vaccine in toddlers: a randomized controlled trial. JAMA *280*, 1685–1689.

MacLennan, J., Kafatos, G., Neal, K., Andrews, N., Cameron, J.C., Roberts, R., Evans, M.R., Cann, K., Baxter, D.N., Maiden, M.C., and Stuart, J.M. (2006). Social behavior and meningococcal carriage in British teenagers. Emerg. Infect. Dis. *12*, 950–957.

MacLennan, J., Obaro, S., Deeks, J., Lake, D., Elie, C., Carlone, G., Moxon, E.R., and Greenwood, B. (2001). Immunologic memory 5 years after meningococcal A/C conjugate vaccination in infancy. J. Infect. Dis. *183*, 97–104.

MacLennan, J., Obaro, S., Deeks, J., Williams, D., Pais, L., Carlone, G., Moxon, R., and Greenwood, B. (1999). Immune response to revaccination with meningococcal A and C polysaccharides in Gambian children following repeated immunisation during early childhood. Vaccine *17*, 3086–3093.

MacLennan, J.M., Shackley, F., Heath, P.T., Deeks, J.J., Flamank, C., Herbert, M., Griffiths, H., Hatzmann, E., Goilav, C., and Moxon, E.R. (2000). Safety, immunogenicity, and induction of immunologic memory by a serogroup C meningococcal conjugate vaccine in infants: A randomized controlled trial. JAMA *283*, 2795–2801.

Maiden, M.C., and Stuart, J.M. (2002). Carriage of serogroup C meningococci 1 year after meningococcal C conjugate polysaccharide vaccination. Lancet *359*, 1829–1831.

Mandrell, R.E., Azmi, F.H., and Granoff, D.M. (1995). Complement-mediated bactericidal activity of human antibodies to poly alpha 2 – >8 N-acetylneuraminic acid, the capsular polysaccharide of *Neisseria meningitidis* serogroup B.J. Infect. Dis. *172*, 1279–1289.

Mandrell, R.E., and Zollinger, W.D. (1982). Measurement of antibodies to meningococcal group B polysaccharide: low avidity binding and equilibrium binding constants. J. Immunol. *129*, 2172–2178.

Mandrell, R.E., and Zollinger, W.D. (1989). Human immune response to meningococcal outer membrane protein epitopes after natural infection or vaccination. Infect. Immun. *57*, 1590–1598.

Massari, P., Henneke, P., Ho, Y., Latz, E., Golenbock, D.T., and Wetzler, L.M. (2002). Cutting edge: Immune stimulation by neisserial porins is toll-like receptor 2 and MyD88 dependent. J. Immunol. *168*, 1533–1537.

Massari, P., Ram, S., Macleod, H., and Wetzler, L.M. (2003). The role of porins in neisserial pathogenesis and immunity. Trends Microbiol. *11*, 87–93.

Massari, P., Visintin, A., Gunawardana, J., Halmen, K.A., King, C.A., Golenbock, D.T., and Wetzler, L.M. (2006). Meningococcal porin PorB binds to TLR2 and requires TLR1 for signaling. J. Immunol. *176*, 2373–2380.

McChesney, D., Tramont, E.C., Boslego, J.W., Ciak, J., Sadoff, J., and Brinton, C.C. (1982). Genital antibody response to a parenteral gonococcal pilus vaccine. Infect. Immun. *36*, 1006–1012.

McGuinness, B., Barlow, A.K., Clarke, I.N., Farley, J.E., Anilionis, A., Poolman, J.T., and Heckels, J.E. (1990). Deduced amino acid sequences of class 1 protein (PorA) from three strains of *Neisseria meningitidis*. Synthetic peptides define the epitopes responsible for serosubtype specificity. J. Exp. Med. *171*, 1871–1882.

Mesa, C., de Leon, J., and Fernandez, L.E. (2006). Very small size proteoliposomes derived from *Neisseria meningitidis*: An effective adjuvant for generation of CTL responses to peptide and protein antigens. Vaccine *24*, 2692–2699.

Milagres, L.G., Gorla, M.C., Rebelo, M.C., and Barroso, D.E. (2000). Bactericidal antibody response to *Neisseria meningitidis* serogroup B in patients with bacterial meningitis: effect of immunization with an outer membrane protein vaccine. FEMS Immunol. Med. Microbiol. *28*, 319–327.

Milagres, L.G., Gorla, M.C., Sacchi, C.T., and Rodrigues, M.M. (1998). Specificity of bactericidal antibody response to serogroup B meningococcal strains in Brazilian children after immunization with an outer membrane vaccine. Infect. Immun. *66*, 4755–4761.

Mond, J.J., Lees, A., and Snapper, C.M. (1995a). T-cell-independent antigens type 2. Annu. Rev. Immunol. *13*, 655–692.

Mond, J.J., Vos, Q., Lees, A., and Snapper, C.M. (1995b). T-cell independent antigens. Curr. Opin. Immunol. *7*, 349–354.

Moore, K.W., de Waal Malefyt, R., Coffman, R.L., and O'Garra, A. (2001). Interleukin-10 and the interleukin-10 receptor. Annu. Rev. Immunol. *19*, 683–765.

Mueller, J.E., Borrow, R., and Gessner, B.D. (2006). Meningococcal serogroup W135 in the African meningitis belt: epidemiology, immunity and vaccines. Expert Rev. Vaccines *5*, 319–336.

Munkley, A., Tinsley, C.R., and Heckels, J.E. (1991). Blocking of bactericidal killing of *Neisseria meningitidis* by antibodies directed against class 4 outer membrane protein. Microbial Pathogen. *11*, 447–452.

Naess, L.M., Oftung, F., Aase, A., Wetzler, L.M., Sandin, R., and Michaelsen, T.E. (1998). Human T-cell responses after vaccination with the Norwegian group B meningococcal outer membrane vesicle vaccine. Infect. Immun. *66*, 959–965.

Norheim, G., Aseffa, A., Yassin, M.A., Mengistu, G., Kassu, A., Fikremariam, D., Tamire, W., Merid, Y., Hoiby, E.A., Caugant, D.A., Fritzsonn, E., Tangen, T., Alebel, T., Berhanu, D., Harboe, M., and Rosenqvist, E. (2007). Serum antibody responses in Ethiopian meningitis patients infected with *Neisseria meningitidis* serogroup A sequence type 7. Clin. Vaccine Immunol. *14*, 451–463.

Norheim, G., Aseffa, A., Yassin, M.A., Mengistu, G., Kassu, A., Fikremariam, D., Tamire, W., Merid, Y., Hoiby, E.A., Caugant, D.A., Fritzsonn, E., Tangen, T., Melak, B., Berhanu, D., Harboe, M., Kolberg, J., and Rosenqvist, E. (2008). Specificity of the sub-capsular antibody responses following serogroup A meningococcal disease in Ethiopian patients. Clin. Vaccine Immunol. *15*, 863–871.

O'Brien, J.P., Goldenberg, D.L., and Rice, P.A. (1983). Disseminated gonococcal infection: a prospective analysis of 49 patients and a review of the pathophysiology and immune mechanisms. Medicine *62*, 395–406.

Oftung, F., Naess, L.M., Wetzler, L.M., Korsvold, G.E., Aase, A., Hoiby, E.A., Dalseg, R., Holst, J., Michaelsen, T.E., and Haneberg, B. (1999). Antigen-specific T-cell responses in humans after intranasal immunization with a meningococcal serogroup B outer membrane vesicle vaccine. Infect. Immun. *67*, 921–927.

Ohga, S., Aoki, T., Okada, K., Akeda, H., Fujioka, K., Ohshima, A., Mori, T., Minamishima, I., and Ueda, K. (1994). Cerebrospinal fluid concentrations of interleukin-1 beta, tumour necrosis factor-alpha, and interferon gamma in bacterial meningitis. Arch. Dis. Child. *70*, 123–125.

Olsen, S.F., Djurhuus, B., Rasmussen, K., Joensen, H.D., Larsen, S.O., Zoffman, H., and Lind, I. (1991). Pharyngeal carriage of *Neisseria meningitidis* and *Neisseria* lactamica in households with infants within areas with high and low incidences of meningococcal disease. Epidemiol. Infect. *106*, 445–457.

Patel, M., and Lee, C.K. (2005). Polysaccharide vaccines for preventing serogroup A meningococcal meningitis. Cochrane Database Syst Rev. CD001093.

Paulnock, D.M. (1992). Macrophage activation by T-cells. Curr Opin Immunol 4, 344–349.

Peltola, H., Makela, H., Kayhty, H., Jousimies, H., Herva, E., Hallstrom, K., Sivonen, A., Renkonen, O.V., Pettay, O., Karanko, V., Ahvonen, P., and Sarna, S. (1977). Clinical efficacy of meningococcus group A capsular polysaccharide vaccine in children three months to five years of age. N. Engl. J. Med. 297, 686–691.

Perez-Melgosa, M., Ochs, H.D., Linsley, P.S., Laman, J.D., van Meurs, M., Flavell, R.A., Ernst, R.K., Miller, S.I., and Wilson, C.B. (2001). Carrier-mediated enhancement of cognate T-cell help: the basis for enhanced immunogenicity of meningococcal outer membrane protein polysaccharide conjugate vaccine. Eur. J. Immunol. *31*, 2373–2381.

Pettersson, A., Kortekaas, J., Weynants, V.E., Voet, P., Poolman, J.T., Bos, M.P., and Tommassen, J. (2006). Vaccine potential of the *Neisseria meningitidis* lactoferrin-binding proteins LbpA and LbpB. Vaccine *24*, 3545–3557.

Pihlgren, M., Friedli, M., Tougne, C., Rochat, A.F., Lambert, P.H., and Siegrist, C.A. (2006). Reduced ability of neonatal and early-life bone marrow stromal cells to support plasmablast survival. J. Immunol. *176*, 165–172.

Plummer, F.A., Chubb, H., Simonsen, J.N., Bosire, M., Slaney, L., Nagelkerke, N.J., Maclean, I., Ndinya-Achola, J.O., Waiyaki, P., and Brunham, R.C. (1994). Antibodies to opacity proteins (Opa) correlate with a reduced risk of gonococcal salpingitis. JClinInvest *93*, 1748–1755.

Pollard, A.J., and Frasch, C. (2001). Development of natural immunity to *Neisseria meningitidis*. Vaccine *19*, 1327–1346.

Pollard, A.J., Galassini, R., Rouppe van der Voort, E.M., Hibberd, M., Booy, R., Langford, P., Nadel, S., Ison, C., Kroll, J.S., Poolman, J., and Levin, M. (1999a). Cellular immune responses to *Neisseria meningitidis* in children. Infect. Immun. *67*, 2452–2463.

Pollard, A.J., Galassini, R., van der Voort, E.M., Booy, R., Langford, P., Nadel, S., Ison, C., Kroll, J.S., Poolman, J., and Levin, M. (1999b). Humoral immune responses to *Neisseria meningitidis* in children. Infect. Immun. *67*, 2441–2451.

Pollard, A.J., and Granoff, D.M. (2007). Reconsideration of the use of meningococcal polysaccharide vaccine. Paed. Infect. Dis. J. *26*, 716 – 722.

Pollard, A.J., and Levin, M. (2000). Production of low-avidity antibody by infants after infection with serogroup B meningococci. Lancet *356*, 2065–2066.

Pollard, A.J., Ochnio, J., Ho, M., Callaghan, M., Bigham, M., and Dobsong, S. (2004). Disease susceptibility to ST11 complex meningococci bearing serogroup C or W135 polysaccharide capsules, North America. Emerg. Infect. Dis. *10*, 1812–1815.

Prins, J.M., Lauw, F.N., Derkx, B.H., Speelman, P., Kuijper, E.J., Dankert, J., and van Deventer, S.J. (1998). Endotoxin release and cytokine production in acute and chronic meningococcaemia. Clin. Exp. Immunol. *114*, 215–219.

Raghunathan, P.L., Jones, J.D., Tiendrebeogo, S.R., Sanou, I., Sangare, L., Kouanda, S., Dabal, M., Lingani, C., Elie, C.M., Johnson, S., Ari, M., Martinez, J., Chatt, J., Sidibe, K., Schmink, S., Mayer, L.W., Konde, M.K., Djingarey, M.H., Popovic, T., Plikaytis, B.D., Carlone, G.M., Rosenstein, N., and Soriano-Gabarro, M. (2006). Predictors of immunity after a major serogroup W-135 meningococcal disease epidemic, Burkina Faso, 2002. J. Infect. Dis. *193*, 607–616.

Ram, S., Mackinnon, F.G., Gulati, S., McQuillen, D.P., Vogel, U., Frosch, M., Elkins, C., Guttormsen, H.K., Wetzler, L.M., Oppermann, M., Pangburn, K., and Rice, P.A.(1999). The contrasting mechanisms of serum resistance of *Neisseria gonorrhoeae* and group B *Neisseria meningitidis*. Mol. Immunol. *36*, 915–928.

Ramsay, M.E., Andrews, N.J., Trotter, C.L., Kaczmarski, E.B., and Miller, E. (2003). Herd immunity from meningococcal serogroup C conjugate vaccination in England: database analysis. BrMJ *326*, 365–366.

Reller, L.B., MacGregor, R.R., and Beaty, H.N. (1973). Bactericidal antibody after colonization with *Neisseria meningitidis*. J. Infect. Dis. *127*, 56–62.

Rennels, M., King, J., Jr., Ryall, R., Manoff, S., Papa, T., Weddle, A., and Froeschle, J. (2002). Dose escalation, safety and immunogenicity study of a tetravalent meninogococcal polysaccharide diphtheria conjugate vaccine in toddlers. Pediatr. Infect. Dis. J. *21*, 978–979.

Rennels, M., King, J., Jr., Ryall, R., Papa, T., and Froeschle, J. (2004). Dosage escalation, safety and immunogenicity study of four dosages of a tetravalent meninogococcal polysaccharide diphtheria toxoid conjugate vaccine in infants. Pediatr. Infect. Dis. J. *23*, 429–435.

Rice, P.A., and Goldenberg, D.L. (1981). Clinical manifestations of disseminated infection caused by *Neisseria gonorrhoeae* are linked to differences in bactericidal reactivity of infection strains. Ann. Intern. Med. *95*, 175–178.

Rice, P.A., Vayo, H.E., Tam, M.R., and Blake, M.S. (1986). Immunoglobin G antibodies directed against protein III block killing of serum resistant *Neisseria gonorrhoeae* by immune sera. J. Exp. Med. *164*, 1735–1748.

Richmond, P., Kaczmarski, E., Borrow, R., Findlow, J., Clark, S., McCann, R., Hill, J., Barker, M., and Miller, E. (2000). Meningococcal C polysaccharide vaccine induces immunologic hyporesponsiveness in adults that is overcome by meningococcal C conjugate vaccine. J. Infect. Dis. *181*, 761–764.

Robbins, J.B., Myerowitz, L., Whisnant, J.K., Argaman, M., Schneerson, R., Handzel, Z.T., and Gotschlich, E.C. (1972). Enteric bacteria cross-reactive with *Neisseria meningitidis* groups A and C and Diplococcus pneumoniae types I and 3. Infect. Immun. *6*, 651–656.

Robinson, K., Neal, K.R., Howard, C., Stockton, J., Atkinson, K., Scarth, E., Moran, J., Robins, A., Todd, I., Kaczmarski, E., Gray, S., Muscat, I., Slack, R., and Ala'Aldeen, D.A. (2002). Characterization of humoral and cellular immune responses elicited by meningococcal carriage. Infect. Immun. *70*, 1301–1309.

Robinson, K., Wooldridge, K.G., Wells, D.B., Hasan, A., Todd, I., Robins, A., James, R., and Ala'Aldeen, D.A. (2005). T-cell-stimulating protein A elicits immune responses during meningococcal carriage and human disease. Infect. Immun. *73*, 4684–4693.

Rosenqvist, E., Hoiby, E.A., Wedege, E., Bryn, K., Kolberg, J., Klem, A., Ronnild, E., Bjune, G., and Nokleby, H. (1995). Human antibody responses to meningococcal outer membrane antigens after three doses of the Norwegian group B meningococcal vaccine. Infect. Immun. *63*, 4642–4652.

Ross, S.C., and Densen, P. (1984). Complement deficiency states and infection: epidemiology, pathogenesis and consequences of neisserial and other infections in an immune deficiency. Medicine (Baltimore) *63*, 243–273.

Salit, I.E. (1981). Meningococcemia caused by serogroup W135. Association with hypogammaglobulinemia. Arch Intern Med *141*, 664–665.

Santos, G.F., Deck, R.R., Donnelly, J., Blackwelder, W., and Granoff, D.M. (2001). Importance of complement source in measuring meningococcal bactericidal titers. Clin. Diagn. Lab. Immunol. *8*, 616–623.

Schneerson, R., Barrera, O., Sutton, A., and Robbins, J.B. (1980). Preparation, characterization, and immunogenicity of *Haemophilus influenzae* type b polysaccharide–protein conjugates. J. Exp. Med. *152*, 361–376.

Shapiro, E.D., Capobianco, L.A., Berg, A.T., and Zitt, M.Q. (1989). The immunogenicity of *Hemophilus influenzae* type B polysaccharide-*Neisseria meningitidis* group B outer membrane protein complex vaccine in infants and young children. J. Infect. Dis. *160*, 1064–1067.

Simpson, S.D., Ho, Y., Rice, P.A., and Wetzler, L.M. (1999a). T lymphocyte response to *Neisseria gonorrhoeae* porin in individuals with mucosal gonococcal infections. J. Infect. Dis. *180*, 762–773.

Simpson, S.D., Ho, Y., Rice, P.A., and Wetzler, L.M. (1999b). T lymphocyte response to *Neisseria gonorrhoeae* porin in individuals with mucosal gonococcal infections. J. Infect. Dis. *180*, 762–773.

Sjursen, H., Wedege, E., Rosenqvist, E., Naess, A., Halstensen, A., Matre, R., and Solberg, C.O. (1990). IgG subclass antibodies to serogroup B meningococcal outer membrane antigens following infection and vaccination. APMIS *98*, 1061–1069.

Slavik, I., Kuzemenska, P., Kozuch, O., Matoska, J., Pokorny, J., and Sekeyova, M. (1991). The complex of flavivirus envelope polypeptide with meningococcal proteosomes elicits formation of virus-neutralizing antibodies. Acta Virol *35*, 313–321.

Snape, M.D., Kelly, D.F., Salt, P., Green, S., Snowden, C., Diggle, L., Borkowski, A., Yu, L.-m., Moxon, E.R., and Pollard, A.J. (2006). Serogroup C meningococcal glycoconjugate vaccine in adolescents: persistence of bactericidal antibodies and kinetics of the immune response to a booster vaccine more than 3 years after immunization. Clin Infect Dis *43*, 1387–1394.

Snape, M.D., Perrett, K.P., Ford, K., John, T., Pace, D., Yu, L.M., Langley, J.M., McNeil, S., Dull, P.M., Ceddia, F.C., Anemona, A., Halperin, S.A., Dobson, S., and Pollard, A.J. (2008). Immunogenicity of a tetravalent meningococcal glycoconjugate vaccine in infants. JAMA *299*, 173 - 184.

Snapper, C.M., Rosas, F.R., Moorman, M.A., and Mond, J.J. (1997). Restoration of T-cell-independent type 2 induction of Ig secretion by neonatal B-cells *in vitro*. J. Immunol. *158*, 2731–2735.

Southern, J., Deane, S., Ashton, L., Borrow, R., Goldblatt, D., Andrews, N., Balmer, P., Morris, R., Kroll, J.S., and Miller, E. (2004). Effects of prior polysaccharide vaccination on magnitude, duration, and quality of immune responses to and safety profile of a meningococcal serogroup C tetanus toxoid conjugate vaccination in adults. Clin. Diagn. Lab. Immunol. *11*, 1100–1104.

Sparling, P.F., Holmes, K.K., Mardh, P.A., Sparling, P.F., Wiesner, P.J., Cates, J.W., Lemon, S.M., and Stamm, W.E. (1990). Biology of *Neisseria gonorrhoeae* In Sexually Transmitted Diseases (New York, McGraw-Hill), pp. 131–148.

Stephens, D.S. (1999). Uncloaking the meningococcus: dynamics of carriage and disease. Lancet *353*, 941–942.

Stuart, J.M., Cartwright, K.A., Robinson, P.M., and Noah, N.D. (1989). Effect of smoking on meningococcal carriage. Lancet *2*, 723–725.

Swanson, J.L., Robbins, K., Barrera, O., and Koomey, J.M. (1987). Gene conversion variations generate structurally different pilin polypeptides in *Neisseria gonorhoeae*. J. Exp. Med. *165*, 1016–1025.

Taams, L.S., Smith, J., Rustin, M.H., Salmon, M., Poulter, L.W., and Akbar, A.N. (2001). Human anergic/suppressive CD4(+)CD25(+) T-cells: a highly differentiated and apoptosis-prone population. Eur. J. Immunol. *31*, 1122–1131.

Thiesen, B., Greenwood, B., Brieske, N., and Achtman, M. (1997). Persistence of antibodies to meningococcal

IgA1 protease versus decay of antibodies to group A polysaccharide and Opc protein. Vaccine *15*, 209–219.

Thompson, M.J., Ninis, N., Perera, R., Mayon-White, R., Phillips, C., Bailey, L., Harnden, A., Mant, D., and Levin, M. (2006). Clinical recognition of meningococcal disease in children and adolescents. Lancet *367*, 397–403.

Toropainen, M., Saarinen, L., Wedege, E., Bolstad, K., Michaelsen, T.E., Aase, A., and Kayhty, H. (2005). Protection by natural human immunoglobulin M antibody to meningococcal serogroup B capsular polysaccharide in the infant rat protection assay is independent of complement-mediated bacterial lysis. Infect. Immun. *73*, 4694–4703.

Trinchieri, G. (2001). Regulatory role of T-cells producing both interferon gamma and interleukin 10 in persistent infection. J. Exp. Med. *194*, F53–57.

Trotter, C., Borrow, R., Andrews, N., and Miller, E. (2003). Seroprevalence of meningococcal serogroup C bactericidal antibody in England and Wales in the pre-vaccination era. Vaccine *21*, 1094–1098.

Trotter, C., Findlow, J., Balmer, P., Holland, A., Barchha, R., Hamer, N., Andrews, N., Miller, E., and Borrow, R. (2007). Seroprevalence of bactericidal and anti-outer membrane vesicle antibodies to *Neisseria meningitidis* group B in England. Clin. Vaccine Immunol. *14*, 863–868.

Trotter, C.L., Andrews, N.J., Kaczmarski, E.B., Miller, E., and Ramsay, M.E. (2004). Effectiveness of meningococcal serogroup C conjugate vaccine 4 years after introduction. Lancet *364*, 365–367.

Trotter, C.L., and Edmunds, W.J. (2006). Reassessing the cost-effectiveness of meningococcal serogroup C conjugate (MCC) vaccines using a transmission dynamic model. Med Decis Making *26*, 38–47.

Trotter, C.L., and Ramsay, M.E. (2007). Vaccination against meningococcal disease in Europe: review and recommendations for the use of conjugate vaccines. FEMS Microbiol. Rev. *31*, 101–107.

Tsirpouchtsidis, A., Hurwitz, R., Brinkmann, V., Meyer, T.F., and Haas, G. (2002). Neisserial immunoglobulin A1 protease induces specific T-cell responses in humans. Infect. Immun. *70*, 335–344.

Twumasi, P.A., Jr., Kumah, S., Leach, A., O'Dempsey, T.J., Ceesay, S.J., Todd, J., Broome, C.V., Carlone, G.M., Pais, L.B., Holder, P.K., Plikaytis, B.D., and Greenwood, B.M. (1995). A trial of a group A plus group C meningococcal polysaccharide–protein conjugate vaccine in African infants. J. Infect. Dis. *171*, 632–638.

Urwin, R., Holmes, E.C., Fox, A.J., Derrick, J.P., and Maiden, M.C. (2002). Phylogenetic evidence for frequent positive selection and recombination in the meningococcal surface antigen PorB. Mol. Biol. Evol. *19*, 1686–1694.

van der Ley, P., Heckels, J.E., Virji, M., Hoogerhout, P., and Poolman, J.T. (1991). Topology of outer membrane porins in pathogenic *Neisseria* spp. Infect. Immun. *59*, 2963–2971.

van der Voort, E.R., van Dijken, H., Kuipers, B., van der Biezen, J., van der Ley, P., Meylis, J., Claassen, I., and Poolman, J. (1997). Human B- and T-cell responses after immunization with a hexavalent PorA meningococcal outer membrane vesicle vaccine. Infect. Immun. *65*, 5184–5190.

van Deuren, M., van der Ven-Jongekrijg, J., Bartelink, A.K., van Dalen, R., Sauerwein, R.W., and van der Meer, J.W. (1995). Correlation between proinflammatory cytokines and antiinflammatory mediators and the severity of disease in meningococcal infections. J. Infect. Dis. *172*, 433–439.

Vaughan, T.E., Skipp, P.J., O'Connor, C.D., Hudson, M.J., Vipond, R., Elmore, M.J., and Gorringe, A.R. (2006). Proteomic analysis of *Neisseria lactamica* and *Neisseria meningitidis* outer membrane vesicle vaccine antigens. Vaccine *24*, 5277–5293.

Vella, P.P., Marburg, S., Staub, J.M., Kniskern, P.J., Miller, W., Hagopian, A., Ip, C., Tolman, R.L., Rusk, C.M., Chupak, L.S., and Ellis, R.W. (1992). Immunogenicity of conjugate vaccines consisting of pneumococcal capsular polysaccharide types 6B, 14, 19F, and 23F and a meningococcal outer membrane protein complex. Infect. Immun. *60*, 4977–4983.

Vidarsson, G., van Der Pol, W.L., van Den Elsen, J.M., Vile, H., Jansen, M., Duijs, J., Morton, H.C., Boel, E., Daha, M.R., Corthesy, B., and van De Winkel, J.G. (2001). Activity of human IgG and IgA subclasses in immune defense against *Neisseria meningitidis* serogroup B.J. Immunol. *166*, 6250–6256.

Vieusseaux, G. (1805). Memoire sur le maladie qui a régné a Genêve au printemps de 1805.. Journal de Médecine, Chirurgie et Pharmacie *II*, 163–165.

Virji, M., and Heckels, J. (1988). Nonbactericidal antibodies against *Neisseria gonorrhoeae*: evaluation of their blocking effect on bactericidal antibodies directed against outer membrane antigens. J. Gen. Microbiol. *134*, 2703–2711.

Waage, A., Halstensen, A., Shalaby, R., Brandtzaeg, P., Kierulf, P., and Espevik, T. (1989). Local production of tumor necrosis factor alpha, interleukin 1, and interleukin 6 in meningococcal meningitis. Relation to the inflammatory response. J. Exp. Med. *170*, 1859–1867.

Wedege, E., Hoiby, E.A., Rosenqvist, E., and Bjune, G. (1998). Immune responses against major outer membrane antigens of *neisseria meningitidis* in vaccinees and controls who contracted meningococcal disease during the norwegian serogroup B protection trial. Infect. Immun. *66*, 3223–3231.

Wedege, E., Kuipers, B., Bolstad, K., van Dijken, H., Froholm, L.O., Vermont, C., Caugant, D.A., and van den Dobbelsteen, G. (2003). Antibody specificities and effect of meningococcal carriage in icelandic teenagers receiving the Norwegian serogroup B outer membrane vesicle vaccine. Infect. Immun. *71*, 3775–3781.

Weinberg, G.A., and Granoff, D.M. (1988). Polysaccharide–protein conjugate vaccines for the prevention of *Haemophilus influenzae* type b disease. J. Pediatr. *113*, 621–631.

Weis, N., Berthelsen, L., Wachmann, H., and Lind, I. (2005). The meningococcal antibody test: how useful in the diagnosis of meningococcal disease? Epidemiol. Infect. *133*, 217–227.

Welsch, J.A., and Granoff, D. (2007). Immunity to *Neisseria meningitidis* group B in adults despite lack of serum bactericidal antibody. Clin. Vaccine Immunol *14*, 1596–1602.

Wetzler, L.M., Blake, M.S., Barry, K., and Gotschlich, E.C. (1992). Gonococcal porin vaccine evaluation: comparison of Por proteosomes, liposomes and blebs isolated from rmp deletion mutants. J. Infect. Dis. *166*, 551–555.

Wetzler, L.M., Blake, M.S., and Gotschlich, E.C. (1989a). Protein I (Por) of *Neisseria gonorrhoeae* as an immunogen: liposomes, proteosomes and the lack of blocking antibodies. In Transactions of the American Association of Physicians (Baltimore, MD, Waverly Press), pp. 78–90.

Wetzler, L.M., Gotschlich, E.C., Blake, M.S., and Koomey, J.M. (1989b). The construction and characterization of *Neisseria gonorrhoeae* lacking protein III in its outer membrane. J. Exp. Med. *169*, 2199–2209.

Wetzler, L.M., Ho, Y., and Reiser, H. (1996). Neisserial porins induce B lymphocytes to express costimulatory B7–2 molecules and to proliferate. J. Exp. Med. *183*, 1151–1159.

WHO (2002). Meningococcal Vaccines: polysaccharide and polysaccharide conjugate vaccines. WHO position paper. Weekly Epidemiological Record *40*, 331 – 339.

Wiertz, E.J., Delvig, A., Donders, E.M., Brugghe, H.F., van Unen, L.M., Timmermans, H.A., Achtman, M., Hoogerhout, P., and Poolman, J.T. (1996). T-cell responses to outer membrane proteins of *Neisseria meningitidis*: comparative study of the Opa, Opc, and PorA proteins. Infect. Immun. *64*, 298–304.

Wiertz, E.J., van Gaans-van den Brink, J.A., Gausepohl, H., Prochnicka-Chalufour, A., Hoogerhout, P., and Poolman, J.T. (1992). Identification of T-cell epitopes occurring in a meningococcal class 1 outer membrane protein using overlapping peptides assembled with simultaneous multiple peptide synthesis. J. Exp. Med. *176*, 79–88.

Wiertz, E.J., van Gaans-van den Brink, J.A., Schreuder, G.M., Termijtelen, A.A., Hoogerhout, P., and Poolman, J.T. (1991). T-cell recognition of *Neisseria meningitidis* class 1 outer membrane proteins. Identification of T-cell epitopes with selected synthetic peptides and determination of HLA restriction elements. J. Immunol. *147*, 2012–2018.

Williams, J.N., Jones, G.R., Christodoulides, M., and Heckels, J.E. (2003). Serological correlates of protection against meningococci in a cohort of university students, before and during an outbreak of serogroup C infection. J. Infect. Dis. *187*, 1433–1441.

Zandvoort, A., and Timens, W. (2002). The dual function of the splenic marginal zone: essential for initiation of anti-TI-2 responses but also vital in the general first-line defense against blood-borne antigens. Clin. Exp. Immunol. *130*, 4–11.

Zangwill, K.M., Stout, R.W., Carlone, G.M., Pais, L., Harekeh, H., Mitchell, S., Wolfe, W.H., Blackwood, V., Plikaytis, B.D., and Wenger, J.D. (1994). Duration of antibody response after meningococcal polysaccharide vaccination in US Air Force personnel. J. Infect. Dis. *169*, 847–852.

Zhang, Q., Choo, S., Everard, J., Jennings, R., and Finn, A. (2000). Mucosal immune responses to meningococcal group C conjugate and group A and C polysaccharide vaccines in adolescents. Infection and Immunity *68*, 2692–2697.

Zhang, Q., and Finn, A. (2004). Mucosal immunology of vaccines against pathogenic nasopharyngeal bacteria. J. Clin. Pathol. *57*, 1015–1021.

Zollinger, W.D., and Mandrell, R.E. (1983). Importance of complement source in bactericidal activity of human antibody and murine monoclonal antibody to meningococcal group B polysaccharide. Infect. Immun. *40*, 257–264.

Part IV

Clinical Correlations

Difficulty in Developing a Neisserial Vaccine

11

Kate L. Seib and Rino Rappuoli

Abstract

Diseases caused by *Neisseria meningitidis* and *Neisseria gonorrhoeae* are a significant health problem worldwide, the control of which is largely dependent on the availability and widespread use of comprehensive meningococcal and gonococcal vaccines. Unfortunately, the development of neisserial vaccines has been challenging due to the nature of these organisms, in particular the heterogeneity, variability and/or poor immunogenicity of their outer surface components. As strictly human pathogens, they are highly adapted to the host environment but have evolved several mechanisms to remain adaptable to changing microenvironments and avoid elimination by the host immune system. Currently, serogroup A, C, Y and W-135 meningococcal infections can be prevented by vaccines. However there is no comprehensive serogroup B vaccine, and the prospect of developing a gonococcal vaccine is remote.

Introduction

The World Health Organization (WHO) attributes vaccination, along with access to clean water, as the public health interventions that have had the greatest impact on world health. Since the origin of vaccination in 1796 with the discovery of the smallpox vaccine, there have been numerous breakthroughs in the fight against infectious disease (Levine and Lagos, 2004). The use of vaccines has led to the eradication of smallpox and the near eradication of polio, has significantly diminished childhood disease, and is believed to save more than three million lives a year. However, despite decades of research and continuing advances in the field of vaccinology, vaccines against some of the most serious infectious diseases remain elusive. There is still no vaccine for the three main causes of human mortality: tuberculosis, malaria and human immunodeficiency virus (HIV). Likewise, there is currently no vaccine available for the sexually transmitted infection (STI) gonorrhoea caused by *Neisseria gonorrhoeae*, nor is there a comprehensive vaccine to protect against disease caused by *Neisseria meningitidis*. Although pathogens such as *N. meningitidis* and *N. gonorrhoeae* cause significant disease worldwide, their burden of disease exists primarily in developing countries and hence the economic incentive for vaccine development is often diminished.

The majority of successful vaccines have been developed using classical methods of vaccinology that involve the use of whole organisms, either attenuated or inactivated. Furthermore, they are aimed at pathogens with very little intra-species antigenic diversity or variability. Consequently, natural infection or vaccination with a single agent typically results in a robust, long-lived protective immune response against all strains of the organism. Successful vaccines exist for smallpox, measles, mumps, rubella, diphtheria and yellow fever, all of which have little or no diversity in their antigens and are not geographically or temporally variable (features of ideal vaccines are discussed in Levine and Sztein, 2004). However, the more common situation, and one of the biggest obstacles to vaccine development for the pathogenic *Neisseria*, is that multiple antigenically different strains/types exist for a particular pathogenic species (Lipsitch and O'Hagan, 2007). For these pathogens, infection typically leads to an immune response against the infecting agent that provides differing degrees of cross-immunity to other variants within the species, being more effective against the same (homologous) strain than against other (heterologous) strains. The development of vaccines for the pathogenic *Neisseria* has been further hampered by the highly variable nature of their antigens, as well as difficulties in the investigation and understanding of the host immune responses to these antigens in the absence of ideal animal models. *N. meningitidis* and *N. gonorrhoeae* have an array of mechanisms for developing antigenic variation, predominantly mediated by inherently unstable repetitive DNA sequences (e.g. phase variation, DNA uptake and recombination). A single bacterium can, therefore, give rise to a large number of offspring expressing a diverse range of outer membrane structures, from which organisms with increased fitness (i.e. able to avoid natural or vaccine-induces immune responses, or adapt to changing microenvironment) are selected. Therefore, if a vaccine is to be successful against the pathogenic *Neisseria*, either an antigen must be identified that is stable and conserved between strains, or a vaccine must be developed that incorporates several protective antigens that are common to the majority of disease isolates. Then, once potential vaccines are discovered there are additional regulatory, logistical, economic and social barriers that must be considered.

Many barriers to vaccine development for *N. meningitidis* have been overcome, driven by advances in technology (e.g. genetic engineering techniques, genome sequencing, chemistry techniques) that have increased the understanding of its antigenic variability, epidemiology, immunology, pathogenicity and host–pathogen interactions. At present, there are a number of different *N. meningitidis* vaccines licensed and in use around the world, including simple polysaccharide, conjugate polysaccharide and outer membrane vesicle (OMV) vaccines, all of which have particular benefits in certain situations (Table 11.1). However, the available vaccines also have certain limitations that mean the need for vaccine development continues (Cartwright, 2006; Frasch, 1995). Unfortunately, there is still no licensed universal vaccine against serogroup B meningococcal (MenB) disease or a gonococcal vaccine. This chapter will address the need for vaccines against the pathogenic *Neisseria*, the current state of neisserial vaccines, followed by a closer look at the difficulties associated with neisserial vaccine development, with a particular emphasis on MenB vaccines.

The need for neisserial vaccines

Neisseria meningitidis

Vaccination to prevent meningococcal disease is considered the best control strategy due to its worldwide distribution, rapid onset and progression, devastating effects, and the possibility of treatment failure. Meningococcal disease is thought to account for approximately 1.2 million cases a year and 135,000 deaths. The bulk of the meningococcal disease burden is taken by the developing world, which suffers from epidemic disease caused by serogroups of *N. meningitidis* that are not common in the developed world. A region south of the Saharan desert known as the 'meningitis belt' is prone to large epidemics of meningococcal disease, which occur every 5–12 years with disease rates as high as 10–1000 per 100,000. Endemic (sporadic) cases of meningococcal disease are the usual pattern of disease in the developed world and result in a case frequency of <1–20 cases per 100,000. However, small regions of the developed world also suffer from epidemic disease.

Meningococcal disease is characterized by its rapid onset and progression to meningitis and/or sepsis. A hospital emergency room study revealed that one-third of deaths occurred within 6 hours of admission and two-thirds within 18 hours (van Deuren *et al.*, 2000). Prompt modern medical intervention and antibiotic therapy has reduced the mortality rate from 70–80% to approximately 10% (Cartwright, 1995; Goldacre *et al.*, 2003; van Deuren *et al.*, 2000). However, up to 20% of survivors of meningococcal disease suffer persistent neurological sequelae (mental retardation, hearing loss, speech disorders), loss of limbs or paralysis. A compounding factor is that the early stages of meningococcal disease are hard to distinguish from other febrile illnesses, which often

Table 11.1 Features of currently available meningococcal vaccines

Vaccine type/features[a]	Target	Name (manufacturer[b])	Active constituents	Status and Features
Capsular polysaccharide (CPS)				
T-cell independent response Immunogenic in adults Not immunogenic in young infants No immunological memory Poor response to booster	Mono/ bivalent AC	Menomune or Mengivac A+C® (SP) AC Vax® (GSK)	Serogroup A, C CPS	Licensed in the 1960s, now used in high risk groups
	Trivalent ACW	(GSK, WHO)	Serogroup A, C, W-135 CPS	Licensed in 2003 for epidemic control in Meningitis belt countries
	Tetravalent ACWY	Menomune® (GSK)	Serogroup A, C, W-135, Y CPS	Licensed in 1989 in the US for >2 year olds
CPS-conjugate				
T-cell dependent response Immunogenic in infants, adults Decrease carriage Induce herd immunity Long-term immune memory	Monovalent C	Meningitec™ (Wyeth) Menjugate® (Novartis) NeisVac-C™ (Baxter)	Serogroup C CPS conjugated to tetanus toxoid or CRM_{197}, alum adjuvant	Introduced in UK in 1999, now in use throughout Europe
	Monovalent A	MVA (SI)	Serogroup A CPS	Anticipated to be in use by 2008–9
	Tetravalent ACWY	MCV4, Menactra™ (SP) Menveo™ (Novartis)	Serogroup A, C, Y, W-135 CPS each conjugated to diphtheria toxoid or CRM_{197}	Licensed in US in 2005, for >11 year olds Phase III in infants
Combination CPS-Conjugate				
Similar features to conjugate vaccines	MenC, Hib	Menitorix™ (GSK)	Hib and serogroup C CPS each conjugated to tetanus toxoid	Licensed in UK in 2005
Outer membrane vesicle (OMV)				
Strain-specific 'tailor-made' Mix of OMPs and LPS PorA is immunodominant antigen Low efficacy in young infants	Monovalent B	MenBvac (NIPH, Novartis) MeNZB (NZMoH, Novartis)	Strain 44/76 (B:15:P1.7,16) OMV Strain NZ98/254 (B:4:P1.7b,4) OMV	Used for epidemic control in Norway, NZ since 2004, respectively
	Bivalent BC	VA-MENGOC-BC (Finlay, GSK)	Strain CU385 (B:4P1.19,15) OMV, serogroup C CPS	Used for epidemic control in Cuba since 1989, also used in Brazil

a Vaccine abbreviations: CPS, capsular polysaccharide; OMV, outer membrane vesicle; OMP, outer membrane protein; Hib, *Haemophilus influenzae* type B.

b Company abbreviations: SP, Sanofi Pasteur; GSK, GlaxoSmithKline; WHO, World Health Organization; Novartis, Novartis Vaccines and Diagnostics (ex-Chiron); SI, Serum Institute India; NIPH, National Institute for Public Health Norway; NZMoH, Institute for Public Health New Zealand.

leads to treatment delays. Furthermore, concerns exist that widespread antibiotic resistance could emerge in *N. meningitidis*, as has been the case with *N. gonorrhoeae*. Resistance to sulphonamides emerged in the 1950s, followed by decreased susceptibility to penicillins and chloramphenicol in the 1980s and 1990s, respectively. Strains with decreased susceptibility to rifampicin, ciprofloxacin and tetracycline have also been reported (reviewed in Chapter 13).

Chemoprophylaxis remains effective in the control of small outbreaks by the treatment of known contacts (in schools, barracks, etc). However, issues of resistance, access to and availability of drugs prevents this from being a realistic approach to control large epidemic outbreaks. Given the delays and limited access to medical care in developing countries for patients with meningococcal meningitis or sepsis, major reductions in morbidity and mortality will be impossible in these areas without effective prevention strategies. Although several vaccines for meningococcal disease are licensed, no universal highly efficacious vaccine is available, and the need for research and development continues.

Neisseria gonorrhoeae

Vaccination is considered the best approach for appropriate control of gonococcal disease due to the high incidence of *N. gonorrhoeae* infection, the association of gonococcal disease and increased HIV infection, and the complications associated with asymptomatic cases and antibiotic resistance. Disease caused by *N. gonorrhoeae* is a significant health problem worldwide, with an estimated global incidence of 62 million cases a year, and an average of 22 million cases at any given time, most occurring in South and South-East Asia, sub-Saharan Africa and South and Central America (Gerbase *et al.*, 1998). However, reported numbers only represent a fraction of the actual incidence of infection due to under-reporting and the high number of asymptomatic cases. Gonococcal infections are the second most common reportable infectious disease in the United States, accounting for annual medical costs exceeding $1.1 billion (Aledort *et al.*, 2005). *N. gonorrhoeae* is typically characterized by a symptomatic localized inflammatory response of the urethra in men (urethritis) and the endocervix in women (endocervicitis). However, ascending reproductive tract infections in women can lead to complications such as pelvic inflammatory disease (PID; inflammation of the uterus, fallopian tubes, ovaries or other pelvic structures) in 10–40% of infected women. Sequelae of PID include infertility, ectopic pregnancy, fetal wastage, neonatal blindness and disseminated gonococcal infections (DGI) (Hook and Holmes, 1985). Infection with *N. gonorrhoeae* also significantly increases the risk of infection with HIV, with increased susceptibility to infection in HIV-negative patients and increased transmission by HIV-positive patients (Cohen *et al.*, 1997; Levine *et al.*, 1998).

While a single-dose treatment of antibiotics is able to cure gonorrhoea, antibiotic resistance is an increasing problem and continual surveillance is needed (Chapter 13). Penicillins and tetracyclines are no longer effective, and decreased susceptibility has appeared to newer macrolides, spectinomycin, quinolones and the third-generation cephalosporins. The ability of *N. gonorrhoeae* to rapidly acquire antibiotic resistance means that effective treatment is becoming unaffordable in communities with the highest burden of disease, and even newly developed antibiotics will only provide a short-term solution to control this disease (Tapsall, 2006).

N. gonorrhoeae is transmitted by direct sexual contact, has no reservoir other than the human host and survives poorly outside the human body. Therefore, in theory, it could be eradicated by vaccination. However, development of a vaccine has not been considered a major health priority by funding agencies, presumably due to the fact that infection is treatable by antibiotics, is not typically life threatening, and due to public opinion and the controversy that arises regarding vaccination against STIs. The link between *N. gonorrhoeae* and HIV, as well as the successful development of the human papillomavirus (HPV) vaccine (Ferenczy and Franco, 2007), has stimulated increased interest in a gonococcal vaccine. However, a recent expert opinion paper stated that even if a vaccine was developed, it would be unlikely to receive priority funding for production, trial, distribution and use within 5 years (Tapsall, 2006). To be maximally effective, a gonococcal vaccine should be given to adolescents before they become sexually active. As such, several social barriers and worries, such as non consent from parents and the fear of increased sexual risk-taking behaviours following STI vaccination, must be considered (Stanberry and Rosenthal, 2005).

Difficulties in conventional neisserial vaccine development

Failure of early classical whole cell and toxoid vaccinology approaches

The first barriers to vaccine development were realized with the failure of initial attempts to generate meningococcal and gonococcal vaccines by classical whole cell methods. Killed whole cell vaccines against *N. meningitidis* were first used in the 1930s, but showed only low levels of protection (believed to be serogroup-specific and due to capsule polysaccharide) and had generally unacceptable levels of adverse reactions (reviewed in Cartwright, 1995). Live attenuated vaccines for *N. meningitidis* were never seriously pursued as a valid approach due to safety concerns over the testing and use of live meningococci. *N. meningitidis*, like *N. gonorrhoeae*, are highly competent for DNA uptake (discussed below), providing a potential mechanism for reversion of attenuated strains by uptake of virulence-associated genes. It was also feared that attenuated strains could cause disease in individuals with a higher susceptibility to meningococcal infection due to complement deficiencies or other predisposing factors. However, the advent of molecular biology and the genomic era has greatly improved the engineering of bacteria to ensure attenuation, and has resulted in renewed discussion and investigation in recent years on the use of live attenuated strains of *N. meningitidis* (Li *et al.*, 2004; Newcombe *et al.*, 2004).

The search for a meningococcal exotoxin was also initiated in the 1930s (Ferry and Steele, 1935; Kuhns *et al.*, 1938) in an attempt to develop a vaccine similar to the successful tetanus and diphtheria toxoid vaccines. However, pathogenic *Neisseria* do not produce typical exotoxins and any toxin-like clinical effects were most likely a result of the lipid A component of meningococcal lipopolysaccharide (LPS), a potent endotoxin that causes cytopathic damage to endothelial cells (Dunn *et al.*, 1995).

Classical methods of vaccinology also failed to enable the development of a vaccine for *N. gonorrhoeae*. Initial attempts at vaccination involved the use of either whole cells (Eyre and Stewart, 1909) or killed/autolysed bacteria (Greenberg *et al.*, 1974). However, these vaccines did not provide protection in clinical trials and were not pursued further.

Successes and limitations of meningococcal capsular polysaccharide vaccines

The inability and/or reluctance to use classical methods of vaccinology required research to focus on the identification and characterization of specific components that could be used in subunit vaccines against the pathogenic *Neisseria*. The possibility of vaccination against *N. meningitidis* was realized decades ago, with the discovery that carriage or an episode of meningococcemia is an immunizing event leading to protective immunity against the organism (Goldschneider *et al.*, 1969a,b). Carriage of *N. meningitidis*, and possibly *Neisseria lactamica*, strains induces an immunologic response to *N. meningitidis* so that by young adulthood, the majority of people have measurable antibody to the pathogenic serogroups of *N. meningitidis* (Edwards *et al.*, 1977; Goldschneider *et al.*, 1969b; Robbins *et al.*, 1972). In the 1960s it was discovered that serum antibodies directed against the capsule play a major part in protection against meningococcal disease by promoting complement-mediated killing of the organism (Goldschneider *et al.*, 1969a, b; Gotschlich *et al.*, 1969) (Chapter 8). This provided a potential target for vaccine development, and the use of capsular polysaccharides in protection had already been established in the 1940s with the development of a pneumococcal capsular polysaccharide vaccine (MacLeod *et al.*, 1945).

N. meningitidis strains have traditionally been classified into twelve serogroups based on their antigenically distinct capsular polysaccharides. Six of the 12 serogroups (A, B, C, W-135, X and Y) account for more than 90% of meningococcal disease (Pollard, 2004). Therefore, at a minimum, vaccines should aim to protect against these variants. To overcome the antigenic diversity of the meningococcal capsule, vaccines to date have either focused on a particular serogroup for use in a particular region, or have taken a multivalent approach to raise immunity against several clinically significant antigenic (capsular) types (see Table 11.1 for currently available meningococcal vaccines).

The MenB capsular polysaccharide has homology to the human polysialic acid expressed in various tissues. As a result of this host mimicry, it is poorly immunogenic and does not induce a protective response (Finne *et al.*, 1987; Nedelec *et al.*, 1990). This mimicry has also raised significant concerns about autoimmunity in vaccinated individuals, although this has not been observed to date.

N. gonorrhoeae is not capsulated. Consequently, this approach is not relevant for the development of a gonococcal vaccine.

Simple polysaccharide vaccines available for MenA, C, W-135 and Y

Several combinations of serogroup A, C, W-135 and Y polysaccharides have been used in vaccines over the years in different countries, and there are currently bivalent, trivalent and tetravalent polysaccharide vaccines in use (Table 11.1). One of the first meningococcal vaccines, a bivalent AC vaccine, was developed in the late 1960s in response to military outbreaks and epidemics in Africa (Gotschlich *et al.*, 1969). In 1982, the tetravalent A, C, W-135, Y polysaccharide vaccine (Menomune) was licensed for use in the US for individuals 2 years of age and older (Vodopija *et al.*, 1983). A trivalent A, C, W-135 vaccine was recently formulated for use in Africa and is discussed further below (WHO, 2004). The polysaccharide vaccines are safe with only mild local adverse events, and are immunogenic and efficacious (>85%) in children older than 2 years and adults (Al-Mazrou *et al.*, 2005; Gotschlich *et al.*, 1969). However, these vaccines are typically only used for people in high-risk groups and in response to epidemics due to several limitations.

Limitations of simple polysaccharide vaccines

Polysaccharide vaccines are poorly immunogenic in young infants and children less than 2 years old, the major group at risk of meningococcal disease. For example, a recent study of a tetravalent polysaccharide vaccine showed low bactericidal antibody responses to all serogroups in 6- and 12-month-old infants (e.g. 29%, 11%, 13% and 14% of 12-month olds had a SBA titre of ≥ 8 at 1 month post vaccination against serogroup A, C, W-135, and Y, respectively)(Al-Mazrou *et al.*, 2005). Polysaccharide vaccines do not reliably induce immunological memory due to their failure to generate memory T-cells. Protection wanes over time and is short-lived (1–3 years) in children younger than 5 years of age, while antibodies have been detected in adults up to 10 years after vaccination (Reingold *et al.*, 1985; Zangwill *et al.*, 1994). Furthermore, attempts to boost protection with repeated vaccination may result in immunological hyporesponsiveness, i.e. diminished antibody response (Borrow *et al.*, 2000). Another possible drawback is that the polysaccharide vaccines do not prevent mucosal colonization or nasopharyngeal carriage and therefore do not provide herd immunity (Stephens *et al.*, 2007). These limitations are common to other polysaccharide vaccines, including vaccines against *Streptococcus pneumoniae* and *Haemophilus influenzae* type b (Hib). Despite their limitations, the use of these polysaccharide antigens has been the founding step for further development of neisserial vaccines.

Conjugate polysaccharide vaccines available for MenA, C, W-135 and Y

Many of the problems of simple polysaccharide vaccines were overcome with the discovery that the chemical conjugation of the capsular polysaccharide to an immunogenic protein carrier (e.g. CRM197, a genetically detoxified diphtheria toxoid, or tetanus toxoid) greatly increased its immunogenicity. This led to the development of several successful meningococcal monovalent group C and tetravalent A, C, W-135, Y conjugate vaccines (Table 11.1) (Costantino *et al.*, 1992; Granoff *et al.*, 2008). In contrast to the simple polysaccharide vaccines, conjugated polysaccharide vaccines elicit a T-cell-dependent immune response (Lesinski and Westerink, 2001a,b), which enables them to be immunogenic in children under 2 years of age and induce immunological memory. Furthermore, revaccination provides a booster response, with no immune hyporesponsiveness (Richmond *et al.*, 1999; Richmond *et al.*, 2000). These conjugate vaccines have also been shown to decrease nasopharyngeal carriage of meningococci (Ramsay *et al.*, 2003).

The monovalent meningococcal group C (MenC) conjugate vaccines provide long-term protection and are suitable for children as young as 6 weeks of age, older children and adults (Table 11.1) (Lakshman and Finn, 2002; Pollabauer *et al.*, 2005). These vaccines were introduced into the UK in 1999 in response to the rising rates of MenC that had been seen throughout Europe in the mid 1990s and have had a huge impact on the incidence of disease, resulting in a more than 90% decrease in the number of deaths and clinical cases, and a 66% decrease in asymptomatic carriage (Ramsay *et al.*, 2003; Trotter *et al.*, 2004). Similar campaigns have been successful in decreasing the rate of disease in Ireland, Spain, the Netherlands, Belgium, Iceland, Portugal (Trotter and Ramsay, 2007), Canada (De Wals *et al.*, 2007) and Australia (Booy *et al.*, 2007).

The tetravalent A, C, W-135, and Y conjugate vaccine (MCV4 or Menactra, Sanofi Pasteur) contains each of the 4 serogroup polysaccharides conjugated separately to diphtheria toxoid (Table 11.1). This vaccine was approved in 2005 in the US for use in persons from 2–55 years of age (Bilukha and Rosenstein, 2005), and is now recommended for routine vaccination of all persons aged 11–18 years and high risk groups (CDC, 2007). However, this conjugate vaccine is not licensed for infants. This vaccine elicits an initial antibody response that is similar to that of the simple polysaccharide vaccine but is more durable, and revaccination can elicit a rise in antibody levels. The vaccine has shown 100% efficiency at raising bactericidal antibody titres in 11–18 year olds (Pichichero, 2005; Pollard, 2004). A new MenACWY–CRM conjugated vaccine (Menveo™, Novartis Vaccines) has been shown during a Phase II study to be well tolerated and elicit a strong immune response in infants as young as 6 months old (Table 11.1) (Snape *et al.*, 2008). This vaccine is currently in multiple phase III clinical trials and is on track for regulatory submissions in 2008 (Novartis, 2007).

Limitations of conjugate polysaccharide vaccines

While the development of conjugate vaccines overcame many of the limitations of the simple polysaccharide vaccines, several concerns and limitations remain based on the use of the capsule polysaccharide as the sole antigen and the use of serogroup-specific vaccines. The polysaccharide capsule plays an important role in meningococcal fitness, facilitating its transmission, colonization and virulence by protecting the meningococcus from desiccation, phagocytic killing, opsonization, and complement-mediated bactericidal killing (Chapters 7–10) (Frosch and Vogel, 2006; Schneider *et al.*, 2007). However, the capsule is not essential for survival outside the bloodstream, and has variable expression both within a single infection and within the population. While the majority of disease-causing strains are capsulated, up to 30% of meningococcal isolates from carriers are acapsulate and non-serogroupable (Dolan-Livengood *et al.*, 2003; Yazdankhah and Caugant, 2004).

Variable expression of the meningococcal capsule is seen due to phase variation (described below and in Chapter 3) and capsule switching, providing a selective advantage that allows the organism to evade opsonization or neutralization by natural or vaccine-induced protective anticapsular antibodies. Meningococci that cannot be grouped (acapsulate) can be recovered from the nasopharynx of contacts with identical genetic markers to outbreak strains (Dolan-Livengood *et al.*, 2003). Capsule switching between serogroups B and C has been reported during several outbreaks (Stefanelli *et al.*, 2003; Swartley *et al.*, 1997) and has even been seen to occur rapidly within the contact of an index case leading to a secondary case (Vogel *et al.*, 2000). This switching was discovered by molecular analysis of meningococcal strain collections that contain isolates with otherwise identical genetic markers (e.g. ET-type) but that express different capsular polysaccharides (Caugant *et al.*, 1986; Swartley *et al.*, 1997). Reports have emerged suggesting that capsule switching occurred after vaccination campaigns using serogroup C conjugate vaccine in Spain (Alcala *et al.*, 2002) and France (Lancellotti *et al.*, 2006), but there has been no clinically significant evidence of this occurring in the UK after mass vaccination (Trotter *et al.*, 2006). Switching results from DNA transformation and horizontal gene transfer between MenB and C strains co-colonizing the human nasopharynx, resulting in allelic exchange of capsule biosynthesis genes (e.g. the polysialyltransferase gene *siaD*).

Different meningococcal vaccines have been adopted for use in different countries depending on the major serogroup present. However, the changing nature of the predominant strains causing endemic and epidemic disease in different regions highlights the danger associated with using serogroup-specific vaccines. For example, serogroup A is responsible for the majority of epidemic disease in the African meningitis belt. However, since 2000, serogroup W-135 has also emerged as a cause of these outbreaks (Mueller *et al.*, 2006). MenB is the main cause of epidemic disease in industrialized countries; an outbreak of MenB disease has been ongoing in New Zealand since 1991 (Dyet *et al.*, 2005), with previous clonal epidemics of MenB disease occurring in Brazil (Sacchi *et al.*, 1992), Norway (Lystad and Aasen, 1991), Chile (Cruz *et al.*, 1990) and Cuba (Sierra *et al.*, 1991). MenB is also responsible for most of the endemic meningitis in industrialized countries, accounting for 30–40% of cases in North America and for up to 80% in certain European countries, with most of the remaining cases being caused by serogroup C strains. Serogroup A has declined in incidence to very low levels in these regions since the Second World War (Tikhomirov *et al.*, 1997). The incidence of endemic serogroup Y disease in the US has increased significantly from 2% in 1989–1991 to 39% for the period 1996– 2001, whereas this serogroup is an infrequent cause of disease in Europe (Pollard and Scheifele, 2001; Rosenstein *et al.*, 1999; Yazdankhah *et al.*, 2004). There was initial concern about serogroup replacement after the introduction of the monovalent MenC vaccine in the UK (e.g. increased MenB disease in the vacancy created by the elimination of serogroup C organisms). However, this has not emerged as a major issue (Maiden and Spratt, 1999). These problems would be overcome with a universal meningococcal vaccine, but should also be reduced with the increased use of tetravalent vaccines (Table 11.1).

Other concerns over these vaccines include the possibility that the vaccine-induced immunological priming for memory responses may not be sufficient to protect against acute meningococcal infection due to the rapid onset of disease, and it has been suggested that the persistence of circulating antibodies is required (Spoulou *et al.*, 2007). Another interesting issue is that since meningococcal carriage is an immunizing event, vaccines should prevent invasive disease but preserve carriage (Taha and Alonso, 2007). Questions also remain regarding the optimal dose, schedule and boosting for conjugate vaccines, as well as the best conjugates and conjugate methodology to use. These issues are further complicated by the perceived high cost of the conjugate vaccines and the determination of cost-effective schedules (De Wals *et al.*, 2007; Shepard *et al.*, 2005). Although no serious adverse events have been associated with the conjugate vaccines, reports of 17 cases of Guillain–Barré syndrome (a rate within the expected background incidence) temporally related to vaccination in the US has prompted a warning to be included in the package insert for MCV4 and could cause public concern about the use of this vaccine (CDC, 2006).

Vaccines to overcome the burden of disease in the meningitis belt are within reach

A crucial challenge for meningococcal vaccine development is the control of disease in the developing world. The limited duration of protection of simple polysaccharide vaccines, the absence of efficacy in infants and young children, and the cost and limited production capacity of conjugate polysaccharide vaccines, makes large-scale vaccination programs impractical and/or unaffordable in developing countries. Significant support is needed from the international community to ensure that existing and new vaccines are available in developing countries, especially in the meningitis belt of sub-Saharan Africa. Currently, WHO, the Meningitis Vaccine Project (MVP) and the International Coordinating Group on Vaccine Provision for Epidemic Meningitis (ICG) provide valuable 'Push' and 'Pull' mechanisms to increase development and access of meningococcal vaccines for the developing world. The ICG, established by WHO in 1997, coordinates and finances an international vaccine stockpile that is made available to countries in the African meningitis belt during the epidemic season. The MVP launched an initiative in 2001, in partnership with WHO and the Program for

Appropriate Technology in Health (PATH), to develop a monovalent serogroup A conjugate vaccine at an affordable price for the developing world, to be produced by the Serum Institute of India Ltd (Laforce *et al.*, 2007). This serogroup A tetanus toxoid conjugate vaccine (PsA-TT) has been shown to be safe, immunogenic, and able to elicit persistent functional antibody titres in adults (Kshirsagar *et al.*, 2007) and toddlers (Viviani, 2007). MVP plans to produce 25–35 million doses per year at a cost of less than US$0.5 per dose, and to license the vaccine for single-dose mass vaccination campaigns of people from 1 to 29 years of age in African meningitis belt countries, starting in 2008–2009. Another public–private initiative was formed between WHO and GlaxoSmithKline Biologicals after the emergence of a serogroup W-135 epidemic in Burkina Faso in early 2002. A trivalent ACW-135 simple polysaccharide vaccine was licensed, produced and ready for use in the African meningitis belt by the 2003 epidemic season (Table 11.1). With more than 3 million doses used during 2006 (from a stockpile of approximately 6 million), this trivalent vaccine is currently one of the main tools for epidemic control in Africa (Nelson *et al.*, 2007).

Barriers to development of the next generation neisserial vaccines: MenB vaccines

Since the widespread introduction of the conjugate MenC vaccine into infant immunization schedules, and the increased use of tetravalent vaccines in the US, MenB remains the largest challenge in confronting meningococcal disease in the developed world. Serogroup B strains are responsible for the majority of endemic and epidemic meningococcal disease in developed countries; one-third of disease in the US (Rosenstein *et al.*, 1999) and more than half in Europe (Noah and Henderson, 2002). The global incidence has been estimated at between 20,000 and 80,000 cases per year, accounting for 2000–8000 deaths annually. The highest rate of MenB disease occurs in infants under the age of 1 year, as a result of waning maternal antibodies (Cartwright *et al.*, 2001). The need to produce protection in very young infants is a major challenge due to the immaturity of the immune system, the slow development of protective immune responses from vaccines, and the reliance on booster doses to produce lasting protection in infants. Serogroup B has two main epidemiological patterns. Unlike serogroup A and C epidemics that usually resolve in 1–3 years, MenB epidemics evolve slowly and may persist for more than a decade. By contrast, endemic disease is caused by very diverse strains, and understanding this diversity and developing a vaccine that protects against all, or the majority of, strains is an enormous task.

Major outer membrane structures as candidate vaccine antigens

Characterization of non-capsular vaccine targets has been the main focus of MenB vaccine development, due to the inability to use the capsular polysaccharide-based approach that has been adopted for serogroup A, C, W-135 and Y vaccines. A number of surface structures involved in infection and pathogenicity (Merz and So, 2000; Nassif *et al.*, 1999) have been identified as potential vaccine candidates and include LPS and outer membrane proteins (OMPs) such as the porins (PorA, PorB), Opa and type IV pili (Table 11.2). Following infection by MenB, antibodies against LPS and several OMPs, including PorA, PorB and Opa, have been detected, indicating the potential of these structures to be used in subunit vaccines (Griffiss *et al.*, 1984; Pollard and Frasch, 2001; Pollard *et al.*, 1999). However, the pathogenic *Neisseria* are able to avoid the immune response by mimicry of host components, the expression of antigens that cloak or block killing by the immune system and an unusual number of mechanisms to generate antigenic heterogeneity and variability, all of which have made development of OMP or LPS subunit vaccines a difficult task. Nonetheless, several of these antigens have been assessed for use in vaccines, with limited success in providing a comprehensive MenB vaccine. The only currently available MenB vaccines, the strain-specific outer membrane vesicle (OMV) vaccines (Table 11.1), will be described, followed by an overview of the barriers that exist in developing a broadly protective MenB vaccine. The general theme that emerges is the need for a multi-component vaccine to provide comprehensive protection and minimize the possibility of vaccine escape.

Table 11.2 Possible vaccine candidates of *N. meningitidis*

Ag	Function/comments	Diversity, variability, immunomodulation	Vaccine approaches
Major OM structures			
CPS	Involved in resistance to serum Absent in NG and NL	13 serogroups- A, B, C, E-29, H, I, K, L, M, W-135, X, Y, Z PV due to SSR, IS1301 in *siaA*, *oatWY* AV-serogroup switching by horizontal recombination Serogroup B CPS mimics human glycoproteins	Antigen of all NM A, C, W-135, Y vaccines (Table 11.1) N-propionylate MenB CPS (Bruge *et al*., 2004) Peptide mimetics (Lo Passo *et al*., 2007) and anti-idiotypes of non cross-reacting epitopes(Park *et al*., 2005)
LPS	Principal component of OM Lipid A component is a potent endotoxin	12 immunotypes, most strains have 1–6 antigenically different LOS Variable length and composition of cores sugars and terminal epitopes Lacto-N-neotetraose moiety in some strains is similar to host glycoproteins Sialylation masks LOS and nearby Ag	Invariant inner core (Plested *et al*., 2003) Peptide mimetics of conserved epitopes (Ngampasutadol *et al*., 2006)
Pili	Long protruding polymeric proteins involved in colonization, motility, transformation	Class I and II AV- recombination between *pilE* expression locus and *pilS* silent loci PV- post translational modifications- glycosylation, phosphorylcholine Soluble pilin scavenges Ab	Pilus-associated protein Q (PilQ) present in OMV vaccines (Williams *et al*., 2007)
Por	Most abundant OMP, constitutive transmembrane Function as molecular pores for ion exchange	PorA- 10 Serosubtypes, PorB- 20 Serotypes Sequence typing of *porA* VR1 and VR2 PV- polyG between the –10 and –35 domains of promoter of *porA* Deletion by recombination between flanking repetitive regions of *porA*	Immunodominant Ag of OMV vaccines (Holst *et al*., 2005); (Rosenqvist *et al*., 1991) Multivalent Por OMV vaccines (Luijkx *et al*., 2006; van den Dobbelsteen *et al*., 2007; van der Ley *et al*., 1995) Removal from OMV (Weynants *et al*., 2007)
Opa	Involved in host cell adherence and invasion	Opa repertoires vary between strains ~ 3–4 *opa, opaA, opaB, opaD, opaJ* loci throughout the genome Sequence typing based on sequencing of HV1, HV2, SV Recombinational reassortment among opa within and between strains PV- CTCTT repeats in ORF May act as immunomodulators via interaction with immune system	Liposomes with recombinant OpaB, OpaJ (de Jonge *et al*., 2004)
Opc	Involved in host cell adherence and invasion	2 paralogous *opc* genes (*opcA* is transcribed, *opcB* is a pseudogene) PV- C repeats in promoter	Key Ag of OMV vaccines (Holst *et al*., 2005)
Rmp	Abundant, unknown function	Antibodies to Rmp 'block' other functional antibodies	Key Ag of OMV vaccines (Holst *et al*., 2005) Removed from OMV vaccines (Claassen *et al*., 1996)

Ag	Function/comments	Diversity, variability, immunomodulation	Vaccine approaches
Other OMPs of interest			
fHBP	GNA1870/LP2086 Lipoprotein	Variants 1–3, subvariants (SBA activity is variant-specific) Binds complement regulator, factor H, increasing resistance to complement killing	Immunodominant Ag of 5CVMB (Giuliani *et al.*, 2006b) Overexpression in OMV (Koeberling *et al.*, 2007) LP2086 from 2 subfamilies (Zhu *et al.*, 2005)
NadA	Adhesin, invasin, member of OCA family	PV – TAAA repeats in promoter, alter binding of IHF transcription factor PV- IS1301 insertion	Ag of 5CVMB (Giuliani *et al.*, 2006b)
TbpA, TbpB	Transferrin binding protein Regulated by iron and Fur	Heterogeneous, TbpB more so than TbpA	Recombinant TbpA and TbpB (West *et al.*, 2001) Overexpression of TbpA in OMV (Weynants *et al.*, 2007)
LbpA, LbpB	Lactoferrin binding protein Regulated by iron and Fur	Heterogeneous, LbpB more so than LbpA	Overexpression of LbpB in OMV, purified LbpA (Pettersson *et al.*, 2006)
NspA	Neisserial surface protein A	Conserved	Recombinant NspA vaccine (Halperin *et al.*, 2007) Present in OMV vaccines (Holst *et al.*, 2005) Overexpression in OMV vaccines (Weynants *et al.*, 2007)
FetA	Ferric enterobactin transport protein A (was FrpB) Iron-limitation-induced	Highly variable exposed loops, typing into FetA variant families based on variable region Deletion by recombination between flanking repetitive regions	Combination of six PorA, 5 FetA variants to provide broad protection (Urwin *et al.*, 2004) Present in OMV vaccines (Holst *et al.*, 2005)

OM, outer membrane; CPS, capsule polysaccharide; LPS, lipopolysaccharide; Ab, antibody; Ag, antigen; NM, *N. meningitidis*; NG, *N. gonorrhoeae*; NL, *N. lactamica*; PV, phase variation; AV, antigenic variation; SBA, serum bactericidal assay; SSR, simple sequence repeats.

Tailor-made outer membrane vesicle vaccines for MenB

A variety of 'tailor-made' MenB OMP-based OMV vaccines have been developed and licensed to control epidemics dominated by a single clone. OMV vaccines have been used in Norway (Fredriksen *et al.*, 1991), Cuba (Sierra *et al.*, 1991), Chile (Boslego *et al.*, 1995) and New Zealand (Oster *et al.*, 2005). They are able to induce protective antibodies against the homologous strain in all age groups, and have proved successful in curbing epidemic disease (see Table 11.1 for details). OMV vaccines consist of intact outer membrane and contain reduced amounts of LPS and a complex mixture of OMPs. The most abundant antigens are PorA, PorB, RmpM and OpcA, but other antigens present include FetA, FbpA, NspA and TdfH (Ferrari *et al.*, 2006; Holst *et al.*, 2005; Uli *et al.*, 2006; Vipond *et al.*, 2006). OMVs are prepared by detergent extraction, followed by purification to reduce the amount of LPS to non-toxic levels and to remove aggregated/denatured proteins (Frasch *et al.*, 2001).

The main limitation of OMV vaccines is that they are strain specific and do not provide protection against heterologous (non-vaccine) strains due to the antigenic diversity of their protein antigens (Table 11.2, Chapter 3). PorA has been identified as the immunodominant antigen and the main target of serum bactericidal antibodies (Rosenqvist *et al.*, 1991). Due to the antigenic diversity of PorA, cross-protection is not seen between strains with heterologous PorA antigens (Tappero *et al.*, 1999), limiting the use of the available OMV vaccines to clonal disease outbreaks, rather than for prevention of endemic disease caused by diverse strains. This has been particularly evident with the New Zealand vaccine MeNZB™, which has been successful in reducing disease caused by the epidemic strain, but has not affected the incidence of MenB disease caused by other strains (Martin, 2007). It has been estimated that 20 different PorA proteins would have to be included in an OMV vaccine to protect against 80% of the MenB strains circulating in the USA (Tondella *et al.*, 2000). The use of OMV vaccines requires substantial epidemiological data, disease surveillance and strain characterization (Chapter 12).

Antigenic diversity and variability

Population structure of N. meningitidis

An understanding of population diversity and dynamics, and the mechanisms that drive them, is fundamental to vaccine design. Compared to clonal species like Salmonella and *Escherichia coli*, *N. meningitidis* and *N. gonorrhoeae* have non-clonal populations (Maiden and Caugant, 2006; Smith *et al.*, 1993), which undergo frequent horizontal gene transfer and recombination events due to their natural competence for DNA uptake (Maiden, 1993)(see below for details). This has resulted in the existence of numerous diverse strains within the population, which differ with respect to geography and can change over time, posing an enormous problem for the selection of appropriate vaccine antigens.

N. meningitidis strains have traditionally been typed using serological methods based on the antigenic diversity of their surface structures: 12 serogroups based on their distinct capsular polysaccharides (see above for details); 13 immunotypes based on LPS (Scholten *et al.*, 1994); and numerous serotypes and serosubtypes based on the major outer membrane proteins PorB and PorA, respectively (Frasch *et al.*, 1985). A sequence-based method for meningococcal serosubtyping, determined by the sequence of the PorA variable regions, VR1 and VR2, in the surface-exposed loops, has identified 163 variants of VR1 and 444 of VR2 to date (http://*Neisseria*.org/nm/typing/pora). These typing schemes demonstrate the antigenic diversity that exists within this species and the problem of using these structures as vaccine antigens.

Multi-locus sequence typing (MLST) is now considered the gold standard for genetically typing variable bacteria, and has revealed the dynamic and diverse nature of *N. meningitidis*, isolated from both asymptomatic carriage and disease-associated populations (Brehony *et al.*, 2007; Caugant, 1998; Yazdankhah *et al.*, 2004). MLST is based on the similarity of the nucleotide sequence type (ST) of several housekeeping genes that are thought to be selectively neutral. More than 6000 ST profiles have been identified, forming 37 'ST complexes' (http://pubmlst.org/*Neisseria*/), with a few of these complexes (e.g. ST-11, ST-32, ST-8, ST-7, ST-41/44, ST-1, ST-4 and ST-5) causing the majority of disease. The distribution of these hypervirulent complexes, like serogroups, is temporally and geographically variable (Caugant *et al.*, 2007). A limitation of MLST for vaccine development is its reliance on housekeeping genes that are not exposed to the host immune system. Surface-exposed antigens show high levels of sequence diversity among strains due to evolutionary pressure to avoid the host immune response. Such antigens should be included in new typing methods to investigate population structures in a way that is relevant for vaccine development.

Antigenic variation

In the pathogenic *Neisseria*, the major surface structures (e.g. pili, Opa, LPS) are all subject to antigenic variation (Table 11.2 and Chapter 3). This is primarily the result of differential expression of individual genes within a gene family to produce functionally conserved but antigenically distinct antigens. The random switch to expression of previously silent genes enables avoidance of immune clearance by natural or vaccine-induced immune responses. A vaccine must, therefore, be able to cover all possible antigenic variants. For example, *N. meningitidis* strains have 3–4 outer membrane Opa proteins, each of which is antigenically different and under a genetic switch that independently controls expression so that, during infection, bacteria can express none, one or multiple Opas in various combinations (Aho, 1989). Antigenic variation of pili is mediated by a different mechanism. Pilin of *Neisseria* is expressed from the *pilE* locus, but homologous recombination between the *pilE* gene and a number of non-expressed 'silent' *pilS* genes results in a change in the *pilE* sequence. These variants differ in their transformability, adherence and immunogenicity (Virji *et al.*, 1992).

Phase variation

Vaccine design is further complicated by the fact that pathogenic *Neisseria* use mutational 'hotspots' to mediate phase variation, the high frequency reversible on/off or graded switching of gene expression of many surface-exposed virulence factors, including capsule, LPS, pili, Opa, Opc and PorA (see Table 11.2 and Chapter 3 for details). Vaccine-induced immune pressure against a phase variable antigen can result in selection of a subpopulation that has expression of that antigen switched off and is not targeted by the vaccine. Phase variation in *Neisseria* is typically mediated by simple repetitive DNA motifs that exhibit high mutation rates by a slipped-strand mispairing (SSM) mechanism, resulting in the loss or gain of repeat units during replication (Bayliss *et al.*, 2001; Maskell *et al.*, 1993; Power and Moxon, 2006). While phase variation is seen in several pathogenic bacteria (Moxon *et al.*, 2006), the pathogenic *Neisseria* are unique due to the high number of genes, over 100, that are potentially phase variable (Martin *et al.*, 2003; Snyder *et al.*, 2001). In contrast, *H. influenzae* contains less than 20 (Hood *et al.*, 1996), and *E. coli* has less than 10 potentially phase variable genes (Blattner *et al.*, 1997).

Many phase variable genes in *Neisseria* undergo simple on/off switching. For example, changes in the number of CTCTT repeats within the open reading frame of *opa* can introduce frameshift mutations in the protein (Stern *et al.*, 1986). In contrast, Opc contains a homopolymeric tract of cytosine residues in its promoter region and exhibits strong, intermediate or weak expression as the number of repeats affects critical spacing within the promoter (Sarkari *et al.*, 1994). Phase variation of LPS expression, pili expression and pili post-translational modification is mediated by several genes that phase vary independently, enabling the organism to randomly switch between a huge repertoire of immunologically distinct structures (Kahler and Stephens, 1998; Power *et al.*, 2003). Insertion and excision of *IS1301* into genes of *N. meningitidis* also mediates phase variation of capsule expression [*siaA* (Hilse *et al.*, 1996), *siaD* (Sadler *et al.*, 2003)] and OMP expression (*porA* (Newcombe *et al.*, 1998), *nadA* (Comanducci *et al.*, 2002)).

An 'indirect' mechanism of phase variation, the 'phasevarion' (phase variable regulon), has recently been identified in pathogenic bacteria, which poses a new problem for vaccine development (Fox *et al.*, 2007; Srikhanta *et al.*, 2005). Two distinct phase variable methylases of type III restriction modification systems have been described in *N. meningitidis* and *N. gonorrhoeae* that act as phasevarions. The phase variable on/off switching of the methylase alters the methylation pattern of the genome, and can alter the expression of multiple genes by changing the affinity of regulatory factors for the promoter of the gene (Srikhanta *et al.*, 2009). The phasevarion of *N. meningitidis* includes the lactoferrin receptor proteins, LbpA and LbpB, which are vaccine candidates in *N. meningitidis* (Pettersson *et al.*, 2006). These proteins were not previously identified as potentially phase variable due to the lack of typical features of phase variable genes within their open reading frame or promoter.

Transformation, recombination and horizontal gene transfer

DNA uptake and recombination is a common feature of pathogenic *Neisseria*, which leads to antigenic variability due to the addition of new genes, allelic replacement, the creation of new mosaic alleles, or the deletion of alleles. Meningococci are naturally competent and readily take up DNA from the environment by transformation, and undergo frequent recombination events (Table 11.2 and Chapter 3)(Hamilton and Dillard, 2006). Transformation and intra- or inter-species horizontal gene transfer has had a huge impact on the evolution of the pathogenic

Neisseria, as is particularly evident in the spread of antibiotic resistance in certain strains (Spratt *et al.*, 1992). This process also enables *N. meningitidis* to acquire additional pilin genes in their *pilS* loci (Seifert *et al.*, 1988) and switch capsule types (described above) (Swartley *et al.*, 1997). In the genome of *N. meningitidis* Z2491, 60 coding regions (one-third of which were predicted to be surface-exposed and thus potential vaccine candidates) had a significantly lower G+C content than the average G+C content of the genome, a common feature of recently acquired DNA (Parkhill *et al.*, 2000). In addition, strains have been isolated in which potential vaccine candidates have been deleted as a result of recombination at repetitive elements flanking the genes, including *porA* (van der Ende *et al.*, 1999) and *fetA* (Claus *et al.*, 2007). The possibility of a broadly protective vaccine has been proposed based on six PorA and five FetA variant sequences to overcome the variability of these antigens (Urwin *et al.*, 2004). However, the fact that *porA* and *fetA* deletions have been seen supports the need for multivalent vaccines.

DNA repair mechanisms and mutator strains

Increased rates of random mutation and phase variation have been identified in 'mutator clones' of *N. meningitidis* (Giraud *et al.*, 2001; Richardson *et al.*, 2002), representing another mechanism that allows avoidance of the immune system and vaccine-induced immunity. This does not directly affect the selection of vaccine antigens. However, it does increase the need for post vaccine surveillance to monitor for vaccine escape mutants (Chapter 12). Mutator clones increase in frequency towards the end of an epidemic, presumably due to immune pressure leading to selection of these hypermutable bacteria with an increased ability to avoid immune recognition. This may also be relevant to vaccine-induced immune pressure. In a panel of epidemic serogroup A strains, 57% had substantially increased frequencies of missense (up to 10-fold) and phase variation (10- to 100-fold) mutation rates (Achtman, 2004; Richardson *et al.*, 2002). The global increase in mutation of these strains is often the result of mutations within genes of the mismatch repair (MMR) pathway, typically *mutL*, *mutS* and *mutY* (Colicchio *et al.*, 2006; Davidsen *et al.*, 2005; Richardson *et al.*, 2002). Decreased activity of the MMR system has also been associated with increased transformation and DNA recombination in meningococci (Alexander *et al.*, 2004). In addition, the DNA-repair profile of *N. meningitidis* (Davidsen and Tonjum, 2006; Davidsen *et al.*, 2007) and *N. gonorrhoeae* (Kline *et al.*, 2003) is reduced compared to *E. coli*. While the major DNA-repair pathways [base excision repair (BER), MMR, and nucleotide excision repair (NER)] are present, these pathways have reduced complexity and a functional SOS response pathway is absent.

Immune evasion and diversion

The pathogenic *Neisseria* have an array of mechanisms to decrease the immunogenicity of antigens (e.g. host mimicry, antigen masking or shedding, formation of blocking antigens), thereby decreasing their effectiveness as vaccine targets (Table 11.2, Chapters 7–10). Attempts have been made to circumvent the low immunogenicity and/or auto-immunity issues of some surface structures by the chemical alteration or rational redesign of these antigens. Mimicry of human structures by the MenB polysaccharide capsule is discussed above and has so far prevented the use of polysaccharide vaccines for this serogroup. A Phase I clinical trial was conducted with a modified MenB capsular polysaccharide-tetanus toxoid conjugate vaccine, in which the N-acetyl groups of the sialic acid residues of the polysaccharide were replaced with N-propionyl groups (Bruge *et al.*, 2004). The vaccine was considered safe and immunogenic. However, the vaccine-induced antibodies were not functional in bactericidal antibody assays or other assays used to measure correlates of protection. Anti-idiotypic antibodies aimed at specific polysaccharide epitopes have also been used (Park *et al.*, 2005). Peptide mimics of a capsule polysaccharide epitope have been investigated (Moe *et al.*, 1999; Westerink *et al.*, 1995) and have recently been shown to induce capsule-specific non-human cross-reactive serum bactericidal activity, and provide protection in the passive infant rat model (Lo Passo *et al.*, 2007). Capsular-based approaches may gain further momentum after a recent study showed there was no association between MenB disease and autoimmune disease (Howitz *et al.*, 2007).

Different carbohydrate moieties of *N. meningitidis* and *N. gonorrhoeae* LPS (e.g. lacto-N-neotetraose, the digalactoside Galα(1–4)Gal)) mimic glycosphingolipids present on human cells, making them poorly immunogenic (Mandrell and Apicella, 1993). LPS of many strains is also sialylated, mimicking sialic acid structures commonly found in human tissue (Vimr and Lichtensteiger, 2002). Moreover, LPS sialylation masks underlying epitopes, preventing binding of bactericidal antibodies and/or activation of the complement cascade (Elkins *et al.*, 1992; Wetzler *et al.*, 1992). In attempts to overcome these issues, LPS vaccines have focused on non-variable inner core epitopes (Plested *et al.*, 2003) and peptide mimetics (Ngampasutadol *et al.*, 2006; Tiwana *et al.*, 2005).

Several other factors should be considered that can hinder vaccine design and the assessment of the immune response raised against vaccine candidates. Antibodies directed against RmpM (reduction modifiable protein M) block the normal bactericidal activity of other antibodies directed against antigens of both *N. meningitidis* (Munkley *et al.*, 1991) and *N. gonorrhoeae* (Rice *et al.*, 1986). The presence of RmpM in vaccines has thus been avoided in certain situations (e.g. removed from the multivalent OMV vaccine (Claassen *et al.*, 1996)), although no evidence was seen of the induction of blocking antibodies after use of the Norwegian OMV vaccine (Rosenqvist *et al.*, 1999). Shedding of antigens enables scavenging of antibodies before they are able to attach to live cells. For example, alternative posttranslational processing of pilin results in secretion of a smaller soluble pilin (S-pilin) (Haas *et al.*, 1987). *N. meningitidis* (Namork and Brandtzaeg, 2002) and *N. gonorrhoeae* (Pettit and Judd, 1992) both shed outer membrane vesicles or 'blebs' during growth that contain OMPs and LPS. These 'blebs' may also play an immunosuppressive role (Lee *et al.*, 2007). IgA protease produced by pathogenic *Neisseria* leads to antigenic cloaking. IgA protease is secreted by *N. meningitidis* and *N. gonorrhoeae*, which results in cleavage of IgA1 to non-functional fragments. The Fab fragments of specific IgA1 are still able to bind antigens and in doing so make them inaccessible to functional antibodies (Kilian and Reinholdt, 1987).

Identifying, investigating and evaluating protective immune responses

The absence of appropriate animal models for exclusive human pathogens

The lack of an animal model that accurately mimics human disease caused by the pathogenic *Neisseria* has posed a major barrier to the fundamental understanding of host–pathogen interactions and to vaccine development and assessment. Meningococci and gonococci are obligate human bacteria, with no other natural reservoir of infection. This is due to the specificity of an array of proteins for human receptors or compounds, including the interaction of pili with human CD46 (Gill and Atkinson, 2004), Opa and Opc with human CEACAM1 (CD66) (Virji *et al.*, 1996) and the specificity of neisserial iron uptake systems for human iron transport proteins (Genco and Desai, 1996; Rohde and Dyer, 2003). Several *in vitro* assays, cell lines (primary and immortalized), organ culture systems and animal models have been developed to investigate various stages of neisserial colonization and infection (reviewed in Arko, 1989; Gorringe *et al.*, 2005; Nassif and So, 1995). However, none of these models are ideal since they either do not utilize the route of entry of natural infection and/or do not mimic the human disease profile (i.e. bacteraemia, tissue involvement, clinical symptomology). For *N. meningitidis*, there is no model available to study the progression from nasopharyngeal mucosal carriage to septicaemia and/or meningitis. However, the mouse and infant rat systems are considered to be the most suitable (Gorringe *et al.*, 2005). The molecular and genetic tools available to study mouse models, and the development of 'humanized' transgenic (e.g. CD46-expressing) or specially treated (e.g. iron) mouse models, for *N. meningitidis* have been recent advancements.

It is likely that several important interactions between *Neisseria* and the host immune system are not present in the available models (e.g. the vaccine candidate GNA1870 binds to human specific factor H, a key regulator of the alternative complement pathway, (Chapter 8) (Schneider *et al.*, 2007)). Antigens that are constitutively expressed during infection are the best vaccine targets. However, the antigens that elicit protective immunity after colonization with *N. meningitidis* or *N. lactamica* remain unknown, and it is possible that they are only expressed *in vivo* in the human host and may not be seen during *in vitro* study. Studies using conditions that closely reflect the *in vivo* environment are therefore important for the discovery of vaccine candidates. Indeed, the interaction of a MenB strain with human epithelial cells induced changes in the expression of 347 genes (Grifantini *et al.*, 2002), over 200 genes are regulated in response to iron (Grifantini *et al.*, 2004), and several genes are regulated in response to oxygen starvation (Bartolini *et al.*, 2006).

Vaccine evaluation

The evaluation of potential vaccine candidates and the determination of protection provided by formulated vaccines are difficult tasks for *Neisseria*. This is due to the lack of an ideal animal disease model, the relatively low incidence of disease in terms of generating statistically significant data following clinical trials, and the need to assess protection against globally diverse strains. Therefore, correlates and surrogates of protection are required to evaluate the efficacy of vaccine candidates (Borrow and Miller, 2006), as well as large strain panels that represent the global diversity of meningococcal strains (see Chapter 12 for a review of epidemiology in the vaccine era). The established correlate of protection for meningococcal disease is the serum bactericidal antibody (SBA) assay, which measures complement-mediated bacterial killing by serum antibodies. In the 1960s it was established that for serogroup C, a SBA titre of ≥4 (measuring functional antibody activity using human complement) correlates with protection against disease. An inverse relationship was also identified between the age-specific prevalence of SBA activity and the incidence of serogroup A, B, and C disease (Goldschneider *et al.*, 1969a, b; Gotschlich *et al.*, 1969; Holst *et al.*, 2003). Meningococcal serogroup C polysaccharide conjugate vaccines were licensed in the United Kingdom and elsewhere on the basis of safety and immunogenicity data, with efficacy trials based on SBA data carried out post-vaccine introduction due to the low incidence of disease (Table 11.1) (Miller *et al.*, 2006). Another accepted surrogate of protection is the maintenance of serum anti-polysaccharide IgG levels (Peltola *et al.*, 1977).

While the SBA assay is the gold-standard correlate of protection for *N. meningitidis*, several issues have arisen regarding evaluation of vaccine candidates based solely on this correlate, including the source of complement used, its lack of sensitivity and underestimation of protection (Borrow *et al.*, 2006). Human complement is currently the only acceptable source of complement for vaccine evaluation. Nevertheless, due to problems of obtaining sufficient amounts of human serum (lacking anti-meningococcal antibodies) as a source of complement and of assay standardization, infant rat complement is now often used to indicate putative protection (titres ≥ 8) (Jodar *et al.*, 2000). However, rat complement lacks several human-specific factors that interact with *N. meningitidis* (e.g. factor H, as described above). Colostrum-free bovine serum and baby rabbit complement are also being evaluated as possible complement sources.

For MenB OMV vaccines (Table 11.1), ≥ 4-fold rises in SBA pre- to post-vaccination or SBA titres ≥4 have been correlated with clinical efficacy in several immunogenicity and efficacy studies (Borrow *et al.*, 2006). It is expected that immunogenicity studies will be sufficient for licensure of a universal MenB vaccine that produces sufficient SBA titres. However, while SBA titres correlate with protection, the absence of SBA activity does not necessarily equate to susceptibility. For example, no bactericidal activity was seen against the MenB vaccine candidate GNA2132 for some strains. However, anti-GNA2132 antiserum was able to protect against bacteraemia in the infant rat passive protection model (Welsch *et al.*, 2003). Assays that demonstrate opsonic activity (opsonophagocytosis, OPA) or whole blood bactericidal activity have been suggested as more appropriate markers of protection than the SBA assay (Ison *et al.*, 1999; Lehmann *et al.*, 1999; Vermont and van den Dobbelsteen, 2002). Mouse models can also be used to evaluate active protection from bacteraemia after intraperitoneal injection (Oliver *et al.*, 2002) or from nasal colonization (Zhu *et al.*, 2005), following immunization with candidate vaccines. These assays all remain to be standardized between laboratories and have not been validated as correlates of protection. Efficacy and protection studies would therefore be required to license a MenB vaccine that did not generate SBA activity.

Accurate evaluation of a comprehensive MenB vaccine will also depend on the use of large, well-defined strain panels, representative of global diversity. It is important to know the extent to which any potential vaccine antigens are conserved and expressed in disease-causing strains from around the world. For example, a panel of 85 meningococcal strains was recently described as representative of global MenB population diversity and was used for evaluation of a recombinant vaccine (Giuliani *et al.*, 2006b). It will also be important to continue to monitor global epidemiology due to the variability and instability of meningococci.

Advances and novel approaches for MenB vaccine development

The continuing search for a universal MenB vaccine has primarily focused on the identification of new antigens, the combination of new or previously identified antigens, and/or the rational redesign of antigens to overcome problems of antigenic diversity, variability and poor immunogenicity.

Genome-based 'reverse vaccinology'

The availability of complete genome sequences for *N. meningitidis* (Parkhill *et al.*, 2000; Tettelin *et al.*, 2000) and advanced bioinformatics, microarray and proteomics technologies have revolutionized approaches used to understand bacterial pathogenesis and develop vaccines. An *in silico* genome-based approach, termed 'Reverse Vaccinology' (De Groot and Rappuoli, 2004), was applied to *N. meningitidis* serogroup B and led to the identification of approximately 600 potentially surface-exposed antigens (Pizza *et al.*, 2000). Antigen selection then continued based on a number of criteria: the ability of candidates to be expressed in *E. coli*, to be surface-exposed and accessible to antibodies in sera obtained from immunized mice during ELISA and FACs analysis; capacity to induce bactericidal antibodies in mice and/or passive protection in the infant rat or mouse protection assays; conservation and low heterogeneity in a wide range of meningococcal strains. Twenty-eight proteins were shown to induce a protective response, providing a large pool of vaccine candidates for consideration. The top 5 antigens from this screening process were genome-derived Neisseria antigen 1870 (GNA1870), NadA (GNA1994), GNA2132, GNA1030 and GNA2091 (Table 11.2) (Giuliani *et al.*, 2006b).

The reverse vaccinology approach has led to the development of the 5-component vaccine against MenB, 5CVMB. This combines the 5 top antigens, developed for large-scale manufacturing (GNA2132-GNA1030 and GNA2091-GNA1870 fusion proteins), and formulated with the adjuvant aluminium hydroxide. The multivalent approach was taken to increase the breadth of vaccine coverage and avoid selection of escape mutants. When tested against a panel of 85 meningococcal strains representative of the global population, the vaccine induced bactericidal antibodies in mice against 78% (when adjuvanted by aluminium hydroxide) or 90% (with the addition of CpG oligonucleotides or adjuvanted by MF59) of strains (Giuliani *et al.*, 2006b). This vaccine is currently in phase II studies, and initial unpublished results indicate that this vaccine is well tolerated and stimulates bactericidal antibodies against multiple MenB strains (Giuliani *et al.*, 2006a).

GNA1870 is a lipoprotein that binds factor H, a key regulator of the alternative complement pathway, leading to increased serum resistance, and has thus been named factor H binding protein (fHBP)(Madico *et al.*, 2006). This antigen is expressed by all *N. meningitidis* strains studied. However, the level of expression varies between strains (high, intermediate, or low expressors) and the protein can be classified into 3 variants (Masignani *et al.*, 2003). fHBP confers passive protection in the infant rat model and induces high levels of bactericidal antibodies, efficiently killing even the very low expressors. The bactericidal activity induced by fHBP is variant-specific, but it was suggested that vaccination with the three variants alone could induce immunity against all serogroups of meningococcus (Giuliani *et al.*, 2005; Masignani *et al.*, 2003). Variant group 1 is found in approximately 60% of disease-producing group B isolates (Welsch *et al.*, 2004) and was the focus of 5CVMB vaccine development. fHBP has also been investigated by another group (alternatively designated LP2086), and a vaccine containing recombinant proteins from two different subfamilies is currently in clinical trials (reviewed in Borrow and Balmer, 2007; Zhu *et al.*, 2005).

NadA is a member of the Oca (oligomeric coiled-coli adhesin) family and is involved in adhesion and invasion of *N. meningitidis* (Chapter 6) (Capecchi *et al.*, 2005; Comanducci *et al.*, 2002). This antigen clusters into three well-conserved alleles, induces strong bactericidal antibodies (not influenced by allele diversity), is protective in the infant rat model and is a valuable component of a multivalent vaccine. However, while present in the majority of hypervirulent *N. meningitidis* strains, NadA is absent from Lineage III strains, and present only in approximately 50% of a large panel of *N. meningitidis* strains and 5% of carriage strains (Comanducci *et al.*, 2002; Comanducci *et al.*, 2004). A complex interplay of classical and mutational regulation is also seen with NadA. Variation in the number of TAAA repeats within its promoter region affects binding of transcription factor IHF (integration host factor), changing the level of expression and consequently bactericidal activity (Comanducci *et al.*, 2002; Martin *et al.*, 2005). Some strains also contain the insertion sequence IS*1301* in *nadA* (Comanducci *et al.*, 2002). The remaining three antigens, GNA2132, GNA1030 and GNA2091, all induce protective immune responses and are well conserved in *N. meningitidis* strains, but are less well functionally characterized (Pizza *et al.*, 2000; Welsch *et al.*, 2003).

New methods to identify novel vaccine antigens

Several novel antigens have recently been discovered using proteomic- and genomic-based methods (see http://www.sci.sld.cu/*Neisseria*/*Neisseria*.htm for abstracts from the International Workshop on *Neisseria* Vaccines)(Findlow *et al.*, 2007; Martin, 2007; Pajon *et al.*, 2007). For example, analysis of the proteomic composition of the Cuban OMV vaccine, VA-MENGOC-BC (Table 11.1), using two-dimensional gel electrophoresis and mass spectrometry, enabled the identification of new putative protective proteins. One of these proteins, NMB0928, was found to induce bactericidal antibodies and provide protection in the infant rat model (Delgado *et al.*, 2007).

Several candidate antigens, identified by Expression Library Immunization (ELI), have been shown to induce bactericidal antibodies and protection in mice against meningococcal infection (Yero *et al.*, 2007). ELI was based on the immunization of mice with subsets of a genomic expression library containing approximately 6000 sequences from a MenB strain, to enable presentation of all the potentially protective antigens to the host. This was followed by analysis of serum samples for bactericidal activity and the ability to mediate passive protection in infant rats. While this approach successfully used DNA vaccine vectors to induce protective immune responses, the antigens discovered will be further studied for use in recombinant subunit vaccines, since DNA immunization is not yet a feasible approach for meningococcal vaccination.

Vaccines based on OMV technology

There are several new OMV vaccines under investigation, which are based on altered expression of various antigens within the strain(s) used to generate the OMVs, in an attempt to increase vaccine coverage (reviewed by Holst, 2007). While OMV vaccines induce an immune response against several antigens, the variability and immunodominance of PorA has led to limited cross-strain protection. Therefore, OMVs from more than one strain have been combined to give bivalent (Boutriau *et al.*, 2007; Sandbu *et al.*, 2007), hexavalent (HexaMen) (de Kleijn *et al.*, 2000; Luijkx *et al.*, 2006) and nonovalent (NonaMen) (van den Dobbelsteen *et al.*, 2007) vaccines. NonoMen is composed of the nine most frequent subtypes occurring in developed countries, formulated from 3 trivalent PorA strains that lack PorB and capsular polysaccharide, and express altered LPS. NonoMen provides broader protection than HexaMen, although low responses are seen to two subtypes.

Strain H44/76 has been genetically modified to eliminate the immunodominant PorA, the capsular polysaccharide and the α-chain of LPS, and to overexpress several minor conserved OMPs (TbpA, Hsf, NspA, and Omp85) by promoter replacement or by gene delivery (Weynants *et al.*, 2007). These OMPs are surface-exposed, well conserved among MenB strains and are involved in several aspects of the meningococcal infectious process. They have also previously been identified as potential vaccine candidates. A cross-protective response was generated to this OMV vaccine in a limited number of heterologous strains. Use of OMVs from strain H44/76 that had been engineered to overexpress GNA1870 has also resulted in increased immune responses and broader protection against MenB and other *N. meningitidis* strains (Koeberling *et al.*, 2007).

Neisseria lactamica-based vaccines

Carriage of the commensal *Neisseria* species, *N. lactamica,* which shares many common surface structures with *N. meningitidis* but lacks PorA and capsule (Snyder and Saunders, 2006), is believed to provide natural immunity to *N. meningitidis*. Hence, *N. lactamica* live bacteria, killed whole cells, OMVs and OMPs have been considered for use as vaccines against meningococcal disease (Abel *et al.*, 2007; Braun *et al.*, 2004; Finney *et al.*, 2008; Sardinas *et al.*, 2006; Vaughan *et al.*, 2006). Immunization of mice with live *N. lactamica* or *N. lactamica* OMVs protects animals against lethal meningococcal challenge (Li *et al.*, 2006; Oliver *et al.*, 2002). A *N. lactamica* OMV vaccine is currently in phase I trials and preliminary data have shown a good safety and immunogenicity profile, with rises in SBA and OPA activity seen against several meningococcal strains (Gorringe, 2007a,b).

Neisseria gonorrhoeae

Vaccine development for *N. gonorrhoeae* has been extremely difficult, largely due to the same factors that have hampered meningococcal vaccine development, as described above. In particular, antigenic diversity and variability, and the lack of an ideal animal model. These issues and additional barriers specific to *N. gonorrhoeae* are briefly discussed below, and a review of gonococcal vaccine development can be found by Mietzner *et al.* (2004).

Gonococcal antigenic diversity and variability

N. gonorrhoeae, to an even greater extent than *N. meningitidis,* is non-clonal and is usually referred to as a panmictic organism (Smith *et al.*, 1993). A number of typing methods have been developed for *N. gonorrhoeae,* but understanding the nature and global trends of disease-causing strains has been hampered by the fact that no single typing scheme has been widely adopted. Typing methods have been based on variable phenotypic characteristics, including growth requirements for specific nutrients and cofactors (auxotyping), antibiotic sensitivity, as well as serological reactivity to porin (serovar typing) and combined auxotype/serovar (A/S) typing (outlined in Bennett *et al.*, 2007). A worldwide collection of more than 1400 *N. gonorrhoeae* strains was composed of 107 different A/S classes (Knapp *et al.*, 1984). Genotypic typing methods based on sequencing *opa*, *por* and the transferrin-binding protein subunit B gene (*tbpB*) are highly discriminating. The combination of *por* and *tbpB* sequence typing forms the basis of *N. gonorrhoeae* multi-antigen sequence typing (NG-MAST). Typing of 474 gonococcal isolates using NG-MAST, identified 166 different *por* alleles and 86 different *tbpB* sequences, giving 197 different two-locus allelic profiles (Martin *et al.*, 2004). MLST analysis of a panel of *N. gonorrhoeae* isolates has recently been performed, revealing a genetic diversity similar to *N. meningitidis* (Bennett *et al.*, 2007).

N. gonorrhoeae undergoes frequent antigenic variation, phase variation, and DNA transformation and recombination, as described above for *N. meningitidis*. Indeed, gonococci expressing multiple variants of pilin (Seifert *et al.*, 1994; Swanson *et al.*, 1987), LPS (Schneider *et al.*, 1991) and Opa protein (Jerse *et al.*, 1994; Swanson *et al.*, 1988) have been isolated from human male volunteers that were infected with a predominantly homogenous inoculum. This antigenic variation is presumed to play a role in the persistence and recidivism of gonococcal infections. It also explains the poor success of gonococcal vaccines up to this point (Blake and Wetzler, 1995; Mietzner *et al.*, 2004; Russell *et al.*, 1999).

Neisseria gonorrhoeae: infection does not confer protection

The highest incidence of gonorrhoea in both men and women occurs between 15 and 29 years of age. Whereas a major problem with *N. meningitidis* is inducing immunity in infants, the main issue with *N. gonorrhoeae* is inducing any protective immunity at all. The predominant morbidity of *N. gonorrhoeae*, like many other STIs (*Chlamydia trachomatis*, HPV, herpes simplex virus), is a result of mucosal infection and, compared to bacteraemic infections, mucosal infections typically do not induce durable protection against re-infection. Natural gonococcal infection, or local intravaginal immunizations lead to only weak local humoral and cellular immune responses, no apparent protective immunity is developed and high levels of reinfection are seen (Hedges *et al.*, 1999; Plummer and Brunham, 1987; Schmidt *et al.*, 2001). Even though serovar-specific antibodies are produced to Por epitopes, reinfection with the same serovar occurs (Fox *et al.*, 1999; Plummer *et al.*, 1989)

The immunological nature of the genital tract has posed a barrier to vaccine development for *N. gonorrhoeae* and other STIs. The male and female genital tract is distinct from other components of the mucosal immune system due to the lack of inductive mucosal-associated lymphoid sites, areas analogous to intestinal Peyer's patches that are responsible for uptake, processing, and presentation of antigens. Furthermore, IgG is believed to be the predominant isotype, rather than IgA3–5 and secretory IgA (S-IgA), which are considered essential to the humoral mucosal immune system (Mestecky *et al.*, 2005). However, increased levels of IgA1 specific to homologous gonococci in cervical mucosa have been observed (Hedges *et al.*, 1999). Gonococci evade host defences via several of the mechanisms described above for *N. meningitidis*, but are also believed to subvert the natural immune response through unknown mechanisms. It is important to identify and understand these strategies to provide new directions in the development of effective gonococcal vaccines (Stanberry and Rosenthal, 2005).

Conclusions and future trends

Disease caused by the pathogenic *Neisseria* remains a major worldwide cause of morbidity and mortality, even after the development of vaccines to protect against several serogroups of meningococcal disease. The development of Men A, C, W-135, Y simple polysaccharide and the improved conjugate polysaccharide vaccines, as well as tailor made OMV MenB vaccines, has dramatically reduced disease in areas where these vaccines have been widely used. The main difficulties that have arisen during the development of neisserial vaccines are based on the exquisite adaptation of the bacteria to the human host and their incredible array of mechanisms to avoid the immune system. However, continuing development of new technologies has driven the rapid increase in understanding of the nature of these organisms, the host immune system, and vaccine approaches. Progress made in the genomic and post-genomic era has put the realization of a universal MenB vaccine within reach, and research in this field will likely be strongly influenced by the outcome of the promising MenB vaccine that is now in phase II trials (Giuliani *et al.*, 2006b). However, the situation remains less promising for a gonococcal vaccine.

As the number of vaccines given in infancy continues to increase, the idea of a universal or combination vaccine containing protective meningococcal antigens becomes more attractive. Indeed, a combination Hib/group C meningococcal conjugate vaccine (Menitorix™, GlaxoSmithKline) was licensed in the U.K. in 2005, and several other vaccine combinations are in trials, including a conjugate 9-valent *S. pneumoniae*/MenC vaccine (Buttery *et al.*, 2005), a Hib-bivalent MenC and Y-Tetanus Toxoid Conjugate vaccine (Hib-MenCY-TT) (Habermehl *et al.*, 2006), a diphtheria-tetanus-whole cell pertussis-hepatitis B/Hib (DTPw-HB/Hib)/bivalent MenC and A vaccine (Forgor *et al.*, 2006), and a Hib–MenC conjugate vaccine in combination with diphtheria-tetanus-acellular pertussis (DTPa)-hepatitis B virus (HBV)-inactivated poliovirus (IPV) (Schmitt *et al.*, 2007; Tejedor *et al.*, 2007).

The question, 'can a universal meningococcal vaccine be developed?', must be asked in light of the novel strategies to identify vaccine candidates that continue to evolve. After more than a century of vaccine development using traditional approaches established by Louis Pasteur, the use of reverse vaccinology in developing a MenB vaccine has pioneered a new era of vaccinology. The plausible next step is the use of the 'pan-genome' approach to define the total gene repertoire of *N. meningitidis*, or even of the pathogenic *Neisseria*, and identify novel conserved antigens

(Muzzi *et al.*, 2007). This pan-genome approach was successfully applied to identify antigens for a Group B *Streptococcus* (GBS) vaccine, based on the genome sequences of eight strains (Maione *et al.*, 2005). Comparison of a combination of meningococcal genomes from disease-associated strains of several serogroups as well as carriage strains should prove very useful for vaccine antigen selection. Determination of the neisserial proteome exposed to the immune systems, the 'surface-proteome', would also be valuable to the identification of novel protective antigens. This proteomic approach, recently utilized in vaccine antigen discovery for group A *Streptococcus* (Rodriguez-Ortega *et al.*, 2006), could be adapted to *N. meningitidis* and *N. gonorrhoeae* grown in contact with human cells. The next step in the post-genome era of vaccine development has been proposed to be based on 'structural vaccinology', similar to the established structure-based design of antivirals (Serruto and Rappuoli, 2006), for the rational design of peptide mimetics of bactericidal epitopes. For example, the crystal structure of a PorA epitope in complex with a Fab fragment of a bactericidal antibody has been determined (Oomen *et al.*, 2005). The structure of the immunodominant domain of GNA1870, a protective antigen of *N. meningitidis* identified by reverse vaccinology and included in the 5CVMB, has also been solved by NMR spectroscopy (Cantini *et al.*, 2006).

Improvements in adjuvant and vaccine delivery, including mucosal immunization, will play an important role in new vaccine formulations. The increased focus on novel vaccines (e.g. consisting of recombinant purified proteins) has also resulted in the need to develop novel adjuvants and delivery systems to improve antigen immunogenicity (O'Hagan and Rappuoli, 2006). MF59, a new oil-in-water emulsion, was seen to enhance coverage of the 5CVMB (Giuliani *et al.*, 2006b), and the Adjuvant System AS04, consisting of an immunomodulator, 3-O-desacyl-4′-monophosphoryl lipid A (MPL), adsorbed on aluminium salt, has been shown to be effective (Garcon *et al.*, 2007). The development of vaccines based on novel approaches necessitates the co-development, standardization and acceptance of the basic science, to ensure that regulatory agencies license these vaccines, and that policy makers recommend them. This includes new bacterial typing systems to accurately reflect the population diversity of *Neisseria*, assays to use as correlates of protection, and assays to measure novel adjuvants. Research in these areas will undoubtedly play a large role in the future of neisserial vaccine development.

Web resources

Vaccine information

www.sci.sld.cu/Neisseria/Neisseria.htm
www.meningitis.org.uk/health-professionals/2007-conference
Neisseria.org/ipnc/
www.meningvax.org/
www.immunise.moh.govt.nz/
www.vaccines.org/
www.vaccinesafety.edu/package_inserts-bymanuf.htm
www.clinicaltrials.gov/

Meningococcal and gonococcal information

www.Neisseria.org
www.cdc.gov/ncidod/dbmd/diseaseinfo/meningococcal_g.htm
www.meningitis-trust.org/
www.who.int/topics/sexually_transmitted_infections/en/

References

Abel, A., Sanchez, S., Arenas, J., Criado, M.T., and Ferreiros, C.M. (2007). Bioinformatic analysis of outer membrane proteome of *Neisseria meningitidis* and *Neisseria lactamica*. Int. Microbiol. *10*, 5–11.

Achtman, M. (2004). Population structure of pathogenic bacteria revisited. Int. J. Med. Microbiol. *294*, 67–73.

Aho, E.L. (1989). Molecular biology of class 5 outer membrane proteins of *Neisseria meningitidis*. Microb. Pathog. *7*, 249–253.

Al-Mazrou, Y., Khalil, M., Borrow, R., Balmer, P., Bramwell, J., Lal, G., Andrews, N., and Al-Jeffri, M. (2005). Serologic responses to ACYW135 polysaccharide meningococcal vaccine in Saudi children under 5 years of age. Infect. Immun. *73*, 2932–2939.

Alcala, B., Arreaza, L., Salcedo, C., Uria, M.J., De La Fuente, L., and Vazquez, J.A. (2002). Capsule switching among C:2b:P1.2,5 meningococcal epidemic strains after mass immunization campaign, Spain. Emerg. Infect. Dis. *8*, 1512–1514.

Aledort, J.E., Hook, E.W., 3rd, Weinstein, M.C., and Goldie, S.J. (2005). The cost effectiveness of gonorrhea screening in urban emergency departments. Sex. Transm. Dis. *32*, 425–436.

Alexander, H.L., Rasmussen, A.W., and Stojiljkovic, I. (2004). Identification of *Neisseria meningitidis* genetic loci involved in the modulation of phase variation frequencies. Infect. Immun. *72*, 6743–6747.

Arko, R.J. (1989). Animal models for pathogenic *Neisseria* species. Clin. Microbiol. Rev. 2 Suppl., S56–59.

Bartolini, E., Frigimelica, E., Giovinazzi, S., Galli, G., Shaik, Y., Genco, C., Welsch, J.A., Granoff, D.M., Grandi, G., and Grifantini, R. (2006). Role of FNR and FNR-regulated, sugar fermentation genes in *Neisseria meningitidis* infection. Mol. Microbiol. *60*, 963–972.

Bayliss, C.D., Field, D., and Moxon, E.R. (2001). The simple sequence contingency loci of *Haemophilus influenzae* and *Neisseria meningitidis*. J. Clin. Invest. *107*, 657–662.

Bennett, J.S., Jolley, K.A., Sparling, F., Saunders, N.J., Hart, A., Feavers, I.M., and Maiden, M.C. (2007). Species Status of *Neisseria gonorrhoeae*: Evolutionary and Epidemiological Inferences from MLST.BMC Biol. *5*, 35.

Bilukha, O.O., and Rosenstein, N. (2005). Prevention and control of meningococcal disease. Recommendations of the Advisory Committee on Immunization Practices (ACIP). MMWR Recomm Rep. *54*, 1–21.

Blake, M.S., and Wetzler, L.M. (1995). Vaccines for gonorrhea: where are we on the curve? Trends Microbiol. *3*, 469–474.

Blattner, F.R., Plunkett, G., 3rd, Bloch, C.A., Perna, N.T., Burland, V., Riley, M., Collado-Vides, J., Glasner, J.D., Rode, C.K., Mayhew, G.F., Gregor, J., Davis, N.W., Kirkpatrick, H.A., Goeden, M.A., Rose, D.J., Mau, B., and Shao, Y. (1997). The complete genome sequence of *Escherichia coli* K-12. Science *277*, 1453–1474.

Booy, R., Jelfs, J., El Bashir, H., and Nissen, M.D. (2007). Impact of meningococcal C conjugate vaccine use in Australia. Med J Aust *186*, 108–109.

Borrow, R., and Balmer, P. (2007). The future of meningoccal vaccines. Pediatr. Health *1*, 51–61.

Borrow, R., Carlone, G.M., Rosenstein, N., Blake, M., Feavers, I., Martin, D., Zollinger, W., Robbins, J., Aaberge, I., Granoff, D.M., Miller, E., Plikaytis, B., van Alphen, L., Poolman, J., Rappuoli, R., Danzig, L., Hackell, J., Danve, B., Caulfield, M., Lambert, S., and Stephens, D. (2006). *Neisseria meningitidis* group B correlates of protection and assay standardization – international meeting report Emory University, Atlanta, Georgia, United States, 16–17 March 2005. Vaccine *24*, 5093–5107.

Borrow, R., Joseph, H., Andrews, N., Acuna, M., Longworth, E., Martin, S., Peake, N., Rahim, R., Richmond, P., Kaczmarski, E., and Miller, E. (2000). Reduced antibody response to revaccination with meningococcal serogroup A polysaccharide vaccine in adults. Vaccine *19*, 1129–1132.

Borrow, R., and Miller, E. (2006). Surrogates of protection. In Handbook of Meningicoccal Disease Infection Biology, Vaccination, Clinical Management, M. Frosch, and M.C.J. Maiden, eds. (Weinheim, Wiley-VCH Verlag GmbH & Co.), pp. 323–341.

Boslego, J., Garcia, J., Cruz, C., Zollinger, W., Brandt, B., Ruiz, S., Martinez, M., Arthur, J., Underwood, P., and Silva, W. (1995). Efficacy, safety, and immunogenicity of a meningococcal group B (15:P1.3) outer membrane protein vaccine in Iquique, Chile. Chilean National Committee for Meningococcal Disease. Vaccine *13*, 821–829.

Boutriau, D., Poolman, J., Borrow, R., Findlow, J., Domingo, J.D., Puig-Barbera, J., Baldo, J.M., Planelles, V., Jubert, A., Colomer, J., Gil, A., Levie, K., Kervyn, A.D., Weynants, V., Dominguez, F., Barbera, R., and Sotolongo, F. (2007). Immunogenicity and safety of three doses of a bivalent (B:4:p1.19,15 and B:4:p1.7-2,4) meningococcal outer membrane vesicle vaccine in healthy adolescents. Clin. Vaccine Immunol. *14*, 65–73.

Braun, J.M., Beuth, J., Blackwell, C.C., Giersen, S., Higgins, P.G., Tzanakaki, G., Unverhau, H., and Weir, D.M. (2004). *Neisseria meningitidis, Neisseria lactamica* and *Moraxella catarrhalis* share cross-reactive carbohydrate antigens. Vaccine *22*, 898–908.

Brehony, C., Jolley, K.A., and Maiden, M.C. (2007). Multilocus sequence typing for global surveillance of meningococcal disease. FEMS Microbiol. Rev. *31*, 15–26.

Bruge, J., Bouveret-Le Cam, N., Danve, B., Rougon, G., and Schulz, D. (2004). Clinical evaluation of a group B meningococcal N-propionylated polysaccharide conjugate vaccine in adult, male volunteers. Vaccine *22*, 1087–1096.

Buttery, J.P., Riddell, A., McVernon, J., Chantler, T., Lane, L., Bowen-Morris, J., Diggle, L., Morris, R., Harnden, A., Lockhart, S., Pollard, A.J., Cartwright, K., and Moxon, E.R. (2005). Immunogenicity and safety of a combination pneumococcal-meningococcal vaccine in infants: a randomized controlled trial. JAMA *293*, 1751–1758.

Cantini, F., Savino, S., Scarselli, M., Masignani, V., Pizza, M., Romagnoli, G., Swennen, E., Veggi, D., Banci, L., and Rappuoli, R. (2006). Solution structure of the immunodominant domain of protective antigen GNA1870 of *Neisseria meningitidis*. J. Biol. Chem. *281*, 7220–7227.

Capecchi, B., Adu-Bobie, J., Di Marcello, F., Ciucchi, L., Masignani, V., Taddei, A., Rappuoli, R., Pizza, M., and Arico, B. (2005). *Neisseria meningitidis* NadA is a new invasin which promotes bacterial adhesion to and penetration into human epithelial cells. Mol. Microbiol. *55*, 687–698.

Cartwright, K. (1995). Introduction and historical perspectives. In Meningococcal disease, K. Cartwright, ed. (Chichester, Wiley), pp. 1–9.

Cartwright, K. (2006). Historical Aspects. In Handbook of Meningicoccal Disease Infection Biology, Vaccination, Clinical Management, M. Frosch, and M.C.J. Maiden, eds. (Weinheim, Wiley-VCH Verlag GmbH & Co.), pp. 1–13.

Cartwright, K., Noah, N., and Peltola, H. (2001). Meningococcal disease in Europe: epidemiology, mortality, and prevention with conjugate vaccines. Report of a European advisory board meeting Vienna, Austria, 6–8 October, 2000. Vaccine *19*, 4347–4356.

Caugant, D.A. (1998). Population genetics and molecular epidemiology of *Neisseria meningitidis*. APMIS *106*, 505–525.

Caugant, D.A., Froholm, L.O., Bovre, K., Holten, E., Frasch, C.E., Mocca, L.F., Zollinger, W.D., and Selander, R.K. (1986). Intercontinental spread of a genetically distinctive complex of clones of *Neisseria meningitidis* causing epidemic disease. Proc. Natl. Acad. Sci. USA *83*, 4927–4931.

Caugant, D.A., Tzanakaki, G., and Kriz, P. (2007). Lessons from meningococcal carriage studies. FEMS Microbiol. Rev. *31*, 52–63.

CDC (2006). Update: Guillain-Barre syndrome among recipients of Menactra meningococcal conjugate vaccine – United States, October 2005-February 2006. MMWR Morb. Mortal. Wkly Rep. *55*, 364–366.

CDC (2007). Revised recommendations of the Advisory Committee on Immunization Practices to Vaccinate all Persons Aged 11–18 Years with Meningococcal Conjugate Vaccine. MMWR Morb. Mortal. Wkly Rep. *56*, 794–795.

Claassen, I., Meylis, J., van der Ley, P., Peeters, C., Brons, H., Robert, J., Borsboom, D., van der Ark, A., van Straaten, I., Roholl, P., Kuipers, B., and Poolman, J. (1996). Production, characterization and control of a *Neisseria meningitidis* hexavalent class 1 outer membrane protein containing vesicle vaccine. Vaccine *14*, 1001–1008.

Claus, H., Elias, J., Meinhardt, C., Frosch, M., and Vogel, U. (2007). Deletion of the meningococcal fetA gene used for antigen sequence typing in invasive and commensal isolates from Germany: frequencies and mechanisms. J. Cli.n Microbiol. 45, 2960–2964.

Cohen, M.S., Hoffman, I.F., Royce, R.A., Kazembe, P., Dyer, J.R., Daly, C.C., Zimba, D., Vernazza, P.L., Maida, M., Fiscus, S.A., and Eron, J.J., Jr. (1997). Reduction of concentration of HIV-1 in semen after treatment of urethritis: implications for prevention of sexual transmission of HIV-1. AIDSCAP Malawi Research Group. Lancet *349*, 1868–1873.

Colicchio, R., Pagliarulo, C., Lamberti, F., Vigliotta, G., Bruni, C.B., Alifano, P., and Salvatore, P. (2006). RecB-dependent mutator phenotype in *Neisseria meningitidis* strains naturally defective in mismatch repair. DNA Repair (Amst) *5*, 1428–1438.

Comanducci, M., Bambini, S., Brunelli, B., Adu-Bobie, J., Arico, B., Capecchi, B., Giuliani, M.M., Masignani, V., Santini, L., Savino, S., Granoff, D.M., Caugant, D.A., Pizza, M., Rappuoli, R., and Mora, M. (2002). NadA, a novel vaccine candidate of *Neisseria meningitidis*. J. Exp. Med. *195*, 1445–1454.

Comanducci, M., Bambini, S., Caugant, D.A., Mora, M., Brunelli, B., Capecchi, B., Ciucchi, L., Rappuoli, R., and Pizza, M. (2004). NadA diversity and carriage in *Neisseria meningitidis*. Infect. Immun. 72, 4217–4223.

Costantino, P., Viti, S., Podda, A., Velmonte, M.A., Nencioni, L., and Rappuoli, R. (1992). Development and phase 1 clinical testing of a conjugate vaccine against meningococcus A and C. Vaccine *10*, 691–698.

Cruz, C., Pavez, G., Aguilar, E., Grawe, L., Cam, J., Mendez, F., Garcia, J., Ruiz, S., Vicent, P., Canepa, I., Martinez, M., Boslego, J., Zollinger, W., Arthur, J. and Caugant, D. (1990). Serotype-specific outbreak of group B meningococcal disease in Iquique, Chile. Epidemiol. Infect. *105*, 119–126.

Davidsen, T., Bjoras, M., Seeberg, E.C., and Tonjum, T. (2005). Antimutator role of DNA glycosylase MutY in pathogenic *Neisseria* species. J. Bacteriol. *187*, 2801–2809.

Davidsen, T., and Tonjum, T. (2006). Meningococcal genome dynamics. Nat. Rev. Microbiol. *4*, 11–22.

Davidsen, T., Tuven, H.K., Bjoras, M., Rodland, E.A., and Tonjum, T. (2007). Genetic interactions of DNA repair pathways in the pathogen *Neisseria meningitidis*. J. Bacteriol. *189*, 5728–5737.

De Groot, A.S., and Rappuoli, R. (2004). Genome-derived vaccines. Expert Rev. Vaccines 3, 59–76.

de Jonge, M.I., Hamstra, H.J., Jiskoot, W., Roholl, P., Williams, N.A., Dankert, J., van Alphen, L., and van der Ley, P. (2004). Intranasal immunisation of mice with liposomes containing recombinant meningococcal OpaB and OpaJ proteins. Vaccine 22, 4021–4028.

de Kleijn, E.D., de Groot, R., Labadie, J., Lafeber, A.B., van den Dobbelsteen, G., van Alphen, L., van Dijken, H., Kuipers, B., van Omme, G.W., Wala, M., Juttmann, R., and Rumke, H.C. (2000). Immunogenicity and safety of a hexavalent meningococcal outer-membrane-vesicle vaccine in children of 2–3 and 7–8 years of age. Vaccine *18*, 1456–1466.

De Wals, P., Coudeville, L., Trottier, P., Chevat, C., Erickson, L.J., and Nguyen, V.H. (2007). Vaccinating adolescents against meningococcal disease in Canada: a cost-effectiveness analysis. Vaccine *25*, 5433–5440.

Delgado, M., Yero, D., Niebla, O., Gonzalez, S., Climent, Y., Perez, Y., Cobas, K., Caballero, E., Garcia, D., and Pajon, R. (2007). Lipoprotein NMB0928 from *Neisseria meningitidis* serogroup B as a novel vaccine candidate. Vaccine *25*, 8420–8431.

Dolan-Livengood, J.M., Miller, Y.K., Martin, L.E., Urwin, R., and Stephens, D.S. (2003). Genetic basis for nongroupable *Neisseria meningitidis*. J. Infect. Dis. 187, 1616–1628.

Dunn, K.L., Virji, M., and Moxon, E.R. (1995). Investigations into the molecular basis of meningococcal toxicity for human endothelial and epithelial cells: the synergistic effect of LPS and pili. Microb. Pathog. *18*, 81–96.

Dyet, K., Devoy, A., McDowell, R., and Martin, D. (2005). New Zealand's epidemic of meningococcal disease described using molecular analysis: implications for vaccine delivery. Vaccine 23, 2228–2230.

Edwards, E.A., Devine, L.F., Sengbusch, G.H., and Ward, H.W. (1977). Immunological investigations of meningococcal disease. III. Brevity of group C acquisition prior to disease occurrence. Scand J. Infect. Dis. 9, 105–110.

Elkins, C., Carbonetti, N.H., Varela, V.A., Stirewalt, D., Klapper, D.G., and Sparling, P.F. (1992). Antibodies to N-terminal peptides of gonococcal porin are bactericidal when gonococcal lipopolysaccharide is not sialylated. Mol. Microbiol. *6*, 2617–2628.

Eyre, J.W.H., and Stewart, B.H. (1909). The treatment of gonococcus infections by vaccines. Lancet *174*, 76–81.

Ferenczy, A., and Franco, E.L. (2007). Prophylactic human papillomavirus vaccines: potential for sea change. Expert Rev. Vaccines *6*, 511–525.

Ferrari, G., Garaguso, I., Adu-Bobie, J., Doro, F., Taddei, A.R., Biolchi, A., Brunelli, B., Giuliani, M.M., Pizza, M., Norais, N., and Grandi, G. (2006). Outer membrane vesicles from group B *Neisseria meningitidis* delta gna33 mutant: proteomic and immunological comparison with detergent-derived outer membrane vesicles. Proteomics *6*, 1856–1866.

Ferry, N.S., and Steele, A.H. (1935). Active immunisation with meningococcus toxin. JAMA *104*, 983–984.

Findlow, J., Balmer, P., Yero, D., Niebla, O., Pajon, R., and Borrow, R. (2007). *Neisseria* Vaccines 2007. Expert Rev. Vaccines 6, 485–489.

Finne, J., Bitter-Suermann, D., Goridis, C., and Finne, U. (1987). An IgG monoclonal antibody to group B meningococci cross-reacts with developmentally regulated polysialic acid units of glycoproteins in neural and extraneural tissues. J. Immunol. *138*, 4402–4407.

Finney, M., Vaughan, T., Taylor, S., Hudson, M.J., Pratt, C., Wheeler, J.X., Vipond, C., Feavers, I., Jones, C., Findlow, J., Borrow, R., and Gorringe, A. (2008). Characterization of the key antigenic components and pre-clinical immune responses to a meningococcal disease vaccine Based on *Neisseria lactamica* outer membrane vesicles. Hum. Vaccin. 4.

Forgor, A.A., Hodgson, A., Chandramohan, D., Reed, Z., Binka, F., Boutriau, D., and Greenwood, B. (2006). A novel DTPw–HBV/Hib–MenAC conjugate combination vaccine administered to infants in northern Ghana is safe and induces immune memory. Paper presented at: 15th International Pathogenic *Neisseria* Conference (Cairns, Australia).

Fox, K.K., Thomas, J.C., Weiner, D.H., Davis, R.H., Sparling, P.F., and Cohen, M.S. (1999). Longitudinal evaluation of serovar-specific immunity to *Neisseria gonorrhoeae*. Am. J. Epidemiol. *149*, 353–358.

Fox, K.L., Srikhanta, Y.N., and Jennings, M.P. (2007). Phase variable type III restriction-modification systems of host-adapted bacterial pathogens. Mol. Microbiol. 65, 1375–1379.

Frasch, C. (1995). Meningococcal Vaccines: Past, Present and Future. In Meningococcal Disease (Chichester, Wiley), pp. 245–284.

Frasch, C.E., Van Alphen, L., Holst, J., Poolman, J., and Rosenqvist, E. (2001). Preparation of outer membrane protein vaccines against meningococcal disease. In Meningococcal Vaccines Methods and Protocols, A.J. Pollard, and M.C.J. Maiden, eds. (Totowa, NJ, Humana Press), pp. 81–107.

Frasch, C.E., Zollinger, W.D., and Poolman, J.T. (1985). Serotype antigens of *Neisseria meningitidis* and a proposed scheme for designation of serotypes. Rev. Infect. Dis. 7, 504–510.

Fredriksen, J.H., Rosenqvist, E., Wedege, E., Bryn, K., Bjune, G., Froholm, L.O., Lindbak, A.K., Mogster, B., Namork, E., Rye, U., *et al.* (1991). Production, characterization and control of MenB-vaccine 'Folkehelsa': an outer membrane vesicle vaccine against group B meningococcal disease. NIPH Ann *14*, 67–79; discussion 79–80.

Frosch, M., and Vogel, U. (2006). Structure and genetics of the meningococcal capsule. In Handbook of Meningicoccal Disease Infection Biology, Vaccination, Clinical Management, M. Frosch, and M.C.J. Maiden, eds. (Weinheim, Wiley-VCH Verlag GmbH & Co.), pp. 145–162.

Garcon, N., Chomez, P., and Van Mechelen, M. (2007). GlaxoSmithKline Adjuvant Systems in vaccines: concepts, achievements and perspectives. Expert Rev Vaccines 6, 723–739.

Genco, C.A., and Desai, P.J. (1996). Iron acquisition in the pathogenic *Neisseria*. Trends Microbiol. *4*, 179–184.

Gerbase, A.C., Rowley, J.T., and Mertens, T.E. (1998). Global epidemiology of sexually transmitted diseases. Lancet 351 Suppl. 3, 2–4.

Gill, D.B., and Atkinson, J.P. (2004). CD46 in *Neisseria* pathogenesis. Trends Mol. Med. *10*, 459–465.

Giraud, A., Radman, M., Matic, I., and Taddei, F. (2001). The rise and fall of mutator bacteria. Curr. Opin. Microbiol. *4*, 582–585.

Giuliani, M.M., Adu-Bobie, J., Comanducci, M., Arico, B., Savino, S., Pizza, M., and Rappouli, R. (2006a). A Universal Vaccine for Serogroup B Meningococcus. Paper presented at: 15th International Pathogenic *Neisseria* Conference (Cairns, Australia).

Giuliani, M.M., Adu-Bobie, J., Comanducci, M., Arico, B., Savino, S., Santini, L., Brunelli, B., Bambini, S., Biolchi, A., Capecchi, B., Cartocci, E., Ciucchi, L., Di Marcello, F., Ferlicca, F., Galli, B., Luzzi, E., Masignani, V., Serruto, D., Veggi, D., Contorni, M., Morandi, M., Bartalesi, A., Cinotti, V., Mannucci, D., Titta, F., Ovidi, E., Welsch, J.A., Granoff, D., Rappuoli, R., and Pizza, M. (2006b). A universal vaccine for serogroup B meningococcus. Proc. Natl. Acad. Sci. USA *103*, 10834–10839.

Giuliani, M.M., Santini, L., Brunelli, B., Biolchi, A., Arico, B., Di Marcello, F., Cartocci, E., Comanducci, M., Masignani, V., Lozzi, L., Savino, S., Scarselli, M., Rappuoli, R., and Pizza, M. (2005). The region comprising amino acids 100 to 255 of *Neisseria meningitidis* lipoprotein GNA 1870 elicits bactericidal antibodies. Infect. Immun. *73*, 1151–1160.

Goldacre, M.J., Roberts, S.E., and Yeates, D. (2003). Case fatality rates for meningococcal disease in an English population, 1963–98: database study. BMJ *327*, 596–597.

Goldschneider, I., Gotschlich, E.C., and Artenstein, M.S. (1969a). Human immunity to the meningococcus. I. The role of humoral antibodies. J. Exp. Med. *129*, 1307–1326.

Goldschneider, I., Gotschlich, E.C., and Artenstein, M.S. (1969b). Human immunity to the meningococcus. II. Development of natural immunity. J. Exp. Med. *129*, 1327–1348.

Gorringe, A. (2007a). The development of meningococcal disease vaccine based on *Neisseria* lactamica: Preliminary analysis of phase 1 clinical trial data. In Meningitis and Septicaemia in Children and Adults, Meningitis Research Foundation Conference November 2007 (Royal Society of Medicine, London).

Gorringe, A.R. (2007b). An update on the development of a meningococcal disease vaccine based on *Neisseria* lactamica. In *Neisseria* Vaccines 2007, April 15–20 (Varadero, Cuba).

Gorringe, A.R., Reddin, K.M., Funnell, S.G., Johansson, L., Rytkonen, A., and Jonsson, A.B. (2005). Experimental disease models for the assessment of meningococcal vaccines. Vaccine 23, 2214–2217.

Gotschlich, E.C., Goldschneider, I., and Artenstein, M.S. (1969). Human immunity to the meningococcus. IV. Immunogenicity of group A and group C meningococcal polysaccharides in human volunteers. J. Exp. Med. *129*, 1367–1384.

Granoff, D., Harrison, L.H., and Borrow, R. (2008). Meningococcal Vaccines. In Vaccines Fifth Edition, S.A. Plotkin, P.A. Offit, and W.A. Orenstein, eds. (Philadelphia, Saunders).

Greenberg, L., Diena, B.B., Ashton, F.A., Wallace, R., Kenny, C.P., Znamirowski, R., Ferrari, H., and Atkinson, J. (1974). Gonococcal vaccine studies in Inuvik. Can J Public Health *65*, 29–33.

Grifantini, R., Bartolini, E., Muzzi, A., Draghi, M., Frigimelica, E., Berger, J., Randazzo, F., and Grandi, G. (2002). Gene expression profile in *Neisseria meningitidis* and *Neisseria* lactamica upon host-cell contact: from basic research to vaccine development. Ann. NY Acad. Sci. 975, 202–216.

Grifantini, R., Frigimelica, E., Delany, I., Bartolini, E., Giovinazzi, S., Balloni, S., Agarwal, S., Galli, G., Genco, C., and Grandi, G. (2004). Characterization of a novel *Neisseria meningitidis* Fur and iron-regulated operon required for protection from oxidative stress: utility of DNA microarray in the assignment of the biological role of hypothetical genes. Mol. Microbiol. 54, 962–979.

Griffiss, J.M., Brandt, B.L., Broud, D.D., Goroff, D.K., and Baker, C.J. (1984). Immune response of infants and children to disseminated infections with *Neisseria meningitidis*. J. Infect. Dis. 150, 71–79.

Haas, R., Schwarz, H., and Meyer, T.F. (1987). Release of soluble pilin antigen coupled with gene conversion in *Neisseria gonorrhoeae*. Proc. Natl. Acad. Sci. USA *84*, 9079–9083.

Habermehl, P., LeRoux-Roels, G., Saenger, R., Maechler, G., and Boutriau, D. (2006). Immunogenicity and Reactogenicity of a Combined *Haemophilus influenzae* type b and *Neisseria meningitidis* Serogroup C and Y-Tetanus Toxoid Conjugate Hib-MenCY-TT) Vaccine Administered at 2, 3 and 4 Months and as a Booster dose at Second Year of Life. Paper presented at: 15th International Pathogenic *Neisseria* Conference (Cairns, Australia.).

Halperin, S.A., Langley, J.M., Smith, B., Wunderli, P., Kaufman, L., Kimura, A., and Martin, D. (2007). Phase 1 first-in-human studies of the reactogenicity and immunogenicity of a recombinant meningococcal NspA vaccine in healthy adults. Vaccine 25, 450–457.

Hamilton, H.L., and Dillard, J.P. (2006). Natural transformation of *Neisseria gonorrhoeae*: from DNA donation to homologous recombination. Mol. Microbiol. *59*, 376–385.

Hedges, S.R., Mayo, M.S., Mestecky, J., Hook, E.W., 3rd, and Russell, M.W. (1999). Limited local and systemic antibody responses to *Neisseria gonorrhoeae* during uncomplicated genital infections. Infect. Immun. *67*, 3937–3946.

Hilse, R., Hammerschmidt, S., Bautsch, W., and Frosch, M. (1996). Site-specific insertion of IS1301 and distribution in *Neisseria meningitidis* strains. J. Bacteriol. *178*, 2527–2532.

Holst, J. (2007). Strategies for Development of Universal Vaccines against Meningococcal Serogroup B Disease: The Most Promising Options and the Challenges Evaluating Them. Hum Vaccin 3.

Holst, J., Feiring, B., Fuglesang, J.E., Hoiby, E.A., Nokleby, H., Aaberge, I.S., and Rosenqvist, E. (2003). Serum bactericidal activity correlates with the vaccine efficacy of outer membrane vesicle vaccines against *Neisseria meningitidis* serogroup B disease. Vaccine *21*, 734–737.

Holst, J., Feiring, B., Naess, L.M., Norheim, G., Kristiansen, P., Hoiby, E.A., Bryn, K., Oster, P., Costantino, P., Taha, M.K., Alonso, J.M., Caugant, D.A., Wedege, E., Aaberge, I.S., Rappuoli, R., and Rosenqvist, E. (2005). The concept of 'tailor-made', protein-based, outer membrane vesicle vaccines against meningococcal disease. Vaccine *23*, 2202–2205.

Hood, D.W., Deadman, M.E., Jennings, M.P., Bisercic, M., Fleischmann, R.D., Venter, J.C., and Moxon, E.R. (1996). DNA repeats identify novel virulence genes in *Haemophilus influenzae*. Proc. Natl. Acad. Sci. USA 93, 11121–11125.

Hook, E.W., 3rd, and Holmes, K.K. (1985). Gonococcal infections. Ann. Intern. Med. *102*, 229–243.

Howitz, M., Krause, T.G., Simonsen, J.B., Hoffmann, S., Frisch, M., Nielsen, N.M., Robbins, J., Schneerson, R., Molbak, K., and Miller, M.A. (2007). Lack of association between group B meningococcal disease and autoimmune disease. Clin. Infect. Dis. 45, 1327–1334.

Ison, C.A., Anwar, N., Cole, M.J., Galassini, R., Heyderman, R.S., Klein, N.J., West, J., Pollard, A.J., Morley, S., and Levin and the Meningococcal, R. (1999). Assessment of immune response to meningococcal disease: comparison of a whole-blood assay and the serum bactericidal assay. Microb. Pathog. *27*, 207–214.

Jerse, A.E., Cohen, M.S., Drown, P.M., Whicker, L.G., Isbey, S.F., Seifert, H.S., and Cannon, J.G. (1994). Multiple gonococcal opacity proteins are expressed during experimental urethral infection in the male. J. Exp. Med. *179*, 911–920.

Jodar, L., Cartwright, K., and Feavers, I.M. (2000). Standardisation and validation of serological assays for the evaluation of immune responses to *Neisseria meningitidis* serogroup A and C vaccines. Biologicals *28*, 193–197.

Kahler, C.M., and Stephens, D.S. (1998). Genetic basis for biosynthesis, structure, and function of meningococcal lipooligosaccharide (endotoxin). Crit. Rev. Microbiol. *24*, 281–334.

Kilian, M., and Reinholdt, J. (1987). A hypothetical model for the development of invasive infection due to IgA1 protease-producing bacteria. Adv .Exp. Med. Biol. 216B, 1261–1269.

Kline, K.A., Sechman, E.V., Skaar, E.P., and Seifert, H.S. (2003). Recombination, repair and replication in the pathogenic neisseriae: the 3 R's of molecular genetics of two human-specific bacterial pathogens. Mol. Microbiol. *50*, 3–13.

Knapp, J.S., Tam, M.R., Nowinski, R.C., Holmes, K.K., and Sandstrom, E.G. (1984). Serological classification of *Neisseria gonorrhoeae* with use of monoclonal antibodies to gonococcal outer membrane protein I.J. Infect. Dis. 150, 44–48.

Koeberling, O., Welsch, J.A., and Granoff, D.M. (2007). Improved immunogenicity of a H44/76 group B outer membrane vesicle vaccine with over-expressed genome-derived Neisserial antigen 1870. Vaccine *25*, 1912–1920.

Kshirsagar, N., Mur, N., Thatte, U., Gogtay, N., Viviani, S., Preziosi, M.P., Elie, C., Findlow, H., Carlone, G., Borrow, R., Parulekar, V., Plikaytis, B., Kulkarni, P., Imbault, N., and LaForce, F.M. (2007). Safety, immunogenicity, and antibody persistence of a new meningococcal group A conjugate vaccine in healthy Indian adults. Vaccine *25*, A101–107.

Kuhns, D., Kisner, P., Williams, M.P., and Moorman, P.L. (1938). The control of meningococcic epidemics by active immunisation with meningococcus soluble toxin: further studies. JAMA *110*, 484–487.

Laforce, F.M., Konde, K., Viviani, S., and Preziosi, M.P. (2007). The Meningitis Vaccine Project. Vaccine *25*, A97–100.

Lakshman, R., and Finn, A. (2002). Meningococcal serogroup C conjugate vaccine. Expert Opin. Biol. Ther. *2*, 87–96.

Lancellotti, M., Guiyoule, A., Ruckly, C., Hong, E., Alonso, J.M., and Taha, M.K. (2006). Conserved virulence of C to B capsule switched *Neisseria meningitidis* clinical isolates belonging to ET-37/ST-11 clonal complex. Microbes Infect. *8*, 191–196.

Lee, H.S., Boulton, I.C., Reddin, K., Wong, H., Halliwell, D., Mandelboim, O., Gorringe, A.R., and Gray-Owen, S.D. (2007). Neisserial outer membrane vesicles bind the coinhibitory receptor carcinoembryonic antigen-related cellular adhesion molecule 1 and suppress CD4+ T lymphocyte function. Infect. Immun. *75*, 4449–4455.

Lehmann, A.K., Halstensen, A., Aaberge, I.S., Holst, J., Michaelsen, T.E., Sornes, S., Wetzler, L.M., and Guttormsen, H. (1999). Human opsonins induced during meningococcal disease recognize outer membrane proteins PorA and PorB. Infect. Immun. *67*, 2552–2560.

Lesinski, G.B., and Westerink, M.A. (2001a). Novel vaccine strategies to T-independent antigens. J. Microbiol. Methods *47*, 135–149.

Lesinski, G.B., and Westerink, M.A. (2001b). Vaccines against polysaccharide antigens. Curr. Drug Targets Infect Disord *1*, 325–334.

Levine, M.M., and Lagos, R. (2004). Vaccines and Vaccination in Historical Perspective. In New Generation Vaccines, M.M. Levine, J.B. Kaper, R. Rappouli, M.A. Liu, and M.F. Good, eds. (New York, Marcel Dekker, Inc.), pp. 1–10.

Levine, M.M., and Sztein, M.B. (2004). Vaccine development strategies for improving immunization: the role of modern immunology. Nat. Immunol. *5*, 460–464.

Levine, W.C., Pope, V., Bhoomkar, A., Tambe, P., Lewis, J.S., Zaidi, A.A., Farshy, C.E., Mitchell, S., and Talkington, D.F. (1998). Increase in endocervical CD4 lymphocytes among women with nonulcerative sexually transmitted diseases. J. Infect. Dis. 177, 167–174.

Li, Y., Sun, Y.H., Ison, C., Levine, M.M., and Tang, C.M. (2004). Vaccination with attenuated *Neisseria meningitidis* strains protects against challenge with live Meningococci. Infect. Immun. 72, 345–351.

Li, Y., Zhang, Q., Winterbotham, M., Mowe, E., Gorringe, A., and Tang, C.M. (2006). Immunization with live *Neisseria* lactamica protects mice against meningococcal challenge and can elicit serum bactericidal antibodies. Infect. Immun. *74*, 6348–6355.

Lipsitch, M., and O'Hagan J, J. (2007). Patterns of antigenic diversity and the mechanisms that maintain them. J.R. Soc Interface. *4*, 787–802.

Lo Passo, C., Romeo, A., Pernice, I., Donato, P., Midiri, A., Mancuso, G., Arigo, M., Biondo, C., Galbo, R., Papasergi, S., Felici, F., Teti, G., and Beninati, C.(2007). Peptide mimics of the group B meningococcal capsule induce bactericidal and protective antibodies after immunization. J. Immunol. *178*, 4417–4423.

Luijkx, T., van Dijken, H., van Els, C., and van den Dobbelsteen, G. (2006). Heterologous prime-boost strategy to overcome weak immunogenicity of two serosubtypes in hexavalent *Neisseria meningitidis* outer membrane vesicle vaccine. Vaccine *24*, 1569–1577.

Lystad, A., and Aasen, S. (1991). The epidemiology of meningococcal disease in Norway 1975–91. NIPH Ann *14*, 57–65; discussion 65–56.

MacLeod, C.M., Hodges, G.R., Heidelberger, M., and Bernhard, W.G. (1945). Prevention of pneumococcal pneumonia by immunization with specific capsular polysaccharides. J. Exp. Med. *82*, 445–465.

Madico, G., Welsch, J.A., Lewis, L.A., McNaughton, A., Perlman, D.H., Costello, C.E., Ngampasutadol, J., Vogel, U., Granoff, D.M., and Ram, S. (2006). The meningococcal vaccine candidate GNA1870 binds the complement regulatory protein factor H and enhances serum resistance. J. Immunol. *177*, 501–510.

Maiden, M.C. (1993). Population genetics of a transformable bacterium: the influence of horizontal genetic exchange on the biology of *Neisseria meningitidis*. FEMS Microbiol Lett *112*, 243–250.

Maiden, M.C., and Spratt, B.G. (1999). Meningococcal conjugate vaccines: new opportunities and new challenges. Lancet *354*, 615–616.

Maiden, M.C.J., and Caugant, D.A. (2006). The Population Biology of *Neisseria meningitidis*: Implications for Meningococcal Disease, Epidemiology and Control. In Handbook of Meningicoccal Disease Infection Biology, Vaccination, Clinical Management, M. Frosch, and M.C.J. Maiden, eds. (Weinheim, Wiley-VCH Verlag GmbH & Co.), pp. 17–35.

Maione, D., Margarit, I., Rinaudo, C.D., Masignani, V., Mora, M., Scarselli, M., Tettelin, H., Brettoni, C., Iacobini, E.T., Rosini, R., D'Agostino, N., Miorin, L., Buccato, S., Mariani, M., Galli, G., Nogarotto, R., Nardi Dei, V., Vegni, F., Fraser, C., Mancuso, G., Teti, G., Madoff, L.C., Paoletti, L.C., Rappuoli, R., Kasper, D.L., Telford, J.L., and Grandi, G. (2005). Identification of a universal Group B streptococcus vaccine by multiple genome screen. Science *309*, 148–150.

Mandrell, R.E., and Apicella, M.A. (1993). Lipo-oligosaccharides (LOS) of mucosal pathogens: molecular mimicry and host-modification of LOS. Immunobiology *187*, 382–402.

Martin, D. (2007). *Neisseria* vaccines 2007 – international workshop. IDrugs *10*, 450–452.

Martin, I.M., Ison, C.A., Aanensen, D.M., Fenton, K.A., and Spratt, B.G. (2004). Rapid sequence-based identification of gonococcal transmission clusters in a large metropolitan area. J. Infect. Dis. 189, 1497–1505.

Martin, P., Makepeace, K., Hill, S.A., Hood, D.W., and Moxon, E.R. (2005). Microsatellite instability regulates transcription factor binding and gene expression. Proc. Natl. Acad. Sci. USA *102*, 3800–3804.

Martin, P., van de Ven, T., Mouchel, N., Jeffries, A.C., Hood, D.W., and Moxon, E.R. (2003). Experimentally revised repertoire of putative contingency loci in *Neisseria meningitidis* strain MC58: evidence for a novel mechanism of phase variation. Mol. Microbiol. *50*, 245–257.

Masignani, V., Comanducci, M., Giuliani, M.M., Bambini, S., Adu-Bobie, J., Arico, B., Brunelli, B., Pieri, A., Santini, L., Savino, S., Serruto, D., Litt, D., Kroll, S., Welsch, J.A., Granoff, D.M., Rappuoli, R., and Pizza, M. (2003). Vaccination against *Neisseria meningitidis* using three variants of the lipoprotein GNA1870. J. Exp. Med. *197*, 789–799.

Maskell, D., Frankel, G., and Dougan, G. (1993). Phase and antigenic variation – the impact on strategies for bacterial vaccine design. Trends Biotechnol *11*, 506–510.

Merz, A.J., and So, M. (2000). Interactions of pathogenic neisseriae with epithelial cell membranes. Annu. Rev. Cell Dev. Biol. *16*, 423–457.

Mestecky, J., Moldoveanu, Z., and Russell, M.W. (2005). Immunologic uniqueness of the genital tract: challenge for vaccine development. Am. J. Reprod Immunol 53, 208–214.

Mietzner, T.A., Thomas, C.E., Hobbs, M.M., and Cohen, M.S. (2004). Vaccines against gonoccoal infection. In New Generation Vaccines, M.M. Levine, J.B. Kaper, R. Rappouli, M.A. Liu, and M.F. Good, eds. (New York, Marcel Dekker, Inc.), pp. 755–773.

Miller, E., Ramsay, M., and Campbell, H. (2006). Vaccination for the control of meningococcal disease: the use of meningococcal vaccines from the public health perspective. In Handbook of Meningicoccal Disease Infection Biology, Vaccination, Clinical Management, M. Frosch, and M.C.J. Maiden, eds. (Weinheim, Wiley-VCH Verlag GmbH & Co.), pp. 403–423.

Moe, G.R., Tan, S., and Granoff, D.M. (1999). Molecular mimetics of polysaccharide epitopes as vaccine candidates for prevention of *Neisseria meningitidis* serogroup B disease. FEMS Immunol. Med. Microbiol. *26*, 209–226.

Moxon, R., Bayliss, C., and Hood, D. (2006). Bacterial contingency loci: the role of simple sequence DNA repeats in bacterial adaptation. Annu. Rev. Genet. *40*, 307–333.

Mueller, J.E., Borrow, R., and Gessner, B.D. (2006). Meningococcal serogroup W135 in the African meningitis belt: epidemiology, immunity and vaccines. Expert Rev. Vaccines *5*, 319–336.

Munkley, A., Tinsley, C.R., Virji, M., and Heckels, J.E. (1991). Blocking of bactericidal killing of *Neisseria meningitidis* by antibodies directed against class 4 outer membrane protein. Microb. Pathog. *11*, 447–452.

Muzzi, A., Masignani, V., and Rappuoli, R. (2007). The pan-genome: towards a knowledge-based discovery of novel targets for vaccines and antibacterials. Drug Discovery Today *12*, 429–439.

Namork, E., and Brandtzaeg, P. (2002). Fatal meningococcal septicaemia with 'blebbing' meningococcus. Lancet *360*, 1741.

Nassif, X., Pujol, C., Morand, P., and Eugene, E. (1999). Interactions of pathogenic *Neisseria* with host cells. Is it possible to assemble the puzzle? Mol. Microbiol. *32*, 1124–1132.

Nassif, X., and So, M. (1995). Interaction of pathogenic neisseriae with nonphagocytic cells. Clin Microbiol. Rev. *8*, 376–388.

Nedelec, J., Boucraut, J., Garnier, J.M., Bernard, D., and Rougon, G. (1990). Evidence for autoimmune antibodies directed against embryonic neural cell adhesion molecules (N-CAM) in patients with group B meningitis. J Neuroimmunol *29*, 49–56.

Nelson, C.B., Birmingham, M., Costa, A., Daviaud, J., Perea, W., Kieny, M.P., and Tarantola, D. (2007). Preparedness for infectious threats: public-private partnership to develop an affordable vaccine for an emergent threat: the trivalent *Neisseria meningitidis* ACW135 polysaccharide vaccine. Am. J. Public Health. 97 Suppl. *1*, S15–22.

Newcombe, J., Cartwright, K., Dyer, S., and McFadden, J. (1998). Naturally occurring insertional inactivation of the porA gene of *Neisseria meningitidis* by integration of IS1301. Mol. Microbiol. *30*, 453–454.

Newcombe, J., Eales-Reynolds, L.J., Wootton, L., Gorringe, A.R., Funnell, S.G., Taylor, S.C., and McFadden, J.J. (2004). Infection with an avirulent phoP mutant of *Neisseria meningitidis* confers broad cross-reactive immunity. Infect. Immun. 72, 338–344.

Ngampasutadol, J., Rice, P.A., Walsh, M.T., and Gulati, S. (2006). Characterization of a peptide vaccine candidate mimicking an oligosaccharide epitope of *Neisseria gonorrhoeae* and resultant immune responses and function. Vaccine *24*, 157–170.

Noah, N., and Henderson, B. (2002). Surveillance of Bacterial Meningitis in Europe 1997/1998 (Communicable Disease Surveillance Centre, London), pp. 1–16.

Novartis (2007). Press release: New data for Menveo™ vaccine show excellent immune response broad protection for infants against meningococcal meningitis (http://www.novartisvaccines.com/press-room/news/20071115_menveo.shtml).

O'Hagan, D.T., and Rappuoli, R. (2006). Novel approaches to pediatric vaccine delivery. Adv. Drug Deliv. Rev. *58*, 29–51.

Oliver, K.J., Reddin, K.M., Bracegirdle, P., Hudson, M.J., Borrow, R., Feavers, I.M., Robinson, A., Cartwright, K., and Gorringe, A.R. (2002). *Neisseria* lactamica protects against experimental meningococcal infection. Infect. Immun. *70*, 3621–3626.

Oomen, C.J., Hoogerhout, P., Kuipers, B., Vidarsson, G., van Alphen, L., and Gros, P. (2005). Crystal structure of an Anti-meningococcal subtype P1.4 PorA antibody provides basis for peptide-vaccine design. J. Mol. Biol. *351*, 1070–1080.

Oster, P., Lennon, D., O'Hallahan, J., Mulholland, K., Reid, S., and Martin, D. (2005). MeNZB: a safe and highly immunogenic tailor-made vaccine against the New Zealand *Neisseria meningitidis* serogroup B disease epidemic strain. Vaccine *23*, 2191–2196.

Pajon, R., Niebla, O., Yero, D., Perez, O., Cabrera, O., Findlow, J., Balmer, P., and Borrow, R. (2007). On the neisserial vaccine quest: *Neisseria* Vaccines 2007. Expert Rev. Anti. Infect. Ther .5, 545–550.

Park, I.H., Youn, J.H., Choi, I.H., Nahm, M.H., Kim, S.J., and Shin, J.S. (2005). Anti-idiotypic antibody as a potential candidate vaccine for *Neisseria meningitidis* serogroup B. Infect. Immun. 73, 6399–6406.

PParkhill, J., Achtman, M., James, K.D., Bentley, S.D., Churcher, C., Klee, S.R., Morelli, G., Basham, D., Brown, D., Chillingworth, T., Davies, R.M., Davis, P., Devlin, K., Feltwell, T., Hamlin, N., Holroyd, S., Jagels, K., Leather, S., Moule, S., Mungall, K., Quail, M.A., Rajandream, M.A., Rutherford, K.M., Simmonds, M., Skelton, J., Whitehead, S., Spratt, B.G., and Barrell, B.G. (2000). Complete DNA sequence of a serogroup A strain of *Neisseria meningitidis* Z2491. Nature *404*, 502–506.

Peltola, H., Makela, H., Kayhty, H., Jousimies, H., Herva, E., Hallstrom, K., Sivonen, A., Renkonen, O.V., Pettay, O., Karanko, V., Ahvonen, P., and Sarna, S. (1977). Clinical efficacy of meningococcus group A capsular polysaccharide vaccine in children three months to five years of age. N. Engl. J. Med. *297*, 686–691.

Pettersson, A., Kortekaas, J., Weynants, V.E., Voet, P., Poolman, J.T., Bos, M.P., and Tommassen, J. (2006). Vaccine potential of the *Neisseria meningitidis* lactoferrin-binding proteins LbpA and LbpB. Vaccine *24*, 3545–3557.

Pettit, R.K., and Judd, R.C. (1992). The interaction of naturally elaborated blebs from serum-susceptible and serum-resistant strains of *Neisseria gonorrhoeae* with normal human serum. Mol. Microbiol. *6*, 729–734.

Pichichero, M.E. (2005). Meningococcal conjugate vaccine in adolescents and children. Clin. Pediatr. (Phila) 44, 479–489.

Pizza, M., Scarlato, V., Masignani, V., Giuliani, M.M., Arico, B., Comanducci, M., Jennings, G.T., Baldi, L., Bartolini, E., Capecchi, B., Galeotti, C.L., Luzzi, E., Manetti, R., Marchetti, E., Mora, M., Nuti, S., Ratti, G., Santini, L., Savino, S., Scarselli, M., Storni, E., Zuo, P., Broeker, M., Hundt, E., Knapp, B., Blair, E., Mason, T., Tettelin, H., Hood, D.W., Jeffries, A.C., Saunders, N.J., Granoff, D.M., Venter, J.C., Moxon, E.R., Grandi, G., and Rappuoli, R. (2000). Identification of vaccine candidates against serogroup B meningococcus by whole-genome sequencing. Science *287*, 1816–1820.

Plested, J.S., Harris, S.L., Wright, J.C., Coull, P.A., Makepeace, K., Gidney, M.A., Brisson, J.R., Richards, J.C., Granoff, D.M., and Moxon, E.R. (2003). Highly conserved *Neisseria meningitidis* inner-core lipopolysaccharide epitope confers protection against experimental meningococcal bacteremia. J. Infect. Dis. 187, 1223–1234.

Plummer, F.A., and Brunham, R.C. (1987). Gonococcal recidivism, diversity, and ecology. Rev. Infect. Dis. *9*, 846–850.

Plummer, F.A., Simonsen, J.N., Chubb, H., Slaney, L., Kimata, J., Bosire, M., Ndinya-Achola, J.O., and Ngugi, E.N. (1989). Epidemiologic evidence for the development of serovar-specific immunity after gonococcal infection. J. Clin. Invest. *83*, 1472–1476.

Pollabauer, E.M., Petermann, R., and Ehrlich, H.J. (2005). Group C meningococcal polysaccharide-tetanus toxoid conjugate vaccine: a meta-analysis of immunogenicity, safety and posology. Hum. Vaccin. *1*, 131–139.

Pollard, A.J. (2004). Global epidemiology of meningococcal disease and vaccine efficacy. Pediatr. Infect. Dis. J. *23*, S274–279.

Pollard, A.J., and Frasch, C. (2001). Development of natural immunity to *Neisseria meningitidis*. Vaccine *19*, 1327–1346.

Pollard, A.J., Galassini, R., van der Voort, E.M., Booy, R., Langford, P., Nadel, S., Ison, C., Kroll, J.S., Poolman, J., and Levin, M. (1999). Humoral immune responses to *Neisseria meningitidis* in children. Infect. Immun. *67*, 2441–2451.

Pollard, A.J., and Scheifele, D. (2001). Meningococcal disease and vaccination in North America. J Paediatr Child Health *37*, S20–27.

Power, P.M., and Moxon, E.R. (2006). Phase Variation and Adaptive Strategies of *N. meningitidis:* Insights into the Biology of a Commensal and Pathogen. In Handbook of Meningicoccal Disease Infection Biology, Vaccination, Clinical Management, M. Frosch, and M.C.J. Maiden, eds. (Weinheim, Wiley-VCH Verlag GmbH & Co.), pp. 99–118.

Power, P.M., Roddam, L.F., Rutter, K., Fitzpatrick, S.Z., Srikhanta, Y.N., and Jennings, M.P. (2003). Genetic characterization of pilin glycosylation and phase variation in *Neisseria meningitidis*. Mol. Microbiol. *49*, 833–847.

Ramsay, M.E., Andrews, N.J., Trotter, C.L., Kaczmarski, E.B., and Miller, E. (2003). Herd immunity from meningococcal serogroup C conjugate vaccination in England: database analysis. BMJ *326*, 365–366.

Reingold, A.L., Broome, C.V., Hightower, A.W., Ajello, G.W., Bolan, G.A., Adamsbaum, C., Jones, E.E., Phillips, C., Tiendrebeogo, H., and Yada, A. (1985). Age-specific differences in duration of clinical protection after vaccination with meningococcal polysaccharide A vaccine. Lancet *2*, 114–118.

Rice, P.A., Vayo, H.E., Tam, M.R., and Blake, M.S. (1986). Immunoglobulin G antibodies directed against protein III block killing of serum-resistant *Neisseria gonorrhoeae* by immune serum. J. Exp. Med. *164*, 1735–1748.

Richardson, A.R., Yu, Z., Popovic, T., and Stojiljkovic, I. (2002). Mutator clones of *Neisseria meningitidis* in epidemic serogroup A disease. Proc. Natl. Acad. Sci. USA *99*, 6103–6107.

Richmond, P., Borrow, R., Miller, E., Clark, S., Sadler, F., Fox, A., Begg, N., Morris, R., and Cartwright, K. (1999). Meningococcal serogroup C conjugate vaccine is immunogenic in infancy and primes for memory. J. Infect. Dis. 179, 1569–1572.

Richmond, P., Kaczmarski, E., Borrow, R., Findlow, J., Clark, S., McCann, R., Hill, J., Barker, M., and Miller, E. (2000). Meningococcal C polysaccharide vaccine induces immunologic hyporesponsiveness in adults that is overcome by meningococcal C conjugate vaccine. J. Infect. Dis. 181, 761–764.

Robbins, J.B., Myerowitz, L., Whisnant, J.K., Argaman, M., Schneerson, R., Handzel, Z.T., and Gotschlich,

E.C. (1972). Enteric bacteria cross-reactive with *Neisseria meningitidis* groups A and C and Diplococcus pneumoniae types I and 3. Infect. Immun. *6*, 651–656.

Rodriguez-Ortega, M.J., Norais, N., Bensi, G., Liberatori, S., Capo, S., Mora, M., Scarselli, M., Doro, F., Ferrari, G., Garaguso, I., Maggi, T., Neumann, A., Covre, A., Telford, J.L., and Grandi, G. (2006). Characterization and identification of vaccine candidate proteins through analysis of the group A Streptococcus surface proteome. Nat. Biotechnol. *24*, 191–197.

Rohde, K.H., and Dyer, D.W. (2003). Mechanisms of iron acquisition by the human pathogens *Neisseria meningitidis* and *Neisseria gonorrhoeae*. Front. Biosci. *8*, d1186–1218.

Rosenqvist, E., Hoiby, E.A., Bjune, G., Bryn, K., Closs, O., Feiring, B., Klem, A., Nokleby, H., and Frolm, L.O. (1991). Human antibody responses after vaccination with the Norwegian group B meningococcal outer membrane vesicle vaccine: results from ELISA studies. NIPH Ann *14*, 169–179; discussion 180–161.

Rosenqvist, E., Musacchio, A., Aase, A., Hoiby, E.A., Namork, E., Kolberg, J., Wedege, E., Delvig, A., Dalseg, R., Michaelsen, T.E., and Tommassen, J. (1999). Functional activities and epitope specificity of human and murine antibodies against the class 4 outer membrane protein (Rmp) of *Neisseria meningitidis*. Infect. Immun. *67*, 1267–1276.

Rosenstein, N.E., Perkins, B.A., Stephens, D.S., Lefkowitz, L., Cartter, M.L., Danila, R., Cieslak, P., Shutt, K.A., Popovic, T., Schuchat, A., Harrison, L.H., and Reingold, A.L. (1999). The changing epidemiology of meningococcal disease in the United States, 1992–1996. J. Infect. Dis. 180, 1894–1901.

Russell, M.W., Hedges, S.R., Wu, H.Y., Hook, E.W., 3rd, and Mestecky, J. (1999). Mucosal immunity in the genital tract: prospects for vaccines against sexually transmitted diseases – a review. Am. J. Reprod. Immunol. *42*, 58–63.

Sacchi, C.T., Pessoa, L.L., Ramos, S.R., Milagres, L.G., Camargo, M.C., Hidalgo, N.T., Melles, C.E., Caugant, D.A., and Frasch, C.E. (1992). Ongoing group B *Neisseria meningitidis* epidemic in Sao Paulo, Brazil, due to increased prevalence of a single clone of the ET-5 complex. J. Clin. Microbiol. *30*, 1734–1738.

Sadler, F., Fox, A., Neal, K., Dawson, M., Cartwright, K., and Borrow, R. (2003). Genetic analysis of capsular status of meningococcal carrier isolates. Epidemiol. Infect. *130*, 59–70.

Sandbu, S., Feiring, B., Oster, P., Helland, O.S., Bakke, H.S., Naess, L.M., Aase, A., Aaberge, I.S., Kristoffersen, A.C., Rydland, K.M., Tilman, S., Nokleby, H., and Rosenqvist, E. (2007). Immunogenicity and Safety of a Combination of Two Serogroup B Meningococcal OMV Vaccines. Clin Vaccine Immunol.

Sardinas, G., Reddin, K., Pajon, R., and Gorringe, A. (2006). Outer membrane vesicles of *Neisseria* lactamica as a potential mucosal adjuvant. Vaccine *24*, 206–214.

Sarkari, J., Pandit, N., Moxon, E.R., and Achtman, M. (1994). Variable expression of the Opc outer membrane protein in *Neisseria meningitidis* is caused by size variation of a promoter containing poly-cytidine. Mol. Microbiol. *13*, 207–217.

Schmidt, K.A., Schneider, H., Lindstrom, J.A., Boslego, J.W., Warren, R.A., Van de Verg, L., Deal, C.D., McClain, J.B., and Griffiss, J.M. (2001). Experimental gonococcal urethritis and reinfection with homologous gonococci in male volunteers. Sex. Transm. Dis. *28*, 555–564.

Schmitt, H.J., Maechler, G., Habermehl, P., Knuf, M., Saenger, R., Begg, N., and Boutriau, D. (2007). Immunogenicity, reactogenicity, and immune memory after primary vaccination with a novel *Haemophilus influenzae-Neisseria meningitidis* serogroup C conjugate vaccine. Clin. Vaccine Immunol. *14*, 426–434.

Schneider, H., Griffiss, J.M., Boslego, J.W., Hitchcock, P.J., Zahos, K.M., and Apicella, M.A. (1991). Expression of paragloboside-like lipooligosaccharides may be a necessary component of gonococcal pathogenesis in men. J. Exp. Med. *174*, 1601–1605.

Schneider, M.C., Exley, R.M., Ram, S., Sim, R.B., and Tang, C.M. (2007). Interactions between *Neisseria meningitidis* and the complement system. Trends Microbiol. *15*, 233–240.

Scholten, R.J., Kuipers, B., Valkenburg, H.A., Dankert, J., Zollinger, W.D., and Poolman, J.T. (1994). Lipo-oligosaccharide immunotyping of *Neisseria meningitidis* by a whole-cell ELISA with monoclonal antibodies. J. Med. Microbio.l *41*, 236–243.

Seifert, H.S., Ajioka, R.S., Marchal, C., Sparling, P.F., and So, M. (1988). DNA transformation leads to pilin antigenic variation in *Neisseria gonorrhoeae*. Nature *336*, 392–395.

Seifert, H.S., Wright, C.J., Jerse, A.E., Cohen, M.S., and Cannon, J.G. (1994). Multiple gonococcal pilin antigenic variants are produced during experimental human infections. J. Clin. Invest. *93*, 2744–2749.

Serruto, D., and Rappuoli, R. (2006). Post-genomic vaccine development. FEBS Lett. *580*, 2985–2992.

Shepard, C.W., Ortega-Sanchez, I.R., Scott, R.D., 2nd, and Rosenstein, N.E. (2005). Cost-effectiveness of conjugate meningococcal vaccination strategies in the United States. Pediatrics *115*, 1220–1232.

Sierra, G.V., Campa, H.C., Varcacel, N.M., Garcia, I.L., Izquierdo, P.L., Sotolongo, P.F., Casanueva, G.V., Rico, C.O., Rodriguez, C.R., and Terry, M.H. (1991). Vaccine against group B *Neisseria meningitidis*: protection trial and mass vaccination results in Cuba. NIPH Ann *14*, 195–207; discussion 208–110.

Smith, J.M., Smith, N.H., O'Rourke, M., and Spratt, B.G. (1993). How clonal are bacteria? Proc. Natl. Acad. Sci. USA *90*, 4384–4388.

Snape, M.D., Perrett, K.P., Ford, K.J., John, T.M., Pace, D., Yu, L.M., Langley, J.M., McNeil, S., Dull, P.M., Ceddia, F., Anemona, A., Halperin, S.A., Dobson, S., and Pollard, A.J. (2008). Immunogenicity of a tetravalent meningococcal glycoconjugate vaccine in infants: a randomized controlled trial. JAMA *299*, 173–184.

Snyder, L.A., Butcher, S.A., and Saunders, N.J. (2001). Comparative whole-genome analyses reveal over 100 putative phase-variable genes in the pathogenic *Neisseria* spp. Microbiology *147*, 2321–2332.

Snyder, L.A., and Saunders, N.J. (2006). The majority of genes in the pathogenic *Neisseria* species are present in non-pathogenic *Neisseria* lactamica, including those designated as 'virulence genes'. BMC Genomics 7, 128.

Spoulou, V.I., Moschou, C.K., Tzanakaki, G., and Theodoridou, M.C. (2007). Conjugate vaccine-induced immunological priming is not protective against acute meningococcal C infection. Vaccine.

Spratt, B.G., Bowler, L.D., Zhang, Q.Y., Zhou, J., and Smith, J.M. (1992). Role of interspecies transfer of chromosomal genes in the evolution of penicillin resistance in pathogenic and commensal *Neisseria* species. J. Mol. Evol. *34*, 115–125.

Srikhanta, Y.N., Maguire, T.L., Stacey, K.J., Grimmond, S.M., and Jennings, M.P. (2005). The phasevarion: a genetic system controlling coordinated, random switching of expression of multiple genes. Proc. Natl. Acad. Sci. USA *102*, 5547–5551.

Srikhanta, Y.N., Dowideit, S.J., Edwards, J.L., Falsetta, M.L., Wu, H.J., Harrison, O.B., Fox, K.L., Seib, K.L., Maguire, T.L., Wang, A.H., Maiden, M.C., Grimmond, S.M., Apicella, M.A., and Jennings, M.P. (2009). Phasevarions mediate random switching of gene expression in pathogenic *Neisseria*. PLoS Pathog. *5*, e1000400.

Stanberry, L.R., and Rosenthal, S.L. (2005). Progress in vaccines for sexually transmitted diseases. Infect. Dis. Clin. North Am. *19*, 477–490.

Stefanelli, P., Fazio, C., Neri, A., Sofia, T., and Mastrantonio, P. (2003). First report of capsule replacement among electrophoretic type 37 *Neisseria meningitidis* strains in Italy. J. Clin. Microbiol. *41*, 5783–5786.

Stephens, D.S., Greenwood, B., and Brandtzaeg, P. (2007). Epidemic meningitis, meningococcaemia, and *Neisseria meningitidis*. Lancet *369*, 2196–2210.

Stern, A., Brown, M., Nickel, P., and Meyer, T.F. (1986). Opacity genes in *Neisseria gonorrhoeae*: control of phase and antigenic variation. Cell *47*, 61–71.

Swanson, J., Barrera, O., Sola, J., and Boslego, J. (1988). Expression of outer membrane protein II by gonococci in experimental gonorrhea. J. Exp. Med. *168*, 2121–2129.

Swanson, J., Robbins, K., Barrera, O., Corwin, D., Boslego, J., Ciak, J., Blake, M., and Koomey, J.M. (1987). Gonococcal pilin variants in experimental gonorrhea. J. Exp. Med. *165*, 1344–1357.

Swartley, J.S., Marfin, A.A., Edupuganti, S., Liu, L.J., Cieslak, P., Perkins, B., Wenger, J.D., and Stephens, D.S. (1997). Capsule switching of *Neisseria meningitidis*. Proc. Natl. Acad. Sci. USA *94*, 271–276.

Taha, M.K., and Alonso, J.M. (2007). Meningococcal Vaccines: To Eradicate the Disease, not the Bacterium. Hum Vaccin *3*, 149–153.

Tappero, J.W., Lagos, R., Ballesteros, A.M., Plikaytis, B., Williams, D., Dykes, J., Gheesling, L.L., Carlone, G.M., Hoiby, E.A., Holst, J., Nokleby, H., Rosenqvist, E., Sierra, G., Campa, C., Sotolongo, F., Vega, J., Garcia, J., Herrera, P., Poolman, J.T., and Perkins, B.A. (1999). Immunogenicity of 2 serogroup B outer-membrane protein meningococcal vaccines: a randomized controlled trial in Chile. JAMA *281*, 1520–1527.

Tapsall, J. (2006). Antibiotic resistance in *Neisseria gonorrhoeae* is diminishing available treatment options for gonorrhea: some possible remedies. Expert Rev. Anti. Infect. Ther. *4*, 619–628.

Tejedor, J.C., Moro, M., Ruiz-Contreras, J., Castro, J., Gomez-Campdera, J.A., Navarro, M.L., Merino, J.M., Martin-Ancel, A., Roca, J., Garcia-Del-Ri, M., Jurado, A., Diez-Delgado, F.J., Omenaca, F., Garcia-Sicilia, J., Boceta, R., Garcia-Corbeira, P., Collard, A., Boutriau, D., Schuerman, L., and Jacquet, J.M. (2007). Immunogenicity and reactogenicity of primary immunization with a novel combined *Haemophilus influenzae* Type b and *Neisseria meningitidis* Serogroup C-tetanus toxoid conjugate vaccine coadministered with a Diphtheria-tetanus-acellular Pertussis-hepatitis B-inactivated poliovirus vaccine at 2, 4 and 6 months. Pediatr. Infect. Dis. J. *26*, 1–7.

Tettelin, H., Saunders, N.J., Heidelberg, J., Jeffries, A.C., Nelson, K.E., Eisen, J.A., Ketchum, K.A., Hood, D.W., Peden, J.F., Dodson, R.J., Nelson, W.C., Gwinn, M.L., DeBoy, R., Peterson, J.D., Hickey, E.K., Haft, D.H., Salzberg, S.L., White, O., Fleischmann, R.D., Dougherty, B.A., Mason, T., Ciecko, A., Parksey, D.S., Blair, E., Cittone, H., Clark, E.B., Cotton, M.D., Utterback, T.R., Khouri, H., Qin, H., Vamathevan, J., Gill, J., Scarlato, V., Masignani, V., Pizza, M., Grandi, G., Sun, L., Smith, H.O., Fraser, C.M., Moxon, E.R., Rappuoli, R., and Venter, J.C. (2000). Complete genome sequence of *Neisseria meningitidis* serogroup B strain MC58. Science *287*, 1809–1815.

Tikhomirov, E., Santamaria, M., and Esteves, K. (1997). Meningococcal disease: public health burden and control. World Health Stat Q. *50*, 170–177.

Tiwana, H., Clow, K.J., Hall, C., Feavers, I.M., and Charalambous, B.M. (2005). The immunogenicity of a conformationally restricted peptide mimetic of meningococcal lipooligosaccharide. Scand. J. Immunol. *62*, 385–392.

Tondella, M.L., Popovic, T., Rosenstein, N.E., Lake, D.B., Carlone, G.M., Mayer, L.W., and Perkins, B.A. (2000). Distribution of *Neisseria meningitidis* serogroup B serosubtypes and serotypes circulating in the United States. The Active Bacterial Core Surveillance Team. J. Clin. Microbiol. *38*, 3323–3328.

Trotter, C.L., Andrews, N.J., Kaczmarski, E.B., Miller, E., and Ramsay, M.E. (2004). Effectiveness of meningococcal serogroup C conjugate vaccine 4 years after introduction. Lancet *364*, 365–367.

Trotter, C.L., and Ramsay, M.E. (2007). Vaccination against meningococcal disease in Europe: review and recommendations for the use of conjugate vaccines. FEMS Microbiol. Rev. *31*, 101–107.

Trotter, C.L., Ramsay, M.E., Gray, S., Fox, A., and Kaczmarski, E. (2006). No evidence for capsule replacement following mass immunisation with meningococcal serogroup C conjugate vaccines in England and Wales. Lancet Infect Dis. *6*, 616–617; author reply 617–618.

Uli, L., Castellanos-Serra, L., Betancourt, L., Dominguez, F., Barbera, R., Sotolongo, F., Guillen, G., and Pajon Feyt, R. (2006). Outer membrane vesicles of the VA-MENGOC-BC vaccine against serogroup B of *Neisseria meningitidis*: Analysis of protein components by two-dimensional gel electrophoresis and mass spectrometry. Proteomics *6*, 3389–3399.

Urwin, R., Russell, J.E., Thompson, E.A., Holmes, E.C., Feavers, I.M., and Maiden, M.C. (2004). Distribution of surface protein variants among hyperinvasive meningococci: implications for vaccine design. Infect. Immun. *72*, 5955–5962.

van den Dobbelsteen, G.P., van Dijken, H.H., Pillai, S., and van Alphen, L. (2007). Immunogenicity of a combination vaccine containing pneumococcal conjugates and meningococcal PorA OMVs. Vaccine *25*, 2491–2496.

van der Ende, A., Hopman, C.T., and Dankert, J. (1999). Deletion of porA by recombination between clusters of repetitive extragenic palindromic sequences in *Neisseria meningitidis*. Infect. Immun. *67*, 2928–2934.

van der Ley, P., van der Biezen, J., and Poolman, J.T. (1995). Construction of *Neisseria meningitidis* strains carrying multiple chromosomal copies of the porA gene for use in the production of a multivalent outer membrane vesicle vaccine. Vaccine *13*, 401–407.

van Deuren, M., Brandtzaeg, P., and van der Meer, J.W. (2000). Update on meningococcal disease with emphasis on pathogenesis and clinical management. Clin. Microbiol. Rev. *13*, 144–166.

Vaughan, T.E., Skipp, P.J., O'Connor, C.D., Hudson, M.J., Vipond, R., Elmore, M.J., and Gorringe, A.R. (2006). Proteomic analysis of *Neisseria* lactamica and N eisseria *meningitidis* outer membrane vesicle vaccine antigens. Vaccine *24*, 5277–5293.

Vermont, C., and van den Dobbelsteen, G. (2002). *Neisseria meningitidis* serogroup B: laboratory correlates of protection. FEMS Immunol. Med. Microbiol *34*, 89–96.

Vimr, E., and Lichtensteiger, C. (2002). To sialylate, or not to sialylate: that is the question. Trends Microbiol. *10*, 254–257.

Vipond, C., Suker, J., Jones, C., Tang, C., Feavers, I.M., and Wheeler, J.X. (2006). Proteomic analysis of a meningococcal outer membrane vesicle vaccine prepared from the group B strain NZ98/254. Proteomics *6*, 3400–3413.

Virji, M., Alexandrescu, C., Ferguson, D.J., Saunders, J.R., and Moxon, E.R. (1992). Variations in the expression of pili: the effect on adherence of *Neisseria meningitidis* to human epithelial and endothelial cells. Mol. Microbiol. *6*, 1271–1279.

Virji, M., Watt, S.M., Barker, S., Makepeace, K., and Doyonnas, R. (1996). The N-domain of the human CD66a adhesion molecule is a target for Opa proteins of *Neisseria meningitidis* and *Neisseria gonorrhoeae*. Mol. Microbiol. *22*, 929–939.

Viviani, S. (2007). Update of progress on a conjugate MenA vaccine for Africa. In Meningitis and Septicaemia in Children and Adults, Meningitis Research Foundation Conference November 2007 (Royal Society of Medicine, London).

Vodopija, I., Baklaic, Z., Hauser, P., Roelants, P., Andre, F.E., and Safary, A. (1983). Reactivity and immunogenicity of bivalent (AC) and tetravalent (ACW135Y) meningococcal vaccines containing O-acetyl-negative or O-acetyl-positive group C polysaccharide. Infect. Immun. *42*, 599–604.

Vogel, U., Claus, H., and Frosch, M. (2000). Rapid serogroup switching in *Neisseria meningitidis*. N. Engl. J. Med. *342*, 219–220.

Welsch, J.A., Moe, G.R., Rossi, R., Adu-Bobie, J., Rappuoli, R., and Granoff, D.M. (2003). Antibody to genome-derived neisserial antigen 2132, a *Neisseria meningitidis* candidate vaccine, confers protection against bacteremia in the absence of complement-mediated bactericidal activity. J. Infect. Dis. 188, 1730–1740.

Welsch, J.A., Rossi, R., Comanducci, M., and Granoff, D.M. (2004). Protective activity of monoclonal antibodies to genome-derived neisserial antigen 1870, a *Neisseria meningitidis* candidate vaccine. J. Immunol. *172*, 5606–5615.

West, D., Reddin, K., Matheson, M., Heath, R., Funnell, S., Hudson, M., Robinson, A., and Gorringe, A. (2001). Recombinant *Neisseria meningitidis* transferrin binding protein A protects against experimental meningococcal infection. Infect. Immun. *69*, 1561–1567.

Westerink, M.A., Giardina, P.C., Apicella, M.A., and Kieber-Emmons, T. (1995). Peptide mimicry of the meningococcal group C capsular polysaccharide. Proc. Natl. Acad. Sci. USA 92, 4021–4025.

Wetzler, L.M., Barry, K., Blake, M.S., and Gotschlich, E.C. (1992). Gonococcal lipooligosaccharide sialylation prevents complement-dependent killing by immune sera. Infect. Immun. *60*, 39–43.

Weynants, V.E., Feron, C.M., Goraj, K.K., Bos, M.P., Denoel, P.A., Verlant, V.G., Tommassen, J., Peak, I.R., Judd, R.C., Jennings, M.P., and Poolman, J.T.(2007). Additive and synergistic bactericidal activity of antibodies directed against minor outer membrane proteins of *Neisseria meningitidis*. Infect Immun. 75, 5434–5442.

WHO (2004). The use of polysaccharide trivalent ACW vaccine for the control of epidemic meningococcal disease outbreaks in countries of the African meningitis belt (World Health Organisation, Department of Communicable Disease Surveillance and Response).

Williams, J.N., Skipp, P.J., Humphries, H.E., Christodoulides, M., O'Connor, C.D., and Heckels, J.E. (2007). Proteomic analysis of outer membranes and vesicles from wild-type serogroup B *Neisseria meningitidis* and a lipopolysaccharide-deficient mutant. Infect. Immun. 75, 1364–1372.

Yazdankhah, S.P., and Caugant, D.A. (2004). *Neisseria meningitidis*: an overview of the carriage state. J. Med. Microbiol. 53, 821–832.

Yazdankhah, S.P., Kriz, P., Tzanakaki, G., Kremastinou, J., Kalmusova, J., Musilek, M., Alvestad, T., Jolley, K.A., Wilson, D.J., McCarthy, N.D., Caugant, D.A., and Maiden, M.C. (2004). Distribution of serogroups and genotypes among disease-associated and carried isolates of *Neisseria meningitidis* from the Czech Republic, Greece, and Norway. J. Clin. Microbiol. 42, 5146–5153.

Yero, D., Pajon, R., Perez, Y., Farinas, M., Cobas, K., Diaz, D., Solis, R.L., Acosta, A., Brookes, C., Taylor, S., and Gorringe, A.(2007). Identification by genomic immunization of a pool of DNA vaccine candidates that confer protective immunity in mice against *Neisseria meningitidis* serogroup B. Vaccine 25, 5175–5188.

Zangwill, K.M., Stout, R.W., Carlone, G.M., Pais, L., Harekeh, H., Mitchell, S., Wolfe, W.H., Blackwood, V., Plikaytis, B.D., and Wenger, J.D. (1994). Duration of antibody response after meningococcal polysaccharide vaccination in US Air Force personnel. J. Infect. Dis. 169, 847–852.

Zhu, D., Zhang, Y., Barniak, V., Bernfield, L., Howell, A., and Zlotnick, G. (2005). Evaluation of recombinant lipidated P2086 protein as a vaccine candidate for group B *Neisseria meningitidis* in a murine nasal challenge model. Infect. Immun. 73, 6838–6845.

Epidemiology in the Vaccine Era

12

Caroline Trotter, Gwenda Hughes and Catherine Ison

Abstract

The pathogenic species of the genus *Neisseria, Neisseria meningitidis* and *Neisseria gonorrhoeae*, share much of their genome, but differ markedly in many other respects. *N. meningitidis* colonizes the mucosal surfaces primarily of the upper respiratory tract as part of the normal flora and occasionally invades to cause systemic disease with subsequent serious sequelae, morbidity and mortality. In contrast, *N. gonorrhoeae* mainly colonises the mucosa of the anogenital tract, rarely invades to cause serious systemic disease but is always considered a pathogen. *N. meningitidis* and *N. gonorrhoeae* are inherently susceptible to antimicrobial agents but *N. gonorrhoeae* has been particularly adept at acquiring or developing resistance to many therapeutic agents, which has not been such a major problem in *N. meningitidis*. Many of the major antigens are shared between these two species with the exception of the polysaccharide capsule found on the meningococcus, which can elicit a protective immune response. There is a lack of an apparent protective immune response to *N. gonorrhoeae*, with multiple episodes occurring in a single individual which results from the extraordinary ability of the organism for antigenic variation, in addition to the absence of the capsule. This chapter addresses the epidemiology of both of these neisserial infections and consequently allows the comparison of an infection with the potential to be controlled by vaccination with a second which is largely controlled by antimicrobial therapy alone. This is all the more interesting given that genetically these organisms are so closely related and came from the same progenitor.

Introduction: epidemiology of meningococcal disease

In most parts of the world today, meningococcal disease is an important public health issue. The rapid onset of severe disease (meningitis and/or septicaemia), high case fatality, serious after-effects, and propensity for outbreaks ensure that the disease receives a high priority. The incidence and epidemiology of meningococcal disease varies markedly over time and according to geographical location. The epidemics of 'cerebrospinal meningitis', described since the early 1800s (the first clear account of which is credited to Viesseux, a Swiss physician, who described an outbreak of meningitis in the Geneva area in 1805; Vieusseux, 1806) no longer occur in Europe, North America and other industrialized regions, but continue to devastate countries in the African 'meningitis belt' (Lapeyssonnie, 1963; Molesworth *et al.*, 2002). Elsewhere, meningococcal disease is endemic and the incidence of disease is much lower (typically from 1 to 10 per 100,000), although 'hyperendemic' periods of higher incidence associated with a particular clone also occur. Pathogenic meningococci have a polysaccharide capsule, and are classified on this basis into 13 known serogroups, of which five (serogroups A, B, C, W135, Y) are responsible for most disease, with X and 29E/Z contributing to a much lesser extent.

The African meningitis belt undoubtedly bears the highest burden of meningococcal disease. This vast area, stretching from Senegal to Ethiopia is home to over 300 million people (LaForce *et al.*, 2007), and is characterized by large epidemics of meningitis which occur in the dry season, periodically, although unpredictably, every 2–10 years (Greenwood, 1999). During the last major epidemic in 1996, over 200,000 cases and 20,000 deaths were reported (Greenwood, 1999). Smaller, localized epidemics occur more frequently, especially in countries at the heart of the belt, such as Burkina Faso and Niger. Even in non-epidemic years, incidence is much higher than that observed in industrialized nations; for example, in Niger the average incidence of meningococcal meningitis was reported to be 26 per 100,000 between 1981 and 1994 (Campagne *et al.*, 2000). Most epidemics in the African meningitis belt have been caused by serogroup A meningococci (Greenwood, 1999) although serogroup C (Whittle *et al.*, 1975; Broome *et al.*, 1983) and more recently serogroup W-135 (Anon, 2002) outbreaks have been described. Recent serogroup A outbreaks have also occurred elsewhere in sub-Saharan Africa, and in China, India, Nepal and the Philippines (Girard *et al.*, 2006) (http://www.who.int/csr/don/archive/disease/meningococcal_disease/en/) but have not been observed in the USA or Western Europe for many decades (Peltola, 1983). The underlying mechanisms leading to epidemics or large outbreaks are not clear. In the African meningitis belt, climatic factors, in particular the arid, dusty winds of the dry season are likely to be important(Molesworth *et al.*, 2003), though not sufficient, since other factors such as population immunity, and pathogen characteristics are also thought to play a role (Moore, 1992).

In Europe, the European Union invasive bacterial infection surveillance (EU-IBIS) network reported that in 2006 the incidence in all participating countries was 1.1 per 100,000, but incidence varied markedly by country, from 0.3 per 100,000 in Italy to 8.3 per 100,000 in Malta. While the sensitivity and specificity of surveillance systems is likely to vary internationally, there are clearly some true differences in the epidemiology as shown, for example, by the variation in the distribution of pathogenic strains. Most disease in Europe is due to serogroups B and C, but some countries, including the UK, Ireland, Spain and The Netherlands experienced significant increases in serogroup C disease in the late 1990s/early 2000s, whereas other nations were not similarly affected. Around 5% of cases in EU-IBIS were due to serogroups W135 and Y. In Russia, serogroup A disease was still reported in the 1990s(Achtman *et al.*, 2001), whereas this serogroup is very rarely reported among the 27 countries participating in EU-IBIS.

In the USA, reported disease incidence since the early 1990s has varied between 0.5 and 1.5 per 100,000 (Harrison, 2006). In addition to serogroups B and C, serogroup Y is also an important cause of disease, contributing around one third of all cases (Rosenstein *et al.*, 1999). In 2005, the incidence of meningococcal disease in Canada was reported to be 0.56 per 100,000, with serogroup B, C, Y being the most common cause of disease in descending order (Navarro *et al.*, 2007). In Australia, incidence from the 1990s onwards has varied between 1.5 and 4 per 100,000, with serogroups B and C predominating. This is similar to Europe, but incidence is low compared to neighbouring New Zealand, which has been experiencing a prolonged 'hyper-endemic' period due to a particular serogroup B clone (NZ98/254, B:4:P1.7b,4); overall incidence here reached 17.4 per 100,000 in 2001. Such 'hyperendemic' periods of disease due to serogroup B have also been described in Norway (B:15: P1.7,15) (Bjune *et al.*, 1991), Brazil (B:4:P1.15)(Sacchi *et al.*, 1992), Cuba (B:4:P1.15) (Sierra *et al.*, 1991), Chile (B:15:P1.3) (Cruz *et al.*, 1990). The use of outer-membrane vesicle vaccines to counter these epidemics is reviewed in Chapter 11. Pilgrims to the Hajj and Umrah in Saudi Arabia must be immunized against meningococcal disease, following large outbreaks, e.g. in 2000 and 2001(Ahmed *et al.*, 2006). Endemic incidence appears to be very low in some countries, e.g. Japan (Horino *et al.*, 2008), for reasons that are not well understood. The burden of meningococcal disease in other parts of the world is less well defined with limited routine surveillance information, although reviews from some regions are available (Hausdorff *et al.*, 2007; Peltola, 2001).

The incidence of meningococcal disease varies by age. In developed countries, the age-specific incidence is consistently highest in infants then declines through childhood. There is (usually) a smaller secondary peak in disease in teenagers and young adults before incidence declines once more, becoming rare in older adults (Fig. 12.1). In the African meningitis belt, surveillance over many years in Niger showed that the burden of disease is concentrated in 0- to 19-year-olds in both epidemic and inter-epidemic periods (Campagne *et al.*, 2000). Case fatality rates (CFR) vary according to setting (bearing in mind that reporting differences are also likely). In Europe, the average CFR in EU-IBIS participating countries was 7.7% in 2006, while in the USA a CFR of 10–14% was reported by CDC in 2004 (Bilukha and Rosenstein, 2005). In Niger, a CFR of 12% was reported in both epidemic and inter-epidemic years (Campagne *et al.*, 2000). Mortality also varies by age, with older adults experiencing the highest case fatality rates in all settings. The risk of death varies according to the characteristics of the infecting organism. For example in a UK study the odds of death resulting from infection with organisms phenotypically identified as serogroup C serotype 2a (most often associated with sequence type 11) and B:2a were significantly higher than compared to B:1 (the most common phenotype) (Trotter *et al.*, 2002). Survivors of meningococcal disease can suffer from a range of sequelae, including deafness, neurological disabilities, learning difficulties, amputation and scarring (the last two associated particularly with meningococcal septicaemia). The prevalence of reported sequelae varies across studies, with a wide range of estimates (Rashid *et al.*, 2008) and differences in prevalence of after-effects by serogroup have also been reported (Erickson and De Wals, 1998).

Meningococcal disease is highly seasonal, with a winter peak incidence in temperate climates and association with the dry season in the African meningitis belt. Host immunity is important in relation to disease risk, as reviewed in Chapter 10. An additional risk factor that has been consistently reported is having close contact with a case, and chemoprophylaxis of contacts to interrupt further transmission has been a key strategy in disease control in developed countries for many years (Purcell *et al.*, 2004). Meningococcal disease appears to have an affinity with the military, with large outbreaks reported particularly among new recruits and after mobilizations (Brundage and Zollinger, 1987). A similar propensity has also been described in university/college students living in halls of residence/dormitories, who share some of the same characteristics as new military recruits, such as age, diversity of geographic backgrounds, and intensive social mixing (CDC, 2000). Other factors that have been reported to be associated with

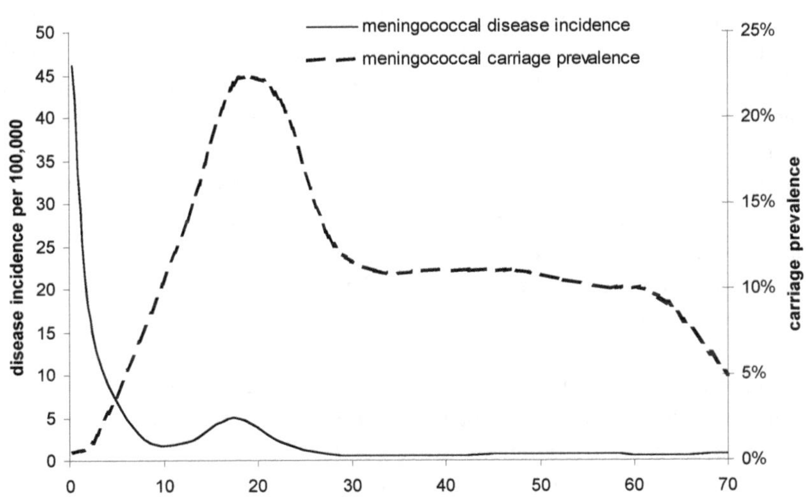

Figure 12.1 Schematic diagram of meningococcal disease incidence and meningococcal carriage prevalence by age in developed countries. Based on laboratory confirmed disease reports in England and Wales and the Stonehouse, Gloucestershire, carriage survey (Cartwright et al., 1987) in the late 1980s.

meningococcal disease in developed countries include lower socioeconomic status (Heyderman *et al.*, 2004), recent or concomitant respiratory infections (particularly influenza) (Cartwright *et al.*, 1991), passive smoking (Coen *et al.*, 2006). In the African meningitis belt, both enteric (Griffiss, 1982) and respiratory co-infections (Mueller *et al.*, 2008) are hypothesized to be important in the occurrence of epidemic disease, although further research is needed to investigate this.

Epidemiology of meningococcal carriage

Despite the public health importance of *N. meningitidis* as a pathogen, the meningococcus is often considered an 'accidental pathogen', whose usual niche is the human nasopharynx. Infection most often results in a period of asymptomatic colonization; disease is a (relatively) rare event. Transmission occurs via respiratory droplets, and most transmission events are from carrier to carrier. Although the relationship between carriage and disease rates is not straightforward(Wenzel, 1973; Trotter *et al.*, 2006), it is necessary to consider carriage as a key component of the natural history of meningococcal infection. Without this understanding, it is difficult to appreciate the epidemiology of infection and the potential impact of control measures which may influence not just disease risk, but also risk of transmission. The population prevalence of carriage is often quoted to be 10%, and while this gives a good indication that carriage is common, it masks the heterogeneity that is observed by both age and setting (Christensen *et al.*, 2008). In general, in developed countries with endemic meningococcal disease, carriage prevalence is low in infants and young children, and rises to a peak in teenagers and young adults. In a large multicentre carriage study in the UK, carriage prevalence in 15- to 17-year-olds was 16–18% (Maiden *et al.*, 2008), although carriage rates of over 50% have been reported in young military recruits (Christensen *et al.*, 2008). In the African meningitis belt, carriage studies have also reported heterogeneous findings, with overall carriage prevalence in the range of 3–30% (Trotter and Greenwood, 2007). The age distribution of carriers was not consistent across studies, probably reflecting both study limitations (such as small sample size) and true differences in population behaviours (such as social mixing) across the meningitis belt (Trotter and Greenwood, 2007).

In industrialized countries, a much wider diversity of strains are detected when asymptomatic carriers are sampled compared to stains isolated from clinical cases of meningococcal meningitis or septicaemia (Yazdankhah and Caugant, 2004). Many carried strains do not have a capsule and are thus non-pathogenic. The risk of disease on exposure is therefore highly dependent on the characteristics of the infecting strain. A long- term study of meningococcal carriage in northern Ghana suggested that there may be important differences in the patterns of carriage in the meningitis belt compared with industrialized countries; the populations of carried meningococci were less diverse, included fewer non-groupable isolates, and were dominated by virulent encapsulated strains (Leimkugel *et al.*, 2007).

The duration of carriage is not robustly defined as most carriage studies are point prevalence studies rather than longitudinal studies, but carriage may be prolonged – lasting for many months. Carriage of *Neisseria lactamica*, a closely related commensal distinguished by its ability to ferment lactose, is also common, particularly in young children (Gold *et al.*, 1978; Bennett *et al.*, 2005). It has been suggested that carriage of *N. lactamica* can protect against meningococcal carriage and disease, but this is not firmly established, particularly as early studies reporting the development of bactericidal antibodies to *N. meningitidis* following *N. lactamica* carriage have not been replicated (Gold *et al.*, 1978).

Impact of meningococcal vaccination

Meningococcal polysaccharide vaccines

Vaccines based on purified high molecular weight meningococcal polysaccharide were developed for initially serogroups A and C in the 1960s (Gotschlich *et al.*, 1969). Early trials showed that these vaccines were safe and immunogenic in adults, but poorly immunogenic in infants and young children, as they do not induce T-cell dependent immunity (thus also offering only short-term protection). The main use of these vaccines has been in outbreak control and epidemic response, rather than as routine vaccines (despite some advocates (Robbins *et al.*, 1997)). The WHO have defined 'alert' (5 per 100,000) and 'epidemic' (15 per 100,000 over 2 consecutive weeks) thresholds for meningococcal meningitis in the African meningitis belt, and recommend that reactive vaccination campaigns be implemented when the epidemic threshold is reached (http://www.who.int/csr/disease/meningococcal/en/index.html). Epidemics continue to occur in the meningitis belt, and while this strategy may be effective in reducing the size of outbreaks (Leake *et al.*, 2002), clearly an effective preventive strategy is preferable to this reactive approach, particularly given the need to (and problems associated with) stockpile vaccines (WHO, 2007). Elsewhere, polysaccharide vaccines have been deployed in response to local outbreaks of disease and in preventative campaigns in high-risk populations such as military recruits.

Meningococcal serogroup C conjugate (MCC) vaccines

In the mid-1990s cases of serogroup C disease increased in the UK, associated with the introduction of a hyper-invasive hyper-virulent serotype 2a ST-11 clone, which had previously caused problems in Canada (Whalen *et al.*, 1995). Although at the time only serogroup C polysaccharide vaccines were licensed, trials of meningococcal serogroup C conjugate (MCC) vaccines were accelerated, and at the end of 1999, newly licensed MCC vaccines were introduced into the UK. Vaccine licensure was based on safety and immunogenicity data; a phase III efficacy trial was not performed (Miller *et al.*, 2001). The UK strategy was to incorporate MCC vaccines into the routine childhood immunization schedule at 2, 3, 4 months of age and run a catch-up campaign targeting all children up to the age of 18 years, which was rolled out as vaccine supply increased. Overall, this strategy was very successful as the number of cases of and deaths from serogroup C disease rapidly declined in the age groups targeted for immunization (Miller *et al.*, 2001; Trotter *et al.*, 2004); in 2007/08 fewer than 30 laboratory confirmed cases were identified in England and Wales (Thomas *et al.*, 2008). MCC vaccines were subsequently introduced into national immunization programmes in Ireland, Spain, The Netherlands, Iceland, Canada, Australia, Portugal, Greece, Germany. Where data are available, a substantial decline in serogroup C disease has been observed(Trotter and Ramsay, 2007). Although short-term effectiveness was reported to be high in all age-groups, in the longer term, vaccine effectiveness wanes, particularly in children immunized in infancy (Trotter *et al.*, 2004; Larrauri *et al.*, 2005). In the UK, the number of serogroup C cases continued to fall despite this (largely unanticipated) rapid waning of direct protection in the youngest age groups. Analysis of surveillance data in the UK and later in The Netherlands showed a marked reduction in the serogroup C attack rate even in unvaccinated individuals (Ramsay *et al.*, 2003; de Greeff *et al.*, 2006), indicating that a large part of the success of the MCC programmes was due to the ability of the vaccines to reduce serogroup C carriage (Maiden and Stuart, 2002), resulting in herd immunity. The importance of targeting teenagers – the most common meningococcal carriers – in order to maximize these indirect effects was also emphasized. The UK immunization schedule was changed in September 2006, so that now two doses of MCC are offered in infancy, with a further dose at 12 months of age, which is intended to improve the duration of direct protection (Chief Medical Officer, 2006). Despite initial concern (Maiden and Spratt, 1999), there is no compelling evidence that selective vaccination has driven capsule replacement (Trotter *et al.*, 2006). However, some important questions, such as the duration of herd immunity effects, remain to be answered and continued high-quality surveillance is essential so that the long-term effects of MCC vaccination can be measured and understood.

Other conjugate vaccines

In early 2005 a quadrivalent meningococcal conjugate vaccine (MCV4) offering protection against serogroups A, C, W135 and Y was licensed for 11- to 55-year-olds in the USA. This vaccine is recommended for routine immunization of adolescents and other high-risk groups (including college freshmen living in dormitories) (Bilukha and Rosenstein, 2005). The inclusion of serogroup Y was of particular benefit to the USA, where a substantial proportion of disease is due to this serogroup. Coverage with MCV4 in the targeted age group is fairly modest; around 32% in 2007(CDC, 2008a). Data on the population impact of MCV4 (including vaccine effectiveness and results of carriage prevalence studies) have not been published to date. Post-licensure surveillance of vaccine safety has revealed a temporal association between MCV4 receipt and onset of Guillain–Barré syndrome, with an estimated risk ratio of 1.78 (95% CI = 1.02–2.85) (CDC, 2006). This finding has not altered the recommendation for universal routine immunization with MCV4 as absolute risk remains low. The current epidemiology of disease in Europe, where few cases are due to W-135 and Y is likely to limit the appeal of the quadrivalent over and above the monovalent serogroup C vaccine, but this may change in the future (Trotter and Ramsay, 2007). Other multivalent and meningococcal conjugate combination vaccines for use in developed countries are likely to become available soon.

Progress is also being made in the development of meningococcal conjugate vaccines for use in the African meningitis belt. The Meningitis Vaccine Project (MVP) is a major initiative which began in 2001 as a partnership between WHO and PATH, funded by the Bill and Melinda Gates Foundation. MVP's aim is to eliminate epidemic meningitis in Africa, through the development, testing, licensure and introduction of an affordable meningococcal serogroup A conjugate vaccine (LaForce *et al.*, 2007). Phase I and II trials of the vaccine, manufactured by the Serum Institute of India, have been successfully completed, and licensure is expected in 2009. The vaccine will be introduced first into Burkina Faso, where 1- to 29-year-olds will be targeted in a mass vaccination campaign, and then rolled across the belt as supply increases.

'Serogroup B' vaccines

Unlike serogroups A, C, W135 and Y, the serogroup B capsule is not an appropriate vaccine target. New approaches for identifying vaccine antigens through genome mining are promising, and hopes are high that a 'universal' vaccine with broad coverage against many meningococcal strains can be developed, as reviewed in Chapter 11. Vaccines that have been used against serogroup B to date have been based on outer membrane proteins, and have targeted specific phenotypes associated with hyperendemic disease. Most recently, a tailor-made vaccine for New Zealand targeting the hyperendemic B:4:P1.4 strain has been used for routine infant immunization and mass catch-up campaigns (O'Hallahan *et al.*, 2005). Although the incidence of disease appeared to be declining before vaccine introduction, observational studies have estimated short-term vaccine effectiveness to be in the region of 73% to 85% (Kelly *et al.*, 2007; Galloway *et al.*, 2008).

Epidemiology of gonococcal infections

N. gonorrhoeae is the causative agent of gonorrhoea, a sexually transmitted infection (STI) which primarily causes a mucosal infection of the urogenital tract. The primary site of infection is the urethra in men and the cervix in women, although infection of the rectum, oropharynx and conjunctiva also occurs. The early signs of gonorrhoea are often mild, and many women and some men show no symptoms and may be unaware of their infection. Men more frequently show signs of infection, most commonly acute urethritis (Hook and Handsfield, 1999). When present, symptoms in women include increased vaginal discharge, dysuria, intermenstrual bleeding and menorrhagia (Hook and Handsfield, 1999).

If undetected or poorly treated, the organism can ascend to the normally sterile upper genital tract causing pelvic inflammatory disease (PID) in women, which can lead to ectopic pregnancy and infertility, and epididymitis and prostatitis in men (Hook and Handsfield, 1999). This occurs more commonly in women than men because it is more often asymptomatic and in developing countries more than industrialized countries because of the lack of good diagnostics and effective therapeutics. Neonatal conjunctivitis may also result when an infected woman gives birth. Rarely, untreated gonorrhoea can spread to the bloodstream or the joints (disseminated gonococcal infection). There is no evidence that the presence of symptoms is associated with the gonococcal serotype or auxotype (Horner *et al.*, 1992); however, the severity of symptoms probably is (Sandstrom *et al.*, 1982). Disseminated gonococcal infection is more common among gonococci of the AHU auxotype

(Noble *et al.*, 1984). Gonorrhoea is treatable with antibiotics, although resistance to some common antimicrobials has become an increasing problem in recent years.

Global burden of gonorrhoea

The World Health Organization (WHO) estimated that, globally, there were 62 million new cases of gonorrhoea in 1999, with 27 million (44%) in South and South-East Asia, 17 million (27%) in sub-Saharan Africa and 7.5 million (12%) in Latin America and the Caribbean (WHO, 2001). Reported rates of infection range considerably between countries, a consequence of the variability in accuracy and completeness of surveillance systems, studies undertaken and diagnostic methods used, as well as in actual population prevalence (De Schryver and Meheus, 1990). The high level of asymptomatic gonorrhoea also leads to under-reporting of infection, especially in women. Consequently, cross-sectional prevalence surveys often provide a better indication of infection burden in populations.

Studies on countries with emerging and developing economies suggest population prevalence can be high, particularly in sub-Saharan Africa. Prevalence of gonorrhoea in women attending family planning and/or antenatal clinics has been found to be about 5% in South Africa (Johnson *et al.*, 2005), 4% in Uganda (Tann *et al.*, 2006), 3% in Kenya (Daly *et al.*, 1994), and 2.5% in Mozambique (Lujan *et al.*, 2008), while in a rural study population in Madagascar 5% of women and 3% of men were infected (Leutscher *et al.*, 2005). In other countries with emerging and developing economies population prevalence varies considerably but tends to be below 2.5%. In South and South-East Asia, gonorrhoea prevalence was 2.3% among postpartum women studied in rural Nepal (Christian *et al.*, 2005) but only 0.2% in women attending antenatal clinics in Thailand (Kilmarx *et al.*, 1998) and Malaysia (WHO, 2008a) and in women attending maternal and child health-care services in Bangladesh (Hawkes *et al.*, 1999). Likewise in the Caribbean, 1.8% of those tested in a population-based sample in Barbados (Adams *et al.*, 2008) and 0.5% of women attending antenatal clinics in Jamaica (Dowe *et al.*, 1998) were infected. A study of women attending gynaecology and family planning clinics in Jordan (Mahafzah *et al.*, 2008) and a community-based study of reproductive age women in Papua New Guinea (Passey *et al.*, 1998) detected gonorrhoea in about 1% of those tested, although WHO has estimated adult population prevalence in Papua New Guinea may be as high as 3.5%(WHO, 1999). Gonorrhoea prevalence among various occupational groups tested in Russia was around 0.3% (Kourbatova *et al.*, 2008) and the adult population prevalence in China has been estimated at 0.15% (WHO, 2008a).

Reported rates of gonorrhoea in Western industrialized countries are mostly derived from established surveillance systems. These data suggest that population prevalence in advanced economies tend to be lower overall, although reported rates of infection will underestimate true prevalence to some extent. The population rate of reported infection is around 0.1% in the USA (CDC, 2007b), 0.05% in men and 0.02% in women in the UK (UK Collaborative Group For HIV and STD Surveillance, 2007) and below 0.01% in most other Western and Central European countries (ESSTI, 2006). Prevalence studies in Western industrialized countries tend to focus on particular high-risk groups but some give an indication of underlying population prevalence. Gonorrhoea prevalence among young women attending family planning clinics was found to be 0.7% in Missouri (Einwalter *et al.*, 2005) and 0.5% in Colorado (Gershman and Barrow, 1996), while among asymptomatic men attending various community-based and clinical venues in four major US cities prevalence ranged between 0% and 1.5% (Gaydos *et al.*, 2006). In Europe, prevalence among young UK women being screened for *Chlamydia trachomatis* at various settings was around 1% in north-west England (Lavelle *et al.*, 2006) and 4% in south London (Rao *et al.*, 2008), but was only 0.3% among young German women participating in human papillomavirus vaccine trials (Barthell *et al.*, 2008).

Current and historical trends

Because gonococcal infection in men usually results in symptoms for which prompt treatment will be sought, trends in diagnoses of gonorrhoea are often considered a reasonable marker of changes in sexual behaviour. As with disease burden, however, interpreting trends in diagnoses of gonococcal infection must be undertaken with care, as it is often hampered by sporadic reporting from surveillance systems (where they exist at all) which are of variable quality. Many countries have no routine surveillance systems, although research studies can sometimes provide an insight. Fluctuating trends in gonococcal infection can therefore be associated with a combination of factors including changes in sexual behaviour, the completeness and coverage of surveillance systems, improvements in diagnostic sensitivity, the effectiveness of treatment and notification of sexual partners, changes in health service provision or infrastructure, and economic and demographic shifts within societies.

Surveillance data from Western industrialized countries are the most comprehensive and give greatest insight into historic and recent trends. Both the USA and the UK have extensive historical data (CDC, 2007b; Hughes and Fenton, 2001). In both countries, the return of the armed forces after the Second World War was followed by a peak in gonorrhoea diagnoses, and more liberal attitudes to sexual behaviour during the 1960s and 1970s heralded a steady rise in diagnoses during this period (CDC, 2007b; Hughes and Fenton, 2001). Most Western industrialized countries experienced a decline in rates of diagnosis during the 1980s and early 1990s which has been associated with large-scale HIV prevention campaigns promoting safer sex (Nicoll *et al.*, 2001; Berglund *et al.*, 2007), but thereafter trends have been more variable. In the USA, gonorrhoea rates fell by 74% between 1975 and 1995 but were relatively stable in the subsequent 10 year period at around 120/100,000 (CDC, 2007b). As in the USA, gonorrhoea rates in Western Europe declined markedly during the 1980s and mid-1990s (Nicoll *et al.*, 2001; Hiltunen-Back *et al.*, 1998; Van der Heyden *et al.*, 2000). Indeed, between 1991 and 1995, numbers of diagnoses declined in all European Union countries with the exception of Ireland (Fenton and Lowndes, 2004). However, rates across Europe have fluctuated in recent years. In the UK, Austria and Sweden, rates rose sharply between 1998 and 2002 before falling again (UK Collaborative Group For HIV and STD Surveillance, 2007; Berglund *et al.*, 2007; ESSTI, 2006). In most other Western European countries rates increased steadily between 1996 and 2005, while in central Europe rates steadily declined over the same period (ESSTI, 2006). Trends in the Western Pacific region over the past two decades have also varied. Numbers of gonorrhoea diagnoses in Hong Kong fell by 10% between 1989 and 1996, but rates in Japan rose by 25% between 1993 and 1997 (WHO, 2008a). In South Australia numbers of diagnoses declined by 35% between 1997 and 2002 but more than doubled between 2002 and 2007 (Anon, 2008).

Few published data are available from emerging and developing economies, especially for the 10-year period prior to 2008, but the limited data available suggest that rates of gonorrhoea have been declining in many areas in recent decades. In sub-Saharan Africa, declining rates of gonorrhoea were observed in Burkina Faso between 1991 and 2001 (Nagot *et al.*, 2004) and in Malawi between 1990 and 1996 (Taha *et al.*, 1998). Gonococcal incidence among young men in Thailand declined 10-fold during the early to mid-1990s (Celentano *et al.*, 1998). In Latin America and the Caribbean, gonorrhoea rates fell in Costa Rica and Chile between 1986 and 1992, but rose in Jamaica over a similar period, although improved reporting may have been responsible for the latter trend (Gerbase *et al.*, 1999). According to WHO, rates of gonorrhoea in Malaysia fell by 25% between 1986 and 1998, and numbers of diagnoses in New Caledonia in the South Pacific fell by over 70% between 1990 and 1997 (WHO, 2008a). As in Western industrialized countries, the introduction of effective HIV/AIDS interventions and the adoption of safer sexual behaviour including increased condom use has likely contributed significantly to the reduction in incidence in emerging and developing economies (Nagot *et al.*, 2004; Celentano *et al.*, 1998; Taha *et al.*, 1998).

Distribution of infection in the population

Gonococcal infection is concentrated in core groups of the population and is geographically localized within countries, often in areas of low socioeconomic status (Hethcote and Yorke, 1984; Rothenberg, 1983; Yorke *et al.*, 1978). In this respect, gonococcal infection differs significantly from another common bacterial STI, *C. trachomatis*. In contrast to gonorrhoea, genital chlamydial infection has only relatively recently become the focus of specific control programmes and is still widely prevalent in sexually active populations (Navarro *et al.*, 2002; Zimmerman *et al.*, 1990).

Core groups are often, but not necessarily, characterized by having higher rates of sexual partner change (Lindan *et al.*, 2005), short intervals between consecutive sexual partnerships (Chen *et al.*, 2008) and/or a high frequency of concurrent sexual partnerships (Gorbach *et al.*, 2005) relative to the general population. However, even in the absence of risky sexual behaviour, a high prevalence of infection may persist in certain population groups. Poor access to treatment services due to inadequate health service infrastructure, low awareness of STIs, and/or high frequency of asymptomatic infection in women perpetuates high levels of undiagnosed and untreated infection (Miller *et al.*, 2003; Leutscher *et al.*, 2005; Gerbase *et al.*, 1999). Furthermore, assortative (like with like) sexual mixing patterns among core group populations maintain high prevalence (Anderson, 2009; Chen *et al.*, 2008).

The characteristics of core groups for gonococcal infection vary globally. Core groups of particular importance include female sex workers (FSWs), men who have sex with men (MSM), and, in western industrialized countries, those of black racial origin. As with most other STIs, young adults, especially adolescent women, are also at particular risk. The considerable variability in reported prevalence and trends within and between countries is strongly influenced by the geographic distribution and migration of core group populations.

Sex workers

Prevalence of gonorrhoea among FSWs can be high, particularly in emerging and developing economies (Cwikel *et al.*, 2008). In countries of sub-Saharan Africa prevalence among FSWs has been found to range between 6% and 34%, with most studies detecting prevalence above 10% (Cwikel *et al.*, 2008; Hoke *et al.*, 2007; Harrison, 2006). Prevalence tends to be even higher in South and South-East Asia, ranging from 9% in Goa to 41% in Indonesia, with about half of recent studies reporting gonorrhoea prevalence in FSWs above 20% (Cwikel *et al.*, 2008; Nguyen *et al.*, 2005; Shahmanesh *et al.*, 2008). Recent studies found prevalence in FSWs in China to be 8% in Kaiyuan City (Wang *et al.*, 2008) and 37% in Gejiu City (Xu *et al.*, 2008), while 14% of FSWs in 3 urban centres in Mongolia were infected (Hagan and Dulmaa, 2007). An average of 8% of FSWs were infected with gonorrhoea in a study of five Central American countries (Soto *et al.*, 2007). Female sex workers in western industrialized countries tend to experience lower levels of gonococcal infection. Recent studies suggest prevalence in FSWs to be about 1% in London (Ward *et al.*, 2004) and Bologna (D'Antuono *et al.*, 2001), 0.6% in Catalonia (Folch *et al.*, 2008) but considerably higher in Athens at about 6% (Papadogeorgaki *et al.*, 2006).

Men who have sex with men

In some Western industrialized countries, a substantial proportion of gonococcal infections in men are homosexually acquired. In Europe in 2005, the percentage of infections among men reported as homosexually acquired ranged between 20% in the Czech Republic to 68% in the Netherlands (ESSTI, 2006), in the USA in 2006 the figure was 22% (CDC, 2008b), while in Australia in 2007, 15% of men with gonorrhoea reported male sexual partners (Anon, 2008). Most studies of gonorrhoea prevalence in MSM are from western industrialized countries. Prevalence among MSM varies considerably according to the setting where the men have been tested. No infections were found in a large community-based population of MSM in Pittsburgh (Cook *et al.*, 2002) and only 0.5% of men attending homosexual entertainment venues in Brisbane were infected (Johansen and Smith, 2002). However, prevalence tends to be higher in MSM attending health care settings, ranging from 2.7% of MSM attending a community health centre in Boston (Mimiaga *et al.*, 2008) to between 7% and 16% of those attending sexually transmitted disease clinics in Canberra, California and London (Gunn *et al.*, 2008; Benn *et al.*, 2007; Currie *et al.*, 2006).

Disparities by race

In western industrialized countries, disparities in the distribution of gonococcal infection by race are striking. In the US in 2006, the rate of gonorrhoea was 18 times higher among African

Americans compared with whites (656 vs. 37/100,000) and African Americans accounted for 69% of all reported cases (CDC, 2007b). Similarly in the UK, rates among urban black ethnic minority populations have been estimated to be at least 10 times higher than in whites, with teenage women and young men being the most heavily affected (Low *et al.*, 2006; Lacey *et al.*, 1997). Over half of gonorrhoea notifications in South Australia are among those of Aboriginal origin (Anon, 2008) and gonococcal infection in New Zealand has been strongly associated with being of Māori or Pacific ethnicity (Azariah and Perkins, 2007). The causes of these disparities are complex, and may relate to group-level social and environmental factors for which race is a marker (Farley, 2006), poor access to healthcare services (Miller *et al.*, 2003) and sexual behaviour and beliefs (Lacey *et al.*, 1997). Assortative sexual mixing among these high prevalence populations also perpetuates the disparities (Anderson, 2009).

Prevention and control

In contrast to meningococcal infections it has not been possible to develop effective vaccines as a control measure. The gonococcus varies most of its surface antigens and hence evades any potentially protective response to an earlier infection. The pilus, as the primary mediator of attachment, was the first vaccine target but this proved unsuccessful (Boslego *et al.*, 1991) and despite considerable research efforts there is no commercially available gonococcal vaccine. There are continuing attempts to produce a gonococcal vaccine (Price *et al.*, 2007) or even a neisserial vaccine (Cornelissen, 2008) with cross-protection against *N. gonorrhoeae* and *N. meningitidis*, which are to be welcomed given the major problems of antimicrobial resistance in gonorrhoea and the potential for gonorrhoea to become an untreatable infection in the foreseeable future. In the absence of a vaccine, control of gonorrhoea can be approached in two main ways: (1) preventing new infections by reducing unsafe sexual behaviour (encouraging consistent and high uptake of condoms and/or reducing rates of sexual partner change), and (2) limiting onward transmission and re-infection through prompt and effective treatment of index cases and their sexual partners. Both of these approaches can be challenging, especially in resource-poor settings.

Because gonorrhoea is concentrated in core group populations it is particularly suitable for targeted interventions (Rothenberg, 1983). Effective marketing to improve use of condoms by sex workers can have a significant impact on gonorrhoea prevalence in resource-poor and more affluent countries (Nagot *et al.*, 2004; Hoke *et al.*, 2007; Hanenberg and Rojanapithayakorn, 1996; Ward *et al.*, 2004) and can even reduce prevalence in lower risk populations (Kilmarx *et al.*, 1998). Other targeted approaches for sex workers which have demonstrated effectiveness include improving the provision of healthcare services (Vu *et al.*, 2007; Wi *et al.*, 2006) and the introduction of presumptive treatment (Wi *et al.*, 2006).

Control programmes for gonorrhoea aimed at broad population groups are not usually considered cost effective (Mahafzah *et al.*, 2008; Kourbatova *et al.*, 2008). However, there is strong evidence that national campaigns aimed at reducing unsafe sexual behaviour in the general population as part of wider STD and HIV/AIDS prevention efforts can be extremely effective at controlling gonorrhoea. The 10-fold decrease in gonorrhoea incidence among young men in northern Thailand during the early to mid-1990s has been associated with the '100% Condom Program' introduced in 1991 which promoted universal condom use for all commercial sex (Celentano *et al.*, 1998). Likewise, declines in gonorrhoea in Malawi and Burkina Faso have been associated with national STD/AIDS prevention programmes (Taha *et al.*, 1998; Nagot *et al.*, 2004). Similar successes in most western industrialized countries have been associated with large-scale HIV prevention campaigns promoting safer sex (Nicoll *et al.*, 2001; Berglund *et al.*, 2007).

Treatment of infected individuals remains the mainstay of control programmes for gonorrhoea but is becoming more challenging as the range of effective agents is reducing. As discussed more fully in Chapter 13 resistance has emerged to the inexpensive and widely available antimicrobial agents including penicillin, tetracycline and ciprofloxacin. The World Health Organization recommends that any first-line therapy for gonorrhoea should result in >95% effective treatment. In many countries resistance is above 5% and hence these agents are not appropriate. The alternatives of the third generation cephalosporins are more expensive and may not always be available. This is particularly problematic in resource poor settings where antibiotics may not be well controlled and may be available over the counter. The choice of therapeutic agents is guided in many countries by surveillance programmes which monitor the trends in antimicrobial resistant gonorrhoea and the emergence of new mechanisms of resistance. There are many such programmes but well established programmes include those in the USA (Wang *et al.*, 2007) Australia (Bates *et al.*, 2008; Tapsall *et al.*, 2008), Western Pacific Region (WHO, 2008b), and in England, Wales and Scotland (GRASP Steering Group, 2007, 2008).

In addition to information on the prevalence of antimicrobial resistant gonorrhoea (GRASP link) some programmes also give enhanced demographic data on the epidemiology of gonorrhoea (Delpech *et al.*, 2008). In England and Wales the Gonococcal Resistance to Antimicrobials Surveillance Programme (GRASP) has shown that the burden of gonorrhoea is found amongst individuals with high-risk sexual behaviour, MSM, particularly those who are HIV positive, young heterosexuals, often with a history of past infection, and those of black Caribbean ethnicity (Delpech *et al.*, 2008). The greatest burden of antimicrobial resistant gonorrhoea is found among MSM, who showed a prevalence of ciprofloxacin resistance of 47% in 2007 compared with 22% among heterosexual men and 8.7% in women (Health Protection Scotland, 2008). This information can be used to target therapy. In the USA, ciprofloxacin remained a therapeutic option for heterosexual gonorrhoea but not for MSM (CDC, 2004) until recently when levels became high in both populations. The US guidelines no longer recommend ciprofloxacin as a first-line therapy for any patient group (CDC, 2007a).

Conclusion and future trends

The epidemiology of these two infections contrasts markedly. While meningococcal disease may be controlled by vaccination, at least for serogroup C disease, gonorrhoea is controlled by effective therapy, at least at the present time. However, despite advances in treatment and intensive care, the case fatality rate for meningococcal disease remains high even in industrialized countries. Because of the rapid onset of disease, opportunities for reducing the burden of disease through treatment may be limited, and primary prevention through the use of effective vaccines is seen as the key to controlling meningococcal disease. Much progress has been made recently, as illustrated in particular by the successful introduction of meningococcal serogroup C conjugate vaccines. New products aimed at preventing meningococcal disease in both developed and developing countries are at advanced stages of development, and offer exciting prospects for the future control of meningococcal disease. In contrast there is still little progress towards a gonococcal vaccine and antimicrobial therapy remains at the forefront of control of gonorrhoea. There is, however, increasing concern that gonorrhoea is becoming less susceptible to the third generation cephalosporins (Matsumoto, 2008), the current treatment of choice, in addition to other agents (Workowski *et al.*, 2008; Tapsall, 2006) and that there are few effective alternative therapies either currently available or on the horizon. It is of considerable concern that gonorrhoea, if not untreatable, could be very difficult to treat in 5–10 years. If that should occur, a safe and effective gonococcal or neisserial vaccine would be essential to control this common sexually transmitted infection.

References

Achtman, M., van der, E.A., Zhu, P., Koroleva, I.S., Kusecek, B., Morelli, G., Schuurman, I.G., Brieske, N., Zurth, K., Kostyukova, N.N., and Platonov, A.E. (2001). Molecular epidemiology of serogroup A meningitis in Moscow, 1969 to 1997. Emerg. Infect. Dis. 7, 420–427.

Adams, O.P., Carter, A.O., Prussia, P., McIntyre, G., and Branch, S.L. (2008). Risk behaviour, healthcare access and prevalence of infection with *Chlamydia trachomatis* and *Neisseria gonorrhoeae* in a population-based sample of adults in Barbados. Sex. Transm. Infect. *84*, 192–194.

Ahmed, Q.A., Arabi, Y.M., and Memish, Z.A. (2006). Health risks at the Hajj. Lancet *367*, 1008–1015.

Anderson, R.M. (2009). Transmission dynamics of sexually transmitted infections. In Holmes, K.K., and Sparling, P.F., eds. Sexually Transmitted Diseases, 3rd edn (New York: McGraw Hill), p. 25.

Anon. (2002). Meningococcal disease, serogroup W135, Burkina Faso. Preliminary report, 2002. Wkly Epidemiol. Rec. 77, 152–155.

Anon. (2008). Sexually Tranmsitted Diseases in South Australia in 2007 – Epidemiologic Report 21. http://www.stdservices.on.net/publications/pdf/annual_report_2007.pdf

Azariah, S., and Perkins, N. (2007). Risk factors and characteristics of patients with gonorrhoea presenting to Auckland Sexual Health Service, New Zealand. N.Z. Med. J. *120*, U2491.

Barthell, E., Woelber, L., Hellner, K., Camerer, B., Gieseking, F., Hauschild, M., Mylonas, I., Friese, K., Sings, H.L., Railkar, R., Gause, C., and Barr, E. (2009). Baseline characteristics and prevalence of HPV *6*, 11, 16, 18 in young German women participating in phase III clinical trials of a quadrivalent HPV (6/11/16/18) vaccine. Arch. Gynecol. Obstet. *279*, 803–807.

Bates, J., Murphy, D., Hicks, V., and Limnios, A. (2008). Annual report of the Australian Gonococcal Surveillance Programme, 2007. Commun. Dis. Intell. *32*, 227–231.

Benn, P.D., Rooney, G., Carder, C., Brown, M., Stevenson, S.R., Copas, A., Robinson, A.J., and Ridgway, G.L. (2007). *Chlamydia trachomatis* and *Neisseria gonorrhoeae* infection and the sexual behaviour of men who have sex with men. Sex. Transm. Infect. *83*, 106–112.

Bennett, J.S., Griffiths, D.T., McCarthy, N.D., Sleeman, K.L., Jolley, K.A., Crook, D.W., and Maiden, M.C. (2005). Genetic diversity and carriage dynamics of *Neisseria* lactamica in infants. Infect. Immun. *73*, 2424–2432.

Berglund, T., Asikainen, T., Grutzmeier, S., Ruden, A.K., Wretlind, B., and Sandstrom, E. (2007). The epidemiology of gonorrhea among men who have sex with men in Stockholm, Sweden, 1990–2004. Sex. Transm. Dis. *34*, 174–179.

Bilukha, O.O. and Rosenstein, N. (2005). Prevention and control of meningococcal disease: recommendations of the advisory committee on immunization practices. MMWR *54*, 1–21.

Bjune, G., Hoiby, E.A., Gronnesby, J.K., Arnesen, O., Fredriksen, J.H., Halstensen, A., Holten, E., Lindbak, A., Nokleby, H., Rosenqvist, E., Solberg, L.K., Closs, O., Eng, J., Froholm, L.O., Lystad, A., Bakketeig, L.S., and Hareide, B. (1991). Effect of outer membrane vesicle vaccine against group B meningococcal disease in Norway. Lancet *338*, 1093–1096.

Boslego, J.W., Tramont, E.C., Chung, R.C., McChesney, D.G., Ciak, J., Sadoff, J.C., Piziak, M.V., Brown, J.D., Brinton, C.C., Jr., Wood, S.W., and. (1991). Efficacy trial of a parenteral gonococcal pilus vaccine in men. Vaccine *9*, 154–162.

Broome, C.V., Rugh, M.A., Yada, A.A., Giat, L., Giat, H., Zeltner, J.M., Sanborn, W.R., and Fraser, D.W. (1983). Epidemic group C meningococcal meningitis in Upper Volta, 1979. Bull. World Health Organ. *61*, 325–330.

Brundage, J.F. and Zollinger, W.D. (1987). Evolution of meningococcal disease epidemiology in the US army. In Evolution of meningococcal disease: Volume I, N.A.Vedros, ed. (Boca Raton: CRC Press Inc.), pp. 6–25.

Campagne, G., Garba, A., Fabre, P., Schuchat, A., Ryall, R., Boulanger, D., Bybell, M., Carlone, G., Briantis, P., Ivanoff, B., Xerri, B., and Chippaux, J.P. (2000). Safety and immunogenicity of three doses of a *Neisseria meningitidis* A+C diptheria conjugate vaccine in infants from Niger. Pediatr. Infect. Dis. J. *19*, 144–150.

Cartwright, K.A., Stuart, J.M., Jones, D.M., and Noah, N.D. (1987). The Stonehouse survey: nasopharyngeal carriage of meningococci and Neisseria lactamica. Epidemiol Infect. *99*, 591–601.

Cartwright, K.A., Jones, D.M., Smith, A.J., Stuart, J.M., Kaczmarski, E.B., and Palmer, S.R. (1991). Influenza A and meningococcal disease. Lancet *338*, 554–557.

CDC (2000). Meningococcal disease and college students. Recommendations of the Advisory committee on Immunization practices (ACI). MMWR *49*, 11–20.

CDC (2004). Increases in fluoroquinolone-resistant *Neisseria gonorrhoeae* among men who have sex with men – United States, 2003, and revised recommendations for gonorrhea treatment, 2004. MMWR Morb. Mortal. Wkly. Rep. *53*, 335–338.

CDC (2006). Update: Guillain-Barre syndrome among recipients of Menactra meningococcal conjugate vaccine – United States, October 2005-February 2006. MMWR Morb Mortal Wkly Rep. *55*, 364–366.

CDC (2007a). Update to CDC's sexually transmitted diseases treatment guidelines, 2006: fluoroquinolones no longer recommended for treatment of gonococcal infections. MMWR Morb. Mortal. Wkly. Rep. *56*, 332–336.

CDC (2007b). Trends in Reportable Sexually Transmitted Diseases in the United States, 2006 – National Surveillance Data for Chlamydia, Gonorrhea, and Syphilis. 1–7. 2007. Centers for Disease Control and Prevention.

CDC (2008a). Vaccination coverage among adolescents aged 13 – 17 Years – United States, 2007. MMWR *57*, 1100–1103.

CDC (2008b). Gonococcal Isolate Surveillance Project (GISP) Annual Report – Surveillance 2006 Supplement. 1–37. 2008. Atlanta, CDC.

Celentano, D.D., Nelson, K.E., Lyles, C.M., Beyrer, C., Eiumtrakul, S., Go, V.F., Kuntolbutra, S., and Khamboonruang, C. (1998). Decreasing incidence of HIV and sexually transmitted diseases in young Thai men: evidence for success of the HIV/AIDS control and prevention program. AIDS *12*, F29-F36.

Chen, M.I., Ghani, A.C., and Edmunds, J. (2008). Mind the gap: the role of time between sex with two consecutive partners on the transmission dynamics of gonorrhea. Sex Transm. Dis. *35*, 435–444.

Chief Medical Officer. Planned changes to the routine childhood immunisation schedule. Department of Health. 8-2-2006.

Christensen, H., Trotter, C L., Bowen, L., and Hickman, M. (2008). *Neisseria meningitidis* carriage by age: a systematic review. International Pathogenic *Neisseria* Conference. 7-9-2008.

Christian, P., Khatry, S.K., Leclerq, S.C., Roess, A.A., Wu, L., Yuenger, J.D., and Zenilman, J.M. (2005). Prevalence and risk factors of chlamydia and gonorrhea among rural Nepali women. Sex. Transm. Infect. *81*, 254–258.

Coen, P.G., Tully, J., Stuart, J.M., Ashby, D., Viner, R.M., and Booy, R. (2006). Is it exposure to cigarette smoke or to smokers which increases the risk of meningococcal disease in teenagers? Int. J. Epidemiol. *35*, 330–336.

Cook, R.L., St George, K., Silvestre, A.J., Riddler, S.A., Lassak, M., and Rinaldo, C.R., Jr. (2002). Prevalence of chlamydia and gonorrhoea among a population of men who have sex with men. Sex Transm. Infect. *78*, 190–193.

Cornelissen, C.N. (2008). Identification and characterization of gonococcal iron transport systems as potential vaccine antigens. Future. Microbiol. 3, 287–298.

Cruz, C., Pavez, G., Aguilar, E., Grawe, L., Cam, J., Mendez, F., Garcia, J., Ruiz, S., Vicent, P., Canepa, I., and. (1990). Serotype-specific outbreak of group B meningococcal disease in Iquique, Chile. Epidemiol Infect *105*, 119–126.

Currie, M.J., Martin, S.J., Soo, T.M., and Bowden, F.J. (2006). Screening for chlamydia and gonorrhoea in men who have sex with men in clinical and non-clinical settings. Sex. Health 3, 123–126.

Cwikel, J.G., Lazer, T., Press, F., and Lazer, S. (2008). Sexually transmissible infections among female sex workers: an international review with an emphasis on hard-to-access populations. Sex. Health 5, 9–16.

D'Antuono, A., Andalo, F., Carla, E.M., and De Tommaso, S. (2001). Prevalence of STDs and HIV infection among immigrant sex workers attending an STD centre in Bologna, Italy. Sex. Transm. Infect. 77, 220.

Daly, C.C., Maggwa, N., Mati, J.K., Solomon, M., Mbugua, S., Tukei, P.M., and Hunter, D.J. (1994). Risk factors for gonorrhoea, syphilis, and trichomonas infections among women attending family planning clinics in Nairobi, Kenya. Genitourin. Med. *70*, 155–161.

de Greeff, S.C., de Melker, H.E., Spanjaard, L., Schouls, L.M., and van, D.A. (2006). Protection from routine vaccination at the age of 14 months with meningococcal serogroup C conjugate vaccine in the Netherlands. Pediatr. Infect. Dis. J. *25*, 79–80.

De Schryver, A. and Meheus, A. (1990). Epidemiology of sexually transmitted diseases: the global picture. Bull. World Health Organ. *68*, 639–654.

Delpech, V., Martin, I., Hughes, G., Nichols, T., James, L., and Ison, C. (2009). Epidemiology and clinical presentation of gonorrhoea in England & Wales: Findings from the Gonococcal Resistance to Antimicrobials Surveillance Programme 2001–2006. Sex Transm. Infect. April 20, epub ahead of print.

Dowe, G., King, S.D., Smikle, M.F., Wynter, H.H., Chout, R., and Klaskala, W. (1998). Prevalence of viral and bacterial sexually transmitted pathogens in Jamaican pregnant women. West Indian Med. J. *47*, 23–25.

Einwalter, L.A., Ritchie, J.M., Ault, K.A., and Smith, E.M. (2005). Gonorrhea and chlamydia infection among women visiting family planning clinics: racial variation in prevalence and predictors. Perspect. Sex. Reprod. Health 37, 135–140.

Erickson, L. and De Wals, P. (1998). Complications and sequelae of meningococcal disease in Quebec, Canada 1990–1994. Clin Infect Dis *26*, 1159–1164.

ESSTI (2006). Sexually Transmitted Infections Surveillance in Europe Annual Report No. 1 – 2006. 1–69.

Farley, T.A. (2006). Sexually transmitted diseases in the Southeastern United States: location, race, and social context. Sex Transm. Dis. *33*, S58-S64.

Fenton, K.A. and Lowndes, C.M. (2004). Recent trends in the epidemiology of sexually transmitted infections in the European Union. Sex Transm. Infect. *80*, 255–263.

Folch, C., Esteve, A., Sanclemente, C., Martro, E., Lugo, R., Molinos, S., Gonzalez, V., Ausina, V., and Casabona, J. (2008). Prevalence of human immunodeficiency virus, *Chlamydia trachomatis*, and *Neisseria gonorrhoeae* and risk factors for sexually transmitted infections among immigrant female sex workers in Catalonia, Spain. Sex. Transm. Dis. *35*, 178–183.

Galloway, Y., Stehr-Green, P., McNicholas, A., and O'Hallahan, J. (2008). Use of an observational cohort study to estimate the effectiveness of the New Zealand group B meningococcal vaccine in children aged under 5 years. Int. J. Epidemiol. *38*, 413–418.

Gaydos, C.A., Kent, C.K., Rietmeijer, C.A., Willard, N.J., Marrazzo, J.M., Chapin, J.B., Dunne, E.F., Markowitz, L.E., Klausner, J.D., Ellen, J.M., and Schillinger, J.A. (2006). Prevalence of *Neisseria gonorrhoeae* among men screened for *Chlamydia trachomatis* in four United States cities, 1999–2003. Sex. Transm. Dis. 33, 314–319.

Gerbase, A.C., Toscano, C., Totan, S., Cuchi, P, Gonalex-Salvatierra, and Zacarias, F. Sexually transmitted diseases in Latin America and the Caribbean. 1999.

Gershman, K.A. and Barrow, J.C. (1996). A tale of two sexually transmitted diseases. Prevalences and predictors of chlamydia and gonorrhea in women attending Colorado family planning clinics. Sex Transm. Dis. *23*, 481–488.

Girard, M.P., Preziosi, M.P., Aguado, M.T., and Kieny, M.P. (2006). A review of vaccine research and development: meningococcal disease. Vaccine *24*, 4692–4700.

Gold, R., Goldschneider, I., Lepow, M.L., Draper, T.F., and Randolph, M. (1978). Carriage of *Neisseria meningitidis* and *Neisseria* lactamica in infants and children. J. Infect. Dis. 137, 112–121.

Gorbach, P.M., Drumright, L.N., and Holmes, K.K. (2005). Discord, discordance, and concurrency: comparing individual and partnership-level analyses of new partnerships of young adults at risk of sexually transmitted infections. Sex. Transm. Dis. *32*, 7–12.

Gotschlich, E.C., Goldschneider, I., and Artenstein, M.S. (1969). Human immunity to the meningococcus: V. The effect of immunisation with meningococcal group C polysaccharide on the carrier state. J. Exp. Med. *129*, 1385–1395.

Greenwood, B. (1999). Meningococcal meningitis in Africa. Trans. Roy. Soc. Trop. Med. Hyg. *93*, 341–353.

Griffiss, J.M. (1982). Epidemic meningococcal disease: synthesis of a hypothetical immunoepidemiologic model. Rev. Infect. Dis. *4*, 159–172.

Gunn, R.A., O'Brien, C.J., Lee, M.A., and Gilchick, R.A. (2008). Gonorrhea screening among men who have sex with men: value of multiple anatomic site testing,

San Diego, California, 1997–2003. Sex Transm. Dis. *35*, 845–848.

Hagan, J.E. and Dulmaa, N. (2007). Risk factors and prevalence of HIV and sexually transmitted infections among low-income female commercial sex workers in Mongolia. Sex Transm. Dis. *34*, 83–87.

Hanenberg, R. and Rojanapithayakorn, W. (1996). Prevention as policy: how Thailand reduced STD and HIV transmission. Aidscaptions. *3*, 24–27.

Harrison, L.H. (2006). Prospects for vaccine prevention of meningococcal infection. Clin. Microbiol. Rev. *19*, 142–164.

Hausdorff, W.P., Hajjeh, R., Al-Mazrou, A., Shibl, A., and Soriano-Gabarro, M. (2007). The epidemiology of pneumococcal, meningococcal, and Haemophilus disease in the Middle East and North Africa (MENA) region – current status and needs. Vaccine *25*, 1935–1944.

Hawkes, S., Morison, L., Foster, S., Gausia, K., Chakraborty, J., Peeling, R.W., and Mabey, D. (1999). Reproductive-tract infections in women in low-income, low-prevalence situations: assessment of syndromic management in Matlab, Bangladesh. Lancet *354*, 1776–1781.

Health Protection Scotland (2008). New variant *Chlamydia trachomatis* in Scotland. HPS Weekly *42*, 39.

Hethcote, H.W. and Yorke, J.A. (1984). Gonorrhea transmission dynamics and control. Berlin: Springer-Verlag), pp. 101–105.

Heyderman, R.S., Ben-Shlomo, Y., Brennan, C.A., and Somerset, M. (2004). The incidence and mortality for meningococcal disease associated with area deprivation: an ecological study of hospital episode statistics. Arch. Dis. Child. *89*, 1064–1068.

Hiltunen-Back, E., Rostila, T., Kautiainen, H., Paavonen, J., and Reunala, T. (1998). Rapid decrease of endemic gonorrhea in Finland. Sex Transm. Dis. *25*, 181–186.

Hoke, T.H., Feldblum, P.J., Damme, K.V., Nasution, M.D., Grey, T.W., Wong, E.L., Ralimamonjy, L., Raharimalala, L., and Rasamindrakotroka, A. (2007). Randomised controlled trial of alternative male and female condom promotion strategies targeting sex workers in Madagascar. Sex Transm. Infect. *83*, 448–453.

Hook, E.W. and Handsfield, H.H. (1999). Gonococcal infections in the adult. In Sexually Transmitted Diseases, 3rd edn. K.K. Holmes, P.F. Sparling, P.-A. Mardh, S.M. Lemon, W.E. Stamm, P. Piot, and J.N. Wasserheit, eds. McGraw-Hill), p. 451.

Horino, T., Kato, T., Sato, F., Sakamoto, M., Nakazawa, Y., Yoshida, M., Onodera, S., Kohda, M., Matsuo, K., Ishiji, T., Takahashi, H., and Watanabe, H. (2008). Meningococcemia without meningitis in Japan. Intern. Med. *47*, 1543–1547.

Horner, P.J., Coker, R.J., Turner, A., Shafi, M.S., and Murphy, S.M. (1992). Gonorrhoea: signs, symptoms and serogroups. Int. J.STD AIDS *3*, 430–433.

GRASP Steering Group (2007). The Gonococcal Resistance to Antimicrobials Surveillance Programme (GRASP) Year 2006 report. (London: Health Protection Agency.)

GRASP Steering Group (2008). The Gonococcal Resistance to Antimicrobials Surveillance Programme (GRASP) Year 2007 report. (London: Health Protection Agency.)

Hughes, G. and Fenton, K. (2001). Epidemiology of STIs:UK. Medicine *29*, 1–2.

Johansen, J.D. and Smith, E. (2002). Gonorrhoea in Denmark: high incidence among HIV-infected men who have sex with men. Acta Derm. Venereol. *82*, 365–368.

Johnson, L.F., Coetzee, D.J., and Dorrington, R.E. (2005). Sentinel surveillance of sexually transmitted infections in South Africa: a review. Sex Transm. Infect. *81*, 287–293.

Kelly, C., Arnold, R., Galloway, Y., and O'Hallahan, J. (2007). A prospective study of the effectiveness of the New Zealand meningococcal B vaccine. Am. J. Epidemiol. *166*, 817–823.

Kilmarx, P.H., Black, C.M., Limpakarnjanarat, K., Shaffer, N., Yanpaisarn, S., Chaisilwattana, P., Siriwasin, W., Young, N.L., Farshy, C.E., Mastro, T.D., and St Louis, M.E. (1998). Rapid assessment of sexually transmitted diseases in a sentinel population in Thailand: prevalence of chlamydial infection, gonorrhoea, and syphilis among pregnant women – 1996. Sex. Transm. Infect. *74*, 189–193.

Kourbatova, E.V., Akovbyan, V.A., Chesson, H.W., Lytkina, I.N., Dmitriev, G.A., Tikhonova, L.I., Koubanova, A.A., Petukhova, I.I., Latypova, M.F., Aboymova, O.A., Lewis, J.S., Ryan, C.A., and Shakarishvili, A. (2008). Assessment of the routine, occupation-based gonorrhea and syphilis screening program in Moscow, Russia: an analysis of sexually transmitted infection prevalence and cost-effectiveness. Sex. Transm. Dis. *35*, 453–460.

Lacey, C.J., Merrick, D.W., Bensley, D.C., and Fairley, I. (1997). Analysis of the sociodemography of gonorrhoea in Leeds, 1989–93. BMJ *314*, 1715–1718.

LaForce, F.M., Konde, K., Viviani, S., and Preziosi, M.P. (2007). The Meningitis Vaccine Project. Vaccine *25*, A97-A100.

Lapeyssonnie, L. (1963). La meningite cerebrospinale en Afrique. Bull. World Health Organ. *28*, 3–114.

Larrauri, A., Cano, R., Garcia, M., and Mateo, S. (2005). Impact and effectiveness of meningococcal C conjugate vaccine following its introduction in Spain. Vaccine 23, 4097–4100.

Lavelle, S.J., Jones, K.E., Mallinson, H., and Webb, A.M. (2006). Finding, confirming, and managing gonorrhoea in a population screened for chlamydia using the Gen-Probe Aptima Combo2 assay. Sex. Transm. Infect. *82*, 221–224.

Leake, J.A., Kone, M.L., Yada, A.A., Barry, L.F., Traore, G., Ware, A., Coulibaly, T., Berthe, A., Mambu Ma, D.H., Rosenstein, N.E., Plikaytis, B.D., Esteves, K., Kawamata, J., Wenger, J.D., Heymann, D.L., and Perkins, B.A. (2002). Early detection and response to meningococcal disease epidemics in sub-Saharan Africa: appraisal of the WHO strategy. Bull. World Health Organ. *80*, 342–349.

Leimkugel, J., Hodgson, A., Forgor, A.A., Pfluger, V., Dangy, J.P., Smith, T., Achtman, M., Gagneux, S., and Pluschke, G. (2007). Clonal Waves of *Neisseria*

Colonisation and Disease in the African Meningitis Belt: Eight-Year Longitudinal Study in Northern Ghana. PLoS Med. *4*, e101.

Leutscher, P., Jensen, J.S., Hoffmann, S., Berthelsen, L., Ramarakoto, C.E., Ramaniraka, V., Randrianasolo, B., Raharisolo, C., Bottiger, B., Rousset, D., Grosjean, P., McGrath, M.M., Christensen, N., and Migliani, R. (2005). Sexually transmitted infections in rural Madagascar at an early stage of the HIV epidemic: a 6-month community-based follow-up study. Sex. Transm. Dis. *32*, 150–155.

Lindan, C., Mathur, M., Kumta, S., Jerajani, H., Gogate, A., Schachter, J., and Moncada, J. (2005). Utility of pooled urine specimens for detection of *Chlamydia trachomatis* and *Neisseria gonorrhoeae* in men attending public sexually transmitted infection clinics in Mumbai, India, by PCR.J. Clin. Microbiol. *43*, 1674–1677.

Low, N., Broutet, N., Adu-Sarkodie, Y., Barton, P., Hossain, M., and Hawkes, S. (2006). Global control of sexually transmitted infections. Lancet *368*, 2001–2016.

Lujan, J., de Onate, W.A., Delva, W., Claeys, P., Sambola, F., Temmerman, M., Fernando, J., and Folgosa, E. (2008). Prevalence of sexually transmitted infections in women attending antenatal care in Tete province, Mozambique. S. Afr. Med. J. *98*, 49–51.

Mahafzah, A.M., Al Ramahi, M.Q., Asa'd, A.M., and El Khateeb, M.S. (2008). Prevalence of sexually transmitted infections among sexually active Jordanian females. Sex. Transm. Dis. *35*, 607–610.

Maiden, M.C. and Spratt, B.G. (1999). Meningococcal conjugate vaccines: new opportunities and new challenges. Lancet *354*, 615–616.

Maiden, M.C., Ibarz-Pavon, A.B., Urwin, R., Gray, S.J., Andrews, N.J., Clarke, S.C., Walker, A.M., Evans, M.R., Kroll, J.S., Neal, K.R., Ala'Aldeen, D.A., Crook, D.W., Cann, K., Harrison, S., Cunningham, R., Baxter, D., Kaczmarski, E., MacLennan, J., Cameron, J.C., and Stuart, J.M. (2008). Impact of meningococcal serogroup c conjugate vaccines on carriage and herd immunity. J. Infect. Dis. 197, 737–743.

Maiden, M.C.J. and Stuart, J.M. (2002). Carriage of serogroup C meningococci one year after meningococcal C conjugate polysaccharide vaccination. Lancet *359*, 1829–1830.

Matsumoto, T. (2008). Trends of sexually transmitted diseases and antimicrobial resistance in *Neisseria gonorrhoeae*. Int. J. Antimicrob. Agents 31 Suppl *1*, S35-S39.

Miller, E., Salisbury, D.M., and Ramsay, M.E. (2001). Planning, registration, and implementation of an immunisation campaign against meningococcal serogroup C disease in the UK: a success story. Vaccine *20*, S58-S67.

Miller, G.C., McDermott, R., McCulloch, B., Fairley, C.K., and Muller, R. (2003). Predictors of the prevalence of bacterial STI among young disadvantaged Indigenous people in north Queensland, Australia. Sex. Transm. Infect. 79, 332–335.

Mimiaga, M.J., Mayer, K.H., Reisner, S.L., Gonzalez, A., Dumas, B., Vanderwarker, R., Novak, D.S., and Bertrand, T. (2008). Asymptomatic gonorrhea and chlamydial infections detected by nucleic acid amplification tests among Boston area men who have sex with men. Sex. Transm. Dis. *35*, 495–498.

Molesworth, A.M., Cuevas, L.E., Connor, S.J., Morse, A.P., and Thomson MC (2003). Environmental risk and meningitis epidemics in Africa. Emerg. Infect. Dis. *9*, 1287–1293.

Molesworth, A.M., Thomson, M.C., Connor, S.J., Cresswell, M.P., Morse, A.P., Shears, P., Hart, C.A., and Cuevas, L.E. (2002). Where is the meningitis belt? Defining an area at risk of epidemic meningitis in Africa. Trans. R. Soc. Trop. Med. Hyg. *96*, 242–249.

Moore, P.S. (1992). Meningococcal meningitis in sub-Saharan Africa: a model for the epidemic process. Clin. Infect. Dis. *14*, 515–525.

Mueller, J.E., Yaro, S., Madec, Y., Somda, P.K., Idohou, R.S., Njanpop Lafourcade, B.M., Drabo, A., Tarnagda, Z., Sangare, L., Traore, Y., Fontanet, A., and Gessner, B.D. (2008). Association of respiratory tract infection symptoms and air humidity with meningococcal carriage in Burkina Faso. Trop. Med. Int. Health. *13*, 1543–1552.

Nagot, N., Meda, N., Ouangre, A., Ouedraogo, A., Yaro, S., Sombie, I., Defer, M.C., Barennes, H., and Van de, P.P. (2004). Review of STI and HIV epidemiological data from 1990 to 2001 in urban Burkina Faso: implications for STI and HIV control. Sex. Transm. Infect. *80*, 124–129.

Navarro, C., Deeks, S.L., Medaglia, A., and Tsang, R. (2007). Enhanced surveillance of invasive meningococcal disease in Canada: 1 January, 2004, through 31 December, 2005. Canada Communicable Disease Report 33.

Navarro, C., Jolly, A., Nair, R., and Chen, Y. (2002). Risk factors for genital chlamydial infection. Can. J. Infect. Dis. *13*, 195–207.

Nguyen, V.T., Nguyen, T.L., Nguyen, D.H., Le, T.T., Vo, T.T., Cao, T.B., and O'Farrell, N. (2005). Sexually transmitted infections in female sex workers in five border provinces of Vietnam. Sex. Transm. Dis. *32*, 550–556.

Nicoll, A., Hughes, G., Donnelly, M., Livingstone, S., De Angelis, D., Fenton, K., Evans, B., Gill, O.N., and Catchpole, M. (2001). Assessing the impact of national anti-HIV sexual health campaigns: trends in the transmission of HIV and other sexually transmitted infections in England. Sex. Transm. Infect. 77, 242–247.

Noble, R.C., Reyes, R.R., Parekh, M.C., and Haley, J.V. (1984). Incidence of disseminated gonococcal infection correlated with the presence of AHU auxotype of *Neisseria gonorrhoeae* in a community. Sex. Transm. Dis. *11*, 68–71.

O'Hallahan, J., Lennon, D., Oster, P., Lane, R., Reid, S., Mulholland, K., Stewart, J., Penney, L., Percival, T., and Martin, D. (2005). From secondary prevention to primary prevention: a unique strategy that gives hope to a country ravaged by meningococcal disease. Vaccine 23, 2197–2201.

Papadogeorgaki, H., Caroni, C., Frangouli, E., Flemetakis, A., Katsambas, A., and Hadjivassiliou, M. (2006). Prevalence of sexually transmitted infections in female sex workers in Athens, Greece – 2005. Eur. J. Dermatol. *16*, 662–665.

Passey, M., Mgone, C.S., Lupiwa, S., Suve, N., Tiwara, S., Lupiwa, T., Clegg, A., and Alpers, M.P. (1998). Community based study of sexually transmitted diseases in rural women in the highlands of Papua New Guinea: prevalence and risk factors. Sex Transm. Infect. *74*, 120–127.

Peltola, H. (1983). Meningococcal disease: still with us. Rev. Infect. Dis. *5*, 71–91.

Peltola, H. (2001). Burden of meningitis and other severe bacterial infections of children in Africa: implications for prevention. Clin Infect Dis *32*, 64–75.

Price, G.A., Masri, H.P., Hollander, A.M., Russell, M.W., and Cornelissen, C.N. (2007). Gonococcal transferrin binding protein chimeras induce bactericidal and growth inhibitory antibodies in mice. Vaccine *25*, 7247–7260.

Purcell B, Samuelsson S, Hahne SJM, Ehrhard I, Heuberger S, Camaroni I, Charlett A, and Stuart JM (2004). Effectiveness of antibiotocs in preventing meningococcal disease after a case: systematic review. BMJ *328*, 1339–1344.

Ramsay, M.E., Andrews, N.J., Trotter, C.L., Kaczmarski, E.B., and Miller, E. (2003). Herd immunity from meningococcal serogroup C conjugate vaccination in England: database analysis. BMJ *326*, 365–366.

Rao, G.G., Bacon, L., Evans, J., Dejahang, Y., Michalczyk, P., and Donaldson, N. (2008). Prevalence of *Neisseria gonorrhoeae* infection in young subjects attending community clinics in South London. Sex. Transm. Infect. *84*, 117–121.

Rashid, H., Booy, R., Christie, D., El Bashir, H., and Viner, R. Outcome of meningitis: a systematic review. 16th International Pathogenic *Neisseria* Conference. 2008.

Robbins JB, Towne DW, Gotslich EC, and Schneerson R (1997). 'Love's labours lost': failure to implement mass vaccination against group A meningococcal meningitis in sub-Saharan Africa. Lancet *350*, 880–883.

Rosenstein, N., Perkins BE, Stephens DS, Lefkowitz L, Cartter ML, Danila R, Cieslak P, Shutt KA, Popovic T, Schuchat A, Harrison LH, Reingold AL, and Active Bacterial Core Surveillance Team (1999). The changing epidemiology of meningococcal disease in the United States, 1992–1996. J. Infect. Dis. 180, 1894–1901.

Rothenberg, R.B. (1983). The geography of gonorrhea. Empirical demonstration of core group transmission. Am. J. Epidemiol. *117*, 688–694.

Sacchi, C.T., Pessoa, L.L., Ramos, S.R., Milagres, L.G., Camargo, M.C., Hidalgo, N.T., Melles, C.E., Caugant, D.A., and Frasch, C.E. (1992). Ongoing group B *Neisseria meningitidis* epidemic in Sao Paulo, Brazil, due to increased prevalence of a single clone of the ET-5 complex. J. Clin. Microbiol. *30*, 1734–1738.

Sandstrom, E., Jonsson, A., Holterman, M., and Wallmark, G. (1982). Detection of repeated gonorrhea by a modified follow-up method. Acta Derm. Venereol. *62*, 419–424.

Shahmanesh, M., Cowan, F.M., Wayal, S.S., Copas, A., Patel, V., and Mabey, D. (2008). The burden and determinants of HIV and sexually transmitted infections in a population based sample of female sex workers in Goa, India. Sex. Transm. Infect. *85*, 50–59.

Sierra, G.V., Campa, H.C., Varcacel, N.M., Garcia, I.L., Izquierdo, P.L., Sotolongo, P.F., Casanueva, G.V., Rico, C.O., Rodriguez, C.R., and Terry, M.H. (1991). Vaccine against group B *Neisseria meningitidis*: protection trial and mass vaccination results in Cuba. NIPH Ann *14*, 195–207.

Soto, R.J., Ghee, A.E., Nunez, C.A., Mayorga, R., Tapia, K.A., Astete, S.G., Hughes, J.P., Buffardi, A.L., Holte, S.E., and Holmes, K.K. (2007). Sentinel surveillance of sexually transmitted infections/HIV and risk behaviors in vulnerable populations in 5 Central American countries. J. Acquir. Immune. Defic. Syndr. *46*, 101–111.

Taha, T.E., Dallabetta, G.A., Hoover, D.R., Chiphangwi, J.D., Mtimavalye, L.A., Liomba, G.N., Kumwenda, N.I., and Miotti, P.G. (1998). Trends of HIV-1 and sexually transmitted diseases among pregnant and postpartum women in urban Malawi. AIDS *12*, 197–203.

Tann, C.J., Mpairwe, H., Morison, L., Nassimu, K., Hughes, P., Omara, M., Mabey, D., Muwanga, M., Grosskurth, H., and Elliott, A.M. (2006). Lack of effectiveness of syndromic management in targeting vaginal infections in pregnancy in Entebbe, Uganda. Sex. Transm. Infect. *82*, 285–289.

Tapsall, J. (2006). Antibiotic resistance in *Neisseria gonorrhoeae* is diminishing available treatment options for gonorrhea: some possible remedies. Expert. Rev. Anti. Infect. Ther. *4*, 619–628.

Tapsall, J.W., Limnios, E.A., and Murphy, D. (2008). Analysis of trends in antimicrobial resistance in *Neisseria gonorrhoeae* isolated in Australia, 1997 2006. J. Antimicrob. Chemother. *61*, 150–155.

Thomas, H.L., Andrews, N., Trotter, C., Ramsay, M., and Miller, E. (2008). Meningococcal C booster not recommended by evidence. BMJ *337*, a1139.

Trotter, C.L. and Greenwood, B.M. (2007). Meningococcal carriage in the African meningitis belt: a review. Lancet Infect. Dis. *7*, 797–803.

Trotter, C.L. and Ramsay, M.E. (2007). Vaccination against meningococcal disease in Europe: review and recommendations for the use of conjugate vaccines. FEMS Microbiol. Rev. *31*, 101–107.

Trotter, C.L., Andrews, N.J., Kaczmarski, E.B., Miller, E., and Ramsay, M.E. (2004). Effectiveness of meningococcal serogroup C conjugate vaccine 4 years after introduction. Lancet *364*, 365–367.

Trotter, C.L., Fox, A.J., Ramsay, M.E., Sadler, F., Gray, S.J., Mallard, R., and Kaczmarski, E.B. (2002). Fatal outcome from meningococcal disease – an association with meningococcal phenotype but not with reduced susceptibility to benzylpenicillin. J. Med. Microbiol. *51*, 855–860.

Trotter, C.L., Gay, N.J., and Edmunds, W.J. (2006). The natural history of meningococcal carriage and disease. Epidemiol. Infect. *134*, 556–566.

Trotter, C.L., Ramsay, M.E., Gray, S., Fox, A., and Kaczmarski, E. (2006). No evidence for capsule replacement following mass immunisation with meningococcal serogroup C conjugate vaccines in England and Wales. Lancet Infec.t Dis. *6*, 616–617.

UK Collaborative Group for HIV and STI Surveillance (2007) Testing Times. HIV and other Sexually

Transmitted Infections in the United Kingdom: 2007. (London: Health Protection Agency, Centre for Infections.) November 2007.

Van der Heyden, J.H., Catchpole, M.A., Paget, W.J., and Stroobant, A. (2000). Trends in gonorrhoea in nine western European countries, 1991–6. European Study Group. Sex Transm. Infect. *76*, 110–116.

Vieusseux, M. (1806). Mémoire sur la maladie qui a régné à Genève au printemps de 1805. J. Méd. Chir. Pharmacol. *11*, 163–182.

Vu, T.N., Van Nghia, K., Phuc, H.T., Thanh, L.N., Thi, B., V, Hoang, D.B., Thu, T.L., Anh, T.N., Thi Kim, T.N., Godwin, P., Fylkesnes, K., and O'Farrell, N. (2007). Impact of a community sexually transmitted infection/HIV intervention project on female sex workers in five border provinces of Vietnam. Sex. Transm. Infect. *83*, 376–382.

Wang, H., Wang, N., Chen, R.Y., Sharp, G.B., Ma, Y., Wang, G., Ding, G., and Wu, Z. (2008). Prevalence and predictors of herpes simplex virus type 2 infection among female sex workers in Yunnan Province, China. Int. J.STD AIDS *19*, 635–639.

Wang, S.A., Harvey, A.B., Conner, S.M., Zaidi, A.A., Knapp, J.S., Whittington, W.L., del Rio, C., Judson, F.N., and Holmes, K.K. (2007). Antimicrobial resistance for *Neisseria gonorrhoeae* in the United States, 1988 to 2003: the spread of fluoroquinolone resistance. Ann. Intern. Med. *147*, 81–88.

Ward, H., Day, S., Green, A., Cooper, K., and Weber, J. (2004). Declining prevalence of STI in the London sex industry, 1985 to 2002. Sex Transm. Infect. *80*, 374–376.

Wenzel, P. (1973). Non-usefulness of carriage rates. Lancet 2, 205.

Whalen, C.M., Hockin, J.C., Ryan, A., and Ashton, F. (1995). The changing epidemiology of invasive meningococcal disease in Canada, 1985 through 1992. Emergence of a virulent clone of *Neisseria meningitidis*. JAMA 273, 390–394.

Whittle, H.C., Evans-Jones, G., Onyewotu, I., Adjukiewicz, A., Turunen, U., Crockford, J., and Greenwood, B.M. (1975). Group C meningococcal meningitis in the northern savanna of Africa. Lancet i, 1377.

WHO (1999). Status and trends of STI, HIV/AIDS at the end of the millennium Western Pacific Region 1999. 1–104.

WHO (2001). Global prevalence and incidence of selected curable sexually transmitted infections:overview and estimates.(WHO/HIV_AIDS/2001.02, WHO/CDS/CSR/EDC/2001.10). 1–43.

WHO (2007). Risk of epidemic meningitis in Africa: a cause for concern. Bull. WHO *82*, 79–87.

WHO (2008a). Regional Strategic Action Plan for the Prevention and Control of Sexually Transmitted Infections 2008–2012. World Health Organization.

WHO (2008b). Surveillance of antibiotic resistance in *Neisseria gonorrhoeae* in the WHO Western Pacific Region, 2006. Commun. Dis. Intell. *32*, 48–51.

Wi, T., Ramos, E.R., Steen, R., Esguerra, T.A., Roces, M.C., Lim-Quizon, M.C., Neilsen, G., and Dallabetta, G. (2006). STI declines among sex workers and clients following outreach, one time presumptive treatment, and regular screening of sex workers in the Philippines. Sex. Transm. Infect. *82*, 386–391.

Workowski, K.A., Berman, S.M., and Douglas, J.M., Jr. (2008). Emerging antimicrobial resistance in *Neisseria gonorrhoeae*: urgent need to strengthen prevention strategies. Ann. Intern. Med. *148*, 606–613.

Xu, J.J., Wang, N., Lu, L., Pu, Y., Zhang, G.L., Wong, M., Wu, Z.L., and Zheng, X.W. (2008). HIV and STIs in clients and female sex workers in mining regions of Gejiu City, China. Sex. Transm. Dis. *35*, 558–565.

Yazdankhah, S.P. and Caugant, D.A. (2004). *Neisseria meningitidis*: an overview of the carriage state. J. Med. Microbiol. *53*, 821–832.

Yorke, J.A., Hethcote, H.W., and Nold, A. (1978). Dynamics and control of the transmission of gonorrhea. Sex. Transm. Dis. *5*, 51–56.

Zimmerman, H.L., Potterat, J.J., Dukes, R.L., Muth, J.B., Zimmerman, H.P., Fogle, J.S., and Pratts, C.I. (1990). Epidemiologic differences between chlamydia and gonorrhea. Am. J. Public Health *80*, 1338–1342.

Molecular Mechanisms of Antibiotic Resistance Expressed by the Pathogenic Neisseriae

13

William M. Shafer, Jason P. Folster and Robert A. Nicholas

Abstract

Diseases caused by the pathogenic *Neisseria* (*N. gonorrhoeae* and *N. meningitidis*) have been successfully treated with antibiotics for the past 70 years. However, a disturbing trend worldwide is the increasing prevalence of strains with resistance to inexpensive and widely available antibiotics (e.g. penicillin, tetracycline and ciprofloxacin) and the emergence of strains exhibiting decreased susceptibility to effective antibiotics that are expensive and not always available (e.g. third-generation cephalosporins and the newer macrolides). Given the global nature of gonococcal and meningococcal diseases, the worldwide distribution of antibiotics, differing social practices in controlling and monitoring antibiotic availability, and geographical differences in treatment regimens, it is likely that the global problem of antibiotic resistance will continue (and worsen) in the foreseeable future. By understanding the mechanisms of antibiotic resistance in gonococci and meningococci, resistance to antibiotics currently in clinical practice can be anticipated and the design of novel antimicrobials to circumvent this problem can be undertaken more rationally. Herein, we review the genetic and physiologic basis by which the pathogenic *Neisseria* developed resistance to historically important antibiotics and how resistance to newer antibiotics is emerging.

Introduction

On the morning of 17 April 2007 the senior author (W.M.S.) stared at a report in the *Atlanta Journal-Constitution* that blared 'Gonorrhea Joins the List of Superbugs.' Apart from confusing the disease and bacterium, this announcement to the good citizens of Georgia was significant because it brought the problem of this sexually transmitted disease to the forefront. In reflection, proclaiming the downfall of an antibiotic in the treatment of gonorrhoea in the popular press served two useful purposes: it reminded the populace of the prevalence of this disease and emphasized the overall problem of antibiotic resistance in the treatment of bacterial infections.

Gonorrhoea is an ancient disease with biblical (Old Testament) references. In contrast, the principal forms of meningococcal disease (meningitis and fulminant septicaemia) appeared much later in the evolution of *homo sapiens*, with the earliest recorded reports from outbreaks in the early 1800s in Europe and the United States (Stephens *et al.*, 2007). Antibiotic resistance expressed by these two strict human pathogens has considerable implications for treatment options and control of disease in the community. For instance, when one considers that there are over 62 million cases of gonorrhoea each year worldwide, the prospect of multidrug-resistant strains gives reason for concern. While effective anti-gonococcal treatment remains, the antibiotics of choice (ceftriaxone and cefixime) are costly and not always available; these are real-life problems for underdeveloped countries that frequently have high incidences of gonorrhoea and many individuals may not have ready access to health-care providers. While considerably less has been published regarding antibiotic resistance trends in the meningococcus, resistance to rifampin, an antibiotic often used to stop the spread of disease during outbreaks, and quinolones occurs frequently (Stephens *et al.*, 2007). Given the close genetic relationship between gonococci and meningococci and their ability to exchange DNA, there are instances where identical mechanisms of resistance (or decreased antibiotic susceptibility) exist in both species. This chapter focuses on the molecular mechanisms of antibiotic resistance expressed by *N. gonorrhoeae* and *N. meningitidis* and, herein, we concentrate on issues dealing with the genetics of antibiotic resistance, the physiological processes that mediate resistance and the regulation of resistance genes.

Historical perspectives and current problems

What path did the gonococcus follow to join the list of bacteria known as 'Superbugs?' Sulphonamides were the first antibiotics employed in the 1930s to treat gonorrhoea, but by the mid-1940s resistance levels were high. Fortunately, the availability of the 'wonder drug' penicillin marked a new era in treatment of this disease, and this antibiotic quickly supplanted the use of sulphonamides to treat gonorrhoea. In addition to its potent activity against the gonococcus (MIC values < 0.01 μg/ml), penicillin was also effective against meningococci and could penetrate the blood brain barrier, allowing for killing of bacteria in the cerebrospinal fluid. The rapid killing of gonococci by penicillin was a significant advance in public health efforts to reduce disease transmission in the community. However, by the late 1940s bad news was at the doorstep for continued efficacy of penicillin in treating patients with bacterial infections due to reports of penicillin-resistant isolates of *Staphylococcus aureus* that produced an enzyme (β-lactamase) that inactivated the antibiotic (Levy, 2002). Despite trends of antibiotic resistance in other bacteria and treatment failures for certain bacterial infections, penicillin remained an effective antibiotic for the treatment of gonorrhoea for the next forty years. However, the epidemic of gonorrhoea in the United States and elsewhere associated with the 'sexual revolution' of the 1960s soon changed this, and by 1972 the level of penicillin required to effectively treat uncomplicated gonococcal infections had increased several-fold and treatment failures were reported in the clinical literature.

The appearance of the β-lactamase plasmid in certain strains of gonococci during the mid-1970s (Ashford *et al.*, 1976; Percival *et al.*, 1976; Phillips, 1976), confirmed the fear that decades-long use of this antibiotic might soon come to an end. In the United States, this prediction was realized ten years later but was not due to high prevalence of β-lactamase-producing gonococci; rather, strains emerged containing chromosomal mutations that collectively rendered gonococci clinically resistant to penicillin. In particular, the appearance of a gonococcal strain that caused an outbreak of penicillin-resistant infections in Durham, NC (Faruki *et al.*, 1985) was a major blow to the continued use of penicillin. Tetracycline, another relatively cheap antibiotic, also was once an effective treatment option, but the appearance of the *tetM* determinant on the conjugative plasmid (Morse *et al.*, 1986) and the emergence of chromosomal mutations that provided resistance to the antibiotic removed tetracycline from CDC-recommended treatment options in the 1980s. The realization that loss of penicillin and tetracycline as treatment options would significantly impact global public health efforts to control gonorrhoea, especially at a time of the continuing HIV/AIDS epidemic, required the use of alternative antibiotics and ciprofloxacin seemed to fit the need. However, while initially effective, resistance soon emerged in Asia (WHO, 1995) and either was exported to other parts of the world (Patrick *et al.*, 1995; Su and Lind, 2001) or emerged independently. In 2000, fluoroquinolone-resistant strains became so prevalent in Hawaii (Iverson *et al.*, 2000) that fluoroquinolones could no longer be used with confidence and were removed from the CDC-recommended treatment regimens (antibiotics are typically removed when prevalence of resistance is ≥ 5%). These resistant strains eventually spread from Hawaii and the Pacific Rim to the US mainland, and in 2007 the fluoroquinolones were removed entirely from the recommended list of antibiotics to treat gonococcal infections with no geographical exceptions.

For continued effective treatment of gonococcal and meningococcal infections, it is essential that active surveillance of isolates be in place so that changes in bacterial susceptibility to antibiotics can be detected and trends in resistance can be monitored. In the United States this is accomplished for gonococci by the CDC and its Gonococcal Isolate Surveillance Project (GISP), which was established in 1986 (Schwarz *et al.*, 1990) to monitor trends in antimicrobial susceptibilities, since most infections are diagnosed by non-culture methods. Active surveillance programmes are ongoing in other parts of the world, largely through the efforts of the WHO, and these have been instrumental in providing data on regional trends and changes in antibiotic susceptibilities; for the sake of brevity, this review will concentrate on epidemiologic data obtained through GISP. Antibiotics currently examined in the GISP system include penicillin, tetracycline, spectinomycin, ciprofloxacin, ceftriaxone, cefixime, and azithromycin. The 27 STD clinics of GISP restrict active surveillance, however, to males 18–45 years of age who visit STD clinics. Isolates from the first 25 men who visit a GISP site each month are cultured and antimicrobial susceptibility tests are performed on isolated gonococci. It is estimated that GISP only covers 2% of all gonococcal cases each year in the USA. While this limited sampling, coupled with the age and gender restrictions, provides a narrow view of gonococci and antibiotic susceptibility, it has been important in tracking both the prevalence and the spread of antibiotic resistance and the emergence of new resistance problems. GISP data are important because they form the basis for year-to-year changes in CDC-recommended treatment options.

The 2007 GISP report, which analysed data from 6199 gonococcal isolates collected and analysed in 2005 (CDC, 2007), highlights the problem of antibiotic resistance expressed by gonococci. Resistance to penicillin and tetracycline is categorized into six main groups: penicillinase-producing *N. gonorrhoeae* (PPNG)/ tetracycline sensitive, plasmid-mediated (*tetM*) tetracycline-resistant (β-lactamase negative) *N. gonorrhoeae* (TRNG), dual PPNG-TRNG strains, chromosomally mediated penicillin-resistant *N. gonorrhoeae* (PenR), chromosomally mediated tetracycline resistance (TetR) and chromosomally mediated resistance to both penicillin and tetracycline (CMRNG); the molecular basis for these resistances are discussed in detail below. In 2005, the percentage of PPNG was 0.5% (from a high of 11% in 1991), the percentage of TTNG was 4.5% and the prevalence of PPNG-TRNG was 0.7%. For 2005, chromosome-mediated resistance was reported to be 1.9% for PenR, 5.9% for TetR and 6.1% for CMRNG. For the entire GISP system, resistance to ciprofloxacin (MIC ≥ 2 μg/ml) was 9.4%, but certain sites, e.g. Honolulu (19.3%), Long Beach (23.5%), and San Francisco (31.3%), reported much higher prevalence values. As highlighted by ciprofloxacin resistance, geographical trends can be monitored by the GISP and region-specific recommendations can be made for treatment based on local changes in antimicrobial susceptibilities. All of the 2005 GISP isolates remained susceptible to spectinomycin, four isolates showed decreased susceptibility to ceftriaxone and separate isolates (both from Los Angeles) exhibited decreased susceptibility to cefixime; it is noteworthy that these latter two strains were resistant to penicillin, tetracycline and ciprofloxacin. The picture for the prevalence of azithromycin-resistant gonococci is less clear due to changes in antibiotic susceptibility-testing media, but nearly 3% had azithromycin MIC values of ≥ 1.0 μg/ml, which is important because treatment failures frequently occur when the MIC values range from 0.125 to 0.5 μg/ml.

Reports from other parts of the world also highlight the critical problem of gonococcal resistance, especially to penicillin, tetracycline and ciprofloxacin. Data from an ongoing active surveillance programme in Western Pacific countries (Fig. 13.1) by the WHO are particularly illuminating (2006), but resistance is also acute in North America, Europe, Asia, South America, the Caribbean and Africa. Penicillin- and/or tetracycline-resistant strains in Cuba (Lianes *et al.*, 2003), China (Ye *et al*, 2002), and certain regions in Africa (Van Dyke *et al.*, 1997) can represent 30–90% of total isolates. Once established in the community, antibiotic-resistant strains can spread rapidly in a region and then establish a global presence. For example, fluoroquinolone resistance emerged quickly in gonococci and its rapid global spread was nothing short of remarkable, resulting in its removal from the CDC-recommended treatment regimen in less than 20 years since its introduction into clinical practice. We can also

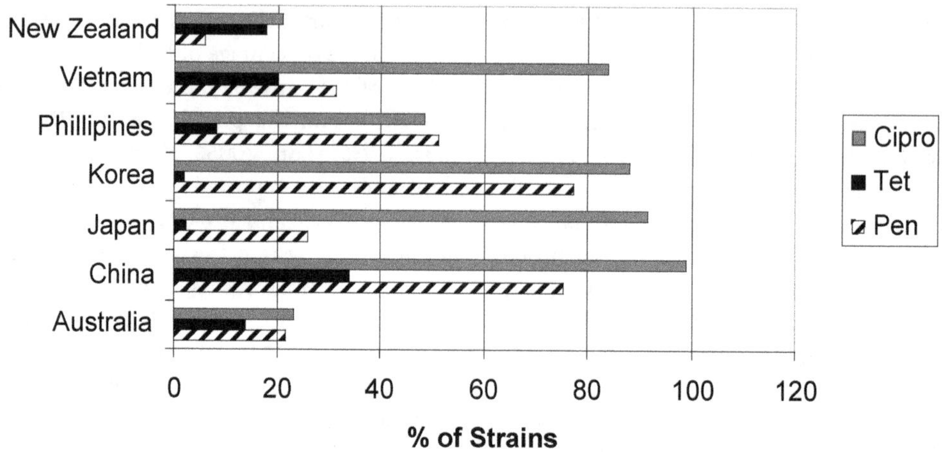

Figure 13.1 The prevalence of gonococci resistant to penicillin, tetracycline and ciprofloxacin in seven selected countries in the Pacific Rim are shown. For both penicillin and tetracycline, the prevalence was determined for both plasmid- and chromosomally mediated mechanisms. Data for resistant isolates in the United States were determined from the 2005 GI-SP report published by the CDC (2007), while the data regarding Pacific Rim countries are from The WHO Western Pacific Gonococcal Antimicrobial Surveillance Programme report of isolates from 2004 published in 2006.

anticipate the emergence of resistance to cefixime and ceftriaxone as gonococcal and meningococcal strains with decreased susceptibility to these cephalosporins have been reported (Ameyama *et al.*, 2002, Wang *et al.*, 2003; Hoffmann *et al.*, 2005; Takahata *et al.*, 2006; Tanaka *et al.*, 2006; Whiley *et al.*, 2007; Bala *et al.*, 2007; Lindberg *et al.*, 2007). This 'MIC creep' is highly reminiscent of the emergence of chromosomally mediated resistance to penicillin and continued reports of strains with decreased susceptibility to these antibiotics should be anticipated; their continued emergence could seriously jeopardize effective treatment options (Tapsall, 2006).

Coordinated, active surveillance programs that monitor changes in meningococcal susceptibility to antibiotics presently do not exist and less definitive data are available regarding resistance trends for *N. meningitidis*. Despite this limitation, it is clear that resistance to antibiotics has also emerged in the meningococcus. In the United States, strains causing invasive meningococcal disease remain sensitive to the most commonly used antibiotics for treating disease or those used in chemoprophylaxis (except for the recent emergence of fluoroquinolone-resistant strains in Minnesota and North Dakota – see below) (Rosenstein *et al.*, 2001). However, in other parts of the world, resistance of meningococci to inexpensive antibiotics that are used for treatment or prophylaxis is a problem (Gallimand *et al.*, 1998; Tapsall *et al.*, 2001). For example, sulphonamides were highly effective in the treatment and prophylaxis of meningococcal disease in the 1940s and 1950s (Kuhns *et al.*, 1943), but sulphonamide-resistant meningococci emerged and caused deadly outbreaks of disease in US military recruits in the 1960s. These outbreaks were the stimulus for the development of the initial capsular polysaccharide vaccine (Gotschlich *et al.*, 1969). Sulphonamide resistance is caused by alterations in the drug target dihydropteroate synthase (DHPS). Two resistance-conferring alleles of *dhps* exist in meningococci and both have successfully spread among meningococci of different serogroups and serotypes. It has been suggested (Fermer *et al.*, 1995) that through acquisition of DNA from other *Neisseria* (possibly the gonococcus), a mosaic *dhps* allele was constructed; this concept of mosaic genes leading to antibiotic resistance is reminiscent of the development of mutations in *penA* that decrease the rate of penicillin acylation of penicillin-binding protein 2 (PBP2) (Spratt, 1988). Rifampin-resistant meningococcal strains also exist (Nolte, 1997) and are likely selected during chemoprophylaxis to eliminate carriage for close contacts of cases (Yagupsky *et al.*, 1993). Meningococcal resistance to quinolones has also been reported (Stephens *et al.*, 2007). Strains with decreased susceptibility to penicillin have been reported throughout the world (Lopardo *et al.*, 1993; Woods *et al.*, 1994; Rosenstein *et al.* 2000; Antignac *et al.*, 2001; Richter *et al.*, 2001; Boras *et al.*, 2001), and these isolates utilize chromosomally mediated resistance mechanisms likely involving *penA* mutations (see below). Penicillinase-producing meningococci were reported in the 1980s in Spain (Fontanals *et al.*, 1989) and Africa (Botha, 1988), but such strains are now apparently rare and have not been isolated in some time. The recent emergence of meningococcal strains in India that are non-susceptible to ceftriaxone and cefotaxime (Manchanda and Bhalla, 2006) is especially alarming. These isolates were associated with potential therapeutic failures in treatment of patients with suspected meningococcemia or meningitis. The molecular basis for this property is not yet clear but the report highlights evolving changes in meningococcal antibiotic susceptibilities.

Ciprofloxacin has been employed for chemoprophylaxis to prevent secondary disease in close contacts of patients with meningococcal disease. Although ciprofloxacin-resistant meningococcal disease has been described in South America, Europe, Asia and Australia, it had not been reported in North America until very recently (CDC, 2008), where it was documented in three cases of serogroup B meningococcal disease in North Dakota and Minnesota. This is a worrisome development and we should anticipate that this resistance will continue to emerge. In the face of this resistance, alternative empiric antimicrobial chemoprophylaxis treatments will be necessary, but it is prudent to monitor changes in meningococcal susceptibility to the alternative antibiotic regimens recommended by the CDC (e.g. ceftriaxone, rifampin and azithromycin).

Mechanisms of resistance

Overview of the physiology and genetics of antibiotic resistance

The physiological mechanisms of antibiotic resistance expressed by gonococci and meningococci are similar to those of other bacteria: antibiotic destruction or modification by enzymatic means, decreased influx of antibiotics, increased efflux of antibiotics, and target modification or protection that reduces affinity for the antibiotic. Fig. 13.2 summarizes mechanisms of gonococcal resistance to ciprofloxacin, penicillin and tetracycline and these are used as examples of the general classes of bacterial resistance mechanisms.

Bacteria develop resistance to antibiotics through mutation or by gene transfer. Both genetic processes have been described in the evolution of antibiotic resistance in gonococci and meningococci. Most resistance determinants in the pathogenic *Neisseria* are chromosomally located and only the TEM-1 β-lactamase (Phillips, 1976) and the *tetM* gene (Morse *et al.*, 1986) are known to be plasmid-borne in gonococci. The following two examples emphasize the importance of spontaneous mutations and the evolution of antibiotic resistance. First, strains expressing the multiple transferable resistance (*mtr*) phenotype (Maness and Sparling, 1973; Guymon *et al.*, 1978; Pan and Spratt, 1994) frequently have a single bp deletion in the *mtrR* promoter or have missense mutations in the region of the *mtrR* gene that encodes the DNA-binding domain of the MtrR repressor (Hagman *et al.*, 1995), and these mutations can influence levels of gonococcal susceptibility to penicillin (Veal *et al.*, 2002) and hydrophobic antibiotics (Hagman *et al.*, 1995). Second, missense mutations in *gyrA* and *parC* result in clinically relevant levels of resistance to fluoroquinolones (Belland *et al.*, 1994). Once established, resistance genes can be spread by transformation or conjugation. Gonococci and meningococci are highly competent for genetic transformation and use a sequence-specific DNA uptake system (Goodman and Scocca, 1988) to recognize DNA from themselves or closely related species. Chromosomal DNA transformation frequencies can be quite high (10^{-1} to 10^{-3} transformants/μg of DNA, depending on the procedure used), while plasmid transformation frequencies are much lower (10^{-6} transformants/μg of DNA). Thus, even though the rate of spontaneous missense mutations that result in resistance is low, horizontal transfer of these alleles by transformation is very effective in disseminating resistance within the community.

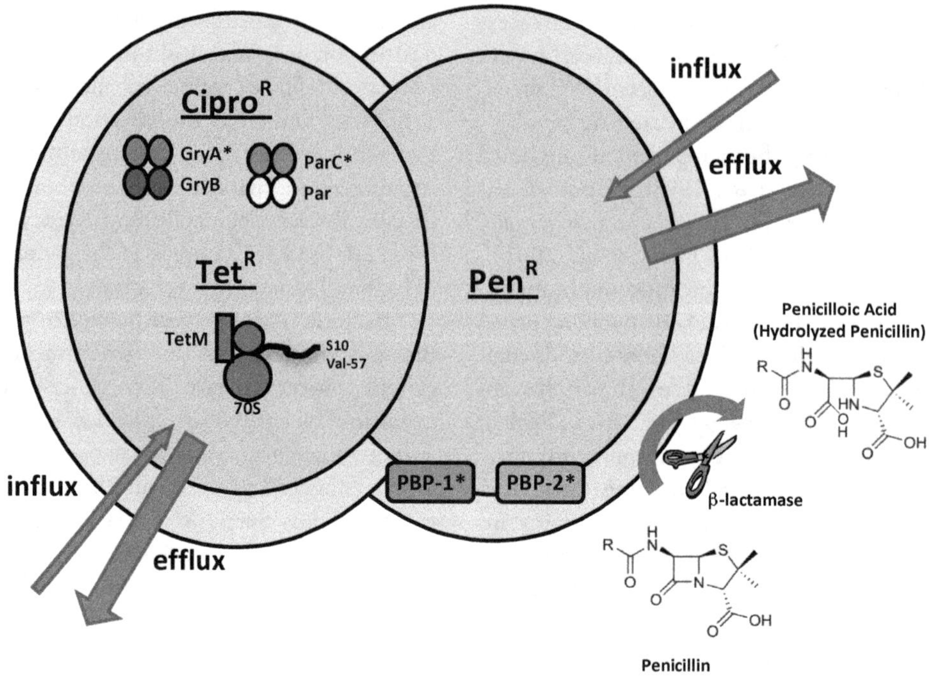

Figure 13.2 Mechanisms used by the pathogenic *Neisseria* to resist penicillin, tetracycline or ciprofloxacin. The mechanisms of penicillin resistance include the action of β-lactamase, decreased influx, increased efflux, and modifications of PBP1 and PBP2. Tetracycline resistance levels can be modulated by changes in influx and efflux. Plasmid-mediated tetracycline levels due to *tetM* are shown by TetM binding to the ribosome. Chromosomally mediated tetracycline resistance involves modification of ribosomal protein S10 at Val-57. Ciprofloxacin resistance is typically due to chromosomal mutations at *gyrA* and *parC*. A colour version of this figure is located in the plate section at the back of the book.

The highly efficient transformation system in both gonococci and meningococci likely accounts for the majority of intra- and inter-species transfer of chromosomal resistance genes in the community; early work by Sarubbi and Sparling (1974) demonstrated transfer of antibiotic resistance determinants during laboratory growth of mixed gonococcal cultures. Conjugal transfer of chromosomal DNA does not occur in *Neisseria,* but the 24.5-megadalton (MDa) plasmid present in certain gonococcal strains can mobilize itself and the β-lactamase plasmid at low frequencies (Eisenstein *et al.,* 1977; Genco *et al.,* 1984). New genetic determinants important for resistance also can be acquired from other microbes that colonize humans. This has occurred in the pathogenic *Neisseria,* as evidenced by the appearance of the TEM-1 β-lactamase-encoding plasmid in gonococci, the *tetM* gene (gonococci and meningococci), which is typically carried in other bacteria on the conjugative transposon Tn*916* (Morse *et al.,* 1986), and macrolide-resistance genes contained on conjugal transposons (Roberts *et al,* 1999); these genetic determinants are discussed in more detail below. Mosaic genes containing DNA sequences from commensal *Neisseria* can be generated by transformation; this seems to have been the case in the construction of mosaic *penA* (Spratt *et al.,* 1992; Bowler *et al.,* 1994) and *dhps* (Fermer *et al.,* 2006) gene sequences that are important in the development of resistance to penicillin and sulphonamides, respectively.

Enzymatic modification of antibiotics

Degradation of penicillin by the TEM-1-type β-lactamase is the best example of this resistance mechanism (Fig. 13.2); this enzyme cleaves the β-lactam ring of penicillin, rendering the antibiotic inactive. β-lactamase-producing *Neisseria gonorrhoeae* were first reported in 1976 from clinical isolates obtained in the Far East (Ashford *et al.,* 1976) and England (Percival *et al.,* 1976). Their rapid global spread was of great concern because they exhibited high levels of resistance to penicillin (MIC>50 µg/ml). Currently, over 50 to 80% of gonococcal isolates in some geographical locations, i.e. the Western Pacific, Latin America, and Africa, produce a β-lactamase, making penicillin therapy unreliable. PPNG possess either a 5.1-kb or 7.1-kb R-plasmid that contains a Tn2 transposable element carrying the *bla* gene, with the smaller plasmid predominating in β-lactamase-producing strains in Africa and Europe. These two plasmids are very closely related and the smaller plasmid was most likely generated by deletions of the larger plasmid during transformation (Sox *et al.*, 1979). The R-plasmids were most likely acquired by conjugal transfer from *Haemophilus parainfluenzae* (Elwell *et al.*, 1977; Roberts *et al.*, 1977), which carries a closely related R-plasmid, RSF0885.

A second example of enzymatic modification of an antibiotic in *Neisseria* was the discovery of chloramphenicol resistance in *N. meningitidis* (Galimand *et al.*, 1998). Chloramphenicol resistance is especially disconcerting because this antibiotic is still used to treat bacterial meningitis in developing countries. Between 1987 and 1996, strains exhibiting high-level chloramphenicol resistance (MIC > 64 µg/ml) were isolated from 12 patients (11 in Vietnam and one in France) and more recently, an additional two isolates have been identified in Australia (Shultz *et al.*, 2003). Resistance expressed by these isolates is due to the activity of chloramphenicol acetyltransferase (CAT), an enzyme that catalyses O-acetylation of chloramphenicol. This reaction prevents the interaction of chloramphenicol with the bacterial ribosome and renders the antibiotic ineffective. CAT is encoded by the *catP* gene, which is located on a truncated transposon, probably derived from Tn*4451*, a mobilizable element from *Clostridium perfringens* (Galimand, 1998). The exact mechanism by which *catP* was obtained by *N. meningitidis* remains unknown. However, the transposon is non-motile due to large deletions of adjacent genes (~80% of the transposon), suggesting that its acquisition was by transformation. Whatever the initial transfer event, the truncations have resulted in a stable element within the meningococcal genome, which, given the high rates of transformation and horizontal transfer in *Neisseria* species raises the likelihood of subsequent spread of chloramphenicol resistance to gonococci and other meningococci.

Decreased influx of antibiotics

The Gram-negative outer membrane (OM) represents an important permeability barrier to antibiotics, particularly hydrophobic compounds (Nikaido, 1994). Antibiotic influx across the OM can be altered by changes in lipopolysaccharide (LPS) structure or porins. Different degrees of truncation of the oligosaccharides of LPS (known as a lipooligosaccharide [LOS] in *Neisseria*) can increase entry of hydrophobic antibiotics. In contrast to the longer LPS produced by enteric bacteria, where deep-rough LPS mutants are hypersusceptible to antibiotics (principally hydrophobic compounds), LOS mutants of gonococci often display only small changes in antibiotic susceptibility and some mutations actually decrease susceptibility (Guyman *et al.*, 1982). However, truncation of gonococcal LOS can impact efflux of hydrophobic antimicrobials (HAs) mediated by the MtrC-MtrD-MtrE efflux system by an unknown mechanism (Lucas *et al.*, 1995).

Certain antibiotics, e.g. penicillin, diffuse into the periplasmic space of Gram-negative bacteria through OM channels formed by porin proteins; thus, loss of porins or subtle amino acid changes can influence antibiotic uptake. The ability of gonococci to modulate influx of penicillin and tetracycline through porins has been the subject of numerous studies. Early studies implicated the major OM protein (originally termed PI) as being important for entry of these antibiotics (Guymon *et al.*, 1978). Gonococci produce one of two allelic forms of porin: $PorB_{1A}$ or $PorB_{1B}$ (also referred to as PIA and PIB). Strains expressing $PorB_{1B}$ are slightly less susceptible than $PorB_{1A}$ strains to penicillin and tetracycline (Sparling *et al.*, 1975; Gill *et al.*, 1998; Olesky *et al.*, 2002). Moreover, certain point mutations that result in amino acid replacements in loop 3 of $PorB_{1B}$ (Gill *et al.*, 1998; Olesky *et al.*, 2002) were found to confer decreased susceptibility of gonococci to penicillin and tetracycline. The most studied mutations (collectively termed *penB*), which result in amino acid replacements at position 120 alone (G120K) or positions 120 and 121 (G120D/A121D), were proposed to decrease antibiotic entry based on studies of porins from other Gram-negative bacteria. Curiously, however, the phenotypic impact of *penB* is only apparent in an *mtrR* mutant strain that over-expresses the *mtrCDE*-encoded efflux pump. Thus, *penB* transformants, in the absence of an *mtrR* determinant, cannot be obtained on penicillin plates in transformation experiments, and replacement of *porB* in a penicillin-susceptible strain with a *penB* allele by allelic exchange does not increase penicillin resistance (Sparling *et al.*, 1975; Olesky *et al.*, 2006). Moreover, permeation studies of liposome-reconstituted wild type and mutant $PorB_{1b}$ porins showed no differences in either pore size or antibiotic-permeation rates. Taken together, these results cast doubt as to whether wild-type and mutant (*penB*) $PorB_{1B}$ proteins differ in their capacity to modulate permeation of penicillin or tetracycline. How *penB* mutations impact levels of gonococcal susceptibility to penicillin and tetracycline is not clear, but there most certainly seems to be some direct or indirect interaction (see below) with an over-expressed *mtr* efflux system.

PilQ is a doughnut-like multimeric structure in the OM composed of 12 identical 75-kDa subunits that is used by gonococci to secrete the growing pilus molecule (Drake and Koomey, 1995; Drake *et al.*, 1997). PilQ also appears to be critical in controlling the intake of antimicrobials by gonococci and meningococci since null or missense mutations in *pilQ* can significantly impact gonococcal susceptibility to structurally distinct antimicrobials (Chen *et al.*, 2004; Zhao *et al.*, 2005). With respect to penicillin, a mutation (*penC*) that decreases gonococcal susceptibility to penicillin mapped to *pilQ* (Zhao *et al.*, 2005), suggesting that PilQ plays a role in modulating penicillin influx. The *penC* missense mutation (later renamed *pilQ2*) appeared to decrease the stability of the PilQ dodecamer, which is normally stable to dissolution in SDS, since Western blots showed only the monomeric protein with no evidence of the high-molecular-mass multimeric form of PilQ. Interestingly, the *pilQ2* mutation had no effect on the MIC of penicillin (or other antibiotics such as tetracycline) in the absence of the *mtrR* and *penB* resistance determinants, suggesting that the rate of antibiotic influx through the PilQ complex in cells lacking *mtrR* and *penB* is only a small fraction of the rate through porins. Only when *mtrR* and *penB* mutations, which together decrease antibiotic levels in the periplasm, are present is the influx through PilQ a significant proportion of the total antibiotic flux. Although PilQ has a role in permeation of antibiotics, it is unlikely that *pilQ2*-like mutations or *pilQ* deletions have any role in clinical resistance, since these mutations disrupt proper piliation that is critical in infection and pathogenesis (see Chapters 5 and 6).

Efflux of antibiotics

The capacity of cells to export drugs from their interior was first described in the field of oncology with the observation that the P-glycoprotein could export anti-tumour agents (reviewed in Gottesman and Ling, 2006). Soon thereafter, the Levy laboratory (McMurray *et al.*, 1980) described an efflux pump in *Escherichia coli* that could export tetracycline and confer high levels of resistance to the antibiotic. It is now recognized (Nikaido, 1996) that most, if not all, bacteria harbour efflux pumps that export antimicrobials of various types and many have the ability to export structurally diverse, seemingly unrelated antimicrobials (Neykahf, 2001). Some bacteria have only a small number of pumps, while others (e.g. *Pseudomonas aeruginosa*) have an astonishingly high number of pumps of different types. Many efflux pumps are constitutively expressed, while others are not produced during bacterial growth in normal laboratory media or conditions, but are activated during growth under stress conditions (or perhaps during infection).

Bacterial efflux pumps can be grouped into the following main categories based on their overall composition and structure of their transporter protein and the entire pump system: (1) the major facilitator (MF) family, (2) the small multidrug resistance (SMR) family, (3) the resistance-nodulation-cell division (RND) family, (4) the multidrug and toxic compound extrusion (MATE) family, and (5) the ATP-binding cassette (ABC) family. MF, SMR and RND transporters are energized by the proton motive force (PMF) of the cytoplasmic membrane, while the ABC family of transporters utilizes ATP hydrolysis for energy. The MATE family of transporters requires an active PMF for efflux of antimicrobials and are Na^+/drug antiporters. Substrate recognition by efflux pumps is mediated by the cytoplasmic membrane transporter protein component of the pump, which likely captures antimicrobials on the cytosolic side of the cytoplasmic membrane, although structural analysis suggests that these pumps can also export compounds from the periplasmic face of the cytoplasmic membrane (see below). Certain efflux pumps are highly selective in substrate recognition, while others recognize a broad range of structurally diverse compounds. The structure of the AcrB transporter of *E. coli*, which on the amino acid sequence level is similar to MtrD in gonococci and meningococci (Hagman *et al.*, 1997; Rouquette-Loughlin *et al.*, 2004), has been solved (Murakami *et al.*, 2002), and more recently AcrB structures with bound substrates were obtained (Murakami *et al.*, 2006). These structures displayed multiple ligand-binding domains, and the presence of a vestibule on AcrB located close to the bilayer on the periplasmic side of the cytoplasmic membrane was also observed. These multiple binding domains provide a mechanism by which transporters can recognize a wide range of antimicrobials. For Gram-negative bacteria that need to export antimicrobials to the extracellular fluid, it is necessary to bridge the transporter to the outer membrane. This function is accomplished by a so-called membrane-fusion protein (AcrA in *E. coli* and MtrC in *Neisseria*) that interacts with an outer membrane protein channel (TolC in *E. coli* and MtrE in the *Neisseria*) (Delahay *et al.*, 1995).

Through classical genetic and gene sequence analysis approaches, four efflux pump systems (*mtrCDE*, *farAB*, *macAB* and *norM*), which appear to be produced by all isolates of *N. gonorrhoeae* and *N. meningitidis*, have been identified (Hagman *et al.*, 1995; Lee and Shafer, 1999; Rouquette-Loughlin *et al.*, 2003, 2005). These four pumps belong to the RND, MF, ABC, and MATE families, respectively (Fig. 13.3). Of these four efflux systems, the *mtrCDE*-, *macAB*- and *norM*-encoded efflux pumps recognize antibiotics relevant to the treatment of gonococcal or meningococcal infections. The *mef*-encoded efflux pump protein can also recognize macrolides (Luna *et al.*, 1999). The *mef* gene is harboured on a conjugative transposon (Luna *et al.*, 2000) but has only been found in a limited number of gonococci.

Based on increases in bacterial susceptibility to antibiotics that result from insertional-inactivation of the gene encoding the cytoplasmic membrane transporter component of the relevant pump, it is now recognized that the MtrC-MtrD-MtrE efflux pump can export structurally diverse hydrophobic antibiotics and penicillin (Hagman *et al.*, 1995; Veal *et al.*, 2002), while the NorM

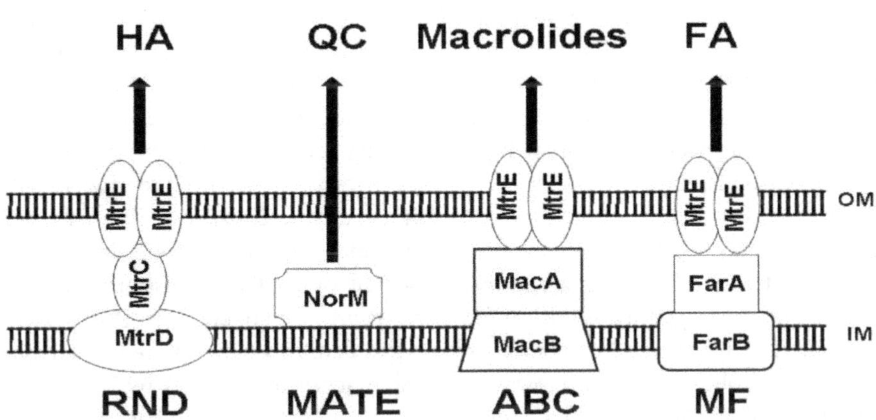

Figure 13.3 Summary of the four conserved efflux pumps of gonococci and meningococci. The classification of the family of pump that they belong to are labeled as RND (resistance/nodulation/division) for the MtrC-MtrD-MtrE system that exports hydrophobic antimicrobial agents (HA), MATE (multidrug and toxic compound extrusion) for the NorM efflux that can exports quinolones and quaternary compound, ABC (ATP-binding cassette) for the MacA–MacB efflux system that exports macrolides and the MF (major facilitator) for the FarA-FarB-MtrE efflux pump that exports long chain fatty acids (FA). The known protein components of each pump are shown. In the case of the MacA-MacB system the OMP channel has not yet been determined but may involve MtrE.

pump exports quinolones (Rouquette-Loughlin *et al.*, 2003) and the MacA–MacB pump exports macrolide antibiotics (Rouquette-Loughlin *et al.*, 2005). The MtrC-MtrD-MtrE and FarA-FarB-MtrE efflux pumps also recognize host-derived antimicrobials, including cationic antimicrobial peptides (Shafer *et al.*, 1998) and long-chain fatty acids (Lee *et al.*, 1999). A functionally intact *mtr* system is required for long term colonization of the mouse vaginal mucosal layer (Jerse *et al.*, 2003) and its capacity to export host antimicrobials has been suggested to be important in virulence (Rouquette-Loughlin *et al.*, 2001).

Bacterial efflux pumps that export antibiotics can have important implications for the efficacy of antibiotic treatment of infections (Nikaido, 1994). By reducing the effective intracellular or periplasmic concentration of an antibiotic, these pumps can render bacteria less susceptible to the antibiotic in question. In addition, if the infecting strain also possesses other mechanisms that decrease susceptibility to the antibiotic, the additive effect of the pump can increase the MIC to a level where antibiotic treatment can fail. This is especially true in isolates that contain mutations that impact efflux pump gene expression (see below). The expression of genes encoding bacterial efflux is subject to multiple layers of regulation and these regulatory processes can become important and relevant to antibiotic therapy when they are lost due to mutation. Indeed, prior to genomic DNA sequencing analysis, most bacterial efflux pumps were discovered by virtue of the impact of loss of regulatory systems that resulted in elevated resistance levels to antibiotics; this topic will be discussed in greater detail below.

Target modification/protection

Antibiotics are effective against bacteria because they often have a specific target that is critical for viability. By binding to their target, they block its function and the microbe succumbs. However, microbes can develop intermediate to high-level resistance through mutations that reduce or abrogate antibiotic binding to the target in question. Because these targets are critical for cell viability, the changes that occur must remodel the active site of the target to lower the affinity for antibiotic without greatly affecting the normal function of the enzyme. Provided that these changes do not alter the function of a target to such an extent that they have a detrimental impact on bacterial physiology, the mutations can persist even in the absence of antibiotic selection. With gonococci and meningococci, antibiotic resistance due to target alteration/protection has been most extensively studied for penicillin, ciprofloxacin, tetracycline and macrolides, which will be discussed below.

Penicillin resistance by target modification

The most extensively studied and perhaps best understood example of how target modification can impact levels of gonococcal susceptibility to penicillin deals with mutations in *penA* that decrease the penicillin acylation rate of PBP2. By themselves, *penA* mutations decrease gonococcal susceptibility to penicillin by four- to six-fold (Sparling *et al.*, 1975). It is now clear from the work of Spratt and co-workers that gonococci and meningococci acquired regions of *penA* sequences from commensal *Neisseria* that possess a PBP2 with a reduced rate of acylation by penicillin (Spratt, 1988; Brannigan *et al.*, 1990; Bowler *et al.*, 1994; Spratt *et al.*, 1992; Antignac *et al.*, 2001). These transformation events resulted in the construction of mosaic *penA* genes. Until recently, the most common alterations were insertion of an additional codon encoding an aspartate residue (Asp345a) that lies just downstream of the penicillin acylation site (Ser310) and 4–8 mutations in the carboxyl terminal region of the protein. Curiously, there is a strict requirement for the inserted amino acid to be an aspartate, as no other amino acid insertion is capable of conferring resistance (Brannigan *et al.*, 1990) These changes are sufficient to re-model PBP2 such that it has a reduced rate of acylation for penicillin but yet retains the transpeptidase activity that is essential for viability. A missense mutation (L421P) in *ponA*, which encodes PBP1, has also been identified; this mutation reduces the rate of penicillin acylation of PBP1 by 3- to 4-fold and can, in an appropriate strain background (Ropp *et al.*, 2002), further decrease gonococcal susceptibility to penicillin. However, *penA* and *ponA* mutations alone are insufficient for clinically significant levels of penicillin resistance in gonococci and require the additive effects of *mtrR*, *penB* and as yet undetermined mutations (see below) to generate a PenR gonococcus.

penA mutations can also influence levels of neisserial susceptibility to third generation cephalosporins (Takahata *et al.*, 2006). Recently, *penA* alleles that are very different than the ones described above have appeared in strains with increased resistance to third-generation cephalosporins such as cefixime and ceftriaxone. These alleles encode up to 60 mutations in PBP2 compared to wild type, and have a mosaic gene organization that appears to have been created following recombination with *penA* genes from the commensal species *N. perflava, N. sicca, N. cinerea,* or *N. flavescens* (Amayama *et al.* 2002; Tanaka *et al.*, 2006). The genetic mechanisms leading to increased cephalosporin resistance have not yet been elucidated; however, while it seems clear that at least a large portion of this resistance is due to these mosaic *penA* alleles, most likely other chromosomal mutations (such as those important for penicillin resistance) are also required to obtain full intermediate-level resistance.

Tetracycline resistance by target modification

Tetracycline resistance in bacteria can occur through efflux or ribosome modification. Ribosome modification can be due to two mechanisms: alterations in ribosomal protein structure that reduce affinity for tetracycline or ribosomal protection by TetM or TetO proteins, which block the capacity of tetracycline to inhibit protein synthesis.

High-level tetracycline-resistance (≥ 16 μg/ml) in gonococci and meningococci is due to the presence of the *tetM* gene. TetM, which resembles elongation factors involved in protein synthesis, confers high-level resistance to tetracycline by binding to ribosomes and causing the release of the antibiotic from ribosomes (Burdett, 1986, 1991), thus permitting protein synthesis to proceed (Fig. 13.2). Indeed, ribosomes isolated from TetM-producing bacteria are resistant to inhibition of protein synthesis by tetracycline *in vitro*. In Gram-positive bacteria, *tetM* is carried on the 16.4-kb conjugative transposon Tn*916* (Clewell and Gawron-Burke, 1986). Conjugative transposons are able to self-mobilize and can be transferred to a broad range of hosts by conjugation. *tetM* initially established itself in gonococci by integrating into the 24.5-MDa conjugative plasmid to produce a 25.2-MDa plasmid (Knapp *et al.*, 1988 and Roberts and Knapp, 1988). Once established, it was stably maintained. It can be transferred to other gonococci and meningococci by conjugation. Many commensal neisseriae naturally harbour a chromosomal copy of *tetM* (Knapp *et al.*, 1988) and it has been suggested that they acquired it from bacteria other than gonococci. The 25.2-MDa plasmid can also be maintained in gonococcal strains with the smaller β-lactamase plasmid, providing high-level resistance to both tetracycline and penicillin; such strains have spread globally. It is interesting to note that while the 25.2-MDa plasmid has been detected in meningococcal strains, the natural 24.5-MDa gonococcal conjugative plasmid has not been detected in this pathogen. Tn*916* can enter *Neisseria* by transformation and plasmids containing Tn*916* have been used to construct random mutations in the meningococcal chromosome (Stephens *et al.*, 1991). Tn*916* can insert into the meningococcal chromosome through two processes (Swartley *et al.*, 1993). Class I insertions utilize a transposition mechanism to integrate the complete Tn*916* into the chromosome, similar to the mechanism utilized by Gram-positive bacteria (Scott *et al.*, 1988; Caparon and Scott, 1989). In contrast, Class II insertions have large deletions at both arms of Tn*916* that frequently eliminate genes necessary for mobilization and conjugation.

High-level chromosomally mediated tetracycline resistance in gonococci (≥ 2 μg/ml) is due to mutations that impact ribosomal protein structure. Early studies by Sparling and colleagues (1975) showed that when the *tet-2* mutation is present in strains with co-resident *mtrR* and *penB* mutations, gonococci are endowed with resistance to tetracycline. Hu *et al.* (2005) cloned *tet-2* and showed that it encoded ribosomal protein S10. This allele of the *rpsJ* gene was found to contain a single point mutation that changed Val-57 to Met in S10; missense mutations that resulted in substitutions of leucine or glutamine were similarly effective in conferring resistance. Val-57 in S10 maps to the vertex of a small loop near the aminoacyl-tRNA region that forms the tetracycline binding site. Hu *et al.* suggested that replacement of the normal Val-57 with large uncharged amino acids would alter rRNA structure, which would thereby reduce the affinity of tetracycline for the ribosome. Importantly, they found the identical mutation in clinical isolates of *N. gonorrhoeae* that expressed tetracycline-resistance and thereby provided a mechanism for tetracycline resistance in TetR and CMRNG gonococci.

Ciprofloxacin resistance by target modification

Quinolone antibiotics target bacterial topoisomerases, which are involved in relaxing supercoiled DNA that forms during replication. The activity of these enzymes is critical for segregation of bacterial chromosomes. Bacteria develop high-level resistance to quinolones through mutations that alter antibiotic recognition of the target enzyme. DNA gyrase consists of a heterotetramer of GyrA and GyrB; mutations in *gyrA* are associated with high-level resistance, whereas mutations in *gyrB* confer lower levels of resistance. The *gyrA* mutations reduce quinolone binding affinity, rendering the enzyme (and the bacteria) resistant to its inhibitory effect. Topoisomerase IV is encoded by the unlinked *parC* and *parE* genes; its enzymatic activity also can be inhibited by quinolone antibiotics, although higher concentrations are required than that needed to inhibit DNA gyrase *in vitro*.

By sequentially passing cells in the presence of increasing amounts of ciprofloxacin, Belland *et al.* (1994) isolated a series of spontaneous mutants of *N. gonorrhoeae* strain FA19 that exhibited different degrees of resistance to ciprofloxacin and nalidixic acid. They observed that mutations in *gyrA* alone provided low- to intermediate-level resistance, but high-level resistance required a co-resident mutation in *parC*. These mutations could easily be transferred to other gonococci by transformation. Levels of resistance were attributed to different amino acid replacements in GyrA and ParC. Thus, a missense mutation at codon 91 in *gyrA* (Ser-91 to Phe-91), which is located within the 'quinolone-resistance determining region' in *E. coli* GyrA, conferred a 100-fold increase in resistance to ciprofloxacin and nalidixic acid. Subsequent mutation at codon 95 (Asp-95 to Asn-95) further increased ciprofloxacin resistance by two-fold. Higher levels of quinolone resistance required mutations in *parC* in addition to those in *gyrA*. These *parC* mutations mapped to codons 88 (Ser-88 to Pro-91) and 91 (Glu-91 to Lys-91). A number of GyrA/ParC amino acid alteration patterns were identified by Trees *et al.* (1999) in ciprofloxacin-resistant strains isolated in the United States and Far East, with the GyrA-91,95/ParC-Asp-86→Asn pattern having the highest prevalence. The important point from their report, however, was that the emergence and spread of ciprofloxacin-resistant strains was not due to the expansion of a single clone that successfully disseminated globally, but rather to the emergence of strains with different mutations and levels of resistance.

Macrolide resistance by target modification

Macrolides, e.g. erythromycin, inhibit protein synthesis in both intact bacteria and cell-free extracts by binding to the 50S ribosomal subunit, preventing translocation of the peptidyl-tRNA, and causing ribosomes to release incomplete polypeptides. rRNA methylases block macrolide binding to 23S rRNA by methylating an adenosine residue at position 2058 (in *E. coli*); this site is in the peptidyl transferase domain. rRNA methylase-encoding genes (collectively referred to as *erm* genes) can be carried by conjugative transposons and certain *erm* genes appear to have entered gonococci in the early 1990s (Roberts *et al.*, 1999). In gonococci, *erm* genes can confer high-level resistance to erythromycin (4–16 μg/ml) or decreased susceptibility to azithromycin (1–4 μg/ml) in the absence of *mtrR* mutations (see below). Gonococcal isolates from Seattle, WA, and Uruguay were reported by Roberts *et al.* (1999) to contain rRNA methylase genes *ermF* or *ermB* and *ermF*; commensal macrolide-resistant *Neisseria* strains can harbour *ermB*, *ermC* and/or *ermF* genes. *ermF* was reported to be part of a complete conjugative element in gonococci and its nucleotide sequence was 97% identical over 374 bp to *ermF* of *Bacteroides fragilis*. The gonococcal-maintained *ermF* could be conjugally transferred to other gonococci, meningococci and *Enterococcus faecalis*. The presence of *ermF* in clinical isolates of gonococci supports the notion that this strict human pathogen can acquire/donate resistance determinants to/from other bacteria that exist at various sites in the human host. Moreover, the capacity of gonococci to conjugally transfer *ermF* to meningococci emphasizes how gene exchange mechanisms can expand the repertoire of antibiotic resistance expressed by these pathogens.

Interplay of resistance mechanisms (or how penicillin met its demise)

Certain mutations by themselves can increase antibiotic resistance to a level that result in treatment failure for the antibiotic in question. In other instances, however, the increase in resistance to a single determinant or mutation is small and is of less clinical significance. However, whereas acquisition of a single resistance determinant may confer only an incremental increase in resistance, it is the cumulative effects of these determinants and their complex interactions that ultimately result in clinically relevant levels of antibiotic resistance. This is precisely the scenario that ultimately resulted in the demise of penicillin as an effective (and cheap) treatment for gonorrhoea. The cumulative effect of chromosomal mutations on levels of gonococcal susceptibility to penicillin was evident from the pioneering work of Sparling and coworkers (Sparling *et al.*, 1975; Faruki and Sparling, 1986), long before gene products of the wild type alleles of the mutated genes were known. More recent studies have identified these genes and associated mutations, but even today the mechanisms of high-level penicillin resistance remain unclear.

The genetics of the stepwise transfer of chromosomally mediated resistance determinants from a resistant clinical gonococcal isolate to a penicillin-susceptible laboratory strain is studied by using DNA isolated from a high level penicillin-resistant CMRNG strain to sequentially transform a penicillin-susceptible laboratory strain (e.g. FA19) to higher levels of resistance (Table 13.1). When FA19 is transformed with DNA from a CMRNG strain and transformants are selected on penicillin concentrations slightly above the MIC, the first resistance determinant transferred is *penA* (Fig. 13.2). The fact that *penA* is the first gene transferred is consistent with the observation that in penicillin-susceptible strains, PBP2 has a ~10-fold higher rate of acylation than PBP1 and thus is the killing target of penicillin at concentrations near the MIC (these two PBPs are the only two essential PBPs of *Neisseria* species). The next determinant transferred, *mtrR*, results in overexpression of the MtrC-MtrD-MtrE efflux pump, which confers resistance to a wide range of hydrophobic compounds and antibiotics. *mtrR* is usually selected for by plating in the presence of erythromycin or detergents such as Triton X-100. The level of penicillin resistance conferred by *mtrR* is at most two-fold, but it is required for high-level resistance (see below). The third step in resistance to penicillin is mediated by *penB*, which encodes amino acid changes in $PorB_{1B}$. *penB* mutations also increase resistance to a broad range of antibiotics that depend on porins to diffuse into the periplasmic space. As described above, *penB* mutations can only be selected for in strains also carrying the *mtrR* determinant (Shafer and Folster, 2006).

While the first three determinants are transferred at high efficiency, transformation of a third-level transformant (MIC_{pen} = 0.75–1 µg/ml) up to the level of resistance of clinical isolates (MIC_{pen} = 4 µg/ml) is very difficult to achieve in the laboratory, and as a consequence the genes involved have not been identified (Foruki and Sparling, 1986). One possibility for a candidate gene involved in high-level penicillin resistance is *ponA*, which encodes the other essential PBP (PBP1) in gonococci. Indeed, Ropp *et al.* (2002) identified a mutant *ponA* allele containing a single missense mutation in a set of high-level penicillin-resistant strains that lowered the rate of acylation for penicillin, but introduction of this determinant into a *penA mtrR penB* transformant had no effect on resistance. However, it was shown that replacement of a mutant *ponA* allele (encoding mutant PBP1) in CMRNG strains with the wild type gene reduced the MIC two-fold. Thus, mutation of PBP1 appears to have a role in resistance, but clearly other unknown determinant(s) are necessary for *ponA* to increase resistance up to levels found in penicillin-resistant clinical isolates. The difficulty in transforming strains to higher level penicillin resistance perhaps suggests that multiple genes are required; if so, this would suggest that the individual determinants have no effect on resistance, and that these genes, much like *mtrR* and *penB*, have complex interactions that require one other to exert their function effects.

Table 13.1 Stepwise transfer of resistance determinants from a clinical penicillin-resistant isolate (FA6140) to a penicillin-susceptible laboratory strain (FA19). The strains go from most susceptible (FA19, bottom) to least susceptible (FA6140, top). DNA from FA6140 was used to sequentially transform FA19 to higher resistance. The first three determinants are readily transferred and well characterized, although the mechanism behind the functional interaction between *mtrR* and *penB* is not clear. Transformation of a third step transformant to higher resistance with FA6140 DNA does not occur (–) and the MIC of the third step transformant, FA19 *penA mtrR penB*, is still 4- to 5-fold lower than the donor strain. Thus, other determinants must be present that increase the MIC to high level as is indicated by the ? marks under the MIC heading. The *ponA* gene is required for full resistance, but evidently requires these other unknown determinants to exert its functional effect.

Strain	Description	Mechanism of resistance determinant	MIC (μg/ml)
FA6140	Clinical isolate from Durham, NC, outbreak	–	4.0
FA6140 *ponA*$_{WT}$	Reversion of mutant ponA allele to a wild-type allele	Mutation in PBP1 that lowers the rate of acylation by penicillin	2.0
–	–	Undefined determinant(s) that modulate resistance; not transformable in the laboratory	???
FA19 *penA mtrR penB*	Third step transformant	Mutations in the constriction loop of the major outer membrane porin (PorB$_{1B}$); functional only with a co-resident *mtrR* mutation	0.75
FA19 *penA mtrR*	Second step transformant	Disregulation/overexpression of the MtrC-MtrD-MtrE efflux pump	0.12
FA19 *penA*	First step transformant	Mutations in PB 2, the lethal target of penicillin, that decrease the ability of penicillin to covalently inactivate the enzyme	0.06
FA19	Penicillin-susceptible laboratory strain	–	0.01

Other antibiotic resistance properties also can be influenced by the interplay of multiple mutations. For instance, as with penicillin, resistance to tetracycline can result from decreased entry due to *penB*, increased efflux due to *mtrR* mutations and decreased target recognition due to missense mutations in *rpsJ* (Sparling *et al.*, 1975; Gill *et al.*, 1998; Olesky *et al.*, 2002; Hu *et al.*, 2005). Mutations that impact efflux pump gene expression (see below) in conjunction with other resistance mechanisms can influence neisserial susceptibility to quinolones and macrolides. The important point is that lessons learned with penicillin should allow us to anticipate that with currently used antibiotics (e.g. cephalosporins), gonococci and meningococci are likely to develop multiple mutations that by themselves may not be clinically significant but their combined action can be of great consequence.

Regulation of resistance genes

Bacteria often regulate expression of their antibiotic resistance genes. Control of resistance gene expression may help to counteract a possible fitness cost, in the absence of antibiotics, observed with certain resistance mechanisms. Regulation can be constitutive or inducible, and both *trans*- and *cis*-acting factors have been identified in gonococci and meningococci as being important in gene regulation. Perhaps the best understood example of how regulatory systems can impact antibiotic resistance levels expressed by gonococci deals with *trans*- and *cis*-acting elements that modulate expression of *mtrCDE* (Fig. 13.4).

The *mtr* system was originally proposed to modulate the influx of structurally diverse hydrophobic antimicrobial compounds into gonococci by altering the permeability property of the OM (Guymon *et al.*, 1978). It was identi-

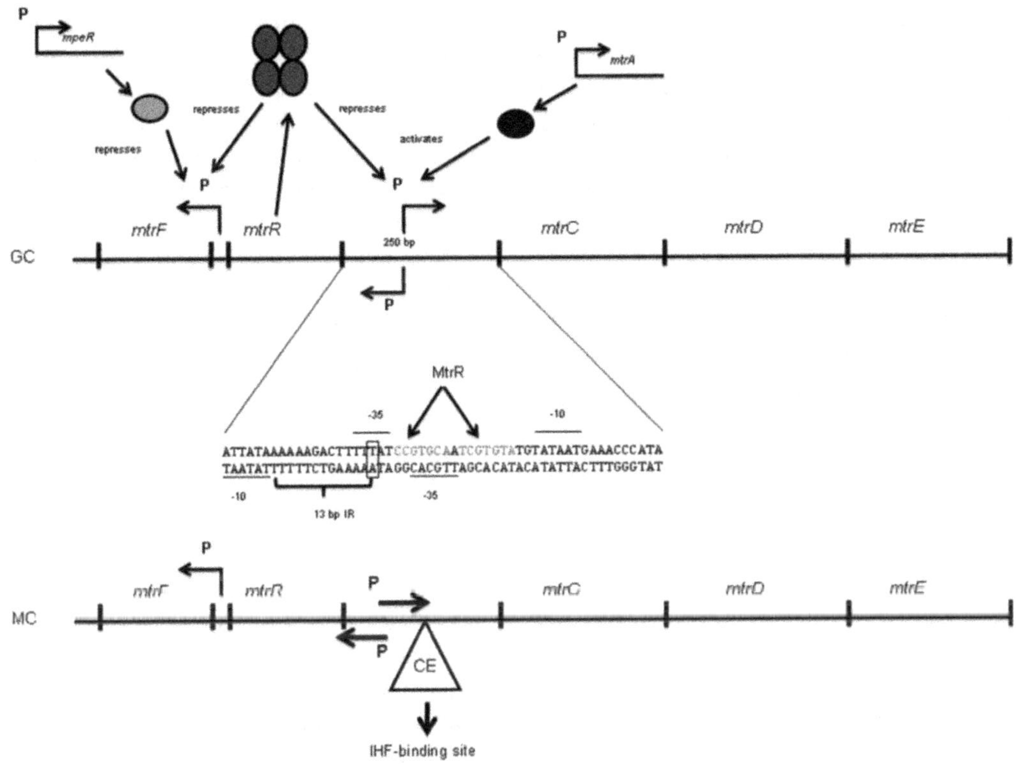

Figure 13.4 Shown is a summary of the mechanisms used by gonococci (GC) and meningococci (MC) to control expression of the *mtrCDE*-encoded efflux system. The position of genes is shown with promoter elements identified by a P and the direction of transcription is shown by a bent arrow. Two dimers of the MtrR repressor bind to the promoter region of *mtrC* at indicated sites (shown in red) in the expanded nucleotide sequence. MtrR and MpeR can also repress transcription of *mtrF* as indicated. The *cis*-acting 13-bp inverted repeat element within the *mtrR* promoter is shown and the site of the single bp deletion that results in high-level expression of *mtrCDE* is indicated by a box. Expression of *mtrCDE* in GC can also be enhanced by MtrA as indicated and binding of MtrA to this region has been inferred. Expression of *mtrCDE* in MC is independent of MtrR and MtrA and involves a 155–159 by Correia element (CE). Integration host factor (IHF) binds to a site within the CE and this negatively controls *mtrCDE* expression. A colour version of this figure is located in the plate section at the back of the book.

fied as such because a single step mutation event increased gonococcal resistance to structurally diverse hydrophobic drugs, dyes and detergents (Maness and Sparling, 1973). Mutations that reversed this resistance and actually rendered gonococci hypersusceptible to these hydrophobic agents (HA) were subsequently identified and such mutants were termed *env* (Sarubbin *et al.*, 1975; Guyman *et al.*, 1975). Pan and Spratt (1994) cloned and sequenced a region of the chromosome from an *mtr* strain that conferred HA-resistance when introduced into a wild type, HA-susceptible strain and found that resistance was due to a missense mutation in a gene that encoded a DNA-binding protein (MtrR) similar to the TetR family of repressors. MtrR contains a helix–turn–helix (HTH) domain (residues 32–53) in the *N*-terminal half and mutations in the HTH-encoding domain were subsequently found to inhibit its DNA-binding activity (Lucas *et al.*, 1997). MtrR acts as two homodimers (Fig. 13.4) (Hoffman *et al.*, 2005) and binds to a DNA sequence that lies just upstream of the *mtrCDE* operon (Fig. 13.4) (Lucas *et al.*, 1997). Mutations in the *mtrR* HTH-coding region as well as missense mutations in the C-terminal-encoding region, which likely impact dimer formation, can enhance transcription of *mtrCDE* and elevate gonococcal resistance to HAs. MtrR binds to the *mtrCDE* promoter in a specific manner and a 15 bp binding site containing two pseudo-direct repeats has been identified (Hoffmann *et al.*, 2005).

In addition to regulating *mtrCDE*, MtrR can positively or negatively regulate other gonococcal genes, including *ponA* and *pilMNOPQ* (Folster *et al.*, 2007) and its regulatory activity modulates gonococcal fitness in a murine vaginal infection model (see below). The *mtrRCDE* locus is also possessed by *N. meningitidis* and the efflux pump is functional (Roquette-Loughlin *et al.*, 2004). However, all meningococci examined to date encode an MtrR protein that is predicted to be non-functional due to nonsense mutations that prematurely truncate the protein or missense mutations that abrogate DNA-binding; as described below, meningococci use an insertion sequence between *mtrR* and *mtrCDE* to control expression of the efflux pump genes.

Another transcription factor (MpeR) is important in regulating expression of HA-resistance through its ability to negatively regulate *mtrF* expression (Fig. 13.4) (Folster *et al.*, 2005). *mtrF* encodes a putative cytoplasmic membrane protein that is required by the MtrC-MtrD-MtrE efflux pump to express high levels of HA-resistance (Veal and Shafer, 2003). MpeR contains an HTH at its C-terminal end and is similar to other DNA-binding proteins in the AraC family of regulators. MtrR and MpeR both regulate *mtrF* expression by independent processes (Fig. 13.4).

Integration host factor (IHF) is a well-described DNA-binding protein in bacteria that bends DNA. IHF negatively regulates *mtrCDE* efflux pump gene expression in meningococci (Rouquette-Loughlin *et al.*, 2004) by binding to a sequence within a Correia element (CE) (Correia *et al.*, 1988) that is positioned upstream of *mtrCDE* (Fig. 13.4); a CE upstream of *mtrCDE* is absent in most gonococci (see exception below). IHF also binds and negatively regulates the *farAB* efflux-encoding operon in gonococci and acts with another transcriptional repressor (FarR) to control expression levels of this efflux pump, which exports antibacterial long chain fatty acids (Lee *et al.*, 2006).

The *mtrCDEF* efflux pump locus can be inducibly expressed when gonococci are grown in the presence of a sublethal level of an antimicrobial agent recognized by the pump (Rouquette *et al.*, 1999). This induction is independent of MtrR but requires the presence of an AraC/XylS-like protein termed MtrA (Fig. 13.4). Transcriptional activation of efflux pump genes is common in bacteria. For example, environmental and antimicrobial stimuli have been shown to modulate expression of the *acrAB* efflux pump system in *E. coli* through the action of several DNA-binding proteins (MarR, Rob and SoxS) (Ma *et al.*, 1996). This inducible resistance process is typically transient and allows for a quick response of bacteria to antimicrobials in their local environment. A high percentage of gonococci bear an 11 bp deletion at the 5′-end of the *mtrA*-coding sequence that prematurely truncates the protein, and such strains are non-inducible. Meningococci possess an intact *mtrA* gene but fail to activate *mtrCDE*, for unknown reasons, when grown in the presence of antimicrobials. MtrA, like MtrR, can directly or indirectly regulate (positively or negatively) several genes in gonococci (Johnson and Shafer, unpublished) and an intact *mtrA* gene in gonococci enhances fitness in a murine infection model (Warner *et al.*, 2007).

While *mtrR*-coding mutations can enhance transcription of *mtrCDE* and elevate HA-resistance in gonococci, high-level resistance requires a *cis*-acting mutation within the overlapping *mtrR* and *mtrCDE* promoters (Hagman and Shafer, 1995). The *mtrR* promoter contains a 13 bp inverted repeat sequence between the −10 and −35 hexamers (Fig. 13.4). A single bp deletion (Hagman and Shafer, 1995) or a dinucleotide insertion (Zarantonelli *et al.*, 2001) within this inverted repeat is sufficient to dramatically reduce transcription of *mtrR*; these *cis*-acting promoter mutations are observed in clinical isolates expressing high levels of HA-resistance. It has been proposed (Hagman and Shafer, 1995; Lucas *et al.*, 1997) that because the *mtrR* and *mtrCDE* promoters overlap on opposite strands at their −35 hexamer region (Fig. 13.4), the single bp deletion or the dinucleotide insertion enhances *mtrCDE* expression by both reducing *mtrR* expression and making the *mtrCDE* promoter more available for interacting with RNA polymerase.

Transcription of *mtrCDE* in meningococci utilizes the same promoter as in gonococci (Rouquette-Loughlin *et al.* 2004). Since all meningococci examined thus far are natural *mtrR* mutants, we expected that transcription of the efflux pump operon would be similar to *mtrR* mutants of gonococci. However, this was not found to be the case and instead, a novel regulatory process was uncovered. Central to the regulation of *mtrCDE* in meningococci is a 155–159 bp CE (Fig. 13.4). This element is positioned numerous times in the gonococcal and meningococcal chromosomes (Correia *et al.*, 1988). In meningococci, but not most gonococci (exceptions discussed below), it is positioned between the *mtrR* and *mtrCDE* genes just downstream of the *mtrCDE* promoter (Rouquette-Loughlin *et al.*, 2004). It appears that it is the major mechanism by which the *mtrCDE* operon in meningococci is regulated. This regulation requires an intact IHF-binding site since its deletion from the CE enhances *mtrCDE* transcription in meningococci. The CE also modulates expression post-transcriptionally through RNaseIII cleavage at sites located within inverted repeats.

A group of gonococcal isolates from Kansas City, MO, were identified (Johnson *et al.*, 2003) that expressed resistance to azithromycin (MIC of 2–4 µg/ml) and erythromycin (4–8 µg/ml) by a mechanism that required an active MtrC-MtrD-MtrE efflux pump. These strains are of particular interest because, unlike most gonococci, they contain a CE between *mtrR* and *mtrCDE*, much like what is observed in meningococci (Rouquette-Loughlin *et al.*, 2004). Some of these macrolide-resistant gonococci also had nonsense or missense mutations in the *mtrR*-coding region that would result in truncation of MtrR, but transformation experiments revealed that resistance was conferred by the CE alone. It is not yet clear how levels of azithromycin resistance mediated by the MtrC-MtrD-MtrE pump are regulated by the CE (D. Trees, personal communication). However, the emergence of these isolates suggests that either a gonococcal CE element located elsewhere on the chromosome was re-positioned to the *mtr* locus or that meningococcal CE DNA sequences containing flanking *mtr* DNA was imported and recombined at this site.

Mutations positioned upstream of the *macAB*- and *norM*-encoded efflux pump genes have been shown to modulate levels of neisserial susceptibility to antibiotics by altering gene expression (Rouquette-Loughlin *et al.*, 2003, 2005). The MacA-MacB efflux system in gonococci and meningococci is similar to the ABC-type MacA-MacB efflux system previously recognized in *E. coli* (Kobayashi *et al.*, 2001), which exports macrolides. The *macA* and *macB* genes in gonococci and meningococci are organized as an operon and the start point of *macAB* transcription lays 37 nucleotides upstream of the translational start codon. Inspection of the putative −10 and −35 hexamer sequences, which are separated by an optimal 17 nucleotides, revealed that the −10 sequence (5′-TA<u>G</u>AAT-3′) contained a point mutation (the underlined G instead of the normal T) in strain FA19 as well as eight other gonococci (including the sequenced strain FA1090). This natural T→G mutation was found to dampen transcription of *macAB*, because transcription was enhanced when it was replaced by the normal T. Importantly, the G→T mutation also increased gonococcal resistance to macrolides by 10–30 fold when it was placed into a strain that lacks the MtrC-MtrD-MtrE efflux pump.

The NorM efflux pump is a Na^+-drug antiporter in the MATE family. It was observed that a point mutation in the −35 hexamer (5′-<u>C</u>TGACG-3′ to 5′-<u>T</u>TGACG-3′) in the gonococcal and meningococcal *norM* promoter could enhance transcription of *norM*, resulting in decreased neisserial susceptibility to norfloxacin and ciprofloxacin (Rouquette-Loughlin *et al.*, 2003). Resistance levels could be further increased by a second mutation that mapped to the ribosome binding site (5′-TG<u>A</u>A-3′ to 5′-TG<u>G</u>A-3′). While the promoter and ribosome binding site mutations could decrease neisserial susceptibility to fluoroquinolones, the levels of resistance were insufficient by themselves to provide clinical resistance; however, they could be significant in strains expressing a level of ciprofloxacin-resistance that is near the MIC breakpoint.

Antibiotic resistance and neisserial fitness

Although strains expressing resistance to an antibiotic have an advantage both *in vitro* and *in vivo* over antibiotic-susceptible strains in the presence of the antibiotic in question, they are frequently less fit in the absence of the antibiotic.

However, compensatory mutations that restore fitness but do not change antibiotic resistance occur frequently in the laboratory and probably in the microbe's natural environment. It is noteworthy that several years after antibiotics are removed from the gonorrhoea treatment guidelines, resistant strains continue to be isolated; this is particularly true in the case of penicillin- and tetracycline-resistant strains in the United States (CDC, 2007). This suggests that mechanisms of resistance are unlikely to have a negative impact on bacterial fitness in human hosts.

The antibiotic resistance literature is replete with examples of how resistance can decrease fitness *in vivo* in the absence of antibiotic pressure. There are a few examples, however, where resistance can actually enhance fitness *in vivo*; one of these is the gonococcal *mtrCDE* efflux pump. Using an experimental murine vaginal infection model, Jerse and co-workers (2003) determined that null mutations in *mtrD* or *mtrE* decrease the capacity of strain FA19 to cause a sustained (12 day) infection. This was likely the result of the inability of the infecting strain to export host-derived HA (such as antimicrobial peptides and/or progesterone) recognized by the pump (Hagman *et al.*, 1997; Shafer *et al.*, 1998). Interestingly, recent results (Warner *et al.*, 2007) from competition experiments that employed wild type strain FA 19 and an isogenetic derivative with an *mtrR* null mutation revealed that loss of MtrR, which results in over-expression of *mtrCDE* and enhanced antimicrobial resistance, greatly increased gonococcal fitness *in vivo*. In contrast, loss of *mtrA*, which would abrogate the HA-inducible resistance phenotype, decreased fitness *in vivo*; however, this decrease in fitness could be reversed by second site mutations that mapped to *mtrR*. While additional genetic experiments using mutant efflux pumps documented the importance of the pump in these changes in fitness, other MtrR- and/or MtrA-regulated genes may also contribute. In collaboration with D. Dyer and L. Jackson (Oklahoma Health Sciences Center), we have determined that these DNA-binding proteins can control numerous gonococcal genes and some of these may be important in fitness. Indeed, MtrR has been shown to bind to DNA sequences upstream of *ponA* and *pilMNOPQ* (Folster *et al.*, 2007), as well as *rpoH* (Folster *et al.*, 2009).

There is suggestive clinical evidence that efflux pumps likely enhance the ability of gonococci to survive during human infection. First, *mtrR* mutants are frequently isolated from the rectum of infected patients (Morse *et al.*, 1982; Shafer *et al*, 1995; Xia *et al.*, 2000), presumably because this environment is rich in HAs such as long chain fatty acids and bile salts. The MtrC-MtrD-MtrE efflux pump is also likely to help gonococci evade innate immune responses involving antimicrobial peptides, since the pump can recognize the human cathelicidin LL-37 (Shafer *et al.*, 1998). Bloodstream isolates from patients with disseminated gonococcal infections (DGI) are often hypersusceptible to HAs and penicillin (Eisenstein and Sparling, 1978) due to small deletions in *mtrC* or *mtrD* (Veal *et al.*, 1998) that render the pump inactive. However, these strains also frequently contain phenotypically suppressed *mtrR* mutations and can donate HA-resistance by transformation to other strains (Eisenstein and Sparling, 1978; Shafer *et al.*, 1995; Veal *et al.*, 1998).

Conclusions and future trends

Since the mid-1980s, mechanisms of antibiotic resistance discussed in this chapter have significantly impacted options for effective treatment (and, by inference, control) of gonococcal infections. As inexpensive and available antibiotics (penicillin, tetracycline, and ciprofloxacin) have been replaced due to resistance, more expensive and not always available antibiotics such as ceftriaxone and cefixime have taken their place and have proven to be effective. However, based on recent reports of strains with decreased susceptibility to these cephalosporins, it is not clear how long these antibiotics will remain in the clinician's arsenal. We should anticipate and plan for the worst case scenario. However, this is not an easy task given the dearth of new antibiotics coming into clinical practice. Fortunately, at present the situation with meningococci is not as gloomy, but there is reason for concern given a clear trend towards decreasing antibiotic susceptibility in this pathogen. In some instances, full resistance of isolates to certain antibiotics used in therapy or chemoprophylaxis has been reported, and the continued emergence and evolution of such meningococci to higher levels of resistance should be expected.

Based on our current understanding of the molecular mechanisms of antibiotic resistance utilized by gonococci and meningococci and their recent trends towards decreased susceptibility to cephalosporins, we predict that both pathogens will develop chromosomal mutations in genes that render these antibiotics less reliable or ineffective during the next 10–20 years. We imagine that these mutations will impact cephalosporin uptake, target recognition and/or efflux. How can this be countered? As we learn more about the structural biology of bacterial targets for antibiotics or those that participate in resistance, new opportunities for antibiotic development should become available. Genome sequencing and analysis projects coupled with insights gained from physiological experiments may identify novel bacterial targets and provide opportunities for rational design of drugs. Advances in drug chemistry will also be important and some progress in this area has been made. For instance, the capacity of the non-β-lactam lactivicins, which inhibit penicillin-binding proteins (Macheboeuf *et al.*, 2007), is an example of a drug that could circumvent mutations that impact bacterial susceptibility to β-lactams. As efflux of antibiotics is an important way that bacteria develop decreased susceptibility to antimicrobials, it may be possible to develop inhibitors of these pumps (Lomovskaya and Watkins, 2002) allowing for a return of certain antibiotics to clinical practice; this would also have the added benefit of decreasing bacterial resistance to mediators of innate host defence mechanisms that are also substrates of efflux pumps. Clearly, new approaches are needed if we are to meet the challenge of having reliable chemotherapeutic agents for the treatment of gonococcal or meningococcal infections. It may be helpful to remember that 'necessity is the mother of invention.' We hope that the prospect of returning to the pre-antibiotic age for many bacterial infections is reason enough that governments, industry and academia will marshal their forces to combat this important problem in public health.

Web resources

Neisseria gonorrhoeae strain FA1090 genome sequence: http://www.genome.ou.edu/

Neisseria meningitidis strains FAM18 and Z2491 genome sequences: http://www.sanger.ac.uk/Projects/N_*meningitidis*

Neisseria meningitidis strain MC58 genome sequence: http://cmr.jcvi.org/tigr-scripts/CMR/GenomePage.cgi?database=gnm

Gonococcal Isolation Surveillance Project (GISP): http://www.cdc.gov/std/gisp/

References

Ameyema, S., Onodera, M., Takahata, S., Minami, N., Maki, K., Endo, H., Goto, H. Suzuki, H., and Oishi, Y. (2002). Mosaic-like structure of penicillin-binding protein 2 gene (*penA*) in clinical isolates of *Neisseria gonorrhoeae* with reduced susceptibility to cefexime. Antimicrob. Agents Chemother. *46*, 3744–3749.

Antignac, A., Kriz, P., Tzanakaki, G., Alonso, J.-M., and Taha, M.-K. (2001). Polymorphism of *Neisseria meningitidis penA* gene associated with reduced susceptibility to penicillin. J. Antimicrob. Chemother. *47*, 285–296.

Ashford, W.A., Golash, R.G., and Henning, V.G. (1976). Penicillinase producing *Neisseria gonorrhoeae*. Lancet *2*, 657–658.

Bala, M., Ray, K., Gupta, S.M., Muralidhar, S., Jain, R.K. (2007). Changing trends of antimicrobial susceptibility patterns of *Neisseria gonorrhoeae* in India and the emergence of ceftriaxone less susceptible *N. gonorrhoeae* strains. J. Antimicrob. Chemother. *60*, 582–586.

Belland, R.J., Morrison, S.G., Ison, C., and Huang, W.M. (1994). *Neisseria gonorrhoeae* acquires mutations in analogous regions of *gyrA* and *parC* in fluroroquinolone-resistant isolates. Mol. Microbiol. *14*, 371–380.

Boras, A., Bozinovic, D., Tenover, F.C., and Popovic, T. (2001). First report of *Neisseria meningitidis* intermediately resistant to penicillin in Croatia. J. Clin. Microbiol. 39, 823.

Botha, P. (1988). Penicillin-resistant *Neisseria meningitidis* in southern Africa. Lancet *i*:54.

Bowler, L.D., Zhang, Q.Y., Riou, J.Y., and Spratt, B.G. (1994). Interspecies recombination between the *penA* genes of *Neisseria meningitidis* and commensal *Neisseria* species during the emergence of penicillin resistance in *Neisseria meningitidis*: natural events and laboratory stimulation. J. Bacteriol. *176*, 333–337.

Brannigan, J.A., Tirodimos, I.A., Zhang, Q.Y., Dowson, C.G., and Spratt, B.G. (1990). Insertion of an extra amino acid is the main cause of the low affinity of penicillin-binding protein 2 in penicillin-resistant strains of *Neisseria gonorrhoeae*. Mol. Microbiol. *4*, 913–919.

Burdett, V. (1986). Streptococcal tetracycline resistance mediated at the level of protein synthesis. J. Bacteriol. *165*, 564–569.

Burdett, V. (1991). Purification and characterization of Tet(M), a protein that renders ribosomes resistant to tetracycline. J. Biol. Chem. *266*, 2872–2877.

Caparon, M.G., and Scott, J.R. (1989). Excision and insertion of the conjugative transposon Tn*916* involves a novel recombination mechanism. Cell. *59*, 1027–1034.

Centers for Disease Control and Prevention (2007). Sexually transmitted disease surveillance 2005 supplement. Gonococcal Isolate Surveillance Project (GISP) Annual Report 2005. pp. 1-110.

Centers for Disease Control and Prevention (2008). Emergence of Fluoroquinolone-resistant *Neisseria meningitidis*—Minnesota and North Dakota, 2007—2008. MMWR. *57*, 173–175.

Chen, C.J., Tobiason, D.M., Thomas, C.E., Shafer, W.M., Seifert, H.S., and Sparling, P.F. (2004). A mutant form of the *Neisseria gonorrhoeae* secretin protein PilQ allows increased entry of heme and antimicrobial compounds. J. Bacteriol. *186*, 730–739.

Clewell, D.B., and Gawron-Burke, C. (1986). Conjugative transposons and dissemination of antibiotic resistance in streptococci. Annu. Rev. Microbiol. *40*, 635–639.

Correia, F.F., Inouye, S., and Inouye, M. (1988). A family of small repeated elements with some transposon-like properties in the genome of *Neisseria gonorrhoeae*. J. Biol. Chem. *263*, 12914–12916.

Delahay, R.M., Robertson, B.D., Balthazar, J.T., Shafer, W.M., and Ison, C. (1997). Involvement of the gonococcal MtrE in the resistance of *Neisseria gonorrhoeae* to toxic hydrophobic compounds. Microbiol *143*, 2127–2133.

Drake, S.L. and Koomey, M. (1995). The product of the *pilQ* gene is essential in the biogenesis of type IV pili in *Neisseria gonorrhoeae*. Mol. Microbiol. *18*, 975–986.

Drake, S.L., Sandstedt, S.A., and Koomey, M. (1997). PilP, a pilus biogenesis lipoprotein in *Neisseria gonorrhoeae*, affects expression of PilQ as a high- molecular-mass multimer. Mol. Microbiol. *23*, 657–668.

Eisenstein, B.I. and Sparling, P.F. (1978). Mutations to increased antibiotic susceptibility in naturally-occurring gonococci. Nature *271*, 242–244.

Eisenstein, B.I., Sox, T., Biswas, G., Blackman, E., and Sparling, P.F. (1977). Conjugal transfer of the gonococcal penicillinase plasmid. Science. *195*, 998–1000.

Elwell, L.P., Roberts, M., Mayer, L.W., and Falkow, S. (1977). Plasmid-mediated beta-lactamase production in *Neisseria gonorrhoeae*. Antimicrob. Agents Chemother. *11*, 528–533.

Faruki, H., and Sparling, P.F. (1986). Genetics of resistance in a non-beta-lactamase-producing gonococcus with relatively high-level penicillin resistance. Antimicrob. Agents Chemother. *30*, 856–860.

Faruki, H., Kohmescher, R.N., McKinney, W.P., and Sparling, P.F. (1985). A community-based outbreak of infection with penicillin-resistant *Neisseria gonorrhoeae* not producing penicillinase (chromosomally-mediated resistance). N. Engl. J. Med. *313*, 607–611.

Fermer, C., Kristiansen, B.-E., Sköld, O., and Swedberg, G. (1995). Sulfonamide resistance in *Neisseria meningitidis* as defined by site-directed mutagenesis could have its origin in other species. J. Bacteriol. *177*, 4669–4675.

Folster, J.P., and Shafer, W.M. (2005). Regulation of *mtrF* expression in *Neisseria gonorrhoeae* and its role in high-level antimicrobial resistance. J. Bacteriol. *187*, 3713–3720.

Folster, J.P., Dhulipali, V., Nicholas, R.A., and Shafer, W.M.. (2007). Differential regulation of *ponA* and *pilMNOPQ* expression by the *Neisseria gonorrhoeae* MtrR transcriptional regulatory protein. J. Bacteriol.*189*, 4569–4577

Folster, J.P., Johnson, P.J., Dhalipali, V., Dyer, D.W., and Shafer, W.M. (2009). MtrR modulates *rpoH* expression and levels of antimicrobial resistance in *Neisseria gonorrhoeae*. J. Bacteriol. *191*, 287–297.

Fontanals, D., Pineda, V., Pons, I., and Rojo, J.C. (1989). Penicillin-resistant beta-lactamase-producing *Neisseria meningitidis* in Spain. Eur. J. Clin. Microbiol. Infect. Dis. *8*, 90–91.

Galimand, M., Gerbaud, G., Guibourdenche, M., Riou, J.-Y., Courvalin, P. (1998). High-level chromosomal chloramphenicol resistance in *Neisseria meningitidis*. N. Engl. J. Med. *339*, 868–918.

Genco, C.A., Knapp, J.S., and Clark, V.L. (1984). Conjugation of plasmids of *Neisseria gonorrhoeae* to other *Neisseria* species: potential reservoirs for the beta-lactamase plasmid. J. Infect. Dis. *150*, 397–401.

Gill, M.J., Simjee, S., Al-Hattawi, K., Robertson, B.D., Easmon, C.S., and Ison, C.A. (1998). Gonococcal resistance to beta-lactams and tetracycline involves mutation in loop 3 of the porin encoded at the *penB* locus. Antimicrob. Agents Chemother. *42*, 2799–2803.

Goodman, S.D., and Scocca, J.J. (1988). Identification and arrangement of the DNA sequence recognized in specific transformation of *Neisseria gonorrhoeae*. Proc. Natl. Acad. Sci. USA. *85*, 6982–6986.

Gotschlich, E.C., Goldschneider, I., and Artenstein, M.S. (1969). Human immunity to the meningococcus. IV. Immunogenicity of group A and group C meningococcal polysaccharides in human volunteers. J. Exp. Med. *129*, 1367–1384.

Gottesman, M.M., and Ling, V. (2006). The molecular basis of multidrug resistance in cancer: the early years of P-glycoprotein research. FEBS Lett, *580*, 998–1009.

Guymon, L.F., and Sparling, P.F. (1975). Altered crystal violet permeability and lytic behavior in antibiotic-resistant and –sensitive strains of *Neisseria gonorrhoeae*. J. Bacteriol. *124*, 757–763.

Guymon, L.F., D.L. Walstad, and Sparling, P.F. (1978). Cell envelope alterations in antibiotic-sensitive and –resistant strains of *Neisseria gonorrhoeae*. J. Bacteriol. *136*, 391–401.

Guymon, L.F., Esser, M., and Shafer, W.M. (1982). Pyocin-resistant lipolysaccharide mutants of *Neisseria gonorrhoeae*: alterations in sensitivity to normal human serum and polymyxin B. Infect. Immun. *36*, 541–547.

Hagman, K.E., Lucas, C.E.,Balthazar, J.T., Snyder, L., Nilles, M., Judd, R.C., and Shafer, W.M.. (1997). The MtrD protein of *Neisseria gonorrhoeae* is a member of the Resistance/Nodulation/Division protein family constituting part of an efflux system. Microbiology. *143*, 2117–2125.

Hagman, K.E., Pan, W. Spratt, B.G., Balthazar, J.T., Judd, R.C., and Shafer, W.M. (1995). Resistance of *Neisseria gonorrhoeae* to antimicrobial hydrophobic agents is modulated by the *mtrCDE* efflux system. Microbiol. *141*, 611–622.

Hoffman, K.M., Williams, D., Shafer, W.M., and Brennan, R.G.. (2005). Characterization of the multiple transferrable repressor, MtrR, from *Neisseria gonorrhoeae*. J. Bacteriol. *187*, 5008–5012

Hoffmann, S.H., Lambertsen, L., Berthelsen, L., Cowan, S. (2005). *Neisseria gonorrhoeae* with increased ceftriaxone MIC in Denmark 2004: bi-locus sequence typing, serotyping, and sexual orientation, abstract WP-035: ISSTDR 16th Meeting, Amsterdam, The Netherlands.

Hu, M., Nandi, S., Davies, C., and Nicholas, R.A. (2005). High-level chromosomally mediated tetracycline resistance in *Neisseria gonorrhoeae* results from a point mutation in the *rpsJ* gene encoding ribosomal protein S10 in combination with the *mtrR* and *penB* resistance determinants. Antimicrob. Agents Chemother. *49*, 4327–4334.

Iverson, C.J., Wang, S.A., Lee, M.V., Ohye, R.G, Trees, D.L., Knapp, J.S., Effler, R.N., O'Conner, N.P., and Levine, W.C. (2004). Fluoroquinolone-resistance among *Neisseria gonorrhoeae* isolates in Hawaii, 1990–2000: role of foreign importation and endemic spread. Sex. Transm. Dis. *31*, 702–708.

Jerse, A.E., Sharma, N.D., Bodner, A.N.B., Snyder, L.A., and Shafer, W.M. (2003). A gonococcal efflux pump system enhances bacterial survival in a female mouse model of genital tract infection. Infect. Immun. *71*, 5576–5582.

Johnson, S.R., Sandul, A.L., Parekh, M., Wang, S.A., Knapp, J.S., and Trees, D.L. (2003). Mutations causing *in vitro* resistance to azithromycin in *Neisseria gonorrhoeae*. Internatl. J. Antimicrob. Agents. *21*, 414–419.

Knapp, J.S., Johson, S.R., Zenilman, J.M., Roberts, M.C., and Morse, S.A. (1988). High-level tetracycline resistance resulting from TetM in strains of *Neisseria* spp., *Kingella denitrificans*, and *Eikenella corrodens*. Antimicrob. Agents Chemother. 32, 765–767.

Kobayshi, N., Nishino, K., Yamaguchi, A. (2001). Novel macrolide-specific ABC-type transporter in *Escherichia coli*. J. Bacteriol. *183*, 5639–5644.

Kuhns, D.M., Nelson, C.T., Feldman, H.A., and Kuhn, L.R. (1943). The prophylactic value of sulfadiazine. JAMA *123*, 335–339.

Lee, E.-H., Hill, S.A., Napier, R., and Shafer, W.M. (2006). Integration Host Factor is required for FarR repression of the *farAB*-encoded efflux pump of *Neisseria gonorrhoeae*. Mol. Microbiol. *60*, 1381–1400

Lee, E.H. and Shafer,W.M. (1999). The *farAB*-encoded efflux pump mediates resistance of gonococci to long-chained antibacterial fatty acids. Mol. Microbiol. *33*, 839–845.

Levy, S.B. (2002). The Antibiotic Paradox (Cambridge, MA: Perseus Publishing).

Lianes, R., Sosa, J., Guzman, D., Liop, A., Valdes, E.A., Martinez, I., Palma, S., and Lantero, M.I. (2003). Antimicrobial susceptibility of *Neisseria gonorrhoeae* in Cuba (1995–1999): implications for treatment of gonorrhea. Sex. Transm. Dis. *30*, 10–14.

Lindberg, R., Fredlund, H. Nicholas, R., and Unemo, M. (2007). *Neisseria gonorrhoeae* isolates with reduced susceptibility to to cefixime and ceftriaxone: association with genetic polymorphisms in *penA*, *mtrR*, *porbIb*, and *ponA*. Antimicrob. Agents Chemother. *51*, 2117–2122.

Lomovskaya, O., and Watkins, W. (2002). Inhibition of efflux pumps as a novel approach to combat drug resistance in bacteria. In: *Microbial Multidrug Efflux* (I.T. Paulsen and K. Lewis, editors), pp. 201–231. (Wymondham: Horizon Scientific Press)

Lopardo, H.A, Santander, A.C., Ceinos, M.D.C., Rubeglio, E.A. (1993). Isolation of moderately penicillin-susceptible strains of *Neisseria meningitidis* in Argentina. Antimicrob. Agents Chemother. 37, 1728–1729.

Lucas, C.E., Balthazar, J.T., Hagman, K.E., and Shafer, W.M.. (1997). The MtrR repressor binds the DNA sequence between the *mtrR* and *mtrC* genes of *Neisseria gonorrhoeae*. J. Bacteriol. *179*, 4123–4128.

Lucas, C.E., Hagman, K.E., Levin, J.C., Stein, D.C., Shafer, W.M. (1995). Importance of lipooligosaccharide structure in determining gonococcal resistance to hydrophobic antimicrobial agents resulting from the *mtr* efflux system. Mol. Microbiol. *16*, 1001–1009.

Luna, V.A., Coates, P., Cove, J.H., Eady, A.E., Nguyen, T.T.H., and Roberts, M.C. (1999). A variety of Gram-positive bacteria carry mobile *mef* genes. J. Antimicrob. Chemother. *44*, 19–25.

Luna, V.A., Cousin, S., Jr., Whittington, W.L.H., and Roberts, M.C. (2000). Identification of the conjugative *mef* gene in clinical *Acinetobacter junii* and *Neisseria gonorrhoeae* isolates. Antimicrob. Agents Chemother. *44*, 2503–2506.

Ma, D., Alberti, M., Lynch, C., Nikaido, H., and Hearst, J.E. (1996). The local repressor AcrR plays a modulating role in regulation of the *acrAB* genes of *Eschericihia coli* by global stress signals. Mol. Microbiol. *19*, 101–112.

Macheboeuf, P., Fischer, D.S., Brown, T., Jr., Zervosen, A., Luxen, A., Joris, B., Dessen, A., and Schodield. (2007). Structural and mechanistic basis or penicillin-binding protein inhibition by lactivicins. Nature Chem. Biol. 3, 563–569.

Manchanida, V. and Bhalla, P. (2006). Emergence of non-ceftiaxine-susceptible *Neisseria meningitidis* in India. J. Clin. Microbiol. *44*, 4290–4291.

Maness, M.J., and Sparling, P.F. (1973). Multiple antibiotic resistance due to a single mutation in *Neisseria gonorrhoeae*. J. Infect. Dis. *128*, 321–330.

McMurray, L., Petrucci, R.E., Jr., and Levy, S.B. (1980). Active efflux of tetracycline encoded by genetically different tetracycline determinants in *Escherichia coli*. Proc. Natl. Acad. Sci. USA. *77*, 3974–3977.

Morse, S.A., Johnson, S.R., Biddle, J.W., Roberts, M.C. (1986). High-level tetracycline resistance in *Neisseria gonorrhoeae* is result of acquisition of streptococcal *tetM* determinant. Antimicrob. Agents Chemother. *30*, 664–670.

Morse, S.A., Lysko, P.G., McFarland, L., Knapp, J.S., Sanstrom, E., Critchlow, C., and Holmes, K.K. (1982). Gonococcal strains from homosexual men have outer membranes with reduced permeability to hydrophobic molecules. Infect. Immun. 37, 432–438.

Murakami, S., Nakashima, R., Yamashita, E., and Yamaguichi, A. (2002). Crystal structure of bacterial multidrug efflux transporter AcrB. Nature. *419*, 587–593.

Murakami, S., Nakashima, R., Yamashita, E., Matsumoto, T., and Yamaguchi, A. (2006). Crystal structures of a multidrug transporter reveal a functionally rotating mechanism. Nature *443*, 173–179.

Neykahf, A.A. (2001). The ostensible paradox of multidrug recognition. J. Mol. Microbiol. Biotechnol. *3*, 151–154.

Nikaido, H. (1994). Prevention of drug access to bacterial targets: permeability barriers and active efflux. Science. *264*, 382–388.

Nikaido, H. (1996). Muiltidrug efflux pumps of Gram-negative bacteria. J. Bacteriol. *178*, 5853–5859.

Nolte, O. (1997). Rifampicin resistance in *Neisseria meningitidis*: evidence from a study of sibling strains, description of new mutations and notes on population genetics. J. Antimicrob. Chemother. *39*, 747–755.

Olesky, M., Hobbs, M., and Nicholas, R.A. (2002). Identification and analysis of amino acid mutations in porin IB that mediate intermediate-level resistance to penicillin and tetracycline in *Neisseria gonorrhoeae*. Antimicrob. Agents Chemother. *46*, 2811–2820.

Olesky, M., Zhao, S., Rosenberg, R.L., and Nicholas, R.A. (2006). Porin-mediated antibiotic resistance in *Neisseria gonorrhoeae*: ion, solute and antibiotic permeation through PIB proteins with *penB* mutations. J. Bacteriol. *188*, 2300–2308.

Pan, W., and Spratt, B.G. (1994). Regulation of the permeability of the gonococcal cell envelope by the *mtr* system. Mol. Microbiol. 11:769–775.

Patrick, D, Shaw, C., and Rekart, M.L. (1995). *Neisseria gonorrhoeae* with decreased susceptibility to ciprofloxacin in British Columbia: an imported phenomenon. Can. Common Dis. Rep. *21*, 137–139.

Percival, A., Corkill, J.E., Arya, P.O., Rowlands, S., Alergani, and Annels, E.H. (1976). Penicillinase producing gonococci in Liverpool. Lancet 2, 137–1382.

Phillips, I. (1976). Beta-lactamase producing penicillin-resistant gonococcus. Lancet. ii, 656–657

Richter, S.S., Gordon, K.A., Rhomberg, P.R., Pfaller, M.A., and Jones, R.N. (2001). *Neisseria meningitidis* with decreased susceptibility to penicillin: report from the SENTRY antimicrobial surveillance program, North America 1998–1999. Diag. Microbiol. Infect. Dis. *41*, 83–88.

Roberts, M. Elwell, L.P., and Falkow, S. (1977). Molecular characterization of two beta-lactamase-specifying plasmids isolated from *Neisseria gonorrhoeae*. J. Bacteriol. *131*, 557–563.

Roberts, M.C., and Knapp, J.S. (1988). Host range of the conjugative 25.2-Megadalton tetracycline resistance plasmid from *Neisseria gonorrhoeae* and related species. Antimicrob. Agents Chemother. *32*, 488–491.

Roberts, M.C., Chung, W.O., Roe, D., Xia, M., Marquez, C., Borthagaray, G., Whittington, W.L., Holmes, K.K. (1999). Erythromycin-resistant *Neisseria gonorrhoeae* and oral commensal *Neisseria* spp. carry known rRNA methylase genes. Antimicrob. Agents Chemother. *43*, 1367–1372.

Ropp, P.A., Hu, M., Olesky, M., Nicholas, R.A. (2002). Mutations in *ponA*, the gene encoding penicillin-binding protein 1, and a novel locus, *penC*, are required for high-level chromosomally mediated penicillin resistance in *Neisseria gonorrhoeae*. Antimicrob. Agents Chemother. *46*, 769–777.

Rosenstein, N.E., Perkins, B.A., Stephens, D.S., Popovic, T., and Hughes, J.M. (2001). Meningococcal disease. N. Engl. J. Med. *344*, 1378–1388.

Rosenstein, N.E., Stocker, S.A., Popovic, T., Tenover, F.C., and Perkins, B.A. (2000). Antimicrobial resistance of *Neisseria meningitidis* in the United States., 1997. Clin. Infect. Dis. *30*, 212–213.

Rouquette-Loughlin, C, Veal, W.L., Lee, E.-H/, Zarantonelli, L. Balthazar, J.T. and Shafer, W.M. (2001). 'Antimicrobial efflux systems possessed by *Neisseria gonorrhoeae* and *Neisseria meningitidis* viewed as virulence factors.' In: *Microbial Drug Efflux*. (Wymondham, UK: Horizon Scientific Press), pp. 187–200.

Rouquette-Loughlin, C., Dunham, S.A., Kuhn, M., Balthazar. J., and Shafer, W.M. (2003). The NorM efflux pump of *Neisseria gonorrhoeae* and *Neisseria meningitidis* recognizes antimicrobial cationic compounds. J. Bacteriol. *185*, 1101–1106.

Rouquette-Loughlin, C.E., Balthazar, J.T., and Shafer, W.M.. (2005). Characterization of the MacA-MacB efflux system in *Neisseria gonorrhoeae*. J. Antimicrob. Chemother.*56*, 856–860.

Rouquette-Loughlin,C.E., Balthazar, J.T., Hill, S.A., and Shafer. W.M. (2004). Modulation of the *mtrCDE*-encoded efflux pump gene complex due to a Correia Element insertion sequence. Mol. Microbiol. *54*, 731–741.

Rouquette, C., Harmon, J.B., and Shafer, W.M. (1999). Induction of the *mtrCDE*-encoded efflux pump system of *Neisseria gonorrhoeae* requires MtrA, an AraC-like protein. Mol. Microbiol. *33*, 651–658.

Sarubbi, F.A., Jr., and Sparling, P.F. (1974). Transfer of antibiotic resistance in mixed cultures of *Neisseria gonorrhoeae*. J. Infect. Dis. *130*, 660–663.

Sarubbi, F.A., Jr., Sparling, P.F., Blackman, E., and Lewis, E. (1975). Loss of low level antibiotic resistance in *Neisseria gonorrhoeae* due to *env* mutations. J. Bacteriol. *124*, 750–756.

Schwarz, S.K., Zenilman, J.M., Schnell, D., Knapp, J.S., Hook, E.W.III, Thompson, S., Judson, F.N., and Holmes, K.K. (1990). National surveillance of antimicrobial resistance in *Neisseria gonorrhoeae*: The Gonococcal Isolate Surveillance Project. JAMA *264*, 1413–1417.

Scott, J.R., Kirchman, P.A., and Caparon, M.G. (1988). An intermediate in transposition of the conjugative transposon Tn*916*. Proc. Natl. Acad. Sci. USA. *85*, 4809–4813.

Shafer, W.M., Qu, X.-D., Waring, A.J., and Lehrer, R.I.. (1998). Modulation of *Neisseria gonorrhoeae* susceptibility to vertebrate antibacterial peptides due to a member of the resistance/nodulation/division efflux pump family. Proc Natl Acad Sci U.S.A. *95*, 1829–1833.

Shafer,W.M., and Folster, J.P. (2006). Towards an understanding of chromosomally mediated penicillin resistance in *Neisseria gonorrhoeae*: evidence for a porin-efflux pump collaboration. J. Bacteriol. *188*, 2297–2299.

Shafer,W.M., Balthazar, J.T., Hagman, K.E. and Morse, S.A.. (1995) Missense mutations that alter the DNA-binding domain of the MtrR protein occur frequently in rectal isolates of *Neisseria gonorrhoeae* that are resistant to faecal lipids. Microbiol. *41*, 907–911.

Shultz, T.R., Tapsall, J.W., White, P.A., Ryan, C.S., Lyras, D., Rood, J.I., Bionotto, E., and Richardson, C.J.L. (2003). Chloramphenicol-resistant *Neisseria meningitidis* containing *catP* isolated in Australia. J. Antimicrob. Chemother. *52*, 856–859.

Sox, T.E., Mohammed, W., and Sparling (1979). Transformation-derived *Neisseria gonorrhoeae* plasmids with altered structure and function. J. Bacteriol. *138*, 510–518.

Sparling, P.F., Sarubbi, F.A., and Blackman, E. (1975). Inheritance of low-level resistance to penicillin, tetracycline, and chloramphenicol in *Neisseria gonorrhoeae.* J. Bacteriol. *124*, 740–749.

Spratt, B.G. (1988). Hybrid penicillin-binding proteins in penicillin-resistant strains of *Neisseria gonorrhoeae.* Nature 332, 173–176.

Spratt, B.G., Bowler, L.D., Zhang, Q.Y., Zhou, J., and Smith, J.M. (1992). Role of interspecies transfer of chromosomal genes in the evolution of penicillin resistance in pathogenic and commensal *Neisseria* species. J. Mol. Evol. *34*, 115–125.

Stephens, D.S., Greenwood, B., and Brandtzaeg, P. (2007). Epidemic meningitis, meningococcaemia, and *Neisseria meningitidis.* Lancet *369*, 2196–2210.

Stephens, D.S., Swartley, J.S., Kathariou, S., and Morse, S.A. (1991). Insertion of Tn*916* in *Neisseria meningitidis* resulting in loss of group B capsular polysaccharide. Infect. Immun. *59*, 4097–4102.

Su, X. and Lind, I. (2001). Molecular basis of high-level ciprofloxacin resistance in *Neisseria gonorrhoeae* strains isolated from Denmark from 1995–1998. Antimicrob. Agents Chemotherap. *45*, 117–123.

Swartley, J.S, McAllister, C.F., Haijeh, R.A., Heinrich, D.W., and Stephens, D.S. (1993). Deletions of Tn*916*-like transposons are implicated in *tetM*-mediated resistance in pathogenic *Neisseria*. Mol. Microbiol. *10*, 299–310.

Takahata, S., Senju, N., Osaki, Y., Yoshida, T., and Ida, T. (2006). Amino acid substitutions in mosaic penicillin-binding protein 2 associated with reduced susceptibility to cefixime in clinical isolates of *Neisseria gonorrhoeae.* Antimicrob. Agents Chemother. *50*, 3638–3645.

Tanaka, M., Nakayama, H., Huruya, K., Konomi, S., Irie, S., Kanayama, A., Saika, T., Kobayashi, I. (2006). Analysis of mutations within multiple genes associated with resistance in a clinical isolate of *Neisseria gonorrhoeae* with reduced ceftriaxone susceptibility that shows a multidrug-resistant phenotype. Int. J. Antimicrob. Agents Chemother. *27*, 20–26.

Tapsall, J. (2006). Antibiotic resistance in *Neisseria gonorrhoeae* is diminishing available treatment options for gonorrhea: some possible remedies. Rev. Anti-Infect. Ther. *4*, 619–628.

Tapsall, J.W., Shultz, T., Limnios, E., Munro, R., Mercer, J., Porritt, R., Griffith, J., Hogg, G., Lum, G., Lawrence, A., *et al.*; National *Neisseria* Nework of Australia (2001). Surveillance of antibiotic resistance in invasive isolates of *Neisseria meningitidis* in Australia 1994–1999. Pathology. *33*, 359–361.

Trees, D.L., Sandul, A.L., Peto-Mesola, V., Aplasca, M-R., Leng, H.B., Whittington, W.L., and Knapp, J.S. (1999). Alterations within the quinolone resistance-determining regions of GyrA and ParC of *Neisseria gonorrhoeae* isolated in the Far East and the United States. Internatl. J. Antimicrob. Agents. *12*, 325–332.

Van Dyck, Crabbe, F., Nazila, N., Bogaerts, J, Munyabikali, J.P., Ghys, P., Diallo, M., and Laga, M. (1997). Increasing resistance of *Neisseria gonorrhoeae* in west and central Africa. Consequence on therapy of gonococcal infection. Sex. Transm. Dis. *24*, 32–37.

Veal, W., Yellen, A., Balthazar, J.T., Pan, W., Spratt, B.G., and Shafer, W.M. (1998). Loss of function mutations in the *mtr* efflux system of *Neisseria gonorrhoeae.* Microbiol. *144*, 621–627.

Veal, W.L., and Shafer, N.M. (2003). Identification of a cell envelope protein (MtrF) involved in hydrophobic antimicrobial resistance in *Neisseria gonorrhoeae.* J. Antimicrob. Chemother. *51*, 27–37.

Veal, W.L., Nicholas, R.A., and Shafer, W.M. (2002). Overexpression of the MtrC-MtrD-MtrE efflux pump due to an *mtrR* mutation is required for chromosomally mediated penicillin resistance in *Neisseria gonorrhoeae.* J. Bacteriol. *184*, 5619–5624.

Wang, S.A., Lee, M.A.C., O'Connor, N., Iverson, C.J., Ohye, R.G., Whiticar, P.M., Hale, J.A., Trees, D.L., Knapp, J.S., Effler, P.V., and Weinstock, H.S. (2003). Multi-resistant *Neisseria gonorrhoeae* with decreased susceptibility to cefixime-Hawaii, 2001. Clin. Infect. Dis. *37*, 849–852.

Warner, D.M., Folster, J.P., Shafer, W.M., and Jerse, A.E. (2007). Regulation of the MtrC-MtrD-MtrE efflux pump system modulates the *in vivo* fitness of *Neisseria gonorrhoeae.* J. Infect. Dis. *196*, 1804–1812.

Whiley, D.M., Limnios, E.A., Ray, S., Sloots, T.P., Tapsall, J.W. (2007). Diversity of *penA* alterations and subtypes in *Neisseria gonorrhoeae* strains from Sydney, Australia, that are less susceptible to ceftriaxone. Antimicrob. Agents Chemother. *51*, 3111–3116.

WHO Western Pacific Gonococcal Antimicrobial Surveillance Programme (2006). Surveillance of antibiotic resistance in *Neisseria gonorrhoeae* in the WHO Western Pacific region, 2004: Commun. Dis. Intell. *30*, 129–132.

Woods., C.R., Smith, A.L., Wasilauskas, B.L., Campos, J., Givner, L.B. (1994). Invasive disease caused by *Neisseria meningitidis* relatively resistant to penicillin in North Carolina. J. Infect. Dis. *170*, 453–456.

World Health Organization Western Pacific Region Gonococcal Surveillance Programme. (1995). 1994 Annual Report. Commun. Dis. Intell. *19*, 495–499.

Xia, M., Whittington, W.L.H., Shafer, W.M. and Holmes, K. (2000). Gonorrhea among men who have sex with men: outbreak caused by a single genotype of erythromycin-resistant *Neisseria gonorrhoeae* with a single basepair deletion in the *mtrR* promoter region. J. Infect. Dis. *181*, 2080–2082.

Yagupsky, P., Ashkenazi, S., and Block, C. (1993). Rifampicin-resistant meningococci causing invasive disease and failure of chemoprophylaxis. Lancet. *341*, 1152–1153.

Ye, S., Su, X., Wang, Q., Yin, Y., Dai, X., and Sun, H. (2002). Surveillance of antibiotic resistance of *Neisseria gonorrhoeae* isolates in China, 1993–1998.

Zarantonelli, L., Borthagary, G., Lee, E.H., Veal, W., and Shafer, W.M.. (2001) Decreased susceptibility to azithromycin and erythromycin mediated by a novel *mtrR* promoter mutation in *Neisseria gonorrhoeae*. J. Antimicrob. Chemother. *47*, 651–654.

Zhao, S., Tobiason, D.M., Hu, M., Seifert, H.S., and Nicholas, R.A. (2005). The *penC* mutation conferring antibiotic resistance in *Neisseria gonorrhoeae* arises from a mutation in the PilQ secritin that interferes with multimer stability. Mol. Microbiol. *57*, 1238–1251.

Index

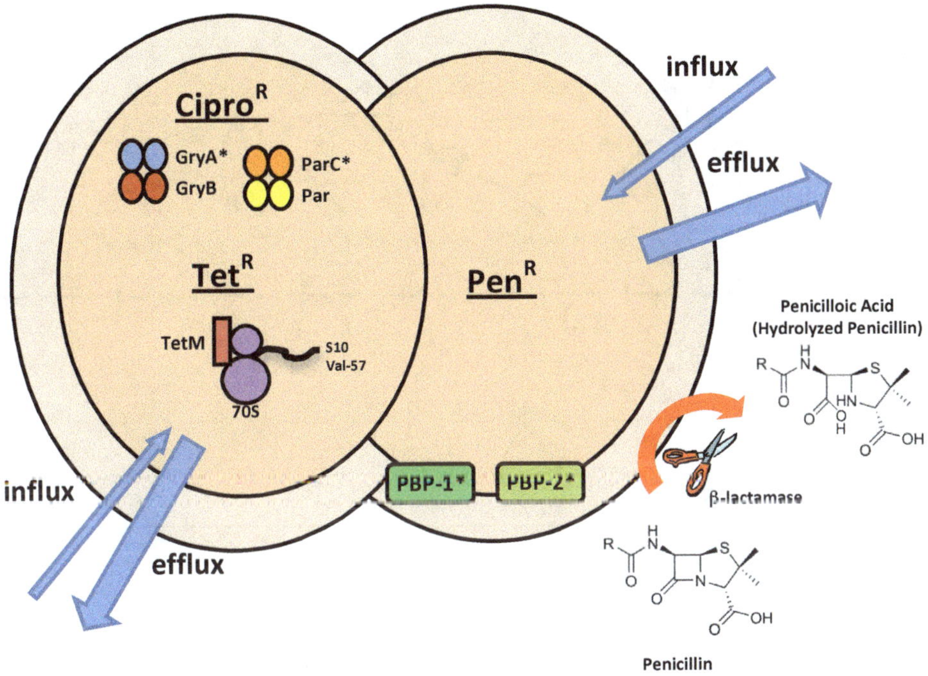

Plate 13.2 Mechanisms used by the pathogenic *Neisseria* to resist penicillin, tetracycline or ciprofloxacin. The mechanisms of penicillin resistance include the action of β-lactamase, decreased influx, increased efflux, and modifications of PBP 1 and PBP 2. Tetracycline resistance levels can be modulated by changes in influx and efflux. Plasmid-mediated tetracycline levels due to *tetM* are shown by TetM binding to the ribosome. Chromosomally mediated tetracycline resistance involves modification of ribosomal protein S10 at Val-57. Ciprofloxacin resistance is typically due to chromosomal mutations at *gyrA* and *parC.*

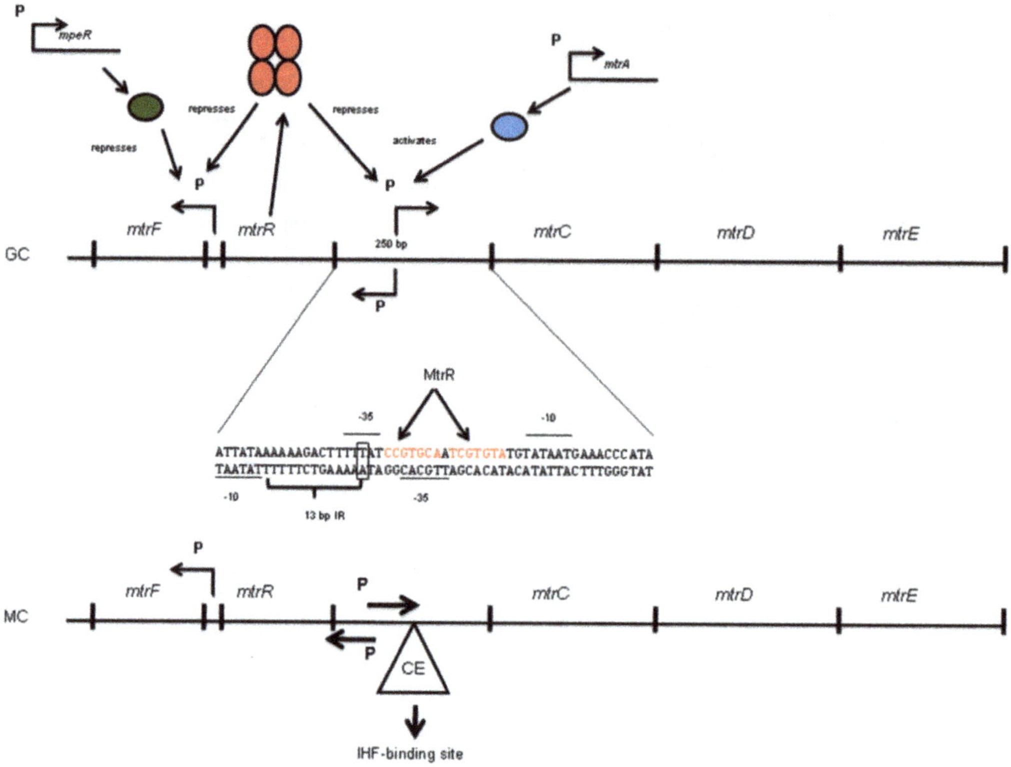

Plate 13.4 Shown is a summary of the mechanisms used by gonococci (GC) and meningococci (MC) to control expression of the *mtrCDE*-encoded efflux system. The position of genes is shown with promoter elements identified by a P and the direction of transcription is shown by a bent arrow. Two dimers of the MtrR repressor bind to the promoter region of *mtrC* at indicated sites (shown in red) in the expanded nucleotide sequence. MtrR and MpeR can also repress transcription of *mtrF* as indicated. The *cis*-acting 13-bp inverted repeat element within the *mtrR* promoter is shown and the site of the single bp deletion that results in high-level expression of *mtrCDE* is indicated by a box. Expression of *mtrCDE* in GC can also be enhanced by MtrA as indicated and binding of MtrA to this region has been inferred. Expression of *mtrCDE* in MC is independent of MtrR and MtrA and involves a 155–159 by Correia element (CE). Integration host factor (IHF) binds to a site within the CE and this negatively controls *mtrCDE* expression.

www.ingramcontent.com/pod-product-compliance
Lightning Source LLC
LaVergne TN
LVHW062227200926
848484LV00006B/1489

* 9 7 8 1 9 0 4 4 5 5 5 1 6 *